KINEMATICS OF MACHINES

By

Dr. SADHU SINGH

Formely, Professor & Head, Mechanical Engineering Department

&

Dean, Faculty of Engineering & Technology

Govind Ballabh Pant University of Agriculture & Technology,

Pantnagar-263145 (Uttarakhand)

Formely, Director (Colleges), Punjab Technical University, Jalandhar

S.K. KATARIA & SONS®

Publisher of Engineering Books

4885/109, Prakash Mahal, Dr. Subhash Bhargav Lane,

Opp. Delhi Medical Association, Daryaganj, New Delhi-110002 (India)

Phone: +91-11-43551243; Mobile: +91-9871775858, 7678425209

Email: katariabook@gmail.com, katariabooks@yahoo.com

Website : www.skkatariaandsons.com

(e-books, subscriptions and bulk buying also available on our website)

Published by :

S.K. Kataria & Sons®

4885/109, Prakash Mahal, Dr. Subhash Bhargav Lane,

Opp. Delhi Medical Association, Daryaganj, New Delhi-110002

Phone: +91-11-43551243; Mobile: +91-9871775858, 7678425209

e-mail: katariabook@gmail.com, katariabooks@yahoo.com

First Edition: 2012; Reprint: 2026

Computer Typesetting by : Stride Computers, Delhi

Printed at : A.S. Offset, Delhi

Preface

Kinematics of Machines has been introduced from the session 2010 – 11 by the Maharishi Dayanand University, Rohtak and Kurukshetra University, Kurukshetra with the revised syllabus (ME: 204 – F). It is offered in the 4th semester to B.Tech./B.E. Mechanical Engineering students. The present book has been written strictly according to the new syllabus.

The book contains eleven chapters on Introduction to Mechanisms; Velocity in Mechanisms; Acceleration in Mechanisms; Kinematic Analysis of Plane Mechanisms; Mechanisms with Lower Pairs; Friction; Belts, Chains and Ropes; Cams; Gears; Gear Trains; and Kinematic Synthesis of Planar Mechanisms. A large number of examples have been solved to illustrate the solution methods. An added feature of the book is that previous questions from MDU and KUK examinations have been solved in each chapter. A large number of problems have been given at the end of each chapter along with Short Answer Type Questions..

The author is highly thankful to Mr. Sanjeev Kataria and Mr. Sharda Singh for their help and cooperation during the preparation of the manuscript of the book and otherwise.

The author expresses his appreciations for his family members for their tolerance and cooperation. My grand children Kanupreet and Amitoej have been a source of great encouragement during the writing of manuscript and proof reading.

Suggestions for the improvement of the book shall be welcome. The author shall appreciate receiving information from readers regarding any errors or misprints left inadvertently.

—AUTHOR

CONTENTS

Chapter 1

Introduction to Mechanisms

Inside This Chapter

1.1 INTRODUCTION

A mechanism is a set of machine elements or components or parts arranged in a specific order to produce a specified motion. First of all we define some of the terms related to the study of mechanisms.

Machine : A machine is a device which transforms energy available in one form to another to do certain type of desired useful work.

Kinematics : It is a subject which deals with the study of relative motion of parts constituting a machine, neglecting forces producing the motion.

Resistant Bodies : These are those bodies which do not suffer appreciable distortion or change in physical form due to forces acting on them, *e.g.*, springs, belts and fluids, etc. Elastic bodies are also resistant bodies if the deformation is not appreciable. It is capable of transmitting the required forces with negligible deformation.

Rigid Bodies : These are those bodies which do not deform under the action of forces. All resistant bodies are considered rigid bodies for the purpose of transmitting motion.

Structure : It is an assemblage of a number of resistant bodies meant to take up loads or subjected to forces having straining actions, but having no relative motion between its members, *e.g.*, a roof truss, a framed railway bridge, machine frame, etc.

Frame : It is a structure which supports moving parts of a machine.

Kinematics of Machines : It is that branch of theory of machines which deals with the study of relative motion of various machine elements neglecting the forces acting on them.

Dynamic of Machines : It is that branch of theory of machines which deals with the study of forces acting on various machine elements.

Theory of Machines : It is the branch of engineering which deals with the study of relative motion of various machine elements and the forces acting on them.

1.2 DIFFERENCES BETWEEN MACHINE AND STRUCTURE

The differences between a machine and structure are given in Table 1.1.

Table 1.1 : Differences between a machine and structure.

Machine	*Structure*
1. There is a relative motion between its members.	There is no relative motion between its members.
2. It transforms available energy into useful work.	There is no transformation of energy.
3. Machine members transmit motion and forces. *Examples* : Car, truck, lathe, etc.	Members take up loads only. *Examples* : Building, bridge, truss, etc.

1.3 KINEMATIC JOINT

Kinematic Joint : A kinematic joint is the connection between two links by a pin. There is clearance between the pin and the hole in the ends of the links being connected so that there is free motion of the links.

1.3.1 Types of Joints

The following types of joints are generally used in a chain :

1. **Binary Joint :** In a binary joint, two links are connected at the same joint by a pin. In the four bar chain shown in Fig. 1.1(a), all the joints are binary joints.
2. **Ternary Joint :** In a ternary joint, three links are connected at the same joint by a pin. It is equivalent to **two** binary joints. In Fig. 1.1(b), joints B and C are ternary joints and others are binary joints.

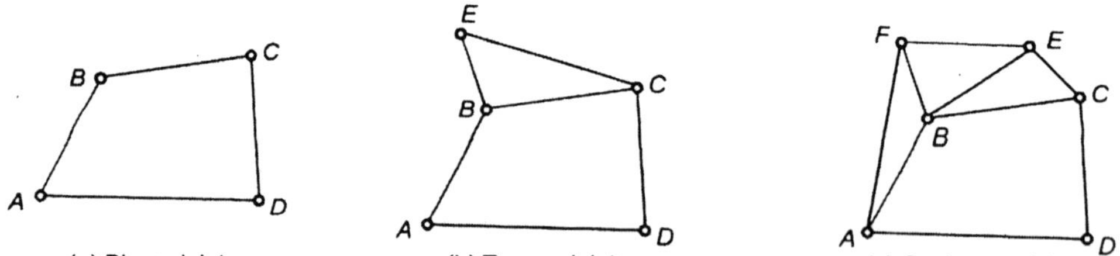

(a) Binary joints (b) Ternary joints (c) Quaternary joints

Fig. 1.1 Types of joints.

3. **Quaternary Joint :** When four links are connected at the same joint by a pin, it is called a quaternary joint. One quarternary joint is equivalent to **four** binary joints. In Fig. 1.1(c), joint *B* is the quaternary joint; *A*, *C*, *E*, *F* are ternary joints; and *D* is the binary joint.

1.4 KINEMATIC LINK

Kinematic Link : A link (or element or kinematic link) is a resistant body (or assembly of resistant bodies), which do not deform under the action of forces and constitute part (or parts) of a machine, connecting other parts, which have motion relative to it.

The links can be classified as : binary, ternary, quaternary, etc. depending upon its ends on which revolute or turning pairs can be placed, as shown in Fig. 1.2. A binary link has two vertices, a ternary link has three vertices, and a quaternary link has four vertices, and so on.

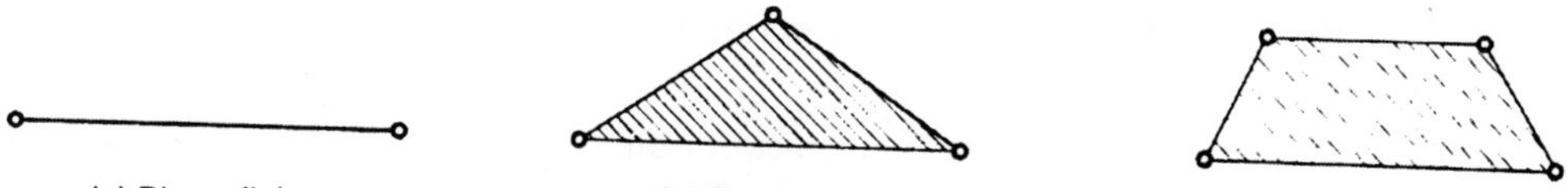

(a) Binary link (b) Ternary link (c) Quaternary link

Fig. 1.2 Types of links.

1.4.1 Types of Links

There are four types of links : rigid, flexible, fluid and floating links.

1. **Rigid Link :** A rigid link is one which does not undergo any deformation while transmitting motion. Links in general are elastic in nature. They are considered rigid if they do not undergo appreciable deformation while transmitting motion, *e.g.*, connecting rod, crank, tappet rod, etc.
2. **Flexible Link :** A flexible link is one which while transmitting motion is partly deformed in a manner not to affect the transmission of motion, *e.g.*, belts, ropes, chains and springs, etc.
3. **Fluid Link :** A fluid link is one which is deformed by having fluid in a closed vessel and the motion is transmitted through the fluid by pressure, as in the case of a hydraulic press, hydraulic jack and fluid brake.
4. **Floating Link :** It is a link which is not connected to the frame.

1.5 KINEMATIC PAIR

The two links of a machine, when in contact with each other, are said to form a pair. A kinematic pair consists of two links which have relative motion between them.

1.5.1 Classification of Kinematic Pairs

Kinematic pairs may be classified according to the following considerations :

1. Type of relative motion
2. Type of contact
3. Type of mechanical constraint.

1. Kinematic Pairs According to the Relative Motion

(i) Sliding Pair : It consists of two elements connected in such a manner that one is constrained to have sliding motion relative to another. For example, a rectangular bar in a rectangular hole [Fig. 1.3(a)], piston and cylinder of an engine, cross-head and guides of a steam engine, ram and its guides in a shaper, tailstock on the lathe bed, etc. all constitute sliding pairs.

(ii) Turning Pair : It consists of two elements connected in such a manner that one is constrained to turn or revolve about a fixed axis of another element. For example, a shaft with collar at both ends revolving in a circular hole [Fig. 1.3(b)], crankshaft turning in a bearing, cycle wheels revolving over their axles, etc. all constitute turning pairs.

(iii) Rolling Pair : When two elements are so connected that one is constrained to roll on another element which is fixed, forms a rolling pair. Ball and roller bearings, belt and pulley surface, a wheel rolling on a flat surface [Fig. 1.3(c)], etc. are examples of rolling pairs.

(iv) Screw (or Helical) Pair : When one element turns about the other element by means of threads, forms a screw pair. The motion in this case is a combination of sliding and turning. The lead screw of a lathe with nut, bolt with a nut [Fig. 1.3(d)], screw with nut of a jack, etc. are some examples of screw pairs.

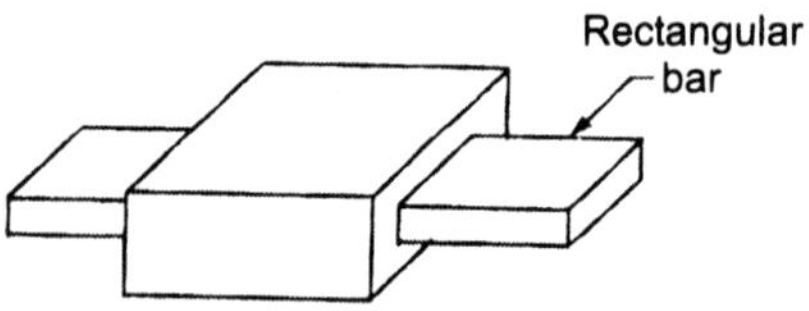

(a) Sliding pair : Rectangular bar in a rectangular hole

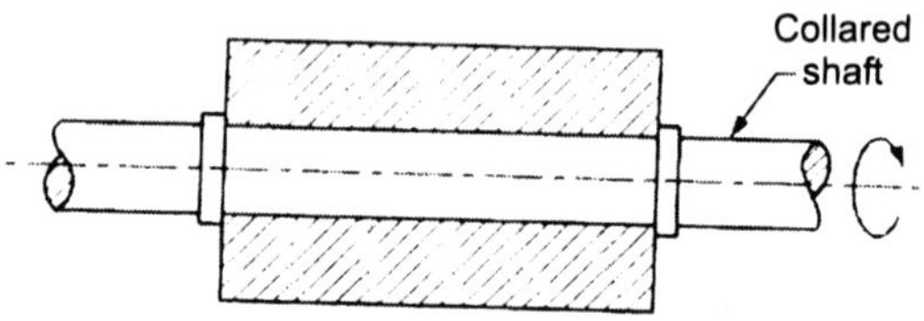

(b) Turning pair : Collared shaft revolving in a circular hole

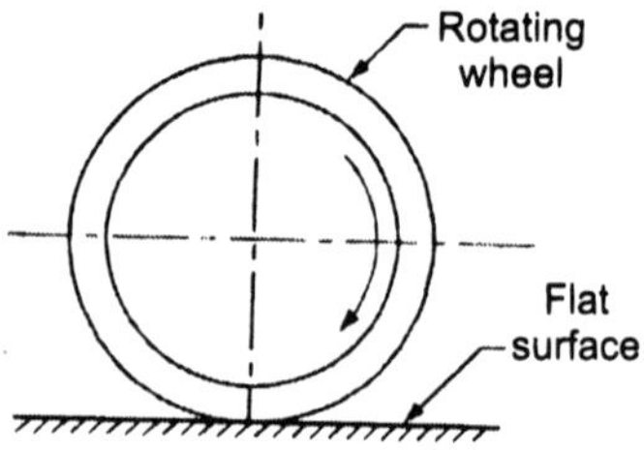

(c) Rolling pair : Wheel rolling on a flat surface

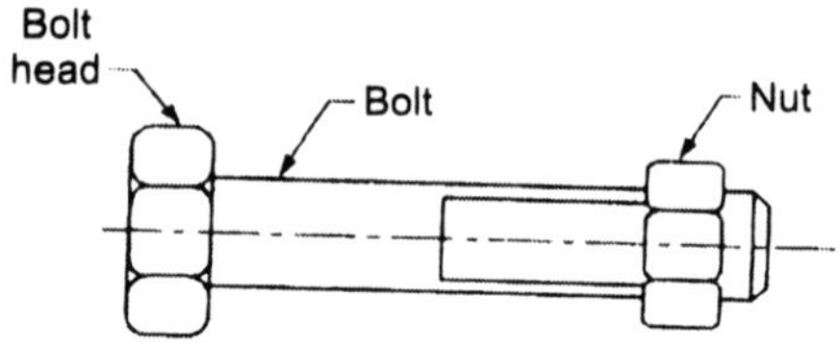

(d) Screw pair Bolt with a nut

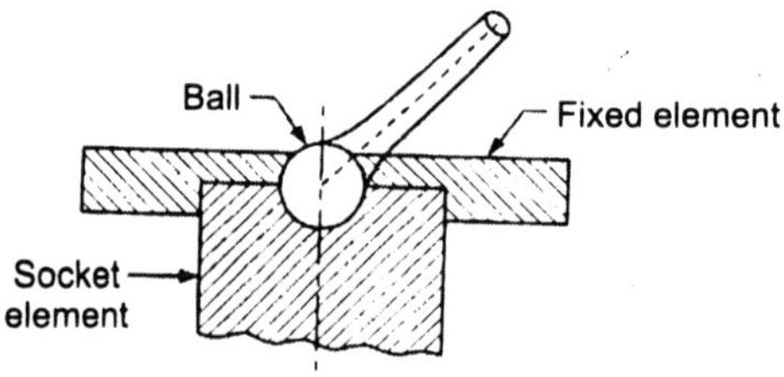

(e) Spherical pair : Ball and socket joint

Fig. 1.3 Types of kinematic pairs according to the type of relative motion.

(v) **Spherical Pair :** When one element in the form of a sphere turns about the other fixed element, it forms a spherical pair. The ball and socket joint [Fig. 1.3(e)], pen stand, mirror attachment of vehicles, etc. are some examples of a spherical pair.

2. Kinematic Pairs According to the Type of Contact

(i) **Lower Pair :** When the two elements have **surface (or area) contact** while in motion and the relative motion being purely **turning or sliding**, is called a lower pair. All sliding pairs, turning pairs and screw pairs form lower pairs. For example, nut turning on a screw, shaft rotating in a bearing, universal joint, all pairs of a slider crank mechanism, pantograph, etc. are lower pairs.

(ii) **Higher Pairs :** When the two elements have **point or line contact** while in motion and the relative motion being the **combination of sliding and turning**, then the pair is known as a higher pair. Belts, ropes and chains drive, gears, cam and follower, ball and roller bearings, wheel rolling on a surface, etc. all form higher pairs.

3. Kinematic Pairs According to the Type of Mechanical Constraint

(i) **Closed Pair :** When the two elements of a pair are held together mechanically in such a manner that only required type of relative motion occurs, is called a closed pair. All lower pairs and some higher pairs, *e.g.,* enclosed cam and follower [Fig. 1.4(a)] are closed pairs.

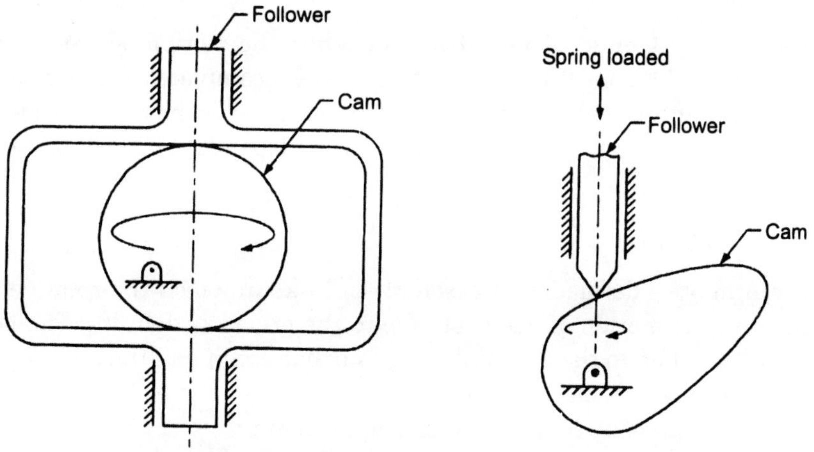

(a) Closed pair (b) Unclosed pair

Fig. 1.4 Closed and unclosed pairs.

(ii) **Unclosed Pair :** When the two elements of a pair are not held mechanically and are held in contact by the action of external forces, are called unclosed pair, *e.g.,* cam and spring loaded follower pair [Fig. 1.4(b)].

1.6 CONSTRAINED MOTION

The three types of constrained motions are :

1. **Completely Constrained Motion :** When the motion between a pair takes place in a definite direction irrespective of the direction of force applied, then the motion is

said to be a completely constrained motion. For example, a square bar in a square hole, a shaft with collars at each end in a circular hole as shown in Fig. 1.3(b), a piston in the cylinder of an internal combustion engine, have all completely constrained motion.

2. **Partially (or successfully) Constrained Motion :** When the constrained motion between a pair is not completed by itself but by some other means, it is said to be successfully constrained motion. For example, the motion of a shaft in a footstep bearing shown in Fig. 1.5 becomes successfully constrained motion when compressive load is applied to the shaft.

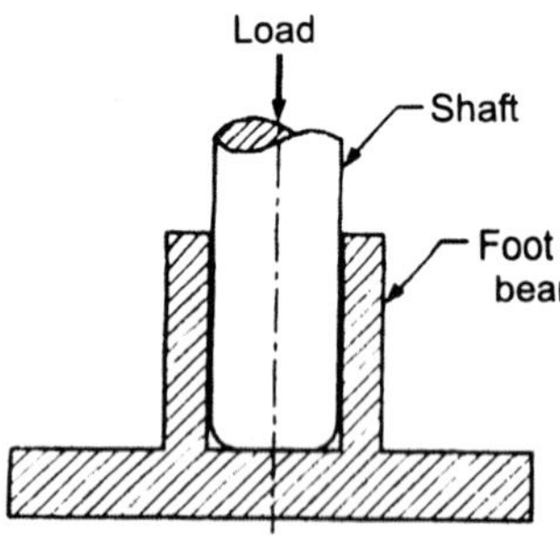

Fig. 1.5 Successfully constrained motion: Loaded shaft in a footstep bearing.

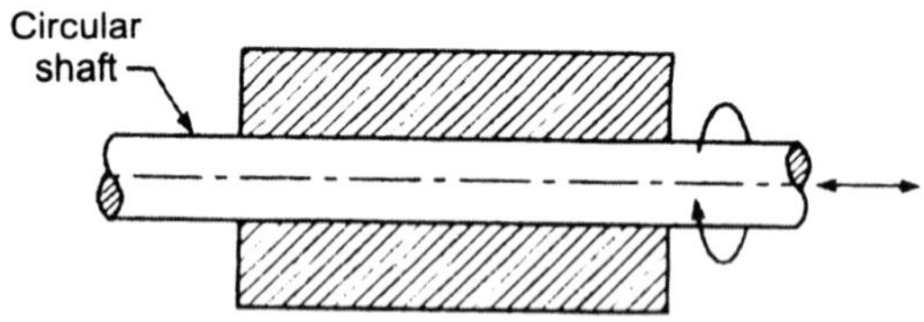

Fig. 1.6 Incompletely constrained motion: Circular shaft in a circular hole.

3. **Incompletely Constrained Motion :** When the motion between a pair can take place in more than one direction, it is said to be incompletely constrained motion, *e.g.*, a circular shaft in a circular hole shown in Fig. 1.6, may have rotary motion as well as reciprocating motion. Both the motions are independent of each other.

1.7 KINEMATIC CHAIN

A kinematic chain may be defined as a assembly of links in which the relative motion of the links is possible and the motion of each relative to the others is definite. The last link of the kinematic chain is attached to the first link. The four-bar chain and the slider-crank chain are some of the examples of a kinematic chain.

The following relationship holds for a kinematic chain :

$$L = 2P - 4 \qquad \text{...(1.1a)}$$

$$J = \left(\frac{3}{2}\right) L - 2 \qquad \text{...(1.1b)}$$

where L = number of binary links
P = number of lower pairs
J = number of binary joints.

If LHS > RHS, then the chain is called a **locked chain** or **redundant chain**
LHS = RHS, then the chain is **constrained**
LHS < RHS, then the chain is **unconstrained**

For a kinematic chain having higher pairs, each higher pair is taken equivalent to two lower pairs and an additional link. In that case,

$$J + \frac{H}{2} = \left(\frac{3}{2}\right) L - 2 \qquad \text{...(1.1c)}$$

■ **Example 1.1** A chain with three links is shown in Fig. 1.7. Prove that the chain is locked.

Solution : Number of binary joints, $J = 3$

Number of binary links, $L = 3$

Number of lower pairs, $P = 3$

Now, $L = 2P - 4$

$$3 = 2 \times 3 - 4 = 2$$

∴ LHS > RHS

Also $J = \left(\frac{3}{2}\right) L - 2$

$$3 = \left(\frac{3}{2}\right) \times 3 - 2 = 2.5$$

∴ LHS > RHS

Therefore, it is a locked chain.

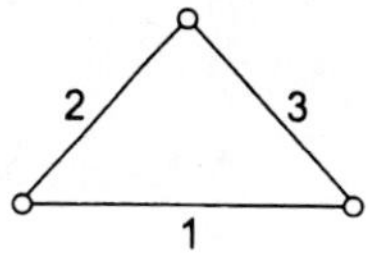

Fig. 1.7

■ **Example 1.2** A four-bar chain is shown in Fig. 1.8. Prove that it is a constrained chain.

Solution : Number of binary joints, $J = 4$

Number of binary links, $L = 4$

Number of lower pairs, $P = 4$

Now, $L = 2P - 4$

$$4 = 2 \times 4 - 4 = 4$$

∴ LHS = RHS

Also $J = \left(\frac{3}{2}\right) L - 2$

$$4 = \left(\frac{3}{2}\right) \times 4 - 2 = 4$$

∴ LHS = RHS

Therefore, it is a constrained chain.

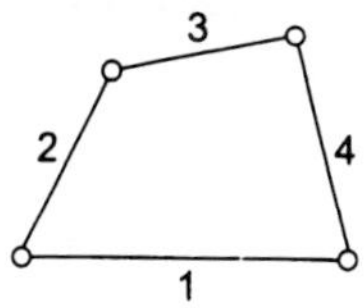

Fig. 1.8

■ **Example 1.3** A five-bar chain is shown in Fig. 1.9. Prove that it is an un-constrained chain.

Solution : Number of binary joints, $J = 5$

Number of binary links, $L = 5$

Number of lower pairs, $P = 5$

Now, $L = 2P - 4$

$$5 = 2 \times 5 - 4 = 6$$

∴ LHS < RHS

Also $J = \left(\frac{3}{2}\right) L - 2$

$$5 = \left(\frac{3}{2}\right) \times 5 - 2 = 5.5$$

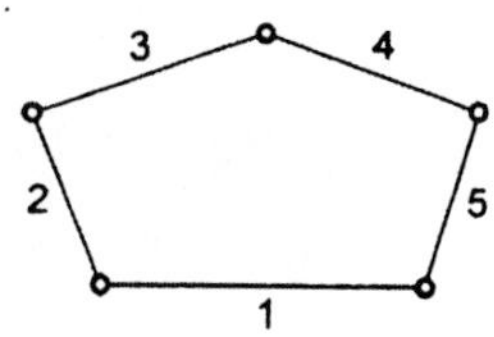

Fig. 1.9

$\therefore$ LHS < RHS

Therefore, it is an un-constrained chain.

■ **Example 1.4** Show that the chain shown in Fig. 1.10 is an un-constrained kinematic chain.

Solution : Number of binary joints,

$$J = 7\,(A = 1, B = 2, C = 2, D = 1, E = 1)$$

Number of binary links, $L = 6$

Number of lower pairs, $P = 9\,(1-2, 2-3, 3-4, 2-4, 4-5, 5-3, 5-6, 3-6, 1-6)$

Now, $L = 2P - 4$

$$6 = 2 \times 9 - 4 = 14$$

$\therefore$ LHS < RHS

Also $J = \left(\frac{3}{2}\right) L - 2$

$$7 = \left(\frac{3}{2}\right) \times 6 - 2 = 7$$

LHS < RHS

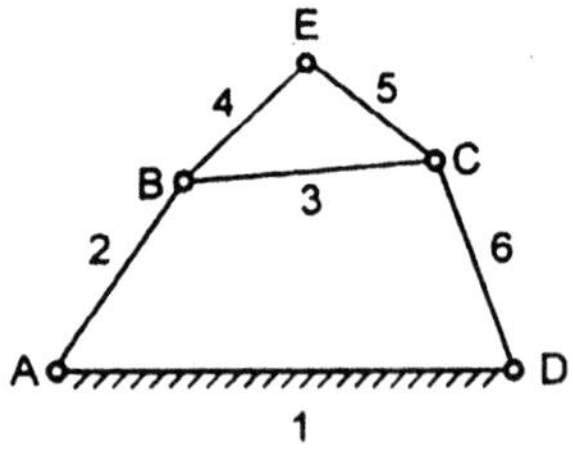

Fig. 1.10

Therefore, it is an un-constrained chain.

■ **Example 1.5** Show that the chain shown in Fig. 1.11 is not a kinematic chain.

Solution : Number of binary joints,

$$J = 13\,(A = 2, B = 4, C = 2, D = 1, E = 2, F = 2)$$

Number of binary links, $L = 9$

Number of lower pairs, $P = 13\,(D = 1, A, C, E, F = 2 \text{ each}, B = 4)$

Now, $L = 2P - 4$

$$9 = 2 \times 13 - 4 = 22$$

$\therefore$ LHS < RHS

Also $J = \left(\frac{3}{2}\right) L - 2$

$$13 = \left(\frac{3}{2}\right) \times 9 - 2 = 11.5$$

LHS < RHS

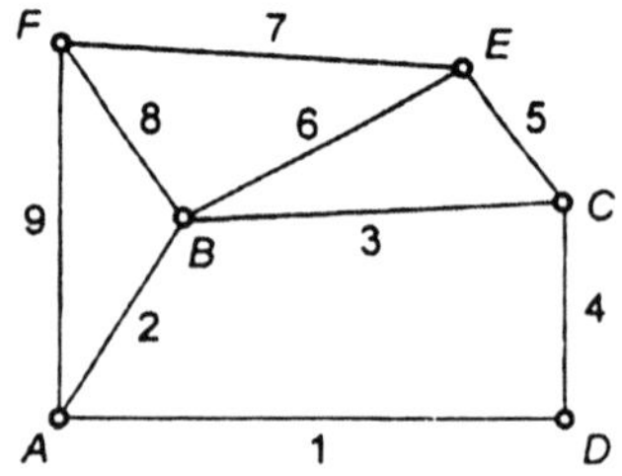

Fig. 1.11

Therefore, it is not a kinematic chain. It is a locked chain or a frame.

1.8 MECHANISM

When one of the links of a kinematic chain is fixed, the chain is called a mechanism.

1.8.1 Types of Mechanisms

The mechanisms are of the following types :

1. **Simple Mechanism :** A mechanism which has four links is called a simple mechanism.
2. **Compound Mechanism :** A mechanism which has more than four links is called a compound mechanism.

3. **Complex Mechanism :** It is formed by the inclusion of ternary or higher order floating link to a simple mechanism.
4. **Planar Mechanism :** When all the links of the mechanism lie in the same plane, it is called a planar mechanism.
5. **Spatial Mechanism :** When the links of the mechanism lie in different planes, it is called a spatial mechanism.

We shall restrict our study to simple and compound planar mechanisms.

1.8.2 Equivalent Mechanisms

Turning pairs of planar mechanisms can be replaced by other type of pairs having one or two degrees of freedom, such as sliding pairs, spring or cam pairs. Such a mechanism is called an

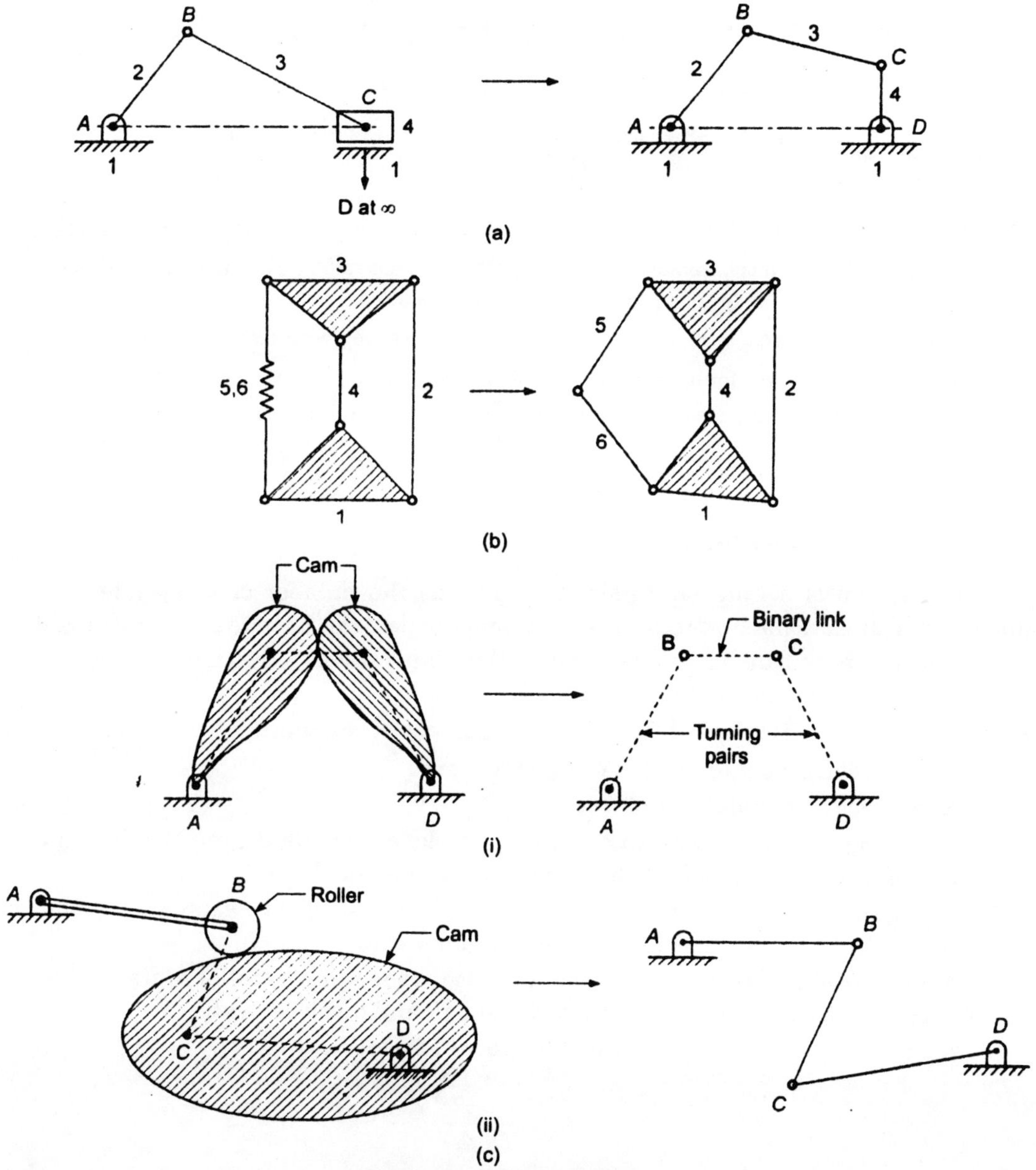

Fig. 1.12 Equivalent mechanisms.

equivalent mechanism. The equivalent mechanism will have same degrees of freedom and shall be kinematically similar.

The following rules may be used to obtain the equivalent mechanism :

1. A sliding pair is equivalent to a turning pair, as shown in Fig. 1.12(a).
2. A spring can be replaced by two binary links, as shown in Fig. 1.12(b).
3. (i) A cam pair can be replaced by one binary link with two turning pairs at each end, as shown in Fig. 1.12(c)-(i).
   (ii) A cam and roller follower can be replaced by a four-bar mechanism as shown in Fig. 1.12(c)-(ii).

1.9 DIFFERENCES BETWEEN MACHINE AND MECHANISM

The differences between a machine and mechanism are given in Table 1.2.

Table 1.2 : Differences between machine and mechanism.

Machine	*Mechanism*
1. It transforms energy into useful work.	It transfers motion between various links.
2. It is the assemblage of mechanisms.	It is the assemblage of links in which one of the links is fixed.
3. It is concerned with transfer of energy.	It is concerned with transfer of motion.
4. It requires external power for operation. *Examples* : Lathe, milling machine, centrifugal pump, etc.	It also requires external power for operation. *Examples* : Engine indicator, steering mechanism, sewing machine mechanism, etc.

1.10 DEGREES OF FREEDOM

Degrees of freedom of a kinematic pair is defined as the number of independent relative motions, both translational and rotational, a kinematic pair can have. An unconstrained rigid body moving in space can have three translations and three rotational motions (*i.e.*, six motions) about the three mutually perpendicular axes.

$$\text{Degrees of freedom} = 6 - \text{number of restraints} \quad \text{...(1.2)}$$

The degrees of freedom of some of the systems are :

- A rigid body has 6 degrees of freedom.
- A rectangular bar sliding in a rectangular hole has one degree of freedom as the motion can be expressed by the linear displacement only.
- The position of the crank of a slider crank mechanism can be expressed by the angle turned through and thus has one degree of freedom.
- A circular shaft rotating in a hole and also translating parallel to its axis has two degrees of freedom, *i.e.*, angle turned through and displacement.
- A ball and a socket joint has three degrees of freedom.

1.10.1 Degrees of Freedom of Planar Mechanisms

Mobility of a Mechanism : The mobility (or masability) of a mechanism is defined as the number of degrees of freedom it possesses. An equivalent definition of mobility is the minimum number of independent parameters required to specify the location of every link within a mechanism.

Gruebler's Criterion :

Let,

n = total number of links in a mechanism

F = degrees of freedom

p = number of lower pairs or simple joints having one degree of freedom

h = number of higher pairs having two degrees of freedom, and so on.

In a mechanism, one link is fixed. Therefore,

$$\text{Number of movable links} = n - 1$$

In a planar mechanism, displacement is possible along two axes (one restraint) and rotation about only one axis (two restraints). Thus, there are three general restraints.

The following relation has been proposed by Gruebler to find the degrees of freedom of a planar mechanism :

$$F = 3(n-1) - 2p - h \qquad \text{...(1.3)}$$

This is known as the *Gruebler's criterion* (or Kutzbach criterion) for degrees of freedom of planar mechanisms.

To determine the degrees of freedom of a mechanism, the presence of a redundant link or redundant pair may also be considered.

(i) A mechanism may have one or more links which do not introduce any extra constraint. Such links are called redundant links (n_r) and should not be taken into account. Similarly redundant joints (p_r) should also not be taken into account.

In Fig. 1.13(a), links 3 and 4 are parallel and are termed as redundant links, as none of them produces extra constraint. By removing one of the two links, the motion remains the same. So one of the two links is considered for calculating the degrees of freedom.

The corresponding kinematic pairs either between links 4 and 2, and 4 and 5; or 3 and 2, and 3 and 5 are considered as redundant pair. Therefore, either of the two links and the corresponding kinematic pair should be considered while calculating the degrees of freedom.

In Fig. 1.13(b), links AB and CD are identical and each leads to same constraint.

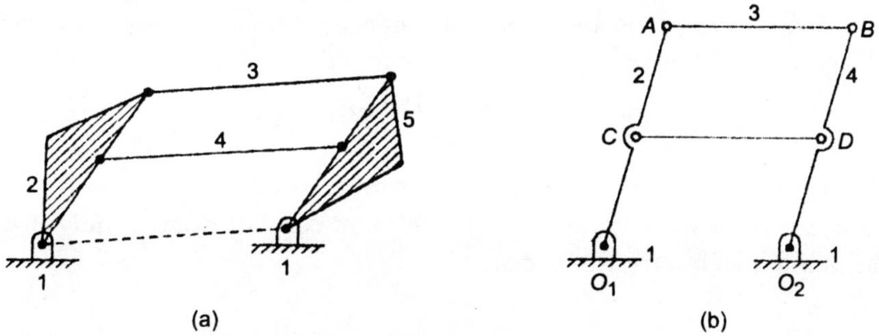

Fig. 1.13

(ii) Sometimes one or more links of a mechanism may have redundant degrees of freedom. If a link can be moved without causing any movement in the rest of the mechanism then the link is said to have redundant degree of freedom (F_r).

In Fig. 1.14(a), the link 3 can slide without causing any movement to the mechanism. The link 3 represents one redundant degree of freedom. In Fig. 1.14(b), roller can rotate without causing any movement in the rest of the mechanism.

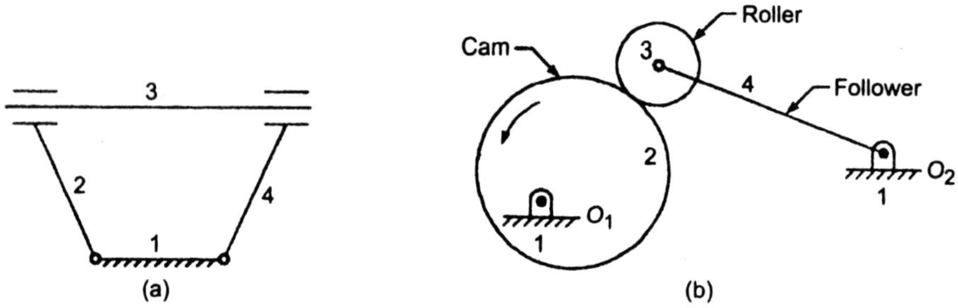

Fig. 1.14

Thus eq. (1.3) can be modified as :

$$F = 3(n - n_r - 1) - 2(p - p_r) - h - F_r \quad \text{...(1.4)}$$

where n_r = redundant number of links

p_r = redundant number of pairs

F_r = redundant degrees of freedom

1.10.2 Planar Mechanism with Lower Pairs Only

For linkages with single degree of freedom only, $h = 0$, and

$$F = 3(n - 1) - 2p \quad \text{...(1.5)}$$

$$2p = 2n_2 + 3n_3 + 4n_4 + \ldots + in_i \quad \text{...(1.5a)}$$

where n_2 = number of binary links

n_3 = number of ternary links

n_4 = number of quaternary links, and so on.

If $F \geq 1$, the device is a mechanism with F degrees of freedom

$F = 0$, the device is a statically determinate structure

$F < -1$, the device is a statically indeterminate structure.

A slider can only move in the specified direction and therefore has only one degree of freedom. Hence, the slider joint also subtracts two degrees of freedom, one translation and one rotation. Hence p in eq. (1.4) also includes slider joints.

Most of the linkages are expected to have one degree of freedom only so that with one input to any of the links, a constrained motion of the other links is obtained. Then

$$1 = 3(n - 1) - 2p$$

or

$$2p = 3n - 4 \quad \text{...(1.5b)}$$

Eq. (1.5b) is similar to eq. (1.5b), where $J = p$ and $L = n$.

As p and n are to be whole numbers, the relation can be satisfied only if n is even. For possible linkages made of binary links only,

$n = 4,\ p = 4$	No excess turning pair
$n = 6,\ p = 7$	One excess turning pair
$n = 8,\ p = 10$	Two excess turning pairs

and so on.

Thus, we find that the number of excess turning pairs increase as the number of links increase. To get the required number of turning pairs from the same number of binary links is not possible. Therefore, the additional pairs or joints can be obtained only from the links having more than two joining points, *i.e.*, ternary or quaternary links, etc.

1.11 FOUR-BAR CHAIN

A four-bar chain has been shown in Fig. 1.15. It consists of four binary links. Link *AD* is fixed (called frame), *AB* is the crank (or driver link), *BC* is the coupler (or connecting rod), and *CD* the lever (or rocker or follower link). θ is the input angle and ϕ the output angle. The coupler *BC* may be a ternary link. The number of degrees of freedom of the four-bar chain is one. The angle between the coupler and output link is called the transmission angle μ.

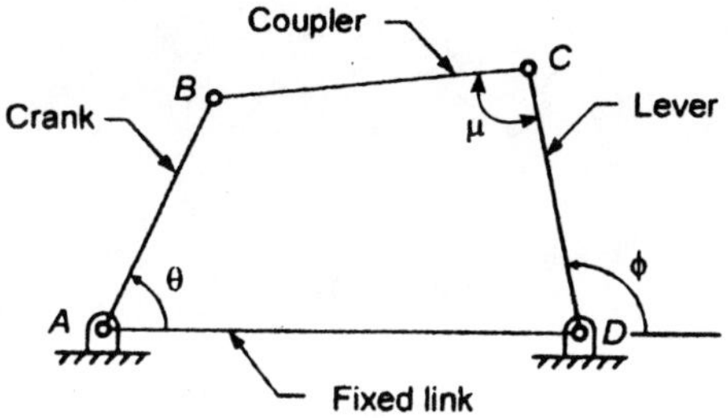

Fig. 1.15 Four-bar chain.

■ **Example 1.6** Calculate the number of degrees of freedom of the linkages shown in Fig. 1.16(a) and (b).

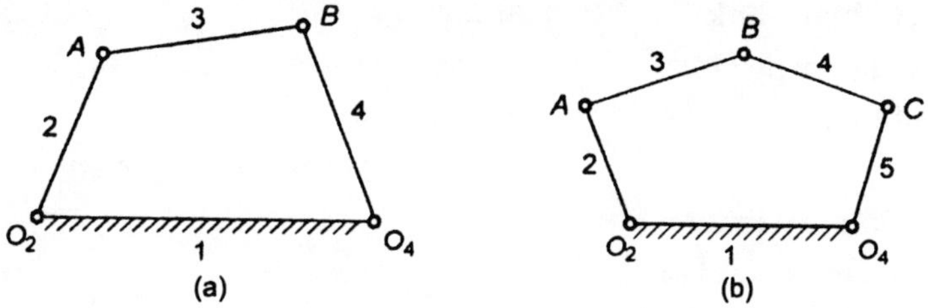

Fig. 1.16

Solution :

(a) Number of binary links, $n = 4$

Number of lower pairs, $p = 4$

Degrees of freedom, $F = 3(n - 1) - 2p$

$$= 3(4 - 1) - 2 \times 4 = 9 - 8 = 1$$

(b) Number of binary links, $n = 5$

Number of lower pairs, $p = 5$

Degrees of freedom, $F = 3(n - 1) - 2p$

$= 3(5 - 1) - 2 \times 5 = 12 - 10 = 2$

■ **Example 1.7** Calculate the number of degrees of freedom of the linkages shown in Fig. 1.17(a) to (c).

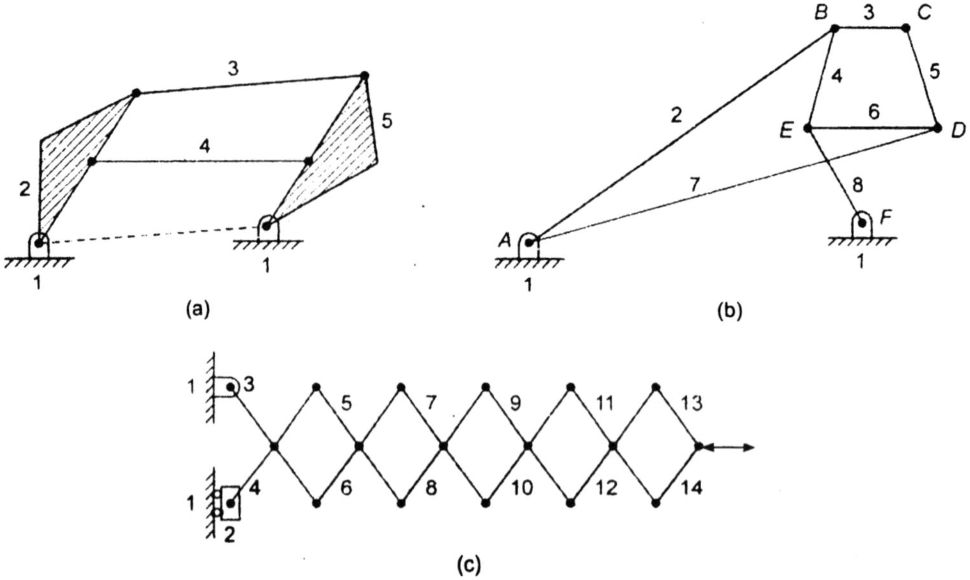

Fig. 1.17

Solution :

(a) In Fig. 1.17(a), links 3 and 4 are parallel and are termed as redundant links, as one of them produces extra constraint. By removing one of these two links, the motion remains same. So one of the two links is considered for calculating the degrees of freedom.

Number of binary links, $n = 4$

Number of lower pairs, $p = 4$

Degrees of freedom $F = 3(n - 1) - 2p$

$= 3(4 - 1) - 2 \times 4 = 9 - 8 = 1$

(b) Number of binary links $n = 8$

Number of simple joints :

Binary joints at C and $F = 2$

Ternary joints at A, B, D and $E = 4$

$p = 2 + 2 \times 4 = 10$

Degrees of freedom, $F = 3(n - 1) - 2p$

$= 3(8 - 1) - 2 \times 10 = 21 - 20 = 1$

(c) Number of binary links, $n = 14$

Number of lower pairs, $p = 18$

Number of higher pairs, $h = 1$ (Slider can rotate and slide)

Degrees of freedom, $F = 3(n - 1) - 2p - h$

$= 3(14 - 1) - 2 \times 18 - 1$

$= 39 - 36 - 1 = 2$

■ **Example 1.8** Determine the number of degrees of freedom of the mechanism shown in Fig. 1.18(a) to (f).

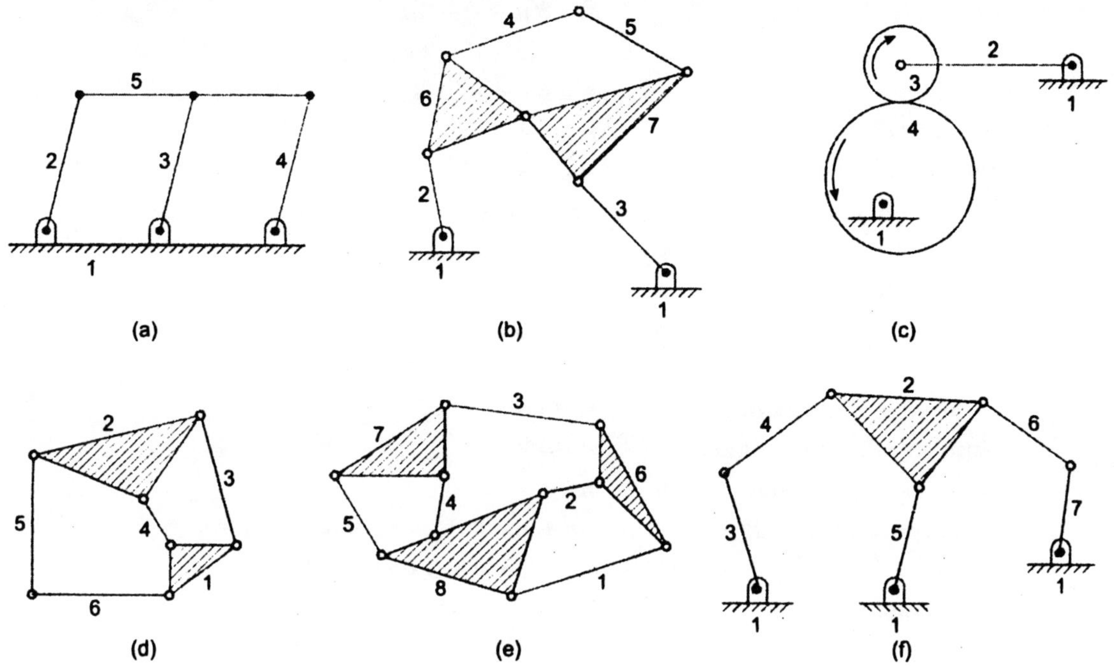

Fig. 1.18

Solution :

(a) Number of binary links, $n_2 = 3$

Number of ternary links, $n_3 = 2$

Total number of links, $n = n_2 + n_3 = 3 + 2 = 5$

$$2p = 2n_2 + 3n_3 = 2 \times 3 + 3 \times 2 = 12$$

Degrees of freedom, $F = 3(n - 1) - 2p$

$= 3(5 - 1) - 12$

$= 12 - 12 = 0$

It is a structure.

(b) Number of binary links, $n_2 = 5$

Number of ternary links, $n_3 = 2$

Total number of links, $n = n_2 + n_3 = 5 + 2 = 7$

$$2p = 2n_2 + 3n_3 = 2 \times 5 + 3 \times 2 = 16$$

Degrees of freedom, $F = 3(n - 1) - 2p$

$= 3(7 - 1) - 16$

$= 18 - 16 = 2$

(c) The roller 3 carried at the end of the output link 2 can be rotated without causing any motion in the rest of the mechanism. Thus, roller 3 is a link with redundant degree of freedom. The roller can thus be considered welded to the output link. Hence, there are three binary links 1, 2 and 4 together with two turning pairs 12, 14 and one higher pair 34. Thus

$$n = 3,\ p = 2,\ h = 1$$

Degrees of freedom,
$$F = 3(n-1) - 2p - h$$
$$= 3(3-1) - 2 \times 2 - 1$$
$$= 6 - 4 - 1 = 1$$

(d) Number of binary links, $n_2 = 4$

Number of ternary links, $n_3 = 2$

Total number of links, $n = n_2 + n_3 = 4 + 2 = 6$

$$2p = 2n_2 + 3n_3 = 2 \times 4 + 3 \times 2 = 14$$

Degrees of freedom,
$$F = 3(n-1) - 2p$$
$$= 3(6-1) - 14$$
$$= 15 - 14 = 1$$

(e) Number of binary links, $n_2 = 5$

Number of ternary links, $n_3 = 2$

Number of quaternary links, $n_4 = 1$

Total number of links, $n = n_2 + n_3 + n_4 = 5 + 2 + 1 = 8$

$$2p = 2n_2 + 3n_3 + 4n_4 = 2 \times 5 + 3 \times 2 + 4 \times 1 = 20$$

Degrees of freedom,
$$F = 3(n-1) - 2p$$
$$= 3(8-1) - 20$$
$$= 21 - 20 = 1$$

(f) Number of binary links, $n_2 = 5$

Number of ternary links, $n_3 = 2$

Total number of links, $n = n_2 + n_3 = 5 + 2 = 7$

$$2p = 2n_2 + 3n_3 = 2 \times 5 + 3 \times 2 = 16$$

Degrees of freedom,
$$F = 3(n-1) - 2p$$
$$= 3(7-1) - 16$$
$$= 18 - 16 = 2$$

■ **Example 1.9** Determine the number of degrees of freedom of the mechanism shown in Fig. 1.19(a) to (f).

Solution :

(a)
$$n_2 = 3,\ n_3 = 2,\ n = 5$$
$$2p = 2 \times 3 + 3 \times 2 = 12$$
$$F = 3(5-1) - 12 = 12 - 12 = 0$$

(b)
$$n_2 = 4,\ n_3 = 2,\ n = 6$$
$$2p = 2 \times 4 + 3 \times 2 = 14$$
$$F = 3(6-1) - 14 = 15 - 14 = 1$$

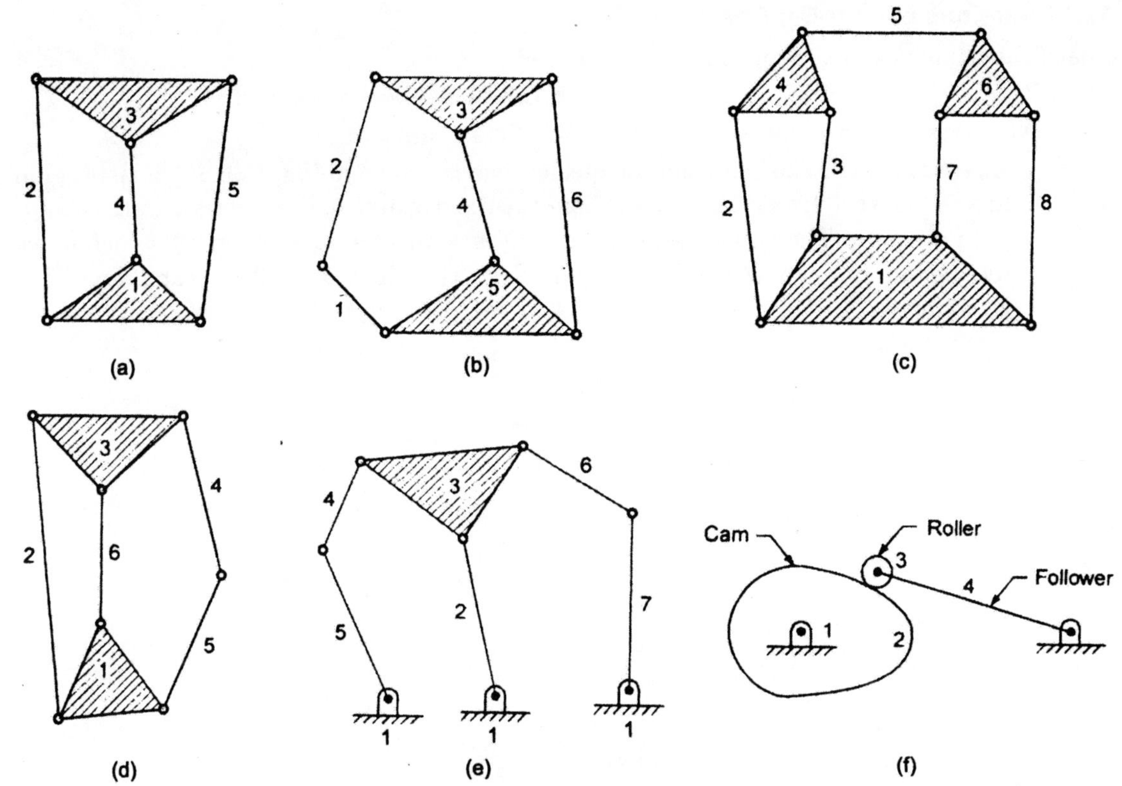

Fig. 1.19

(c)

$$n_2 = 5, n_3 = 2, n_4 = 1, n = 8$$
$$2p = 2 \times 5 + 3 \times 2 + 4 + 1 = 20$$
$$F = 3(8 - 1) - 20 = 21 - 20 = 1$$

(d)

$$n_2 = 4, n_3 = 2, n = 6$$
$$2p = 2 \times 4 + 3 \times 2 = 14$$
$$F = 3(6 - 1) - 14 = 15 - 14 = 1$$

(e)

$$n_2 = 5, n_3 = 2, n = 5 + 2 = 7$$
$$2p = 2n_2 + 3n_3 = 2 \times 5 + 3 \times 2 = 16$$
$$F = 3(n - 1) - 2p$$
$$= 3(7 - 1) - 16 = 18 - 16 = 2$$

(f) The solution has been given in Example 1.8(c).

1.12 INVERSION OF MECHANISMS

A kinematic chain becomes a mechanism when one of its links is fixed. Therefore, as many number of mechanisms can be obtained as many are the links in the kinematic chain. This method of obtaining different mechanisms by fixing different links of a kinematic chain is called inversion of the mechanism. The relative motion between the various links is not altered as a result of inversion, but their absolute motion with respect to the fixed link may alter drastically.

1.12.1 Inversions of Four-Bar Chain

Some of the important inversions of a four-bar chain are :

1. Beam engine
2. Coupled wheel of locomotive
3. Watt's indicator mechanism
4. Slider-crank chain.

1. **Beam Engine :** The beam engine mechanism is shown in Fig. 1.20. It consists of four links. When the crank *AB* rotates about the fixed centre *A*, the lever oscillates about a fixed centre *D*. The end *E* of the lever *CDE* is connected to a piston rod which moves the piston up and down in the cylinder. This is also called crank and lever mechanism.

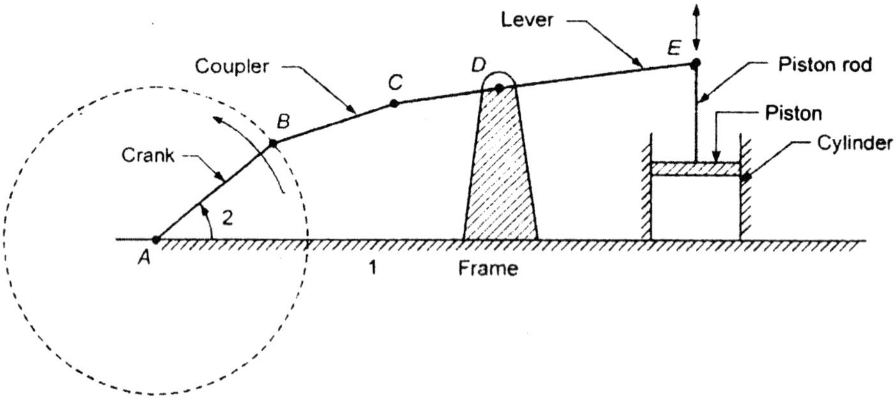

Fig. 1.20 Beam engine.

2. **Coupled Wheel of a Locomotive :** In this mechanism, as shown in Fig. 1.21, the links *AB* and *CD* are of equal lengths and act as cranks. These cranks are connected to the respective wheels. The link *BC* acts as the connecting rod. The link *AD* is fixed

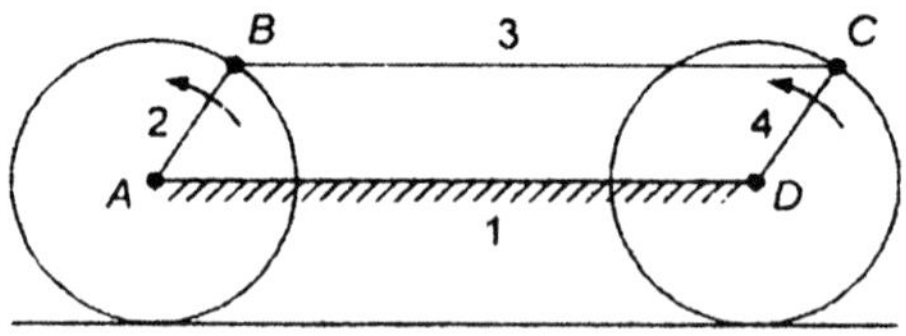

Fig. 1.21 Coupled wheel of a locomotive.

to maintain constant distance between the wheels. This mechanism is used to transmit rotary motion from one wheel to the other. This is also called double crank mechanism.

3. **Watt's Indicator Mechanism :** This mechanism is shown in Fig. 1.22. It consists of four links : fixed link at *A*, link *AC*, link *CE* and link *BFD*. The link *CE* and *BFD* act as levers. The displacement of link *BFD* is directly proportional to the pressure in the indicator cylinder. The point *E* on link *CE* traces out an approximate straight line. It is also called double lever mechanism.

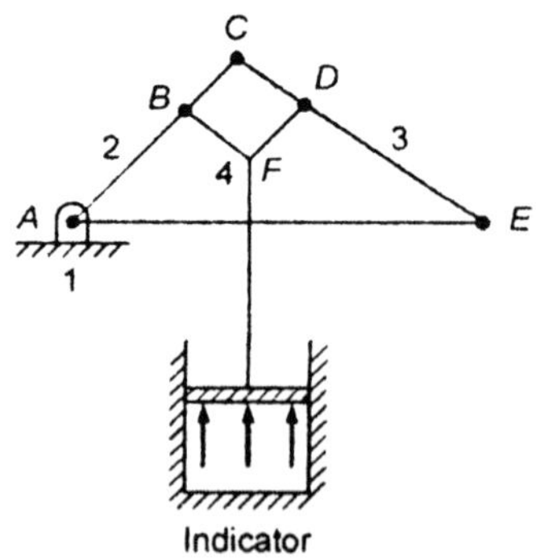

Fig. 1.22 Watt's indicator mechanism.

4. **Single Slider-Crank Chain :** The single slider crank chain shown in Fig. 1.23 consists of three turning pairs

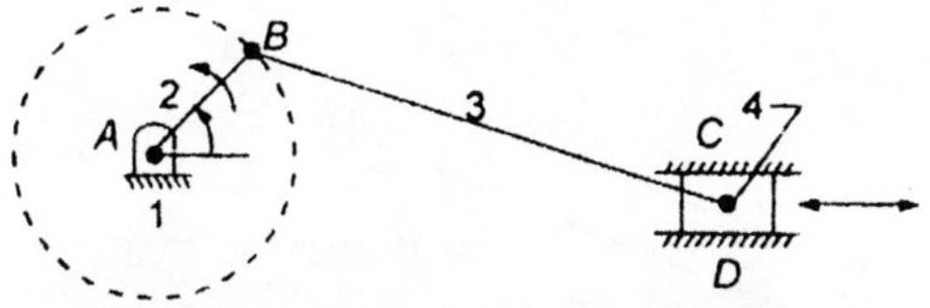

Fig. 1.23 Single-slider-crank mechanism.

and one sliding pair. The link 1 corresponds to the frame of the mechanism, which is fixed. Link 2 is the crank and link 3 the connecting rod. The link 4 is the slider. It is used to convert rotary motion into reciprocating motion and vice-versa. Its important applications are in steam engines, internal combustion engines, reciprocating compressors, etc. If the straight line path of the slider is offset from the fixed point of the crank then it is called offset slider-crank chain. The offset is called the eccentricity.

1.12.2 Inversions of Single-Slider Crank Chain

The inversions of single slider crank chain are :

1. Pendulum pump
2. Oscillating cylinder engine
3. Rotary internal combustion engine
4. Crank and slotted lever quick-return motion mechanism
5. Whitworth quick-return motion mechanism.

1. **Pendulum Pump :** This inversion mechanism is obtained by fixing the link 4, *i.e.,* the sliding pair, as shown in Fig. 1.24. When the link 2 (*i.e.,* crank) rotates, the link 3 (*i.e.,* connecting rod) oscillates about a pin pivoted to the fixed link 4 at *C* and the piston attached to the piston rod (link 1) reciprocates in the cylinder. It is used to supply feed water to a boiler.

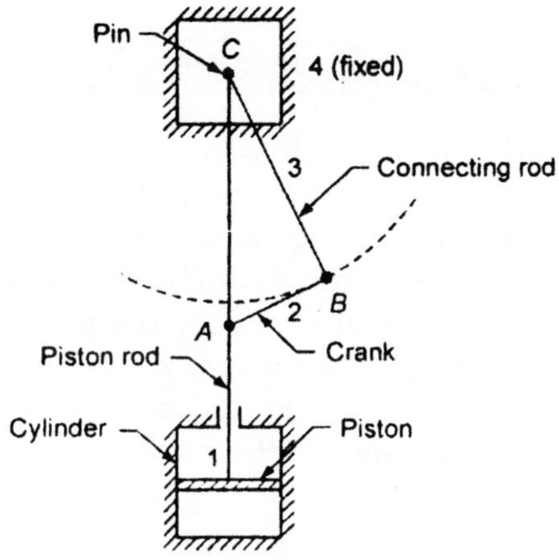

Fig. 1.24 Pendulum pump.

2. **Oscillating Cylinder Engine :** In this mechanism, as shown in Fig. 1.25, link 3 is fixed. When the crank (link 2) rotates, the piston attached to piston rod (link 1)

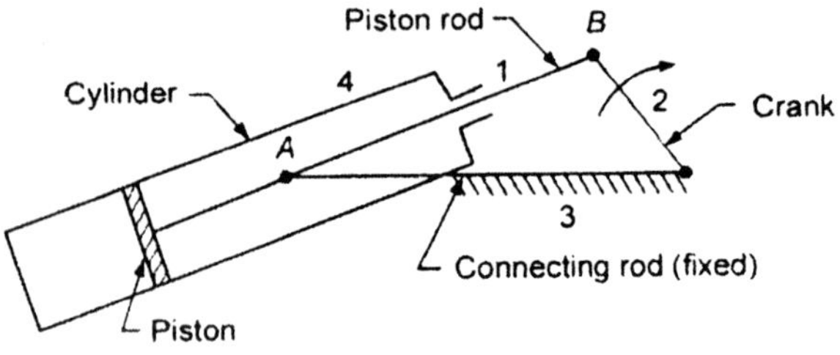

Fig. 1.25 Oscillating cylinder engine.

reciprocates and the cylinder (link 4) oscillates about a pin pivoted to the fixed link at A.

3. **Rotary Internal Combustion Engine (Gnome Engine) :** It consists of several cylinders in one plane and all revolve about fixed centre O, as shown in Fig. 1.26. The crank (link 2) is fixed. When the connecting rod (link 4) rotates, the piston (link 3) reciprocates inside the cylinder forming link 1.

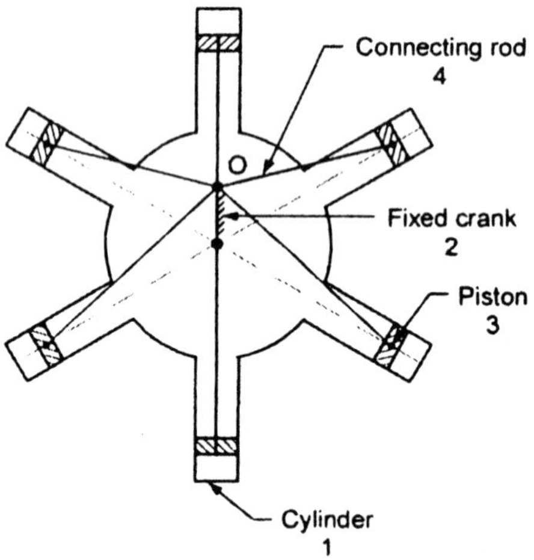

Fig. 1.26 Rotary internal combustion engine.

4. **Crank and Slotted Lever Quick-Return Motion Mechanism :** In this mechanism, as shown in Fig. 1.27, the link AC (link 3) corresponding to the connecting rod is fixed. The driving crank CB (link 2) revolves about centre C. A slider (link 1) attached to the crank pin at B slides along the slotted lever AP (link 4) and make the slotted lever oscillate about the pivoted point A. A short link PQ transmits the motion from AP to the arm which reciprocates with the tool along the line of stroke. The line of stroke is perpendicular to AC produced. This mechanism is mostly used in shaping machines, slotting machine, and rotary internal combustion engines.

Time of cutting stroke/Time of return stroke $= \dfrac{\alpha}{\beta}$ ...(1.6)

Length of stroke $= 2AP\ (CB/AC)$ ...(1.7)

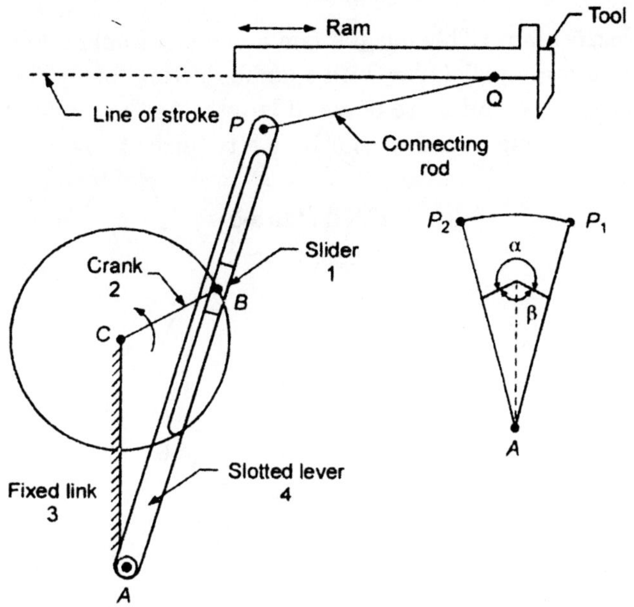

Fig. 1.27 Crank and slotted lever quick-return motion mechanism.

5. **Whitworth Quick-Return Motion Mechanism :** In this mechanism, as shown in Fig. 1.28, the link *CD* (link 2) is fixed. The driving crank *CA* (link 3) rotates about *C*. The slider (link 4) attached to the crank pin at *A* slides along the slotted lever *PA* (link 1) which oscillates about pivot *D*. The connecting rod *PQ* carries the ram at *Q* with cutting tool. The ram reciprocates along the line of stroke. It is used in shaping and slotting machines.

$$\text{Time of cutting stroke/Time of return stroke} = \frac{\alpha}{\beta} \quad \text{...(1.8)}$$

$$\text{Length of stroke} = 2\ PD \quad \text{...(1.9)}$$

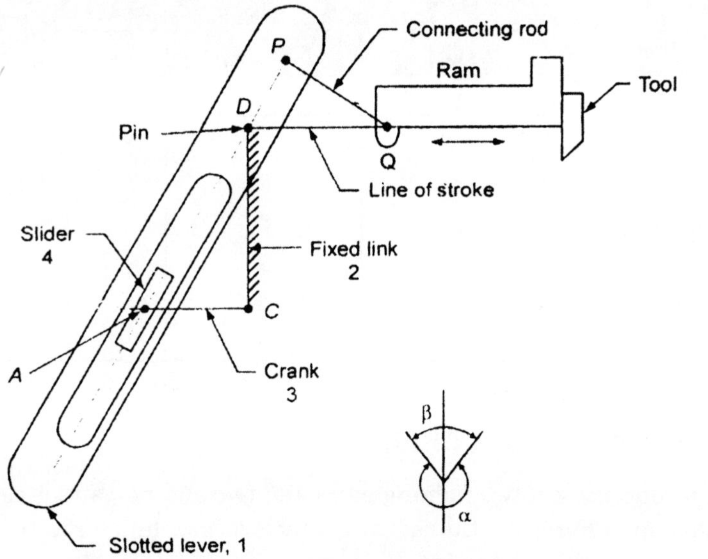

Fig. 1.28 Whitworth quick-return motion mechanism.

6. **Toggle Mechanism :** This mechanism has many applications where it is necessary to overcome a large resistance with a small driving force. Fig. 1.29 shows the toggle mechanism; links 4 and 5 are of equal length. As the angles α decrease and links 4 and 5 approach being collinear, the force F required to overcome a given resistance P decreases as :

$$F = 2\,P \tan \alpha \qquad ...(1.10)$$

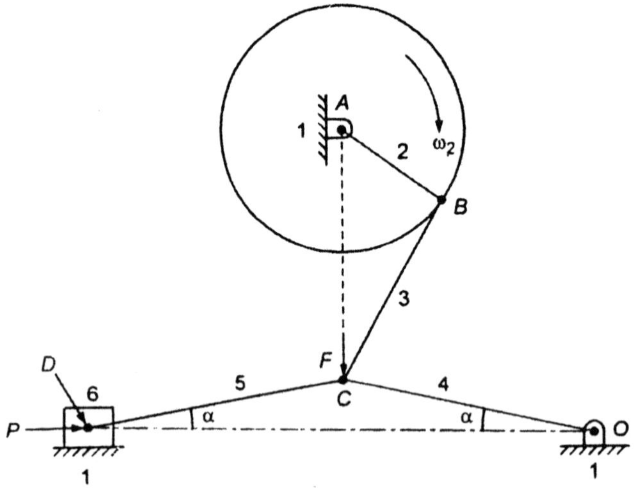

Fig. 1.29 Toggle mechanism.

If α approaches zero, for a given F, P approaches infinity. A stone crusher utilizes this mechanism to overcome a large resistance with a small force. It can be used in numerous toggle clamping devices for holding work pieces.

The summary of single slider crank chain and its inversions is given in Table 1.3.

Table 1.3 : Summary of Single Slider-Crank Chain and its Inversions.

Mechanism	*Links*			
	Fixed	*Rotates*	*Oscillates*	*Reciprocates*
Single slider crank chain	1	2	3	4
Inversions :				
Pendulum pump	4	2	3	1
Oscillating cylinder engine	3	2	4	1
Crank and slotted lever	3	2	4	1
Whitworth mechanism	2	3	1	4
Gnome engine	2	3	1	4

1.13 DOUBLE SLIDER-CRANK CHAIN

A kinematic chain consisting of two turning pairs and two sliding pairs is called double slider – crank chain, as shown in Fig. 1.30. Links 3 and 4 reciprocate, link 2 rotates and link 1 is fixed. Two pairs of the same kind are adjacent.

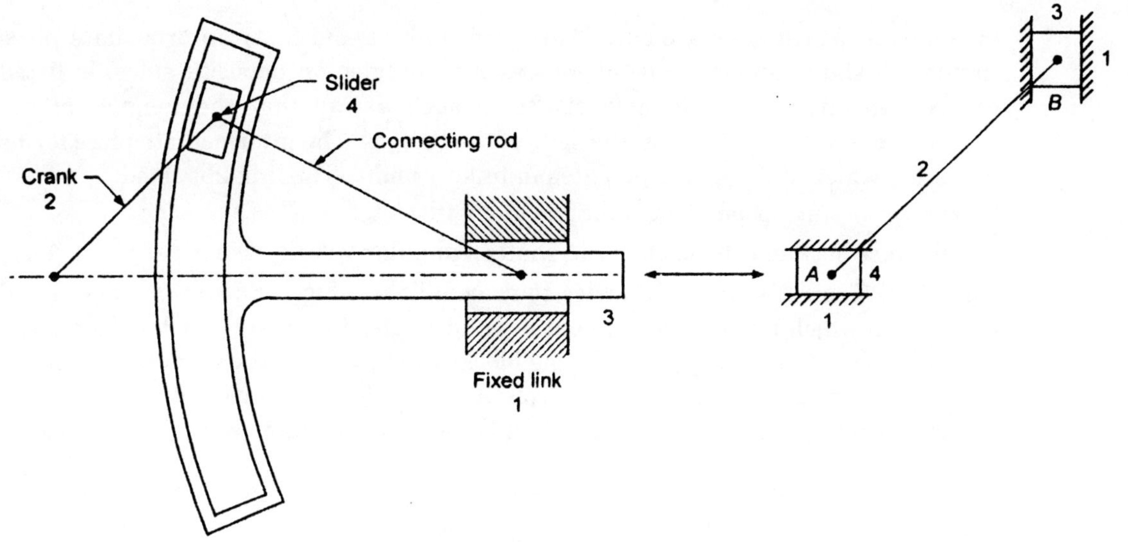

(a) (b)

Fig. 1.30 Double slider-crank chain.

1.13.1 Inversions of Double Slider–Crank Chain

The inversions of double slider–crank chain are :

1. Donkey pump
2. Oldham's coupling
3. Elliptical trammel
4. Scotch yoke.

1. **Donkey Pump :** Figure 1.31 shows a donkey pump, in which link 2 (crank) rotates about point A. One end of the crank is connected to the piston, through the piston rod, which reciprocates vertically in the pump cylinder. This cylinder together with the body of the pump represents the fixed link 1. The other end of the crank is connected to the slider (link 3) which reciprocates horizontally in the cylinders.

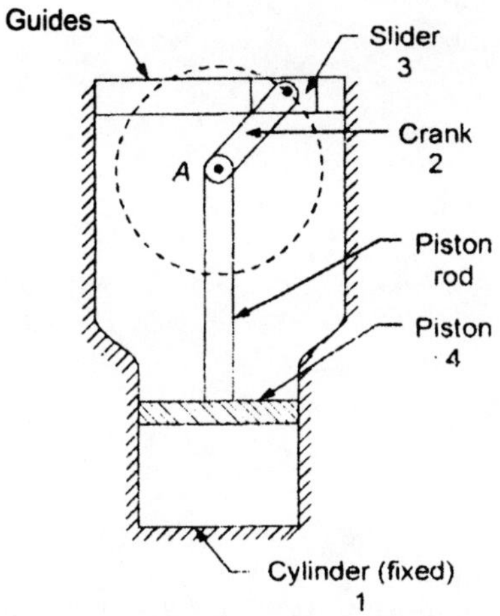

Fig. 1.31 Donkey pump.

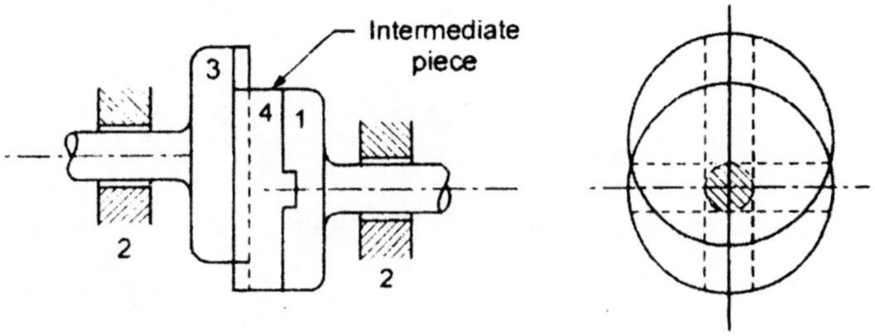

Fig. 1.32 Oldham's coupling.

2. **Oldham's Coupling :** The Oldham's coupling shown in Fig. 1.32, is used to connect two parallel shafts, the distance between whose axes is small and variable. The shafts connected by the coupling rotate at the same speed. The shafts have flanges at

the ends, in which slots are cut. These form links 1 and 3. An intermediate piece circular in shape and having tongues at right angles on opposite sides, is fitted between the flanges of the two shafts in such a way that the tongues of the intermediate piece get fitted in the slots of the flanges. The intermediate piece forms the link 4 which slides or reciprocates in links 1 and 3. The link 2 is fixed.

Maximum sliding speed of each tongue along its slot

= Distance between the axes of the shafts × angular velocity of each shat ...(1.11)

3. **Elliptical Trammel :** It is a device to draw ellipses. Fig. 1.33 shows an elliptical trammel in which two grooves are cut at right angles in a plate which is fixed. The plate forms the fixed link 4. Two sliding blocks are fitted into the grooves. The slides form two sliding links 1 and 3. The link joining slides form the link 2. Any point onthe link 2 or on its extension traces out an ellipse on the fixed plate, when relative motion occurs.

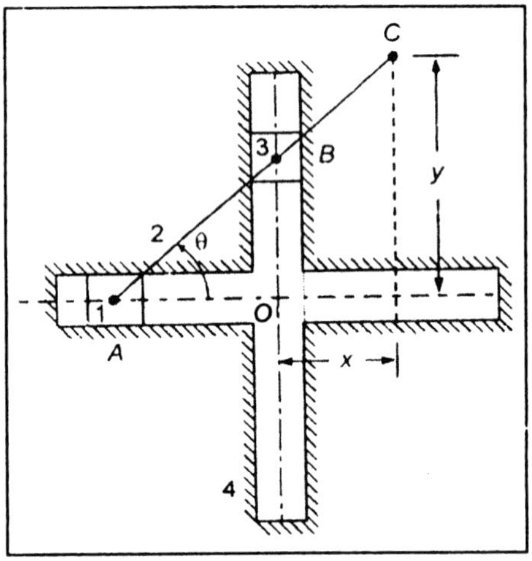

Fig. 1.33 Elliptical trammel.

$$x = BC \cos \theta$$

or $$\frac{x}{BC} = \cos \theta$$

$$y = AC \sin \theta$$

or $$\frac{y}{AC} = \sin \theta$$

Squaring and adding, we get

$$\frac{x^2}{BC^2} + \frac{y^2}{AC^2} = 1 \quad \text{...(1.12)}$$

which is the equation of an ellipse.

4. **Scotch Yoke :** This mechanism gives simple harmonic motion. Its early application was on steam pumps, but it is now used as a mechanism on a test machine to produce vibrations. It is also used as a sine-cosine generator for computing elements. Fig. 1.34 shows a sketch of scoth yoke mechanism.

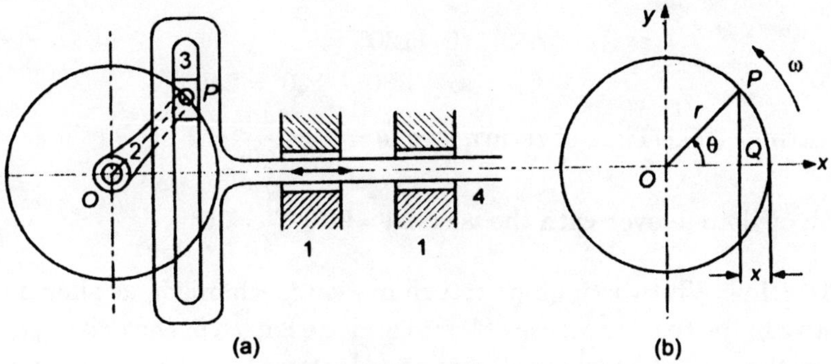

Fig. 1.34 Scotch yoke.

$$x = r - r\cos\theta$$

$$= r(1 - \cos\omega t) \quad \text{...(1.13)}$$

$$v = \frac{dx}{dt} = r\omega\sin\omega t \quad \text{...(1.14)}$$

$$a = \frac{d^2x}{dt^2} = r\,\omega^2\cos\omega t \quad \text{...(1.15)}$$

■ **Example 1.10** **In a crank and slotted lever quick-return mechanism shown in Fig. 1.35, the distance between the fixed centres is 300 mm and the length of the driving crank is 150 mm. Find the inclination of the slotted lever with the vertical in the extreme position and the ratio of time of cutting stroke to return stroke.**

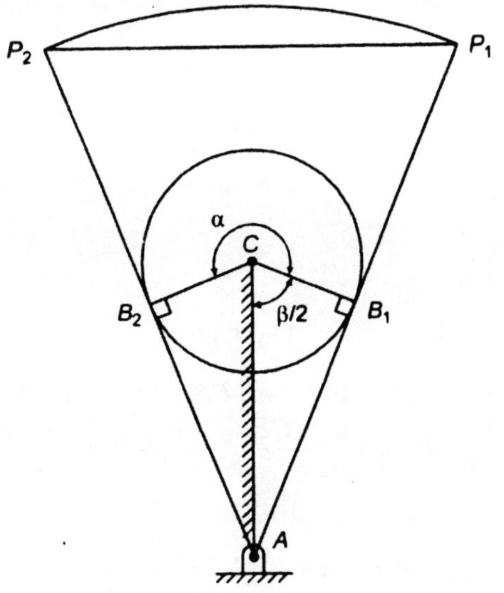

Fig. 1.35

Solution : Given : $AC = 300$ mm, $B_1C = 150$ mm

$$\cos\beta/2 = B_1C/AC$$

$$= \frac{150}{300} = 0.50$$

$$\beta = 120°$$
$$\alpha = 360° - 120° = 240°$$
$$\text{Time of cutting stroke/Time of return stroke} = \frac{\alpha}{\beta} = \frac{240°}{120°} = 2$$
$$\text{Inclination of slotted lever with the vertical} = 90° - \frac{\beta}{2} = 30°$$

■ **Example 1.11** In a Whitworth quick return motion mechanism, as shown in Fig. 1.36, the distance between the fixed centres is 80 mm and the length of the driving crank is 100 mm. The length of the slotted lever is 180 mm and the length of the connecting rod is 150 mm. Calculate the ratio of the time of cutting to return strokes.

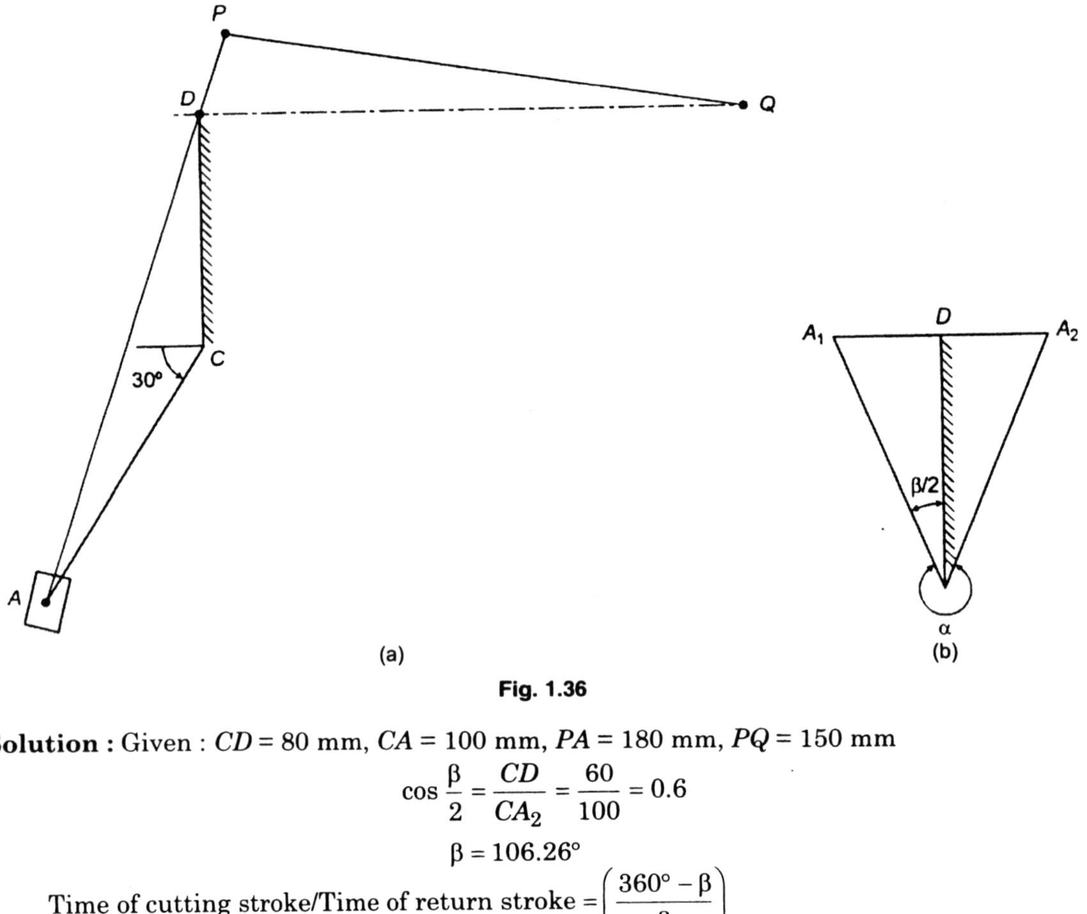

Fig. 1.36

Solution : Given : $CD = 80$ mm, $CA = 100$ mm, $PA = 180$ mm, $PQ = 150$ mm

$$\cos\frac{\beta}{2} = \frac{CD}{CA_2} = \frac{60}{100} = 0.6$$
$$\beta = 106.26°$$
$$\text{Time of cutting stroke/Time of return stroke} = \left(\frac{360° - \beta}{\beta}\right)$$
$$= \frac{253.74°}{106.26°} = 2.388$$

■ **Example 1.12** The distance between two parallel shafts connected by Oldham's coupling is 25 mm. The driving shaft revolves at 240 rpm. Determine the maximum speed of sliding of the tongue of the intermediate piece along its groove.

Solution : $d = 25$ mm, $n = 240$ rpm

$$\omega = 2\pi \times \frac{240}{60} = 25.133 \text{ rad/s}$$

Maximum velocity of sliding $= \omega \times d = 25.133 \times 0.025 = 0.628$ m/s

■ **Example 1.13** In a crank and slotted lever mechanism, the length of crank is 560 mm and the ratio of time of working stroke to return stroke is 2.8. Determine (a) distance between the fixed centres, and (b) the length of the slotted lever, if length of stroke is 250 mm.

Solution : Given : $AB = 560$ mm, stroke = 250 mm, ratio of times = 2.8

The mechanism is shown in Fig. 1.37.

$$\frac{\text{Time of working stroke}}{\text{Time of return stroke}} = 2.8$$

$$\frac{360° - \beta}{\beta} = 2.8$$

$$\beta = 94.737°$$

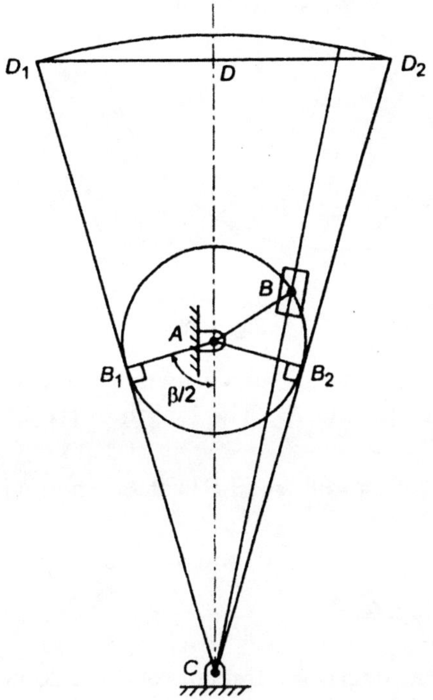

Fig. 1.37

Now
$$\frac{AB_1}{AC} = \cos\left(\frac{\beta}{2}\right)$$

$$AC = \frac{560}{\cos 47.368°} = 826.83 \text{ mm}$$

Length of stroke $= D_1D_2 = 2D_1D$

$$250 = 2CD_1 \sin\left(90° - \frac{\beta}{2}\right)$$

$$= 2CD \sin 47.368°$$

Length of slotted lever, CD_1 = 169.9 mm.

■ **Example 1.14** The configuration of a drag link mechanism is shown in Fig. 1.38. Determine the time ratio and the length of stroke. The crank O_2A rotates clockwise.

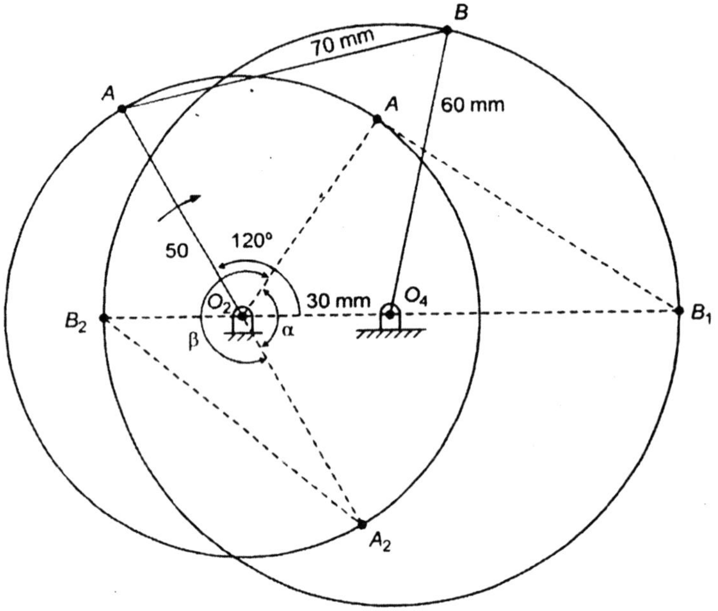

Fig. 1.38

Solution : The drag link mechanism has been drawn as O_2ABO_4 to a scale of 1 cm = 10 mm. The extreme positions of B are B_1 and B_2. The length of stroke is $2 \times O_4B = B_1B_2 = 2 \times 60 =$ 120 mm.

When B is at B_1, then A is at A_1 and when B is at B_2, then A is at A_2. By measurement, we have

$$\alpha = 110° \text{ and } \beta = 250°$$

$$\text{Ratio of times} = \frac{\beta}{\alpha} = \frac{250}{110} = 2.27.$$

■ **Example 1.15** Determine the degrees of freedom of the linkages shown in Fig. 1.39(a) to (e).

Solution :

(a)

$$n_2 = 4, n_3 = 2, n = 4 + 2 = 6$$

$$2p = 2n_2 + 3n_3 = 2 \times 4 + 3 \times 2 = 14$$

$$h = 0 \text{ for pure rolling of roller 3.}$$

$$F = 3(n-1) - 2p - h$$

$$= 3(6-1) - 14 - 0$$

$$= 15 - 14 = 1$$

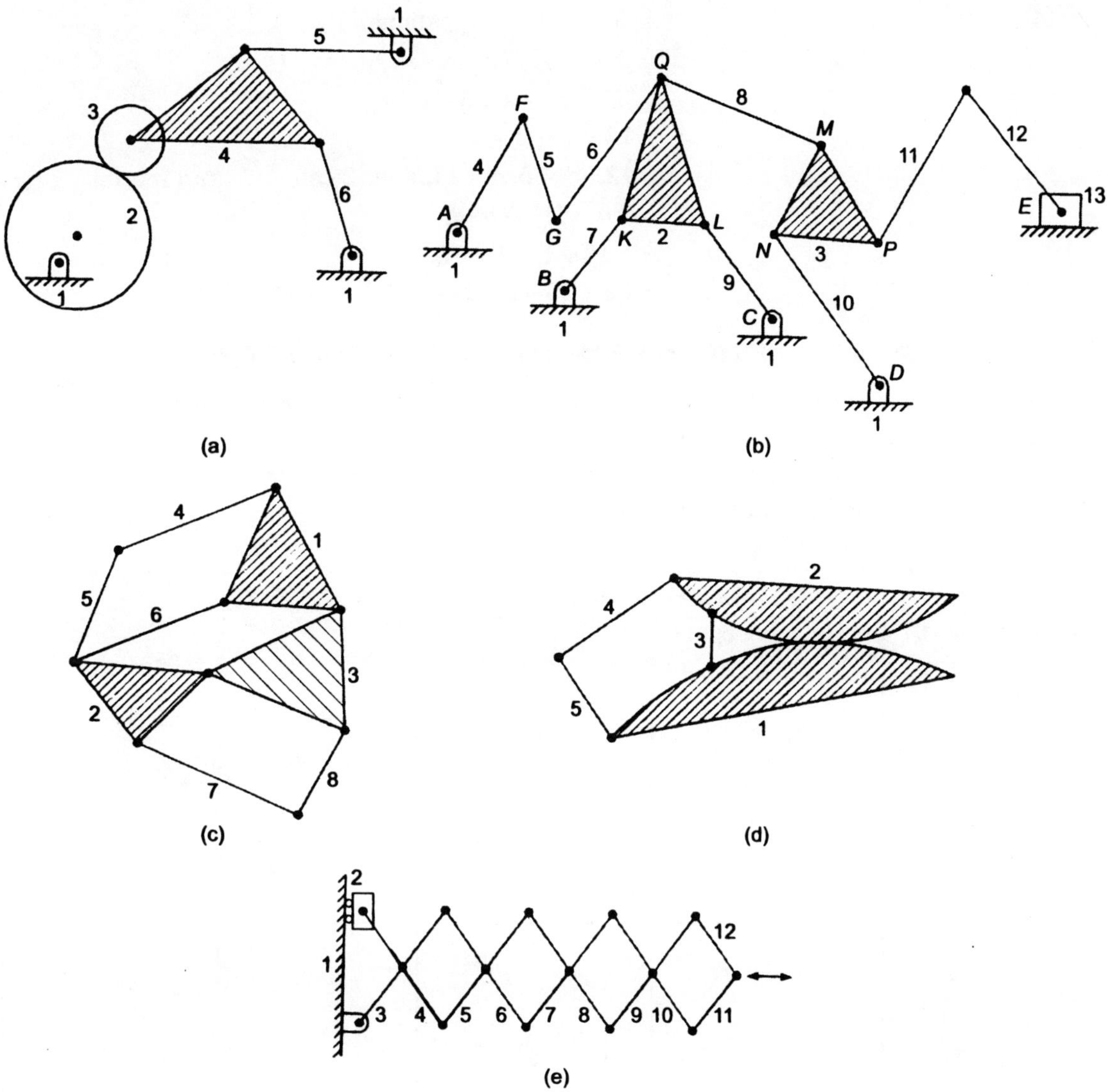

Fig. 1.39

(b) $n = 13$

Binary pairs : $A, B, C, D, E, F, G, H, K, L, M, N, P$, two at Q and one slider.

$p = 16$

$$F = 3(n-1) - 2p$$
$$= 3(13-1) - 2 \times 16$$
$$= 36 - 32 = 4$$

(c)

$$n_2 = 5, n_3 = 3, n = 5 + 3 = 8$$
$$2p = 2 \times 5 + 3 \times 3 = 19$$
$$F = 3(n-1) - 2p$$
$$= 3(8-1) - 19$$
$$= 21 - 19 = 2$$

(d)

$$n = 5,\ p = 5,\ h = 1 \text{ (cam pair)}$$
$$F = 3(n-1) - 2p - h$$
$$= 3(5-1) - 2 \times 5 - 1$$
$$= 12 - 10 - 1 = 1$$

(e)

$$n = 12,\ p = 15,\ h = 1 \text{ (slider 2 can slide and rotate)}$$
$$F = 3(n-1) - 2p - h$$
$$= 3(12-1) - 2 \times 15 - 1$$
$$= 33 - 30 - 1 = 2$$

SOLVED QUESTIONS FROM UNIVERSITY EXAMINATIONS

■ **Example 1.16** Name and sketch a kinematic pair that has the following motions :

(i) Two translations and one rotation

(ii) Three rotations and one translation

(iii) Three rotations and two translations. [KU, Dec., 2006]

Solution :

Name of Kinematic Pair	*Number of Motions*		*Sketch*
	Translatory	*Rotational*	
(i) Prism-plane	2	1	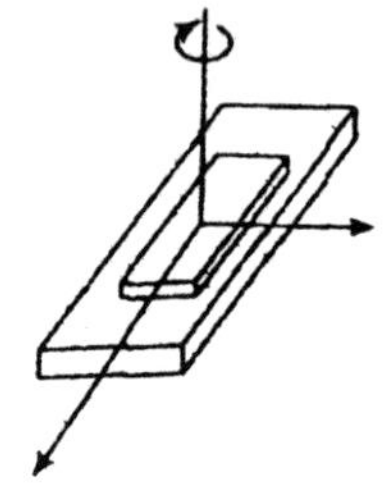
(ii) Sphere-cylinder	1	3	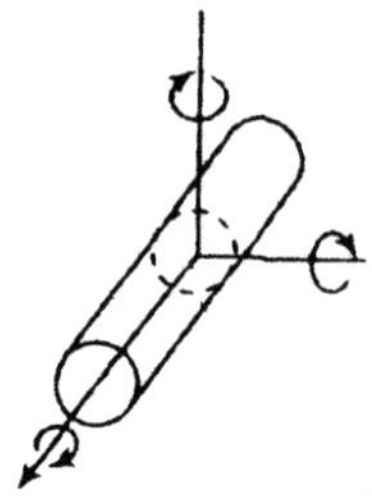
(iii) Sphere-plane	2	3	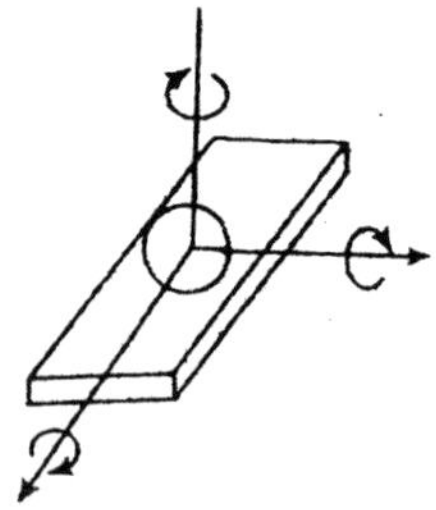

■ **Example 1.17** All possible inversions of a mechanism are mechanisms/are not necessarily mechanisms. Explain. [KU, Dec., 2006]

Solution : When one of the links of a kinematic chain is fixed, it is called a mechanism. An inversion of a mechanism is obtained by fixing various links in turn. Therefore, in a mechanism having *n* links, *n* different types of mechanisms can be obtained. But the relative motion between the various links is not altered in the different inversions. Therefore, it can be concluded that all possible inversions of a mechanism are mechanisms.

■ **Example 1.18** What is meant by equivalent linkages? Explain clearly. [KU, Dec., 2006]

Solution : Refer to art. 1.8.2.

■ **Example 1.19** What is the difference between quick return motion of crank and slotted lever type and that of Whitworth type? [KU. Dec., 2006, 2008]

Solution : Refer to Fig. 1.27 and Fig. 1.28 for crank and slotted lever and Whitworth quick-return motion mechanisms. The differences between them are given in the following table.

Crank and slotted lever mechanism	*Whitworth mechanism*
1. Crank *BC* is smaller than the fixed link *AC*.	Crank *AC* is larger than the fixed link *CD*.
2. The slotted lever moves from one extreme position AP_1 to other extreme position AP_2 during one rotation of crank.	The slotted lever takes complete rotation about pin *D* as crank rotates through one rotation about *C*.
3. The stroke length is adjusted by adjusting the crank length *BC*.	The stroke length is adjusted by adjusting the distance of *P* from *D*.
4. The ratio of cutting time to return time changes with change in stroke length as the crank length changes.	The ratio of cutting time to return time remains constant for different stroke lengths.
5. It is used in those cutting machines which require large stroke length like shapers.	It is used in those cutting machines which require small stroke length like slotters.

■ **Example 1.20** Explain the difference between Mechanism and Machine. [KU, Dec., 2007, 2008]

Solution : Refer to art. 1.9.

■ **Example 1.21** What are different types of kinematic pairs? Give examples. [KU, Dec., 2007, 2009]

Solution : Refer to art. 1.5.1.

■ **Example 1.22** Explain any two inversions of a Double-slider crank chain. [KU, Dec., 2007]

Solution : Refer to art. 1.13.1.

■ **Example 1.23** Determine the degrees of freedom of a minidrafter. [KU, Dec., 2006, 2007]

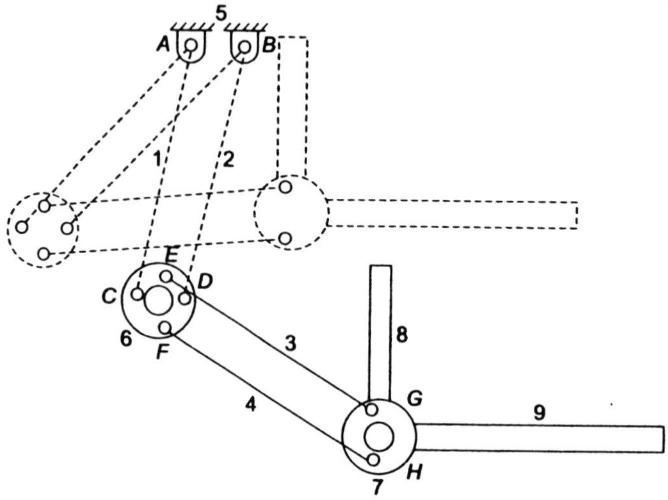

Fig. 1.40 Universal drafting machine (minidrafter).

Solution :

(i) When scales G and H are unclamped. Number of binary links, n :

$$5, 1, 3, 6, 7, 8 = 6$$

Only one link is considered out of 1 and 2, 3 and 4, 8 and 9 as the other link does not produce any constraint and is considered as redundant link.

Number of simple joints, p :

$$A, C, E, K \text{ two sliders} = 6$$

$$F = 3(n - 1) - 2p = 3(6 - 1) - 2 \times 6 = 3$$

(ii) When G and H are clamped.

$$n = 5, \ p = 5$$

$$F = 3(5 - 1) - 2 \times 5 = 2$$

■ **Example 1.24** What do you understand by Constrained motion? What are different types of constrained motions? Explain each type with examples. [KU, Dec., 2009]

Solution : Refer to art. 1.6.

■ **Example 1.25** Explain Whitworth quick return mechanism. [KU, Dec., 2009]

Solution : Refer to art. 1.12.2 part 5.

■ **Example 1.26** What do you understand by inversion of a chain? Describe the mechanisms obtained by inversion of four bar chain. [KU, Dec., 2009]

Solution : Refer to art. 1.12 and art. 1.12.1.

■ **Example 1.27** Briefly discuss the following :

(a) Kinematic links (b) Kinematic pairs

(c) Pantograph (d) One of the straight line mechanisms (Give sketches also)

[MDU, Dec., 2007]

Solution :

(a) Refer to art. 1.4 (b) Refer to art. 1.5

(c) Refer to art 5.2 (d) Refer to art. 5.3.1.

■ **Example 1.28** What do you mean by inversion of a mechanism? Explain with sketches all the inversions of a single slider-crank mechanism. [MDU, May, 2008]

Solution : Refer to art. 1.12 and 1.12.2.

■ **Example 1.29** Differentiate : (i) Lower pair and higher pair (ii) Turning pair and sliding pair (iii) Screw pair and spherical pair (iv) Closed pair and unclosed pair. [MDU, May, 2008]

Solution : Refer to art. 1.5.

■ **Example 1.30** What are binary, ternary and quaternary links? [MDU, Dec., 2008]

Solution : Refer to art. 1.4.

■ **Example 1.31** Explain various inversions of double slider crank chain. [MDU, Dec., 2008, May, 2009]

Solution : Refer to art. 1.13.1.

■ **Example 1.32** Define and explain the terms : Binary joints, ternary joints, quaternary joints, binary links, ternary links, and quarternary links. [MDU, May, 2009]

Solution : Refer to art. 1.3. and 1.4.

■ **Example 1.33** With the help of neat sketch, explain the inversion of slider-crank mechanism which is used in shaper machine. [MDU, Dec., 2009]

Solution : Refer to art. 1.12.2 (4).

■ **Example 1.34** (a) What is meant by equivalent mechanisms?

(b) With the help of neat-sketch explain inversion of double-slider crank mechanism which is used in Oldham's Coupling. [MDU, Dec., 2010]

Solution : (a) Refer to art. 1.8.2. (b) Refer to art. 1.13.1.

SUMMARY

1. Conditions for a chain to be a kinematic chain having lower pairs only are :

$$L = 2P - 4$$

$$J = \left(\frac{3}{3}\right) - 2$$

where L = number of binary links

P = number of lower pairs

J = number of binary joints

If LHS > RHS, then the chain is called a **locked chain or redundant chain**

LHS = RHS, then the chain is **constrained**

LHS < RHS, then the chain is **unconstrained**

2. Gruebler's criterion for degrees of freedom :

$$F = 3(n - 1) - 2p - h$$

where n = number of links

p = number of lower pairs or simple joints having single degree of freedom

h = number of higher pairs

For linkages with single degree of freedom only, $h = 0$, and

$$F = 3(n - 1) - 2p$$

where $2p = 2n_2 + 3n_3 + 4n_4 + ... + in_i$

n_2 = number of binary links

n_3 = number of ternary links

n_4 = number of quaternary links, and so on

$$F = 3(n - n_r - 1) - 2(p - p_r) - F_r$$

where n_r = number of redundant links

p_r = number of redundant simple joints

F_r = redundant degrees of freedom.

If $F \geq 1$, the device is a mechanism with F degrees of freedom

$F = 0$, the device is a statically determinate structure

$F < -1$, the device is a statically indeterminate structure.

3. Inversions of a four-bar chain are :

   Beam engine, coupled wheel of *a* locomotive, Watt's indicator mechanism, single slider crank chain.

4. Inversions of a single slider-crank chain are :

   Pendulum pump, oscillating cylinder engine, rotary internal combustion engine (Gnome engine), crank and slotted lever quick-return motion mechanism, Whitworth quick-return motion mechanism and toggle mechanism.

5. Inversions of double slider-crank chain are :

   Donkey pump, Oldham's coupling, elliptical trammel, and Scotch yoke,

6. In a quick-return motion mechanism,

   Time of cutting stroke/Time of return stroke = angle covered by crank during cutting stroke/angle covered by crank during return stroke.

7. Maximum sliding speed of each tongue along its slot in the Oldham's coupling = distance between the axes of shafts × angular speed of each shaft.

SHORT ANSWER TYPE QUESTIONS

1. Differentiate between a mechanism and a machine.
2. Define kinematic link, kinematic pair and a kinematic chain.
3. What is a resistant body?
4. How do you classify kinematic pairs? Illustrate with examples.
5. Differentiate between (a) Lower and higher pairs,(b) turning and screw pairs,(c) rolling and spherical pairs, and (d) closed and unclosed pairs.
6. Define degrees of freedom of a mechanism. How this is determined?
7. Explain Gruebler's criterion for degrees of freedom for planar mechanisms.
8. What do you understand by an equivalent mechanism?

9. Discuss various types of constrained motions.
10. What is mobility (or movability) of a mechanism?
11. What is a kinematic chain? How this can be ascertained?
12. What do you understand by inversion of a mechanism? List various inversions of a four-bar chain.
13. Describe various inversions of a single slider-crank chain.
14. List the various inversions of a double slider-crank chain and give their applications.

EXERCISES

1. (a) How many types of links you know? Explain with examples.
   (b) Explain at least two cyonstrained motion with suitable examples.
   (c) Sketch and explain any two inversions of a double slider-crank chain.
   (d) What is meant by (i) a resistant body (ii) a lower kinematic pair (iii) a kinematic chain, and (iv) a mechanism.
   (e) Discuss various inversions of double slider crank chain with neat sketches.
2. (a) Explain the difference between machine and structure with examples.
   (b) Explain at least two types of joints which are usually found in chain.
   (c) Sketch and describe the four-bar chain mechanism. Why it is considered to be the basic chain?
   (d) State Gruebler's criterion for ascertaining the degree of freedom of planar mechanisms having turning pairs only.
3. What do you understand by degrees of freedom? Determine the degrees of freedom of the linkages shown in Fig. 1.41.

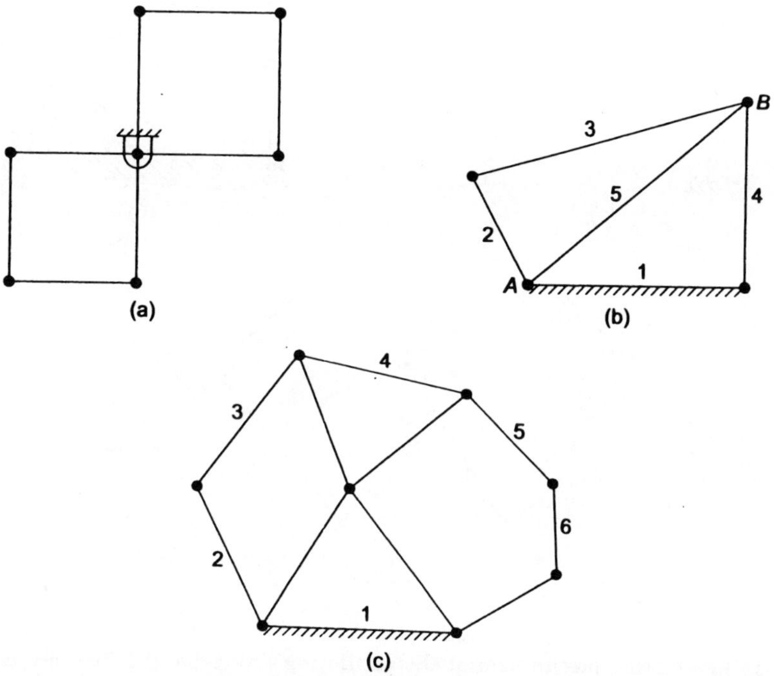

Fig. 1.41

4. Determine the number of degrees of freedom of all the devices shown in Fig. 1.42.

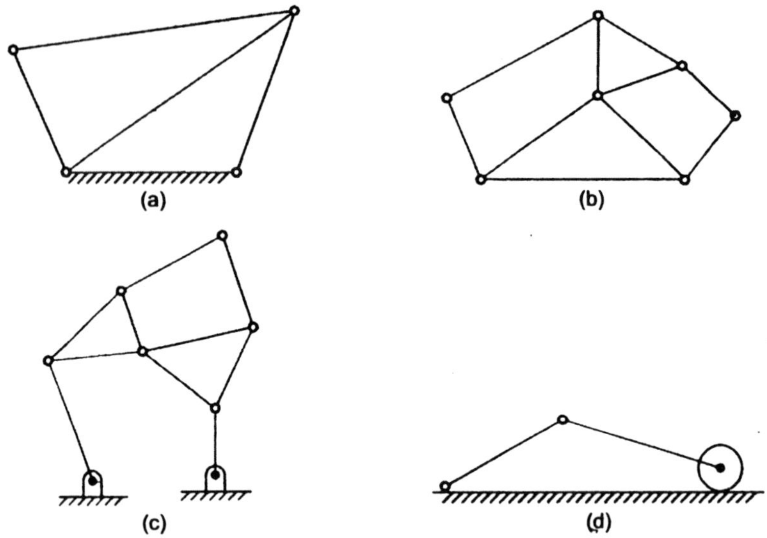

Fig. 1.42

[**Ans.** 0, 1, 2, 2]

5. Draw the equivalent mechanisms of the following and find their degrees of freedom, as shown in Fig. 1.43.

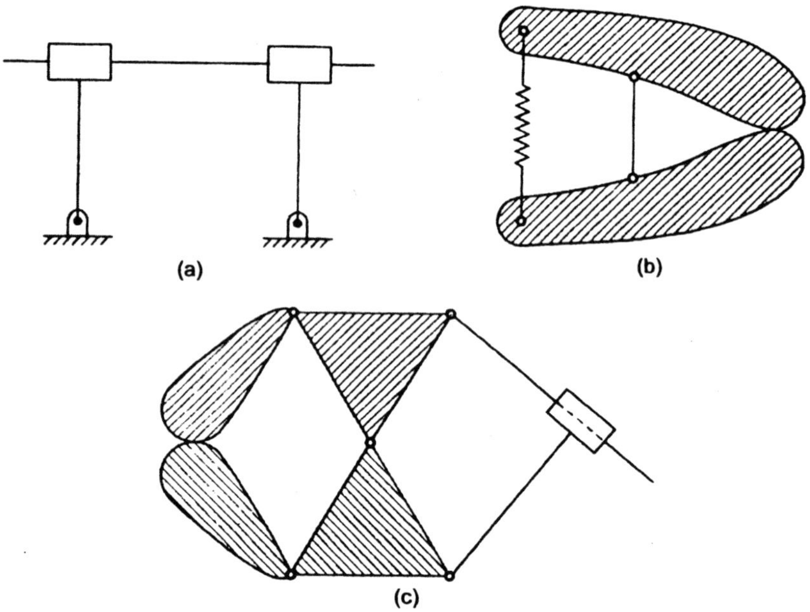

Fig. 1.43

[**Ans.** 1, 1, 2]

6. In a quick-return motion mechanism of the oscillating link type, the distance between the fixed centres is 80 mm and the length of the driving crank is 20 mm. Determine the time ratio of the working stroke to the return stroke. [**Ans.** 1.38]

7. In an off-set slider crank mechanism, the eccentricity is 50 mm, length of crank is 300 mm and length of connecting rod is 500 mm determine the quick return ratio. [**Ans.** 1.29]

8. In a quick-return motion mechanism of the crank and slotted lever type, the ratio of maximum velocities is 2. If the length of stroke is 250 mm, find (a) the length of the slotted lever, (b) the ratio of times of cutting and return strokes, and (c) the maximum cutting velocity in m/s if the crank rotates at 30 rpm. [**Ans.** 375 mm, 1.552, 0.294 m/s]

9. The distance between the parallel shafts of an Oldham's coupling is 15 mm. The driving shaft revolves at 160 rpm. Calculate the maximum speed of sliding of the tongue of the intermediate piece along its groove. [**Ans.** 0.251 m/s]

❑❑❑

Chapter 2

Velocity in Mechanisms

Inside This Chapter

2.1 INTRODUCTION

Kinematics is the study to determine the displacement, velocity and acceleration of the various links of a mechanism. The study of velocity analysis involves the determination of the linear velocities of the various points on the links and the angular velocity of the links. This is important from the point of view that it is used to determine the linear acceleration of the various points on the links and the angular acceleration of the links. This is subsequently used to calculate the forces and torques acting on the links. In this chapter, we shall study the graphical and analytical methods to determine velocity. The various terms related to this study are :

Displacement : The displacement of a body is its change of position with reference to a certain fixed point.

Velocity : It is the rate of change of displacement of a body with respect to time. It is a vector quantity.

Linear Velocity : It is the rate of change of displacement of a body along a straight line with respect to time. Its SI units are m/s.

Angular Velocity : It is the rate of the change of angular position of a body with respect to time. Its units are rad/s.

The relationship between linear velocity v and angular velocity ω is :

$$v = r\omega \qquad \ldots(2.1)$$

where r = distance of the point undergoing displacement from the centre of rotation.

Configuration Diagram : It is the line diagram of a mechanism, drawn on some suitable scale, from which the velocity and acceleration diagrams are drawn.

Velocity Diagram : It is a line diagram of the velocities of the various links of a mechanism drawn on some suitable scale.

2.2 RELATIVE VELOCITY METHOD

The relative velocity of a body A with respect to a body b, v_{ab} is obtained by adding to the velocity of A the reversed velocity of B.

Let v_a = velocity of body A

v_b = velocity of body B.

If $v_a > v_b$, then relative velocity of A w.r.t. B is,

$$\vec{v}_{ab} = \vec{v}_a - \vec{v}_b$$

or

$$\overrightarrow{ba} = \overrightarrow{oa} - \overrightarrow{ob}$$

where $ba = v_{ab}$, $oa = v_a$, $ob = v_b$ to some scale.

Similarly relative velocity of B w.r.t. A is,

$$\vec{v}_{ba} = \vec{v}_b - \vec{v}_a$$

$$\overrightarrow{ab} = \overrightarrow{ob} - \overrightarrow{oa}$$

where $ab = v_{ba}$, $oa = v_a$, $ob = v_b$ to some scale.

Consider two points A and B on a rigid link rotating clockwise about point A, as shown in Fig. 2.1 (a). There can be no relative motion between A and B as long as the distance between A

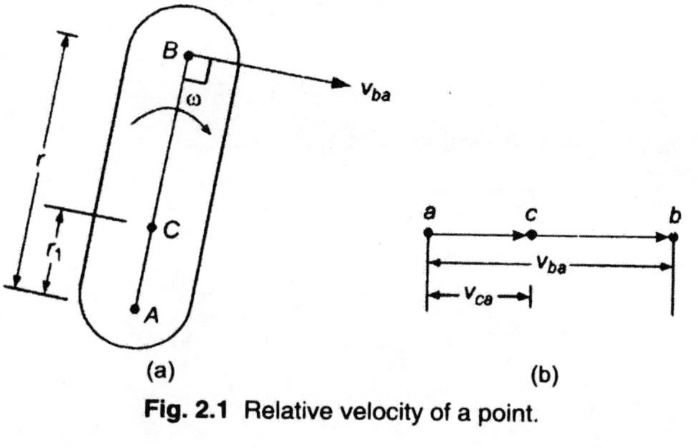

(a) (b)

Fig. 2.1 Relative velocity of a point.

and B with remains the same. Therefore, the relative motion of B with respect to A must be perpendicular to AB. Hence, the direction of relative velocity of two points in a rigid link is always along the perpendicular to the straight line joining these points. Let relative velocity of B with respect to A be represented by $v_{ba} = \omega \cdot AB$, then ab is drawn perpendicular to AB to a convenient scale, as shown in Fig. 2.1 (b). Similarly, the linear velocity of any other point C on AB with respect to A is $v_{ca} = \omega \cdot AC$, and is represented by vector ac. Hence

$$\frac{v_{ba}}{v_{ca}} = \frac{(\omega \cdot AB)}{(\omega \cdot AC)} = \frac{AB}{AC}$$

or $$\frac{ab}{ac} = \frac{AB}{AC} = \frac{r}{r_1} \qquad \text{...(2.2)}$$

Hence the point c divides the vector ab in the same ratio as the point C divides the link AB.

2.3 DETERMINATION OF LINK VELOCITIES

There are two methods to determine the velocities of links of mechanism.

1. Relative velocity method
2. Instantaneous centre method

2.3.1 Relative Velocity Method

Consider a rigid link AB, as shown in Fig. 2.2 (a), such that the velocity of A (v_a) is known in magnitude and direction (vertical) and velocity of B (v_b) is known only in direction (horizontal). To construct the velocity diagram, take a point o. Draw oa representing the magnitude direction of velocity of A. Draw ob along the direction of v_b. From point 'a' draw a line ab perpendicular to AB, meeting ob in b. Then oab is the velocity triangle, as shown in Fig. 2.2 (b). The velocity of point B in magnitude is, $v_b = ob$, and the velocity of B with respect to A, $v_{ba} = ab$. The vector ab is called the **velocity image** of link AB. The velocity of any point C in AB with respect to A is given by,

$$v_{ca} = ac = v_{ba} \cdot \left(\frac{AC}{AB}\right)$$

$$v_c = v_a + \left(\frac{AC}{AB}\right) v_{ba}$$

$$= \overrightarrow{oa} + \left(\frac{ac}{ab}\right) \cdot ab$$

$$= \overrightarrow{oa} + \overrightarrow{ac} = \overrightarrow{oc}$$

Hence, vector oc represents the velocity of point C.

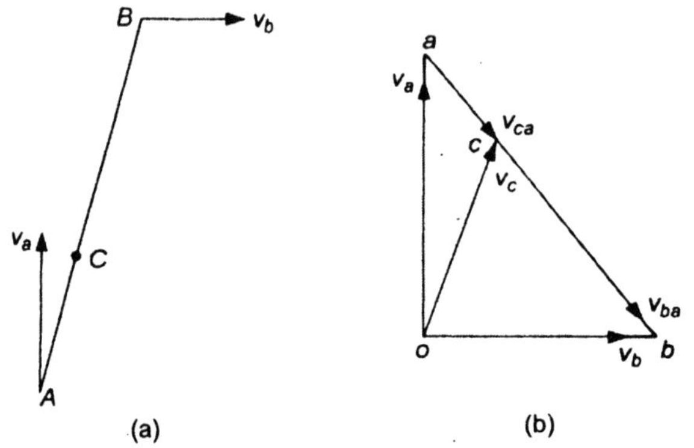

Fig. 2.2 Relative velocity of points in a kinematic link.

2.3.2 Relative Velocity of Points in a Kinematic Link

Consider a link A_1B_1, first undergoing rotation by an amount of $\Delta\theta$ to A', B', and then undergoing translation by an amount Δs_A to occupy the new position A_2B_2, as shown in Fig. 2.3(a). Then

$$\Delta s_B = \Delta s_{BA} + \Delta s_A \quad \text{...(1)}$$

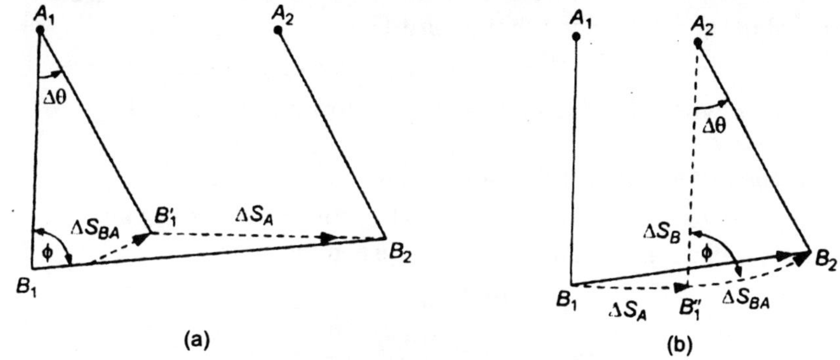

Fig. 2.3 Relative velocity of points in a kinematic link.

Now let the link A_1B_1 first undergo linear translation ΔS_A to A_2B_1'', and then angular rotation of $\Delta\theta$, as shown in Fig. 2.3 (b). Then

$$\Delta s_B = \Delta s_A + \Delta s_{BA} \quad \text{...(2)}$$

Eqs. (1) and (2) are same. Dividing by Δt, we get

$$\Delta s_B / \Delta t = \Delta s_A / \Delta t + \Delta s_{BA} / \Delta t$$

or

$$v_b = v_a + v_{ba}$$

Therefore, the velocity of point B is obtained by adding vectorially the relative velocity of point B w.r.t. point A to the velocity of point A.

Now

$$\Delta s_{BA} = A_1B_1 \,.\, \Delta\theta$$

$$\Delta s_{BA} / \Delta t = A_1B_1 \,.\, \Delta\theta / \Delta t$$

or

$$v_{ba} = A_1B_1 \,.\, w$$

Also

$$\angle\phi = 90°$$

The following conclusions may be drawn from the above analysis :

1. The velocity of any point on the kinematic link is given by the vector sum of the velocity of some other point in the link and the velocity of the first point relative to the other.
2. The magnitude of the velocity of any point on the kinematic link relative to the other point in the kinematic link is the product of the angular velocity of the link and distance between the two points under consideration.
3. The direction of the velocity of any pont on a link relative to any other ponit on the link is perpendicular to the line joining the two points.

2.3.3 Relative Angular Velocities

Consider two links OA and OB connected by a pin joint at O, as shown in Fig. 2.4. Let ω_1 and ω_2 be the angular velocities of the links OA and OB respectively. Relative angular velocity of OA w.r.t. OB is,

$$\omega_{12} = \omega_1 - \omega_2$$

and relative angular velocity of OB w.r.t. OA is,

$$\omega_{21} = \omega_2 - \omega_1$$
$$= -\omega_{12}$$

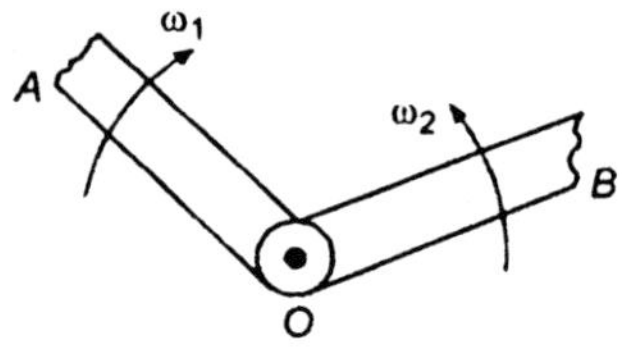

Fig. 2.4 Relative angular velocities.

If r = radius of the pin at joint O, then

Rubbing (or sliding) velocity at the pin joint O

$= (\omega_1 - \omega_2)\,r$, when the links move in the same direction ...(2.3a)

$= (\omega_1 + \omega_2)\,r$, when the links move in the opposite direction ...(2.3b)

2.3.4 Relative Velocity of Points on the Same Link

Consider a ternary link ABC, as shown in Fig. 2.5 (a), such that C is any point on the link. Let v_a and v_b be the linear velocities of points A and B respectively.

Then
$$v_b = v_a + v_{ba}$$
$$\omega_{ab} = \frac{v_{ba}}{AB} = \frac{ab}{AB}$$
$$v_c = v_a + v_{ca}$$
or
$$= v_b + v_{cb}$$

The velocity diagram is shown in Fig. 2.5 (b).

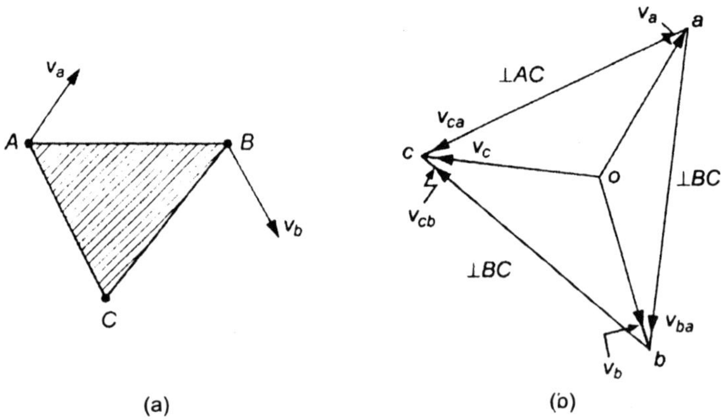

Fig. 2.5 Relative velocity of points on the same link.

Angular velocity of link ABC,

$$\omega_{abc} = \frac{v_{ba}}{AB} = \frac{v_{cb}}{BC} = \frac{v_{ca}}{AC} \quad \text{...(2.4a)}$$
$$= \frac{ab}{AB} = \frac{bc}{BC} = \frac{ac}{AC} \quad \text{...(2.4b)}$$

2.3.5 Forces in a Mechanism

Consider a link AB subjected to the action of forces and velocities as shown in Fig. 2.6. Let A be the driving and B the driven end. When the direction of the forces and velocities is the same, then

Energy at A = Energy at B

$$F_a \times v_a = F_b \times v_b$$

or $$F_b = \frac{F_a v_a}{v_b} \quad \text{...(2.5a)}$$

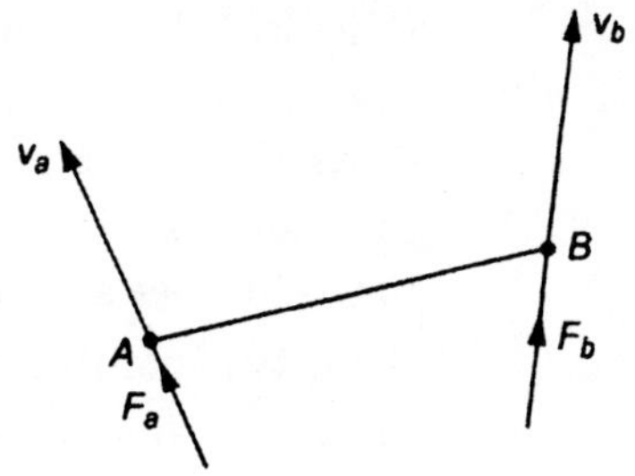

Fig. 2.6 Force and velocity diagram.

Considering the effect of friction, the efficiency of transmission,

$$\eta = \text{output/input}$$
$$= (F_b v_b)/(F_a v_a)$$

or $$F_b = \eta\, F_a v_a / v_b \quad \text{...(2.5b)}$$

When the forces are not in the direction of velocities, then their components along the velocities should be taken.

2.3.6 Mechanical Advantage

Mechanical advantage, MA = Load lifted/Effort applied

$$= \frac{F_b}{F_a}$$

For a mechanism, MA = output torque/Input torque

$$= \frac{T_b}{T_a} = \frac{\omega_a}{\omega_b} \quad \text{...(2.6a)}$$

Considering the effect of friction, $$MA = \eta\, \frac{\omega_a}{\omega_b} \quad \text{...(2.6b)}$$

2.3.7 Four-Bar Mechanism

(a) Simple Four-Bar Mechanism

Consider a simple four-bar mechanism, as shown in Fig. 2.7 (a), in which the crank AB is rotating at angular velocity ω in the anti-clockwise direction. To draw the velocity diagram, adopt the following steps :

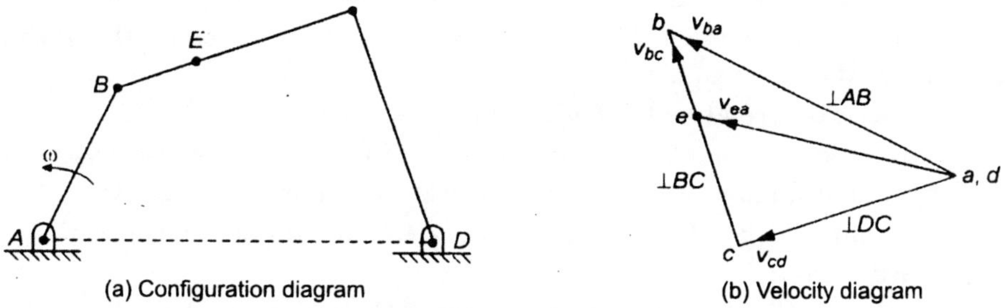

Fig. 2.7 Simple four-bar mechanism.

Procedure :

(i) Velocity Diagram :

1. Calculate the angular velocity of crank AB, $\omega = \frac{2\pi N}{60}$ rad/s

2. Calculate the linear velocity of point B on the crank AB, $v_b = AB \times \omega$
3. The absolute linear velocity of B is perpendicular to AB. Draw $ab \perp AB$ to a convenient scale to represent v_b, as shown in Fig. 2.7 (b).
4. From b draw a line perpendicular to BC and from 'a' draw another line perpendicular to CD to meet each other at point c. Then

   Velocity of point, C, $v_{ca} = ac$

   Velocity of B relative to C, $v_{bc} = cb$
5. Measure cb to obtain the velocity v_{cb}.

(ii) Velocity of intermediate point :

6. To find the velocity of any point E in BC, we have

$$\frac{CE}{CB} = \frac{ae}{cb}$$

or

$$ce = \left(\frac{CE}{CB}\right) . cb$$

Thus locate point e in cb and joint ae. Then

Velocity of E relative to A, $v_{ea} = v_e = ae$

(iii) Angular velocity of links :

7. (a) Velocity of C relative to B, $v_{cb} = bc$

   Point C relative to B moves in the direction – sense given by v_{cb}, *i.e.*, b to c (downwards). Thus C moves in the clockwise direction about B.

$$v_{cb} = \omega_{cb} \times BC = \omega_{cb} \times CB$$

   Angular velocity of link BC, $\omega_{cb} = v_{cb}/CB$ clockwise (cw) ...(1)

   (b) Velocity of B relative to C, $v_{bc} = cb$

   Point B relative to C moves in the direction-sense given by v_{bc}, *i.e.*, c to b (upwards, opposite to bc). Thus B moves in the clockwise direction about C.

$$v_{bc} = \omega_{bc} \times BC = \omega_{bc} \times CB$$

   Angular velocity of link BC, $\omega_{bc} = v_{bc}/BC$ clockwise (cw) ...(2)

   As $CB = BC$ and $v_{cb} = v_{bc}$, therefore, $\omega_{cb} = \omega_{bc}$.

   Thus, we observe that the magnitude of the angular velocities in the same. Further the direction of the angular velocity about one end of the link is the same about the other end.

   (c) Velocity of C relative to D, $v_{cd} = dc$

   It may be seen that C relative to D moves in a direction-sense given by v_{cd}, *i.e.*, d to c. Thus C moves in the counter-clockwise direction about D.

   Angular velocity of link CD, $\omega_{cd} = v_{cd}/CD$ counter clockwise (ccw)

(iv) Rubbing velocities :

8. Two links are joined together by pins passed through holes drilled at the ends of the links. When the pins move during motion the links rub on the pins. The velocity of rubbing will depend upon the relative angular velocity between the links. The rubbing velocity will be equal to the product of the algebraic difference of the angular velocities and the radius of the pin.

The rubbing velocities at pins of joints are :

$A : \omega_{ab} \cdot r_a$

$B : (\omega_{ab} + \omega_{bc})\, r_b$, ω_{ab} is *ccw* and ω_{bc} is cw

$C : (\omega_{bc} + \omega_{dc})\, r_c$, ω_{bc} is *cw* and ω_{dc} ccw

$D : \omega_{cd} \cdot r_d$

where r_a, r_b, r_c and r_d are the radii of pins at *A*, *B*, *C* and *D* respectively.

(b) Mechanism with Ternary Link

Consider the four-bar mechanism with ternary link, as shown in Fig. 2.8 (a) in which the crank *OA* is rotating clockwise with uniform angular speed ω

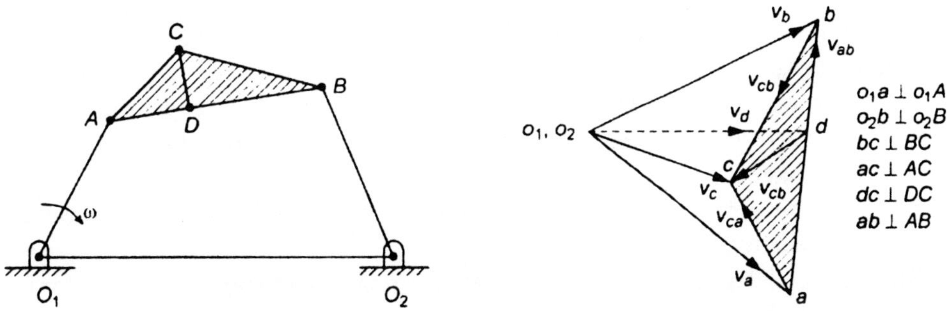

(a) Configuration diagram (b) Velocity diagram

Fig. 2.8 Four-bar mechanism with ternary link.

Procedure :

1. Calculate the angular velocity of crank O_1A, $\omega = \dfrac{2\pi N}{60}$ rad/s
2. Calculate the linear velocity of point *A* on the crank O_1A, $v_c = O_1A \times \omega$
3. The linear velocity of point *A* is perpendicular to O_1A. Therefore, draw $o_1a \perp O_1A$ to a convenient scale, as shown in Fig. 2.8 (b).
4. The velocity of point *B* is perpendicular to O_2B. Therefore, at point o_2, draw $o_2b \perp O_2B$. The relative velocity of *B* w.r.t. *A* is perpendicular to *AB*. Therefore, draw $ab \perp AB$ meeting the line drawn perpendicular to O_2B at *b*. Then
   Velocity of point *b*, $v_b = o_2b$
   Velocity of point *B* relative to point *A*, $v_{ba} = ab$
5. To find the velocity of point *C*, drawn $ac \perp AC$ and $bc \perp BC$ to meet at *c*. Join ac. Then $v_c = o_1c$. Now,

$$v_b = a_a + v_{ba}$$
$$= o_1a + ab = o_1b$$
$$v_c = v_b + v_{cb}$$
$$= o_2b + bc = o_2c$$

and

$$= v_a + v_{ca}$$
$$= o_1a + ac = o_1c$$

Thus, velocity of point *C*, $v_c = o_1c$

6. To find the velocity of any point D in AB, we have

$$\frac{bd}{ba} = \frac{BD}{BA}$$

or

$$v_d = bd = \left(\frac{BD}{BA}\right) \times ba$$

Thus, locate point d in ab and join $o_1 d$.
Then, velocity of point D, $v_d = o_1 d$

7. To find the relative velocity of C relative to D, join cd. Then
Velocity of C relative to D, $v_{cd} = dc$

Velocity Image

In Fig. 2.8, we observe that Δabc is similar to ΔABC in which all the three sides ab, bc and ac are perpendicular to AB, BC and AC respectively. The Δabc is called the **velocity image** of link ABC. The velocity images are very helpful in the velocity analysis of complicated shapes of the linkages. The following points should be kept in mind while using the velocity images :

(i) Velocity image of a link is scaled reproduction of the shape of the link in the velocity diagram from the configuration diagram, rotated, bodily 90° in the direction of the angular velocity.

(ii) The order of the letters in the velocity image is the same as the configuration diagram.

(iii) The ratio of the sizes of different images to the sizes of their respective links is different in the same mechanism.

2.3.8 Slider-Crank Mechanism

Consider the slider-crank mechanism, as shown in Fig. 2.9 (a), in which the crank OC is rotating clockwise with angular speed ω PC is the connecting rod and P is the slider or piston.

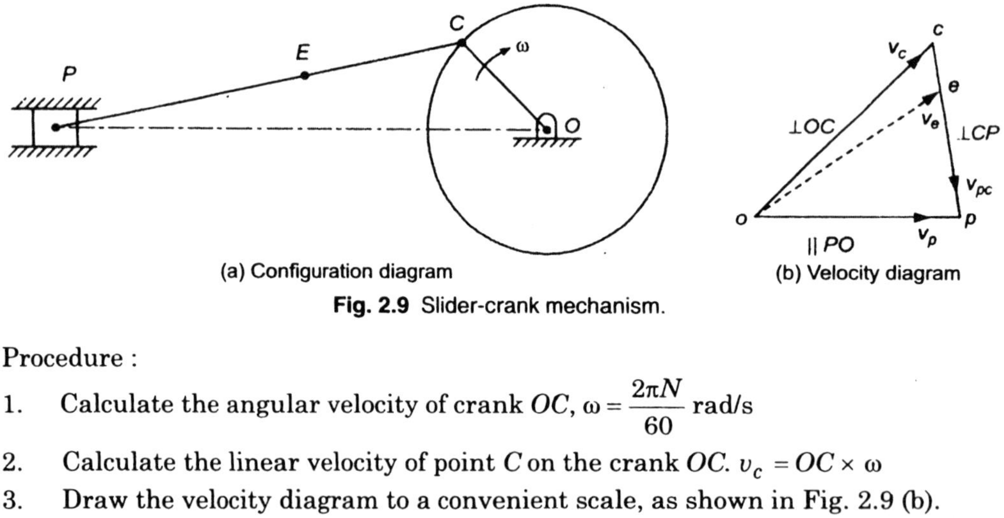

(a) Configuration diagram (b) Velocity diagram

Fig. 2.9 Slider-crank mechanism.

Procedure :

1. Calculate the angular velocity of crank OC, $\omega = \frac{2\pi N}{60}$ rad/s

2. Calculate the linear velocity of point C on the crank OC. $v_c = OC \times \omega$

3. Draw the velocity diagram to a convenient scale, as shown in Fig. 2.9 (b).

4. Draw a line $v_v = oc \perp OC$ from point o, representing the velocity of point C.

5. From point c draw a line perpendicular to CP. The velocity of slider P is horizontal. Therefore, from point o draw a line parallel to OP to intersect the line drawn

perpendicular to CP at p. Then velocity of piston, $v_p = op$ and the velocity of piston P relative to crank pin C is $v_{pc} = cp$.

6. To find the velocity of any point E in CP, we have

$$\frac{CE}{CP} = \frac{ce}{cp}$$

or
$$ce = \left(\frac{CE}{CP}\right) \times cp$$

7. Locate point e in cp, join oe. Then $v_e = oe$.
8. Rubbing velocities at various points are :

$$O : \omega_{oc} \times r_0$$
$$P : \omega_{cp} \times r_p$$
$$C : (\omega_{oc} + \omega_{cp}) \times r_c$$

where r_0, r_p and r_c are the radii of pins at O, P and C respectively.

2.3.9 Crank and Slotted Lever Quick-Return Motion Mechanism

Consider the crank and slotted lower mechanism, as shown in Fig. 2.10 (a). The crank OB is rotating at uniform angular speed ω in the counter clockwise direction.

Let Length of crank $OB = r$

Length of slotted lever $AC = l$

Distance between the fixed centres O and A, $OA = d$

Let the point B be on the crank OB and point P on the slotted lever AC at the same point B.

Procedure :

1. Calculate the angular velocity of crank OB, $\omega = 2\pi N/60$ rad/s
2. Calculate the linear velocity of point B on the crank OB, $v_b = r\omega$
3. Draw $ob = v_b$ and perpendicular to OB to a convenient scale, as shown in Fig. 2.10 (b).
4. The velocity of slotted lever is perpendicular to AC. The velocity of point P on the slider is along the slotted lever. Hence draw a line from b parallel to AC to meet the line perpendicular to AC at p. Then

   Velocity of point P on AC relative to fixed point A, $v_{pa} = v_p = ap$
5. Measure ap.
6. Calculate the velocity of point c. $v_c = ac = (AC/AP) \times ap$
7. Locate point c by extending line ap to c.
8. From point c draw a line perpendicular to CD and from a point o draw a line parallel to the path of slider at D to intersect at point d. Then

   Velocity of cutting tool at D, $v_d = ad$
9. The velocity of point B on crank OB relative to velocity of point P on slotted lever AC is given by,

$$v_{pb} = pb$$

10. The component of velocity of crank perpendicular to slotted lever is zero at positions B_1 and B_2. Thus for these positions of crank, the slotted level reverses its direction of motion.

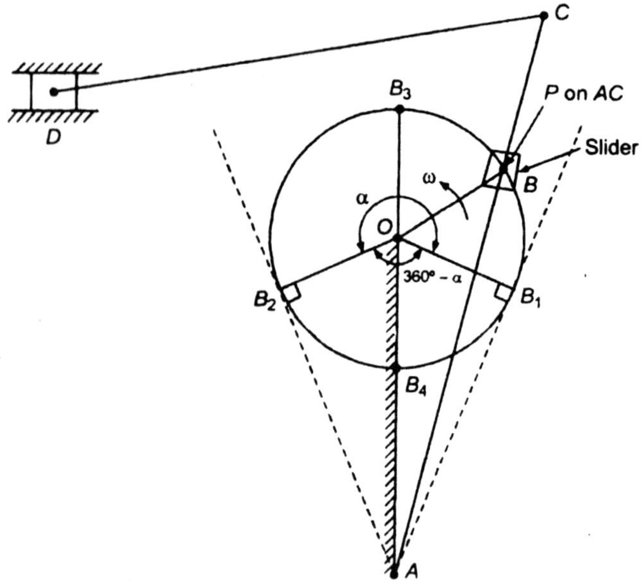

(a) Configuration diagram

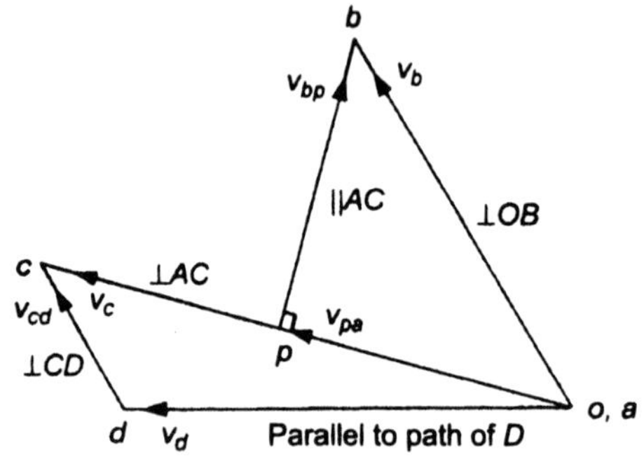

(b) Velocity diagram

Fig. 2.10 Crank and slotted lever mechanism.

$$\frac{\text{Time of cutting stroke}}{\text{Time of return stroke}} = \frac{\alpha}{360° - \alpha} \qquad \text{...(2.7)}$$

11. At positions B_3 and B_4, the component of velocity along the lever is zero, *i.e.*, the velocity of slider at crank pin is zero. Thus the velocity of lever at crankpin is equal to the velocity of crankpin, *i.e.*, $r\omega$ The velocity of lever and slider (tool) at these points is minimum. Maximum cutting velocity occurs at B_3 and maximum return velocity occurs at B_4.

$$\text{Maximum cutting velocity} = (OB \times \omega)\left(\frac{AC}{AB_3}\right)$$

$$= (r\omega)\left[\left(\frac{l}{d + r}\right)\right] \qquad \text{...(2.8a)}$$

$$\text{Maximum return velocity} = (OB \times \omega)\left(\frac{AC}{AB_4}\right)$$

$$= (r\omega)\left[\left(\frac{l}{d-r}\right)\right] \quad \ldots(2.8b)$$

$$\frac{\text{Maximum cutting velocity}}{\text{Maximum return velocity}} = \left[\frac{d-r}{d+r}\right] \quad \ldots(2.9)$$

2.3.10. Whitworth Quick-Return Motion Mechanism

In the Whitworth mechanism,as shown in Fig. 2.11 (a), the crank O_1A rotates with constant angular speed ω in the clockwise direction. The link PB oscillates about pin O_2. The ram C reciprocate on the guides.

The velocity diagram has been drawn in Fig. 2.11 (b).

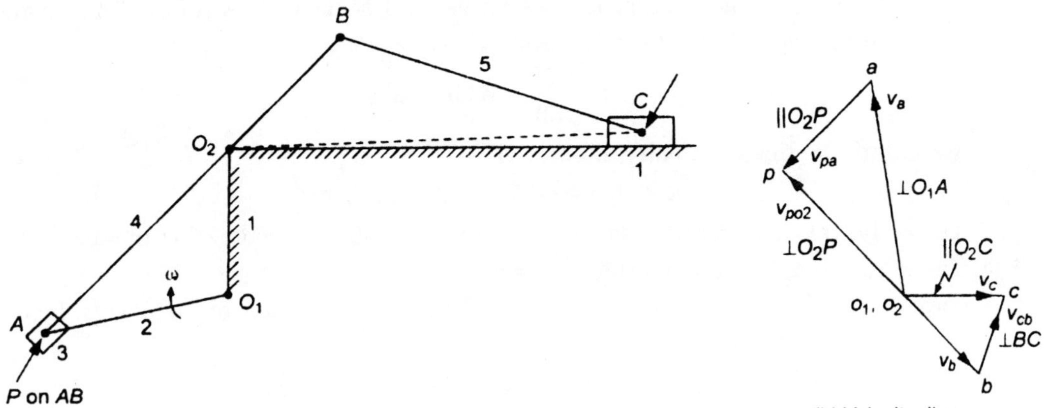

(a) Configuration diagram (b) Velocity diagram

Fig. 2.11 Whitworth quick-return motion mechanism.

Procedure :

1. Calculate the angular velocity of crank O_1A, $\omega = 2\pi N/60$ rad/s
2. Calculate linear velocity of point A on crank O_1A, $v_a = \omega \times O_1A$.
3. Draw $v_a = o_1a \perp O_1A$ to convenient scale.
4. At o_2 draw a line perpendicular to PO_2 and at 'a' draw a line parallel to O_2P to intersect at point p. Then $v_{pa} = ap$ represents the velocity of slider at A and $v_{po2} = o_2p$, the velocity of lever O_2P or PB.
5. Produce po_2 to b so that $\dfrac{po_2}{o_2b} = \dfrac{O_2P}{O_2B}$.
6. Draw $bc \perp BC$ and $o_2c \| O_2C$ to intersect at c.
   Then velocity of the ram C, $v_c = o_2c$.

■ **Example 2.1** In the mechanism shown in Fig. 2.12 (a), the crank O_1A rotates at a uniform speed of 650 rpm. Determine the linear velocity of slider C and the angular speed of the link BC. $O_1A = 30$ mm, $AB = 45$ mm, $BC = 50$ mm, $O_2B = 65$ mm, and $O_1O_2 = 70$ mm.

Solution : Given $N = 650$ rpm, lengths of links.

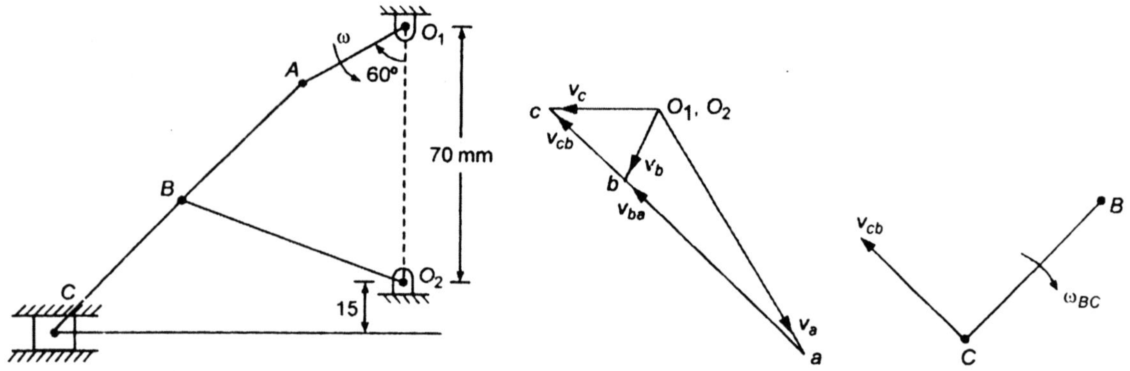

(a) Configuration diagram (b) Velocity diagram (c)

Fig. 2.12

Procedure :

1. Draw the configuration diagram to a convenient scale as shown in Fig. 2.12 (a).
2. Calculate the angular velocity of crank O_1A.

$$\omega = 2\pi \times \frac{650}{60} = 68.07 \text{ rad/s}$$

3. Calculate the linear velocity of point A,

$$v_a = \omega \times O_1A = 68.7 \times 0.03 = 2.04 \text{ m/s}$$

4. Draw the velocity diagram as shown in Fig. 2.12 (b) to a scale of 1 cm = 0.5 m/s.
5. Draw $v_a = o_1a \perp O_1A$ such that $o_1\alpha = 4.08$ cm.
6. Draw a line from o_1 perpendicular to O_2B and another line from 'a' perpendicular to AB to intersect at b. Then

$$ab = v_{ba} \text{ and } o_2b = v_b$$

7. From b draw a line perpendicular to BC and another line from o_2 parallel to the path of motion of the slider to intersect at C. Then

$$\text{Velocity of the slider } C,\ v_c = o_2c$$

By measurement, we get

$$v_c = o_2c = 1.4 \text{ cm} = 0.7 \text{ m/s}$$

$$v_{cb} = bc = 1.3 \text{ cm} = 0.65 \text{ m/s}$$

Angular velocity of link $BC = v_{cb}/BC = 0.65/0.05 = 13$ rad/s clockwise about B. (See Fig. 2.12 (c)).

■ **Example 2.2** The slider C of the toggle mechanism shown in Fig. 2.13 (a), is constrained to move on a horizontal path.The crank O_1A rotates in the counter-clockwise direction at a uniform speed of 180 rpm.

$$O_1A = 200 \text{ mm},\ AB = 400 \text{ mm},\ O_2B = 300 \text{ mm, and } BC = 600 \text{ mm}.$$

Determine (a) velocity of slider C, (b) angular velocity of links AB, O_2B, and BC, (c) rubbing velocities on the pins of 25 mm diameter at A and C, and (d) torque required at the crank O_1A for a force of 2 kN at C.

Solution : Given : $N = 180$ rpm, radius of pin, $r = 12.5$ mm, lengths of links.

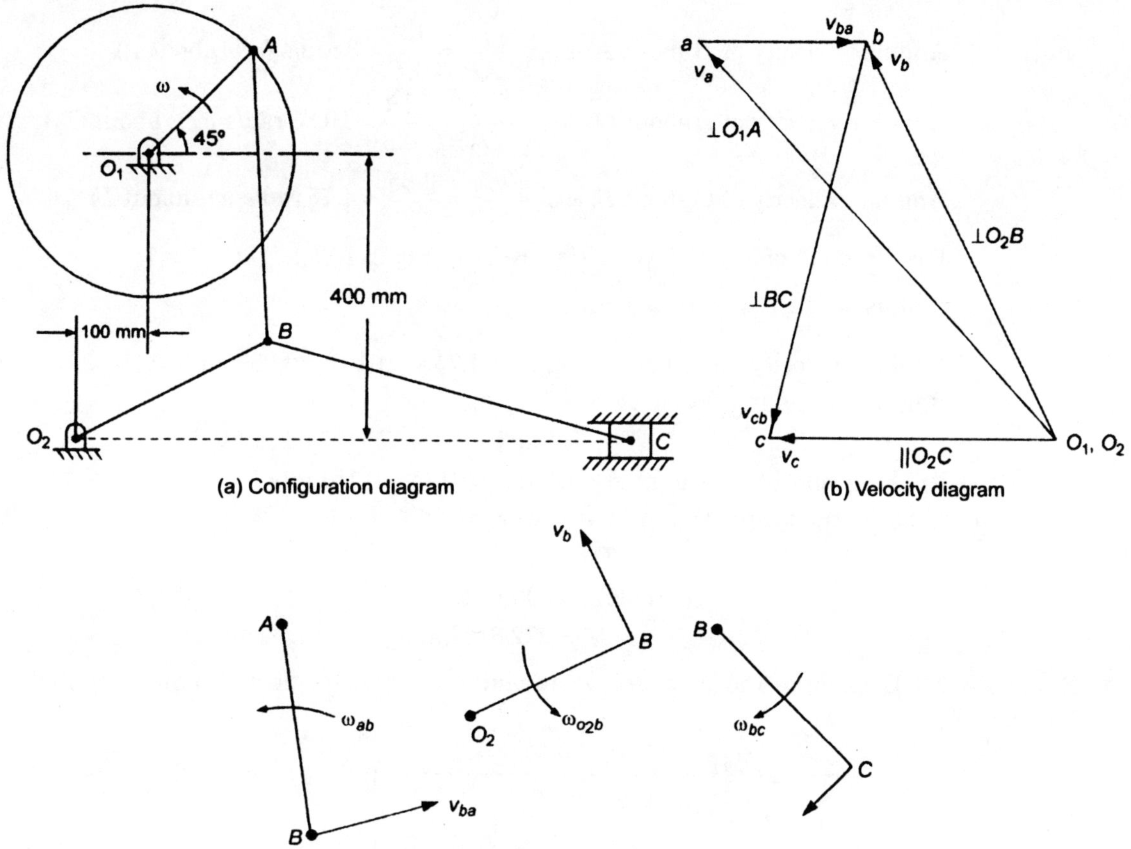

(a) Configuration diagram

(b) Velocity diagram

(c) Finding directions of angular velocities

Fig. 2.13 Toggle mechanism.

Procedure :

1. Draw the configuration diagram to a convenient scale as shown in Fig. 2.13 (a).
2. Calculate the angular velocity of crank O_1A.

$$\omega = 2\pi \times \frac{180}{60} = 18.65 \text{ rad/s}$$

3. Calculate the linear velocity of point A.

$$v_a = \omega \times O_1A = 18.65 \times 0.2 = 3.77 \text{ m/s}$$

4. Draw the velocity diagram as shown in Fig. 2.13 (b) to a scale of 1 cm = 0.5 m/s
5. Draw $v_a = o_1a \perp O_1A$ such that $o_1a = 7.54$ cm.
6. From 'a' draw a line perpendicular to AB and another line from o_1 perpendicular to O_2B to intersect at b. Then $o_2b = v_b$ and $ab = v_{ba}$.
7. From b draw a line perpendicular to BC and another line from o_1 parallel to the path of motion of the slider C to meet at c. Then $o_2c = v_c$. By measurement, we get
   (a) Velocity of the slider, $v_c = o_2c = 4.2$ cm $= 2.1$ m/s
   (b) Velocity of link $AB = v_{ba} = ab = 2.4$ cm $= 1.2$ m/s
   Velocity of point B, $v_b = o_2b = 6.2$ cm $= 3.1$ m/s
   Velocity of link BC, $v_{cb} = bc = 5.7$ cm $= 2.85$ m/s

Angular velocity of B about A, $\omega_{ab} = \frac{v_{ba}}{AB} = \frac{1.2}{0.4} = 3$ rad/s (ccw about A)

Angular velocity of B about O_2, $\omega_b = \frac{v_b}{O_2B} = \frac{3.1}{0.3} = 10.33$ rad/s (ccw about O_2)

Angular velocity of C about B, $\omega_{bc} = \frac{v_{cb}}{BC} = \frac{2.85}{0.6} = 4.75$ rad/s (cw about B)

For direction of angular velocities, refer to Fig. 2.13 (c).

(c) Radius of pin, $r = \frac{25}{2} = 12.5$ mm

Rubbing velocity on pin at $C = \omega_{cb} . r = 4.75 \times 0.0125 = 0.0594$ m/s

Relative angular velocity at $A = \omega_b - \omega_{ba} + \omega_{cb}$

$$= 10.33 - 3 + 4.75 = 12.08 \text{ rad/s}$$

Rubbing velocity on pin at $A = 12.08 \times 0.0125 = 0.151$ m/s

(d) Let T be the torque required at the crank O_1A.

$$T\omega = F_c v_c$$

$$18.85T = 2000 \times 2.1$$

$$T = 222.81 \text{ N.m}$$

■ **Example 2.3** Determine the mechanical advantage of the toggle mechanism shown in Fig. 2.14 (a).

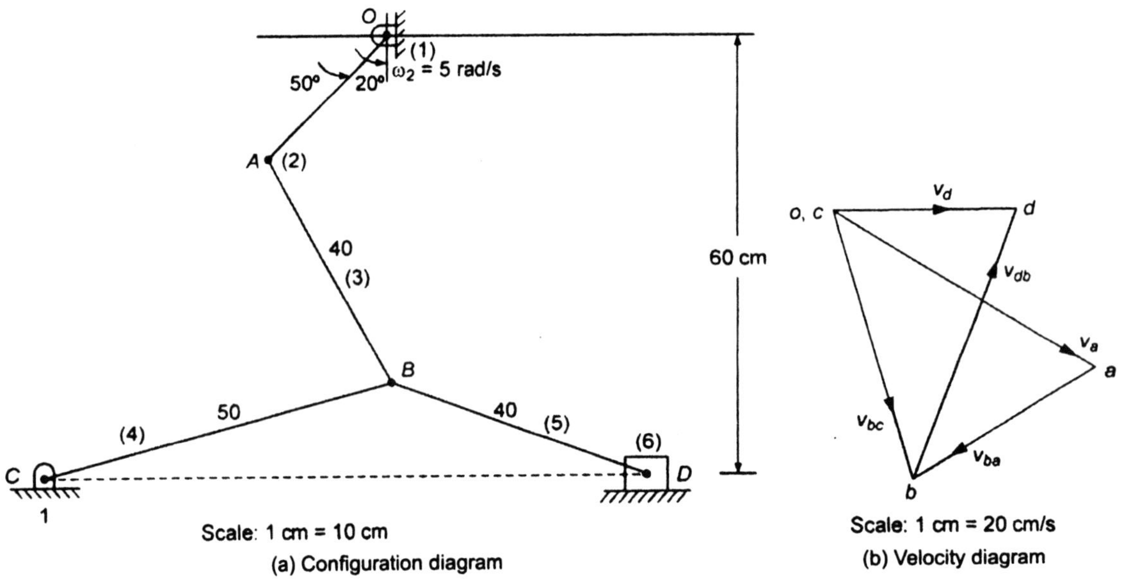

Fig. 2.14

Solution : Given : $\omega = 5$ rad/s, lengths of links.

Procedure :

1. Calculate the linear velocity of point A.

$$v_a = \omega_2 \times OA = 5 \times 20 = 100 \text{ cm/s}$$

2. Draw the configuration diagram to a convenient scale as shown in Fig. 2.14 (a).
3. Draw the velocity diagram to a scale of 1 cm = 20 cm/s, as shown in Fig. 2.14 (b).
4. Draw $v_a = oa \perp OA$, such that oa = 5 cm.
5. Draw $ab \perp AB$ and $cb \perp BC$ to meet at b.
6. Draw $bd \perp BD$ and $cd \parallel CD$ to meet at d.

By measurement, cd = 2.5 cm

Let T_a = Torque input at A

Force, $F_a = \dfrac{T_a}{OA}$

Work input = $F_a \times v_a$

Let F_d = force at D

Work output = $F_d \times v_d$

$$\text{Mechanical advantage} = \frac{F_d}{F_a} = \frac{v_a}{v_d} = \frac{oa}{cd} = \frac{5}{2.5} = 2$$

■ **Example 2.4** The dimensions of various links for the mechanism shown in Fig. 2.15 (a) are : OA = 0.5 m, AB = 1.5 m, $AC = CD$ = 0.9 m. The crank OA has uniform angular speed of 180 rpm. Determine the velocities of sliders B and D. Also calculate the turning moment at O if a force of 500 N acts on B and force of 800N acts on D, as shown.

Solution :

Procedure :

1. Draw configuration diagram as shown in Fig. 2.15 (a) at a scale of 1 cm = 0.2 m, OB = 1.68 m, OE = 0.64 m.
2. Angular speed of OA, $\omega = \dfrac{2\pi \times 180}{60} = 18.85$ rad/s

$$v_a = \omega \times OA = 18.85 \times 0.5 = 9.425 \text{ m/s}$$

3. Draw the velocity diagram as shown in Fig. 2.15 (b) to a scale of 1 cm = 1 m/s.
   (i) Draw $oa \perp OA$, oa = 9.425 cm.
   (ii) Draw $ab \perp AB$ and $ob \parallel$ slider B.
   (iii) Measure ab = 2.8 cm, ob = 10.8 cm so that $v_b = ob$ = 10.8 m/s.
   (iv) Now $\dfrac{ac}{ab} = \dfrac{AC}{AB}$, $ac = \dfrac{0.9}{1.5} \times 2.8 = 1.68$ cm.
   (v) Join oc. Then $oc = v_c$.
   (vi) Draw $cd \perp CD$ and $od \parallel$ slider D. Then $od = v_d$ = 1.5 cm = 1.5 m/s.
   (vii) Turning moment at O,

$$M_o = 500 \sin 45° \times OB - 800 \times OE$$
$$= 500 \times \frac{1}{\sqrt{2}} \times 1.68 - 800 \times 0.64$$
$$= 82.06 \text{ N.m. } (cw)$$

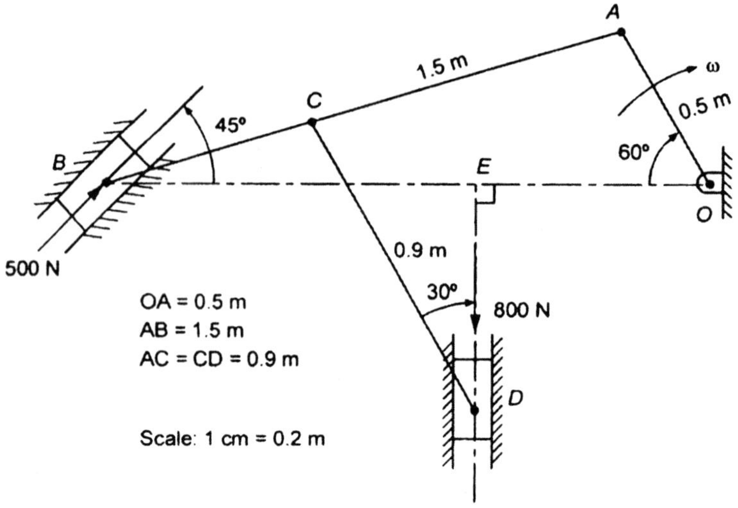

(a) Configuration diagram

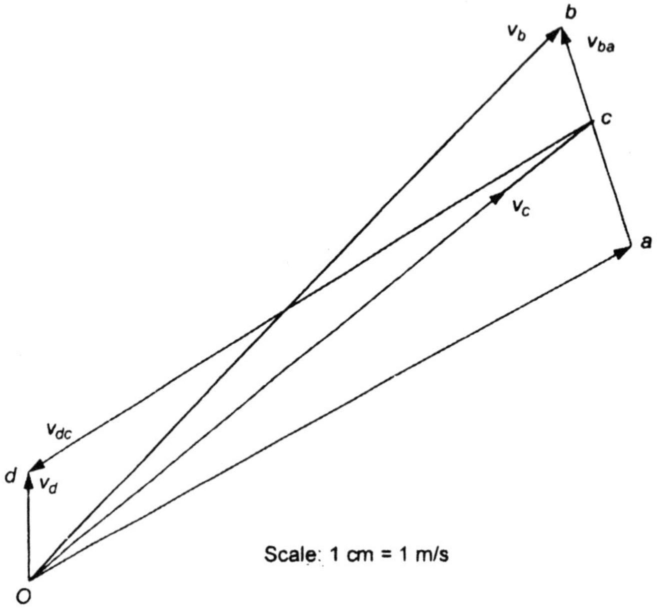

(b) Velocity diagram

Fig. 2.15

■ **Example 2.5** The dimensions of the various links of the mechanism shown in Fig. 2.16 (a) are : $OA = 50$ mm, $AB = 400$ mm, $BC = 150$ mm, $CD = 100$ mm, and $DE = 250$ mm. The crank OA rotates at 60 rpm. Find the velocity of slider E.

Solution :

Procedure :

1. Angular speed of crank OA, $\omega = 2\pi \times \dfrac{60}{60} = 6.28$ rad/s

$$v_a = w \times OA = 6.28 \times 50 = 314.16 \text{ mm/s}$$

2. Draw the configuration diagram to a scale of 1 mm = 5 mm, as shown in Fig. 2.16 (a).

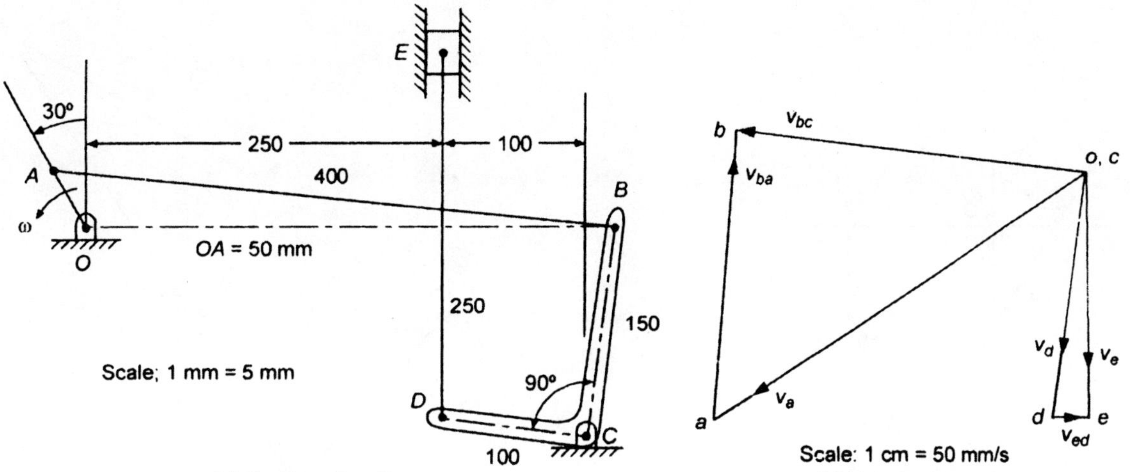

(a) Configuration diagram (b) Velocity diagram

Fig. 2.16

3. Draw the velocity diagram as shown in Fig. 2.16 (b) to a scale of 1cm = 50 mm/s.
   (i) Draw $v_a = oa \perp OA$, $oa = 6.28$ cm
   (ii) Draw $ab \perp AB$ and $cb \perp BC$ meeting at b
   (iii) Measure $cb = v_{bc} = 5$ cm = 250 mm/s
   (iv) Now $\frac{cd}{cb} = \frac{CD}{BC}$, $cd = \frac{100}{150} \times 5 = 3.33$ cm
   (v) Draw $cd \perp CD = 3.33$ cm
   (vi) Draw $de \perp DE$ and $oe \perp OB$. Then $v_e = oe = 3.4 = 117$ mm/s
   Velocity of slider $E = 117$ mm/s

■ **Example 2.6** The dimensions of the various links of the mechanism shown in Fig. 2.19 (a) are : $OA = 30$ mm, $AB = 75$ mm, $BC = 45$ mm, and $BD = 100$ mm. the crank OA rotates at 120 rpm. Determine the velocity of slider D and angular speed of links AB, BC and BD.

Solution :

Procedure :

1. Draw configuration diagram as shown in Fig. 2.17 (a) to a scale of 1 cm= 20 mm.
2. Angular speed of crank OA, $\omega = 2\pi \times \frac{120}{60} = 12.57$ rad/s

$$v_a = \omega \times OA = 12.57 \times 30 = 377 \text{ mm/s.}$$

3. Draw the velocity diagram as shown in Fig.2.17 (b) to a scale of 1 cm = 50 mm/s.
   (i) Draw $v_a = oa \perp OA$, $oa = 7.54$ cm.
   (ii) Draw $ab \perp AB$ and $cb \perp BC$, intersecting at point b.
   (iii) Draw $bd \perp BD$ and $cd \parallel CD$, intersecting at point d.
   (iv) Then $v_{ba} = ab = 7.4$ cm = 370 mm/s

$$v_d = cd = 3.2 \text{ cm} = 160 \text{ mm/s}$$

$$v_{bc} = cb = 3.5 \text{ cm} = 175 \text{ mm/s}$$

$$v_{db} = bd = 2.8 \text{ cm} = 140 \text{ mm/s}$$

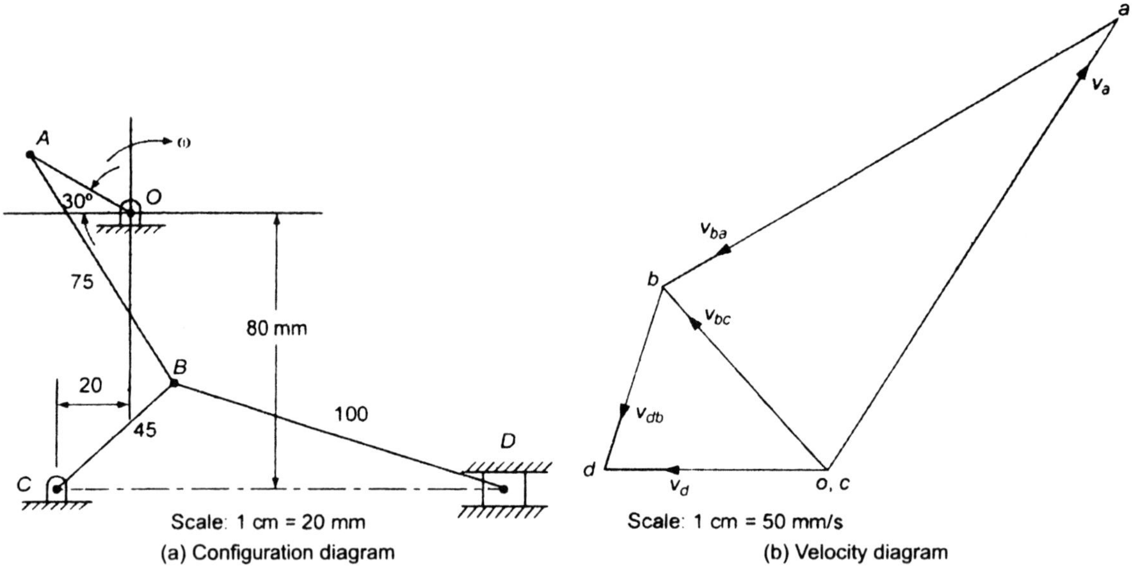

Fig. 2.17

(v) $\omega_{AB} = \dfrac{v_{ba}}{AB} = \dfrac{370}{75} = 4.93$ rad/s, cw

$\omega_{BC} = \dfrac{v_{bc}}{BC} = \dfrac{175}{45} = 3.89$ rad/s, cw

$\omega_{BD} = \dfrac{v_{bd}}{BD} = \dfrac{140}{100} = 1.4$ rad/s, cw.

■ **Example 2.7** The crank OA of the mechanism shown in Fig. 2.18 (a) rotates at120 rpm. The dimensions of the various links are :

$OA = 100$ mm, $AB = CE = 400$ mm, $AC = 125$ mm, and $EF = 300$ mm.The rod CE slides in a slot in trunnion at D. Determine (a) velocity of F, (b) velocity of sliding of CE in D, and (c) angular velocity of CE.

Solution :

Procedure :

(a)

1. Draw the configuration diagram shown in Fig. 2.18 (a) to a scale of 1 cm= 50 mm.
2. Angular speed of crank,OA, $\omega = 2\pi \times \dfrac{120}{60} = 12.567$ rad/s

$$v_a = \omega \times oA = 12.567 \times 100 = 1256.7 \text{ mm/s}$$

3. Draw the velocity diagram as shown in Fig. 2.18 (b) to a scale of 1 cm = 200 mm/s.
   (i) Draw $v_a = oa \perp OA$, $oa = 6.28$ cm.
   (ii) Draw $ab \perp AB$ and $ob \parallel OB$, intersecting at point b. Measure $ab = 3.2$ cm.
   (iii) Now $\dfrac{ac}{ab} = \dfrac{AC}{AB}$, $ac = 3.2 \times \dfrac{125}{400} = 1$ cm.
   (iv) Draw $cd \perp CD$ and $od \parallel CD$, intersecting at point d. Measure $cd = 0.7$ cm.

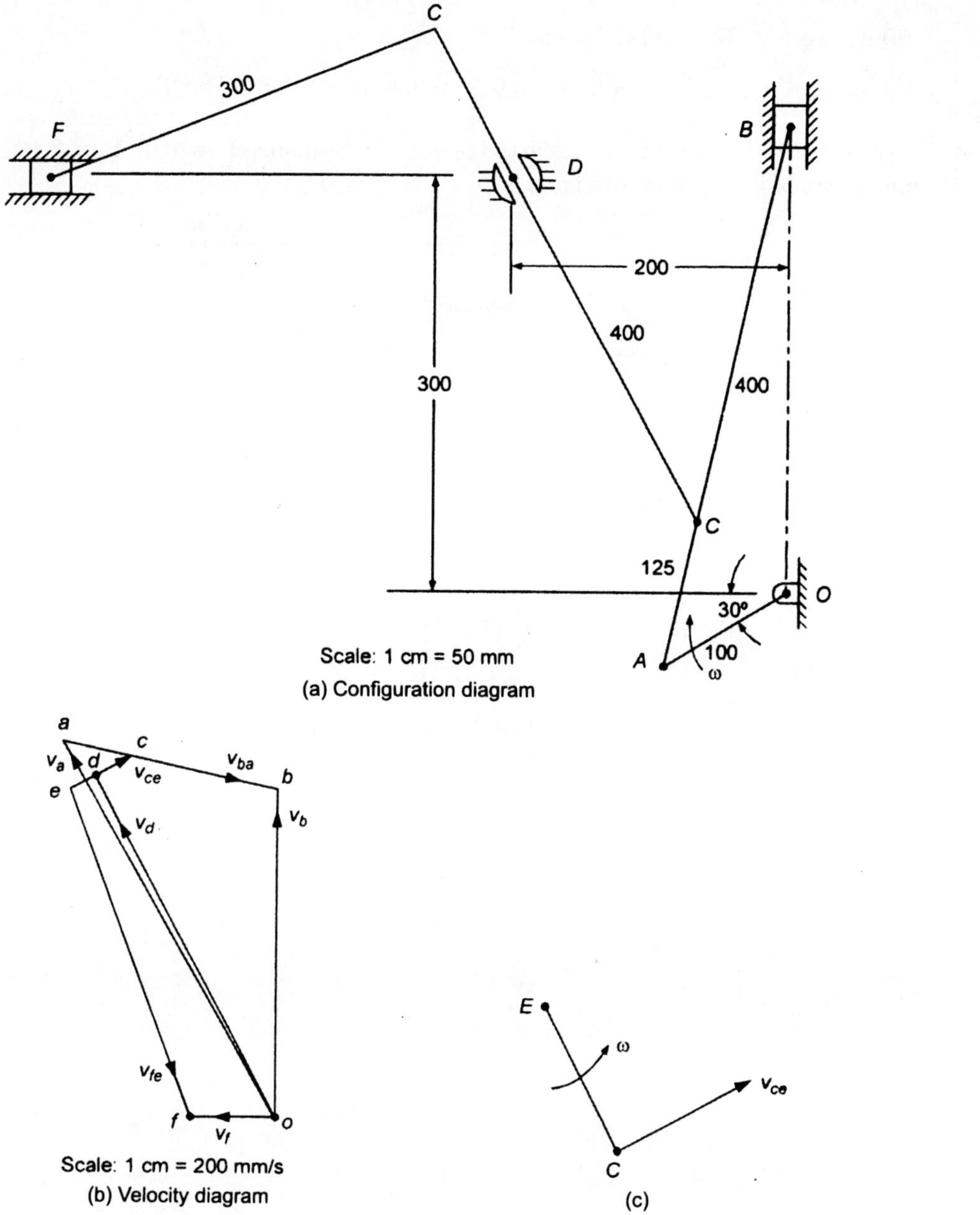

Scale: 1 cm = 50 mm
(a) Configuration diagram

Scale: 1 cm = 200 mm/s
(b) Velocity diagram

(c)

Fig. 2.18

(v) Now $\dfrac{ce}{cd} = \dfrac{CE}{CD}$, $ce = 0.7 \times \dfrac{400}{285} = 0.982\,\text{cm}$, $v_{ce} = 196.4$ mm/s. Produce *cd* to *e* so that $ce = 0.982\,\text{cm}$.

(vi) Draw $ef \perp EF$ and $of \parallel FD$, intersecting at point *f*.
Then $v_f = of = 1.2\,\text{cm} = 240$ mm/s
Thus, velocity of *F*, $v_f = 240$ mm/s

(b) $v_d = od = 5.6 \text{ cm} = 1120 \text{ mm/s}$

(c) $\omega_{ce} = \dfrac{v_{ce}}{CE} = \dfrac{196.4}{400} = 0.491$ rad/s, ccw about E (see Fig. 2.18 (c)).

■ **Example 2.8** The crank and slatted level mechanism shown in Fig. 2.19 (a) has the dimensions of its various links as :

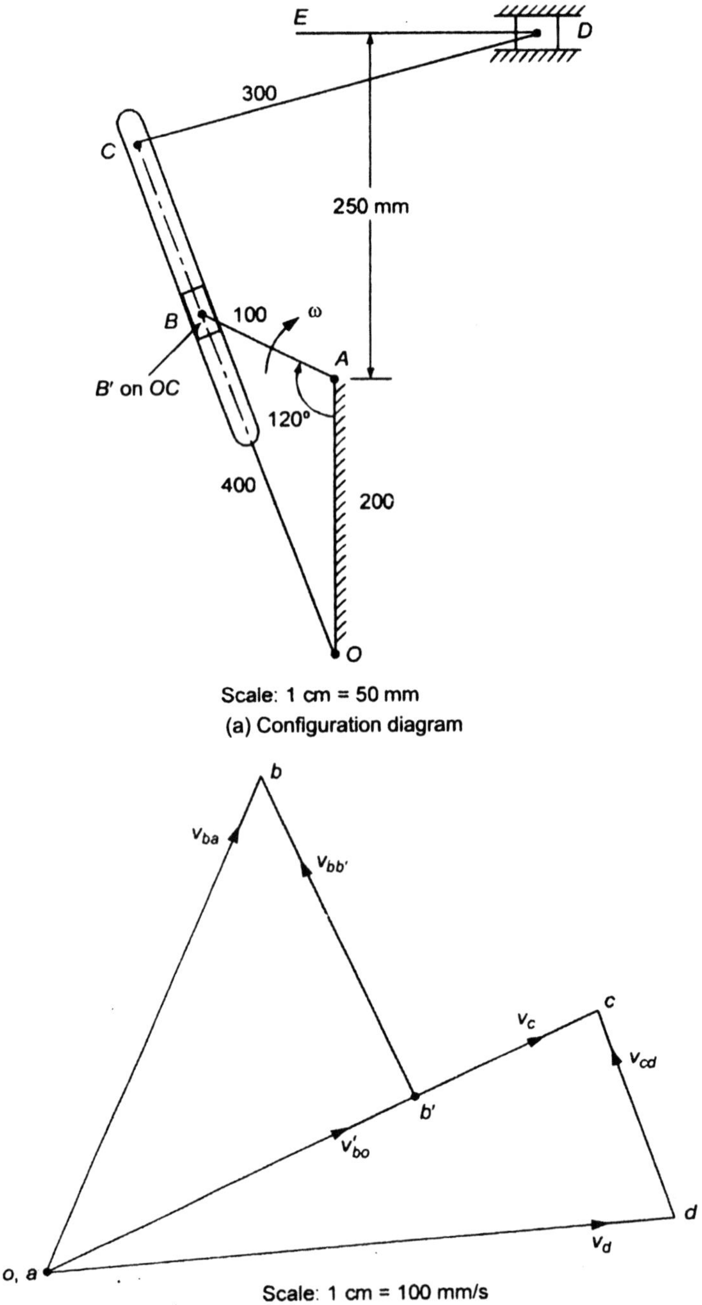

Scale: 1 cm = 50 mm

(a) Configuration diagram

Scale: 1 cm = 100 mm/s

(b) Velocity diagram

Fig. 2.19

$OA = 200$ mm, $AB = 100$ mm, $OC = 400$ mm, and $CD = 300$ mm. The crank AB rotates at 75 rpm. Determine (a) velocity of ram, and (b) angular speed of slotted lever OC.

Solution :

Procedure :

1. Draw the configuration diagram shown in Fig. 2.19 (a) to a scale of 1 cm = 50 mm.
2. Angular speed of crank AB, $\omega = 2\pi \times \dfrac{75}{60} = 7.854$ rad/s

$$v_b = \omega \times AB = 7.854 \times 100 = 785.4 \text{ mm/s.}$$

3. (a) Draw the velocity diagram as shown in Fig. 2.19 (b) to a scale of 1 cm = 100 mm/s
   (i) Draw $v_b = ab \perp AB$, $ab = 7.85$ cm.
   (ii) Draw $bb' \parallel OB$ and $ob' \perp OB$, intersecting at point b'. Measure $ob' = 5.8$ cm.
   (iii) $oc = ob' \times \dfrac{OC}{OB} = 5.8 \times \dfrac{400}{265} = 8.75$ cm. Produce ob' to c so that $oc = 8.75$ cm.
   (iv) Draw $cd \perp CD$ and $od \parallel DE$, intersecting at point d.
   (v) Then $v_d = od = 9$ cm $= 900$ mm/s.

   (b) Angular speed of slotted level OC, (by measurement, $OB' = 265$ mm)

$$\omega_{oc} = \frac{ob'}{OB'} = \frac{5.8 \times 100}{265} = 2.19 \text{ rad/s, cw.}$$

■ **Example 2.9** For the mechanism shown in Fig. 2.20 (a), determine the velocities of points C, E and F. Also calculate the angular velocities of the links BC, CDE and EF. Clank AB rotates at 120 rpm. The dimensions of various links are : $AB = 60$ mm, $BC = 120$ mm, $AD = 50$ mm, $CD = 100$ mm, $DE = 120$ mm, $CE = 50$ mm and $EF = 150$ mm.

Solution :

Procedure :

1. Draw the configuration diagram shown in Fig. 2.20 (a) to scale of 1 cm = 20 mm.
2. Angular speed of crank AB, $\omega = 2\pi \times \dfrac{120}{160} = 12.57$ rad/s

$$v_b = \omega \times AB = 12.57 \times 60 = 752.98 \text{ mm/s.}$$

3. Draw the velocity diagram as shown in Fig. 2.20 (b) to a scale of 1 cm = 200 mm/s
   (i) Draw $v_b = ab \perp AB$, $ab = 3.77$ cm
   (ii) Draw $bc \perp BC$ and $dc \perp DC$ to intersect at point c.
   (iii) Measure $v_{cd} = dc = 4.8$ cm $= 960$ mm/s to get the velocity of point c.
   (iv) Draw $ce \perp CE$ and $de \perp DE$ to intersect at point e. Then $v_{ed} = de = 5.8$ cm $= 1160$ mm/s, the velocity of point E.
   (v) Draw $ef \perp EF$ and $df \perp DF$ to intersect at point f. Then $v_{fd} = df = 9.5$ cm $= 1900$ mm/s, the velocity of point F.
   (vi) $\omega_{EF} = \dfrac{v_{ef}}{EF} = \dfrac{6.7 \times 200}{150} = 8.93$ rad/s, ccw

$$\omega_{BC} = \frac{v_{bc}}{BC} = \frac{2.8 \times 200}{120} = 6.33 \text{ rad/s ccw}$$

$$\omega_{CDE} = \frac{v_{cd}}{CD} = \frac{4.8 \times 200}{100} = 9.6 \text{ rad/s, ccw.}$$

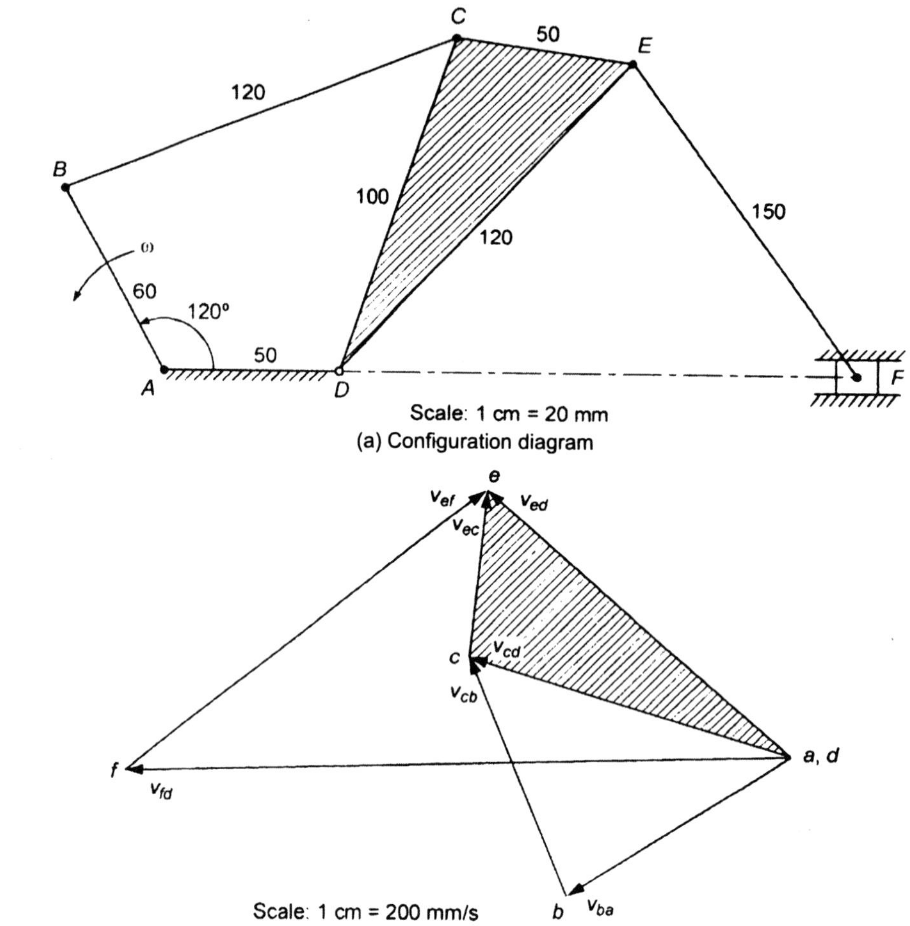

(a) Configuration diagram

(b) Velocity diagram

Fig. 2.20

■ **Example 2.10** For the mechanism shown in Fig. 2.21 (a), determine the angular velocities of links 3 and 4 when link 2 is rotating at 120 rpm. Also find the velocity of point *C* and *D*. $O_2A = 100$ mm, $AB = 250$ mm, $AC = 150$ mm, $BC = 150$ mm, $O_4B = 100$ mm, and $AD = 100$ mm.

Solution :

Procedure:

1. Draw the configuration diagram shown in Fig. 2.21 (a) to a scale of 1 cm = 50 mm.
2. Angular speed of link 2,

$$\omega_2 = 2\pi \times \frac{120}{60} = 12.57 \text{ rad/s}$$

$$v_a = \omega_2 \times O_2A = 12.57 \times 100 = 1257 \text{ mm/s}$$

3. Draw the velocity diagram shown in Fig. 2.21 (b) to scale of 1 cm = 200 mm/s.
   (i) Draw $v_a = o_2a \perp O_2A$. $o_2a = 6.28$ cm.
   (ii) Draw $o_4b \perp O_4B$ and $ab \perp AB$ to intersect at point *b*. Also draw $ac \perp AC$ and $bc \perp BC$ to intersect at point *c*.

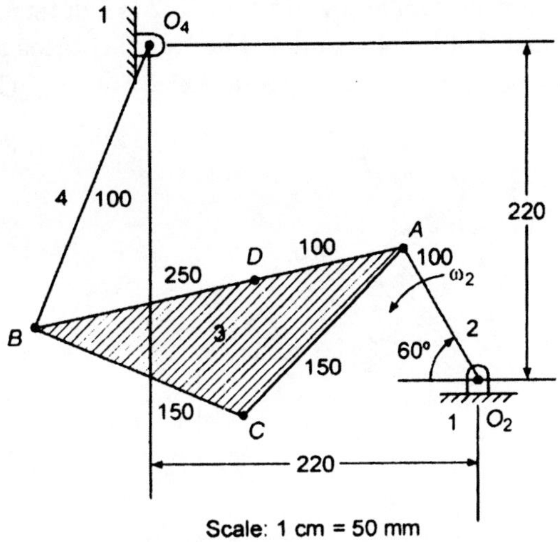

Scale: 1 cm = 50 mm

(a) Configuration diagram

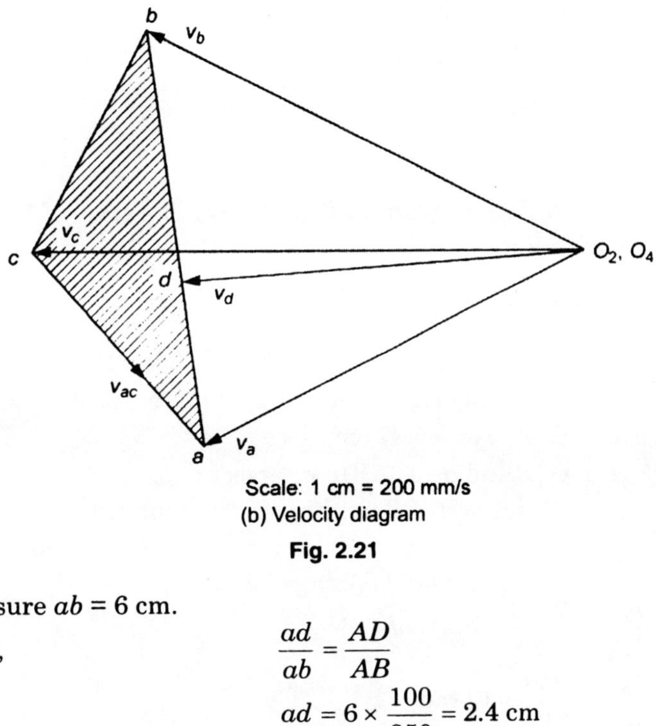

Scale: 1 cm = 200 mm/s

(b) Velocity diagram

Fig. 2.21

(iii) Measure $ab = 6$ cm.

Now,
$$\frac{ad}{ab} = \frac{AD}{AB}$$

$\therefore$
$$ad = 6 \times \frac{100}{250} = 2.4 \text{ cm}$$

Join o_2d . $o_2d = 6\,\text{cm}$

$$v_d = o_2d = 1200\,\text{mm/s}$$
$$v_c = o_2c = 8\,\text{cm} = 1600\,\text{mm/s}$$
$$v_b = o_4b = 7.2 \text{ cm} = 1440\,\text{mm/s}$$

(iv) $w_4 = \dfrac{v_b}{O_4B} = \dfrac{1440}{100} = 14.4$ rad/s, ccw

$$w_3 = \frac{v_{ac}}{AC} = \frac{2.7 \times 200}{150} = 4.93 \text{ rad/s, ccw.}$$

■ **Example 2.11** For the mechanism shown in Fig. 2.22 (a), determine the velocities of points C and A and angular speedy of links 3 and 4. The link 2 rotates at 150 rpm.

$O_2A = 380$ mm, $O_4B = 250$ mm, $AC = 250$ mm, $BC = 400$ mm, $O_2O_4 = 750$ mm.

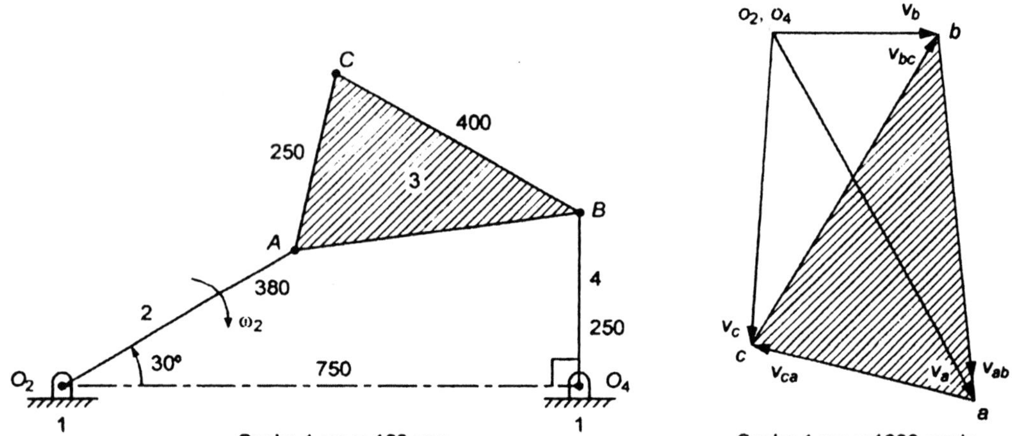

Scale: 1 cm = 100 mm

(a) Configuration diagram

Scale: 1 cm = 1000 mm/s

(b) Velocity diagram

Fig. 2.22

Solution :

Procedure :

1. Draw the configuration diagram in Fig. 2.22 (a) to a scale of 1 cm = 100 m.
2. Angular sped of link 2,

$$\omega_2 = 2\pi \times \frac{150}{60} = 15.7 \text{ rad/s}$$

$$v_a = \omega_2 \times O_2A = 15.7 \times 380 = 5969 \text{ mm/s.}$$

3. Draw the velocity diagram shown in Fig. 2.22 (b) to a scale of 1 cm = 1000 mm/s.
   (i) Draw $o_2a \perp O_2A$, $o_2a = 5.97$ cm.
   (ii) Draw $a_4b \perp O_4B$ and $ab \perp AB$ to intersect at b.
   (iii) Draw $ac \perp AC$ and $bc \perp BC$ to intersect at point C. Join o_2c.
   (iv) Measure $v_c = o_2c = 4.5 \text{ cm} = 4500 \text{ mm/s}$ $v_b = o_2b = 2.3 \text{ cm} = 2300 \text{ mm/s}$

$$v_{ab} = ba = 5.2 \text{ cm} = 5200 \text{ mm/s}$$

   (v) $\omega_4 = \dfrac{v_b}{OB_4} = \dfrac{2300}{250} = 9.2$ rad/s, cw

$$\omega_3 = \frac{v_{ab}}{AB} = \frac{5200}{400} = 13 \text{ rad/s, cw.}$$

■ **Example 2.12** For the mechanism shown in Fig. 2.23 (a), determine the velocity of the slider C. The link 2 rotates at 180 rpm.

$O_2A = 50$ mm, $AB = 100$ mm, $AC = 200$ mm, $BD = 100$ mm, $O_6D = 100$ mm, and $O_2O_6 = 100$ mm.

Solution :

Procedure :

1. Draw the configuration diagram shown in Fig. 2.23 (a) to a scale of 1 cm = 25 cm.

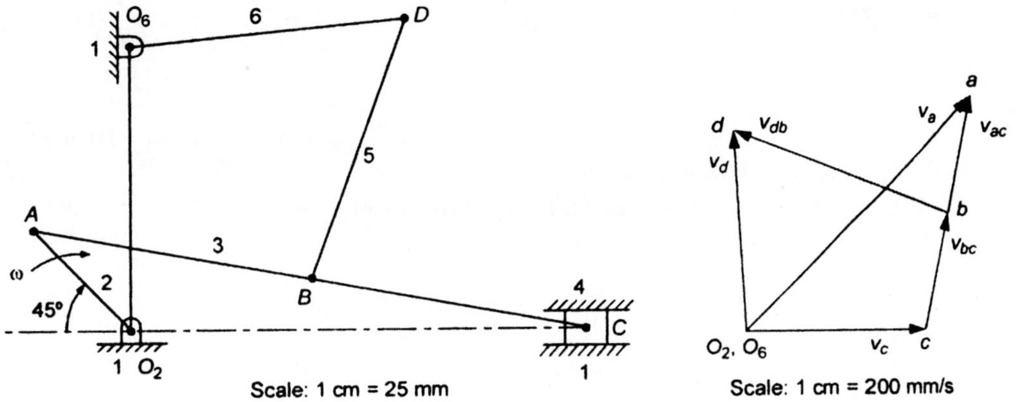

Scale: 1 cm = 25 mm
(a) Configuration diagram

Scale: 1 cm = 200 mm/s
(b) Velocity diagram

Fig. 2.23

2. Angular speed of link 2,

$$\omega_2 = 2\pi \times \frac{180}{60} = 18.85 \text{ rad/s}$$

$$v_a = \omega_2 \times o_2A = 18.85 \times 50 = 942.5 \text{ mm/s.}$$

3. Draw the velocity diagram in Fig. 2.23 (b) to scale of 1 cm= 200 mm/s.
   (i) Draw $v_a = o_2a \perp O_2A$, $o_2a = 4.71$ cm.
   (ii) Draw $ac \perp AC$ and $o_2c \parallel O_2C$ to intersect at c. Measure $ac = 3.5$ cm.
   (iii) Now $\frac{ab}{ac} = \frac{AB}{AC}$, $ab = 3.5 \times \frac{100}{200} = 1.75$ cm.
   (iv) Draw $o_6d \perp O_6D$ and $bd \perp BD$ to intersect at d.
   (v) Velocity of $c = v_c = o_2c = 2.6$ cm= 520 mm/s.

■ **Example 2.13** In the crank-shaper mechanism shown in Fig. 2.24 (a), the link 2 rotates at a constant angular speed of 1 rad/s. Determine the angular speed of link 4, v_{A4}, v_{A2A3}, and v_{A3A4} $O_2A = 50$ mm, $O_4A = 80$ mm, and $O_2O_4 = 120$ mm.

Solution :

Procedure :

1. Draw the configuration diagram shown in Fig. 2.24 (a) to a scale of 1 cm = 25 mm.

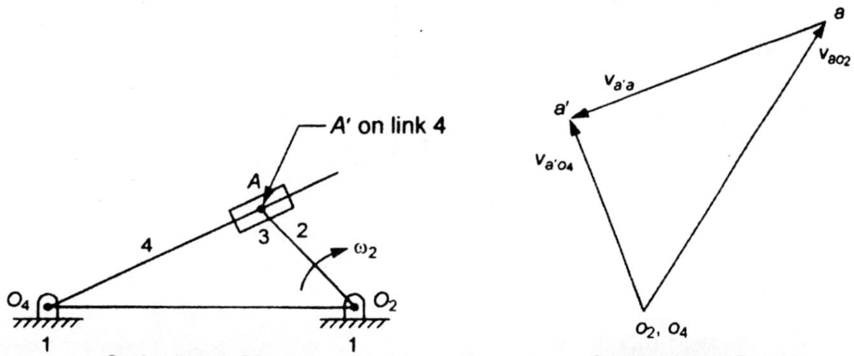

Scale: 1 cm = 25 mm
(a) Configuration diagram

Scale: 1 cm = 10 mm/s
(b) Velocity diagram

Fig. 2.24

2. $\omega_2 = 1$ rad/s (given).

$$v_a = \omega_2 \times o_2A$$
$$= 1 \times 50 = 50 \text{ mm/s}.$$

3. Draw the velocity diagram shown in Fig. 2.24 (b) to a scale of 1 cm = 10 mm/s.
   (i) $v_a = o_2a \perp O_2A$, $o_2a = 5$ cm.
   (ii) Draw $o_4a' \perp O_4A$, and $a'a \parallel O_4A$, to intersect at a'. Then

$$v_{A4} = v_{a'o_4} = o_2a' = 3 \text{ cm} = 30 \text{ mm/s}$$
$$w_4 = \frac{v_{a'o_4}}{o_4B} = \frac{30}{80} = 0.375 \text{ rad/s, ccw}$$
$$v_{A3A4} = aa' = 4 \text{ cm} = 40 \text{ mm/s}$$
$$v_{A2A3} = v_a = 50 \text{ mm/s}.$$

■ **Example 2.14** For the mechanism shown in Fig. 2.25 (a), the link 2 rotates at 160 rad/s. Determine v_b, ω_4 and v_{ba}. $O_2A = 150$ mm, $AB = 200$ mm, and $O_4B = 150$ mm.

Solution :

Procedure :

1. Draw the configuration diagram shown in Fig. 2.25 (a) to a scale of 1 cm = 50 mm.
2. $\omega_2 = 160$ rad/s (given)
   $v_a = \omega_2 \times O_2A = 160 \times 150 = 24{,}000$ mm/s.
3. Draw the velocity of diagram shown in Fig. 2.25 (b) to a scale of 1 cm = 4000 mm/s.
   (i) Draw $v_a = o_2a \perp O_2A, o_2a = 6$ cm.
   (ii) Draw $o_4b \perp O_4B$ and $ab \perp AB$ to intersect at b.
   (iii) Measure $v_b = o_4b = 3 \text{ cm} = 12{,}000$ mm/s.
   (iv) $\omega_4 = \dfrac{v_b}{O_4B} = \dfrac{12{,}000}{150} = 80$ rad/s, ccw

$$v_{ba} = ab = 5.8 \text{ cm} = 23{,}200 \text{ mm/s}.$$

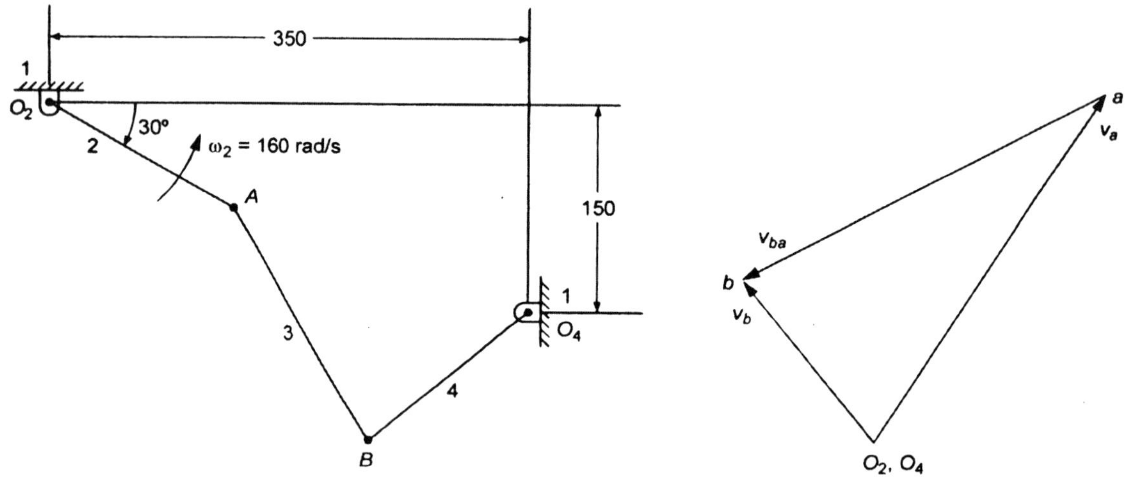

Scale: 1 cm = 50 mm
(a) Configuration diagram

Scale: 1 cm = 4000 mm/s
(b) Velocity diagram

Fig. 2.25

■ **Example 2.15** The driving 2 of the Whitworth quick-return motion mechanism shown in Fig. 2.26 (a) rotates at a constant speed of 6 m/s. Determine the velocity of tool holder.

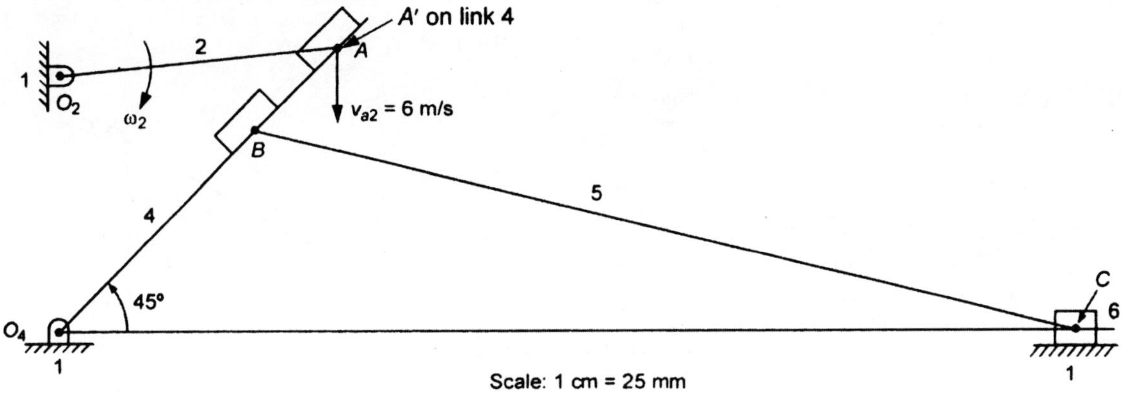

Scale: 1 cm = 25 mm
(a) Configuration diagram

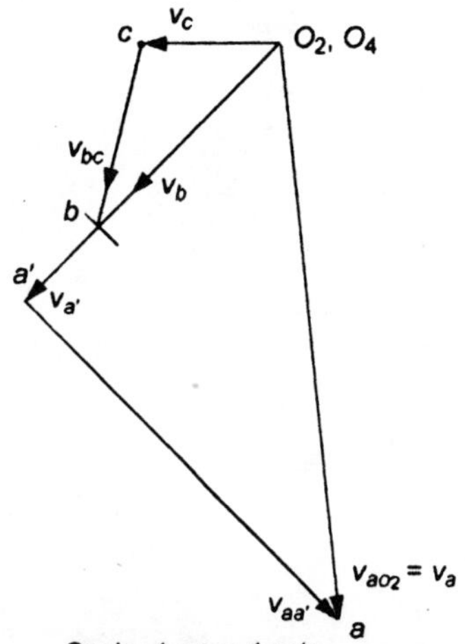

Scale: 1 cm = 1 m/s
(b) Velocity diagram
Fig. 2.26

Solution :

Procedure :

1. Draw the configuration diagram shown in Fig. 2.26 (a) to a scale of 1 cm = 25 mm and velocity shown in Fig. 2.26 (b) to a scale of 1 cm = 1 m/s.
2. Draw $v_{ao_2} = o_2a = v_{a2} \perp O_2A$, $o_2a = 6$ m/s.
3. Draw $o_4a' \parallel O_4A$ and $a'a \perp O_4A$ to intersect at a'.
4. Measure $o_4a' = 3.8$ cm, $O_4A = 5.7$ cm, $O_4B = 4$ cm.
5. Now $\dfrac{o_4b}{o_4a'} = \dfrac{O_4B}{O_4A}$, $o_4b = 3.8 \times \dfrac{4}{5.7} = 2.67$ cm.
6. Draw $bc \perp BC$ and $o_2c \parallel O_4C$ to intersect at point C.
7. Velocity of tool holder, $v_c = o_2c = 1.5$ cm $= 1.5$ m/s.

■ **Example 2.16** For the mechanism shown in Fig. 2.27 (a), determine the velocity of slider and ω_3, ω_4, and ω_5.

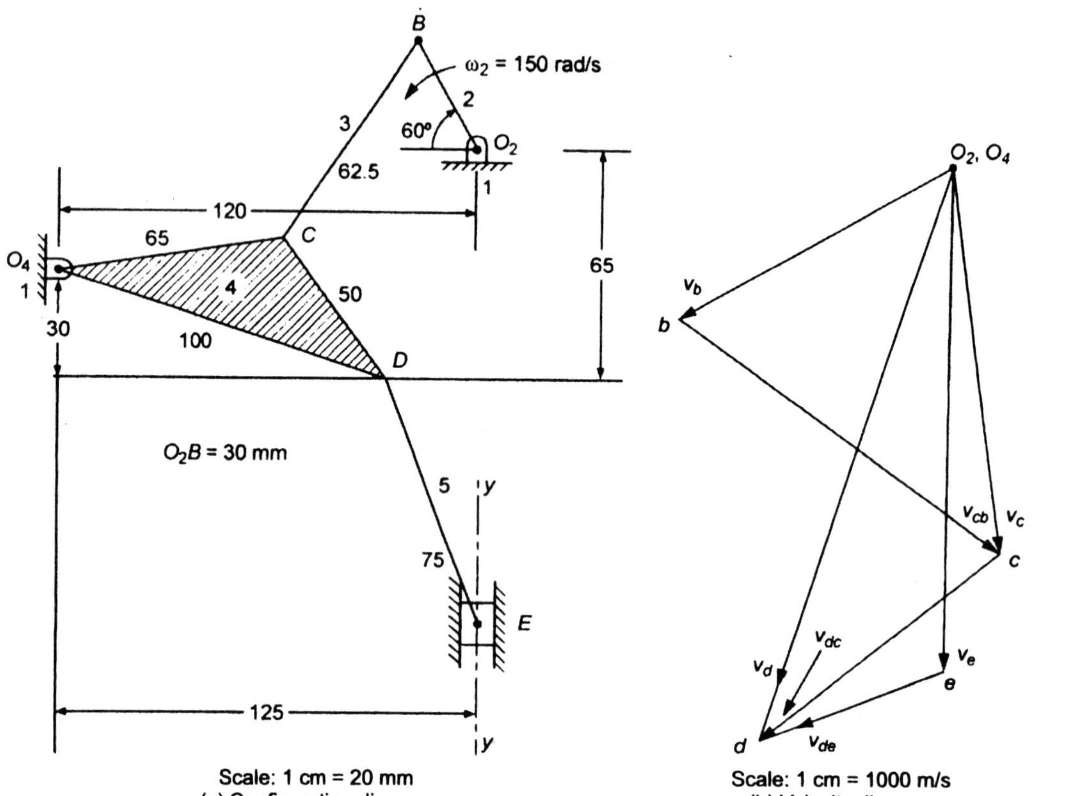

(a) Configuration diagram (b) Velocity diagram

Fig. 2.27

Solution :

Procedure :

1. Draw the configuration diagram shown in Fig. 2.27 (a) to a scale of 1 cm = 20 mm.
2. Angular speed of link 2,

$$\omega_2 = 150 \text{ rad/s}$$

$$v_b = \omega_2 \times O_2B = 150 \times 30 = 4500 \text{ mm/s.}$$

3. Draw the velocity diagram shown in Fig. 2.27 (b) to a scale of 1cm = 1000 mm/s.
   (i) Draw $v_b = o_2b \perp O_2B, o_2b = 4.5$ cm.
   (ii) Draw $bc \perp BC$ and $o_4c \perp O_4C$ to intersect at point c.
   (iii) Draw $cd \perp CD$ and $o_4d \perp O_4D$ to intersect at a.
   (iv) Draw $de \perp DE$ and $o_2e \parallel yy$ to intersect at e.
   (v) $\omega_3 = \dfrac{v_{cb}}{BC} = \dfrac{bc}{BC} = \dfrac{5.8 \times 1000}{62.5} = 92.8$ rad/s, ccw

$$\omega_4 = \frac{v_c}{O_4C} = \frac{o_4c}{O_4C} = \frac{5.5 \times 1000}{65} = 84.6 \text{ rad/s, ccw}$$

$$\omega_5 = \frac{v_{de}}{DE} = \frac{ed}{DE} = \frac{2.8 \times 1000}{75} = 37.3 \text{ rad/s, ccw}$$

Velocity of slider, $v_e = o_4e = 7.2 \times 1000 = 7200$ mm/s or 7.2 m/s.

■ **Example 2.17** For the mechanism shown in Fig. 2.28 (a), $v_E = 5$ m/s. Determine v_d, ω_3 and ω_5.

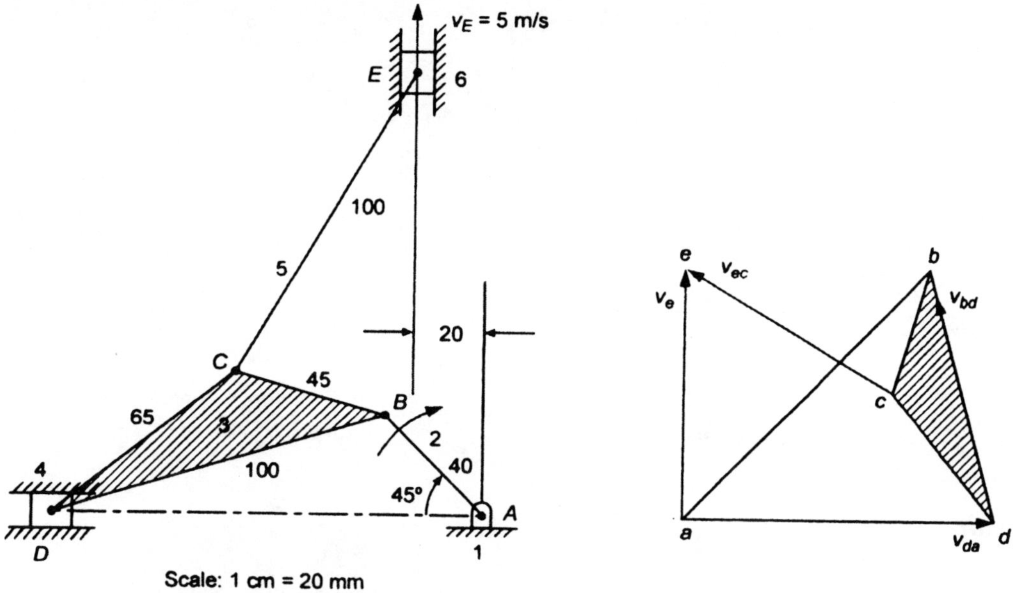

(a) Configuration diagram

(b) Velocity diagram

Fig. 2.28

Solution :

Procedure:

1. Draw the configuration diagram shown in Fig.2.28 (a) to a scale of 1 cm = 20 mm.
2. Draw the velocity diagram shown in Fig. 2.28 (b) by assuming v_b = 5 cm and crank AB rotating clockwise.
3. Draw $v_b = ab \perp AB$, ab = 5 cm.
4. Draw $bd \perp BD$, $ad \parallel AD$ to meet at a point d.
5. Draw $bc \perp BC$ and $cd \perp CD$ to meet at point c.
6. Draw $ae \parallel v_E$ and $ce \perp CE$ to meet at e. Then $ae = v_E$ = 5 m/s = 3.5 cm. Thus scale is : 1 cm = 1.428 m/s.
7. $v_{da} = ad$ = 4.5 cm = 6.45 m/s

   $v_{bd} = db$ = 2.7 cm = 5.28 m/s

$$\omega_3 = \frac{v_{bd}}{BD} = \frac{bd}{BD} = \frac{5.28}{0.1} = 52.8 \text{ rad/s, ccw}$$

$$v_{ec} = ce = 2.4 \text{ cm} = 4.85 \text{ m/s}$$

$$\omega_5 = \frac{v_{ec}}{EC} = \frac{4.85}{0.10} = 48.5 \text{ rad/s, cw}$$

■ **Example 2.18** The crank O_1A of the four-bar linkage shown in Fig. 2.29 (a) is rotating at a uniform angular velocity of 30 rad/s. Draw the velocity polygon and determine the velocity of point B, the angular velocities of the links 3 and 4. $O_1A = 100$ mm, $AB = 200$ mm, $O_2B = 75$ mm, $AC = 100$ mm, $BC = 150$ mm, and $AD = 50$ mm. Also determine the velocity images of all links.

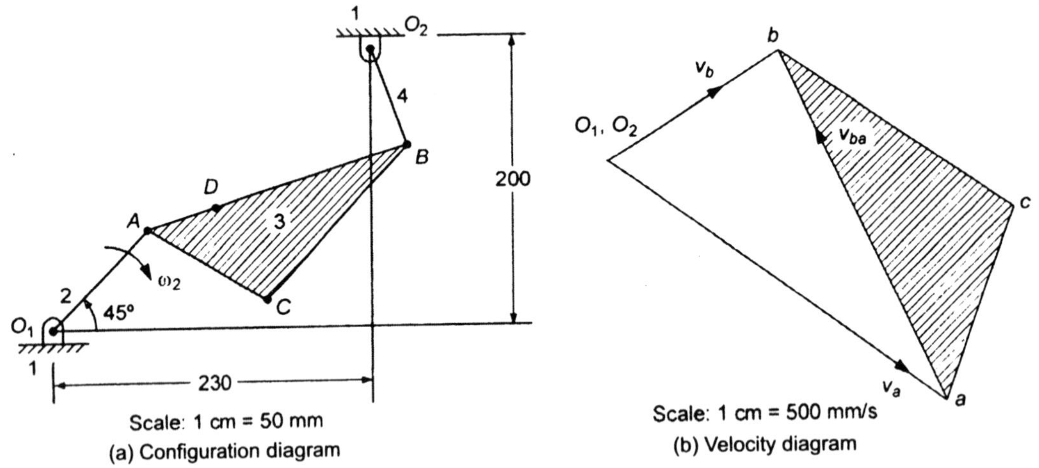

Fig. 2.29

Solution :

Procedure :

1. Draw the configuration diagram shown in Fig.2.29 (a) to a scale of 1 cm= 50 mm.
2. $\omega_2 = 30$ rad/s, $v_a = w_2 \times O_1A = 30 \times 100 = 3000$ mm/s
3. Draw the velocity polygon as shown in Fig. 2.29 (b) to a scale of 1 cm = 500 mm/s.
   (i) Draw $v_a = o_1a \perp O_1A$, $o_1a = 6$ cm.
   (ii) Draw $ab \perp AB$ and $o_2b \perp O_2B$ to intersect at point b.
   (iii) Draw $ac \perp AC$ and $bc \perp BC$ to intersect at point c.
   (iv) $v_{ba} = ab = 5.6$ cm $= 2800$ mm/s
   $v_b = o_1b = 2.9$ cm $= 1450$ mm/s
   (v) $\omega_3 = \dfrac{v_{ba}}{AB} = \dfrac{ab}{AB} = \dfrac{2800}{200} = 14$ rad/s, cw
   $\omega_4 = \dfrac{v_b}{O_2B} = \dfrac{o_2b}{O_2B} = \dfrac{1450}{75} = 19.3$ rad/s, cw.

■ **Example 2.19** In the mechanism shown in Fig. 2.30 (a), the piston D moves in the vertical direction upwards with a velocity of 5 m/s.

$O_1A = 7.5$ cm, $O_1O_2 = 30$ cm, $AB = 25$ cm, $O_2C = O_2B = 10$ cm, $CD = 25$ cm, and $BC = 12.5$ cm. Find the speed in rpm and direction of rotation of crank O_1A.

Solution :

Procedure :

Draw the configuration diagram shown in Fig. 2.30 (a) to a scale of 1 cm = 5 cm. Let us assume that crank O_1A is rotating clockwise. Assume that $v_a = 5$ cm to some scale. Then draw the velocity diagram shown in Fig. 2.30 (b) as follows :

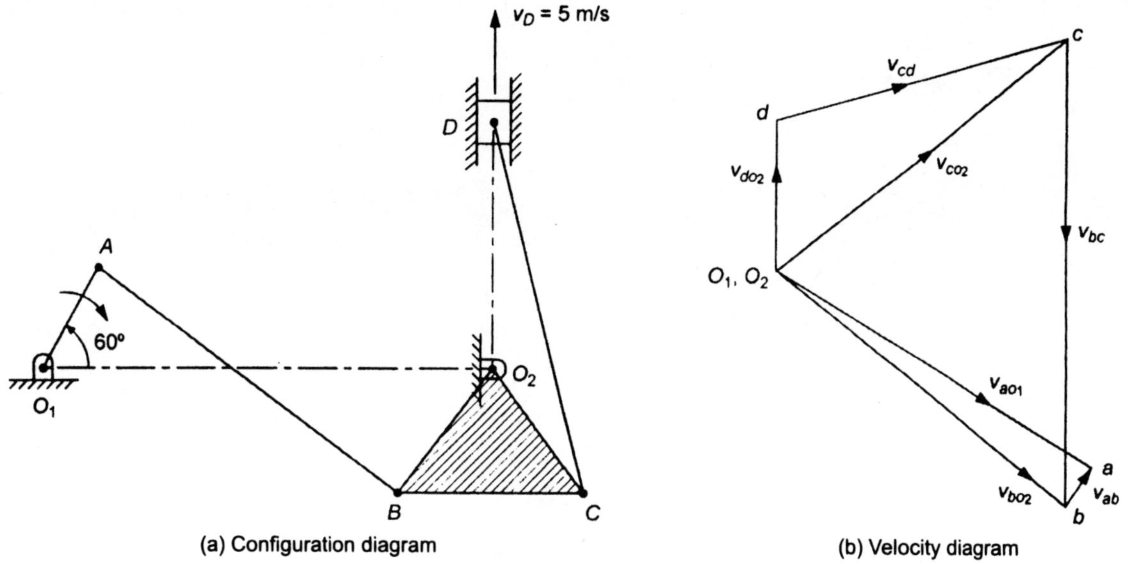

Fig. 2.30

1. Draw $v_a = o_1a \perp O_1A$, $o_1a = 5$ cm.
2. Draw $ab \perp AB$ and $o_2b \perp O_2B$ meeting at point b.
3. Draw $bc \perp BC$ and $o_2c \perp O_2C$ meeting at point c.
4. Draw $o_2d \parallel O_2D$ and $cd \perp CD$ meeting at point d.

Then $$v_D = o_2d = 2 \text{ cm} = 5 \text{ m/s}$$

Hence scale : 1 cm = 2.5 m/s

and $$v_a = 5 \times 2.5 = 12.5 \text{ m/s}$$

$$\omega = \frac{v_a}{O_1A} = \frac{12.5 \times 100}{7.5} = 166.67 \text{ rad/s}$$

$$\frac{2\pi N}{60} = 166.67$$

$$N = 1591.5 \text{ rpm}$$

■ **Example 2.20** The quick-return motion mechanism of the crank and slotted lever type shaping machine is shown in Fig. 2.31 (a).

$$O_1O_2 = 800 \text{ mm}, O_1B = 300 \text{ mm}, O_2D = 1300 \text{ mm}, DR = 400 \text{ mm}.$$

The crank O_1B makes an angle of 45° with the vertical and rotates at 40 rpm in the counter-clockwise direction. Find :

(a) Velocity of ram R, and (b) angular velocity of link O_2D.

Solution :

Procedure :

1. Draw the configuration diagram to a scale of 1 cm = 200 mm.
2. Angular speed of crank O_1B,

$$\omega = \frac{2\pi \times 40}{60} = 4.189 \text{ rad/s}$$

$$v_b = \omega \times O_1B = 4.189 \times 300 = 1256.6 \text{ m/s}$$

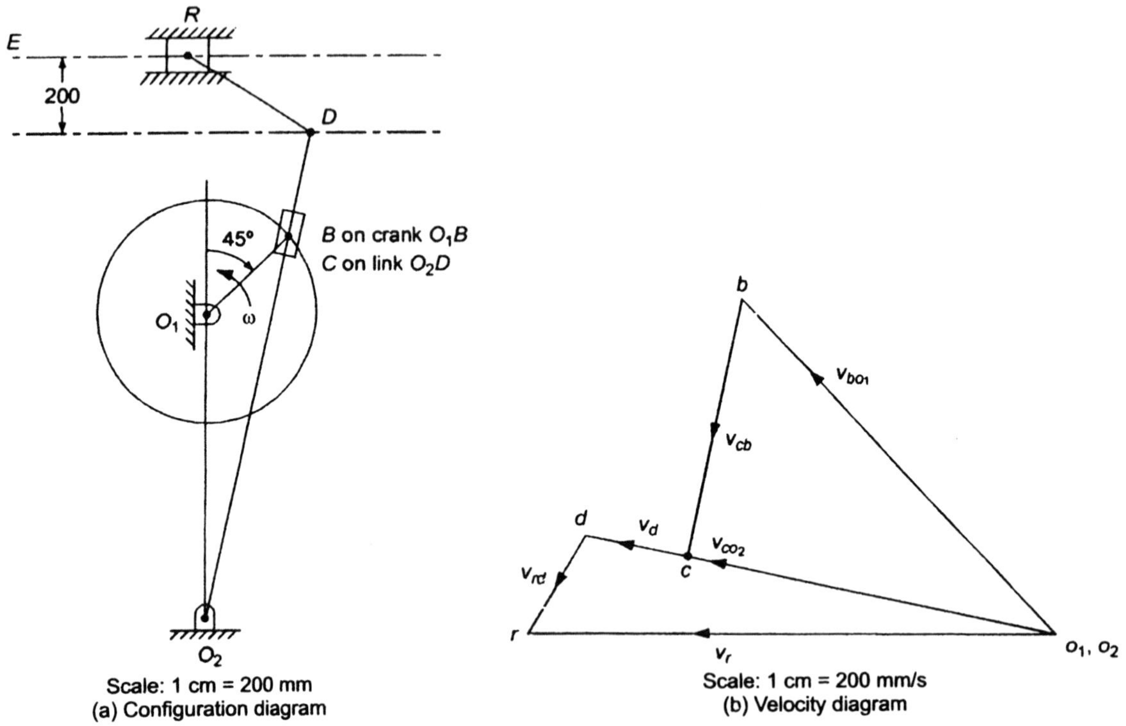

Fig. 2.31

3. Draw the velocity diagram as shown in Fig. 2.31 (b) to a scale of 1 cm= 200 mm/s.
4. Draw $o_1b \perp O_1B$, $o_1b = 6.28$ cm. Also draw be $\| O_2D$ and $o_2d \perp O_2D$.
5. Measure $o_2c = 5.1$ cm.
6. Now $\frac{o_2d}{o_2c} = \frac{O_2D}{O_2C}$, $o_2d = \frac{5.1 \times 6.5}{5.1} = 6.5$ cm.
7. Produce o_2c to d so that $o_2d = 6.5$ cm.
8. Draw $dr \perp DR$ and $o_2r \parallel RE$ to meet at r.
9. Then $v_r = o_1r = 7.1$ cm $= 1420$ mm/s is the velocity of ram.
10. $v_d = o_2d = 6.5$ cm $= 1300$ mm/s, $w_{o_2D} = \frac{v_d}{O_2D} = \frac{1300}{1300} = 1$ rad/s ccw about O_2.

■ **Example 2.21** The dimensions of the mechanism shown in Fig. 2.32 (a) are : $O_1O_2 = 6$ cm, $O_3A = 8$ cm, $O_1B = 20$ cm, $BC = 18$ cm, $O_2C = 20$ cm, $\angle O_1O_2C = 90°$.

The crank O_3A rotates uniformly at 20 rad/s clockwise. Determine the velocity of the slider A and angular velocity of link BC.

Solution :

Procedure :

1. Angular speed of link $O_3A = 20$ rad/s

$$v_{ao_3} = \omega \times O_3A = 20 \times 8 = 160 \text{ cm/s}$$

2. Draw velocity diagram as shown in Fig. 2.32 (b)
3. Draw $v_{ao_3} = o_3a \perp O_3A$ to a scale of 1 cm = 40 cm/s so that $o_3a = 4$ cm
4. Draw $ad \parallel BO_1$ and $o_1b \perp O_1B$. Measure $o_1d = 2.9$ cm and $O_1B = 2.8 \times 4 = 11.2$ cm

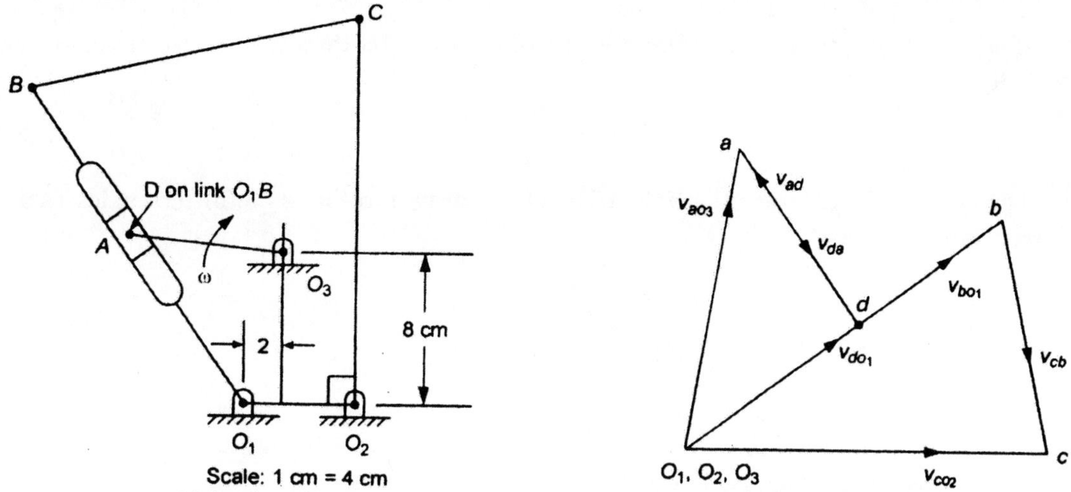

Scale: 1 cm = 4 cm

(a) Configuration diagram

(b) Velocity diagram

Fig. 2.32

$$\text{Now } \frac{o_1 d}{o_1 b} = \frac{O_1 D}{O_1 B}, \; o_1 b = 2.9 \times \frac{20}{11.52} = 5.2 \text{ cm}$$

Produce $o_1 d$ to b so that $o_1 b = 5.2$ cm

5. Draw $bc \perp BC$ and $o_2 c \perp O_2 C$ to meet at c.

Velocity of the slider A, $v_{da} = 2.8 \times 40 = 112$ cm/s

Angular velocity of link $BC = \omega_{BC} = \dfrac{v_{cb}}{BC} = \dfrac{3.1 \times 40}{18} = 6.9$ rad/s, cw.

2.4 INSTANTANEOUS CENTRE METHOD

A line as a whole may be considered to be rotating about an imaginary centre or a given center at a given instant. Such a centre has zero velocity or the link is at rest at this point. This is known as the **instantaneous centre or centre of rotation.** This centre veries from instant to instant for different positions of the link. The locus of these centers is termed the **centrode.**

2.4.1 Velocity of a Point on a Link

Consider two points A and B on a rigid link, having velocities v_a and v_b respectively, as shown in Fig. 2.33 (a). From A and B draw lines perpendicular to the directions of motion and let them meet at I. Then I is the instantaneous centre of rotation of the link AB for its given position.

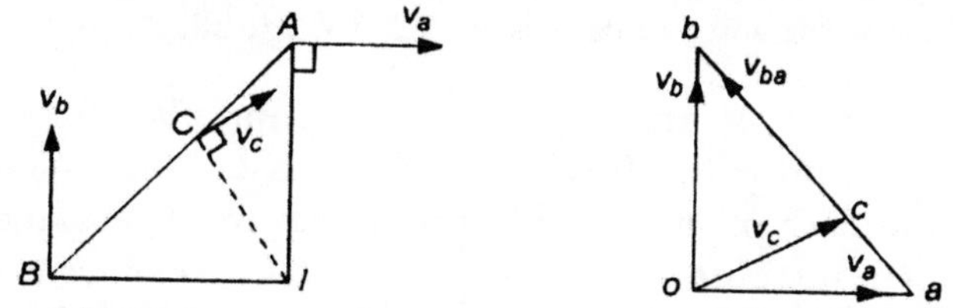

(a) Instantaneous centre

(b) Velocity diagram

Fig. 2.33

If ω is the instantaneous angular velocity of the link AB, then $v_a = \omega \,.\, IA$ and $v_b = \omega \,.\, IB$. Thus

$$v_b = \left(\frac{IB}{IA}\right) \times v_a \qquad \text{...(2.10)}$$

The velocity diagram for the link AB has been drawn in Fig. 2.33 (b). Triangles IAB and oab are similar. Hence

$$\frac{oa}{IA} = \frac{ob}{IB} = \frac{ab}{AB}$$

or

$$\frac{v_a}{IA} = \frac{v_b}{IB} = \frac{v_{ba}}{AB} = \frac{v_c}{IC} = \omega$$

where C is any point on the link AB. ...(2.11)

2.4.2 Properties of Instantaneous Centre

The properties of instantaneous centre are :

1. At the instantaneous centre of rotation, one rigid link rotates instantaneously relative to another for the configuration of the mechanism considered.
2. The two rigid links have no linear velocities relative to each other at the instantaneous centre.
3. The two rigid links have the same linear velocity relative to the third rigid link, or any other link.

2.4.3 Number of Instantaneous Centres

The number of instantaneous centers in a mechanism is equal to the number of possible combinations of the two links. The number of instantaneous centres,

$$N = \frac{n\,(n-1)}{2} \qquad \text{...(2.12)}$$

where n = number of links.

2.4.4 Types of Instantaneous Centres

The instantaneous centres for a mechanism are of the following types :

1. Fixed instantaneous centres
2. Permanent instantaneous centres
3. Neither fixed nor permanent instantaneous centres.

Consider a four-bar mechanism shown in Fig. 2.34. For this mechanism, $n = 4$. Hence $N = \frac{4\,(4-1)}{2} = 6$. The instantaneous centres are : 12, 13, 14, 23, 24, 34.

The instantaneous centres 12 and 14 remain at the same place for the configuration of the mechanism, and are, therefore, called fixed instantaneous centres. The instantaneous centres 23 and 24 move when the mechanism moves, but the joints are of permanent nature. Therefore, they are called permanent instantaneous centres. The instantaneous centres 13 and 24 vary with the configuration of the mechanism and are neither fixed nor perpament instantaneous centres.

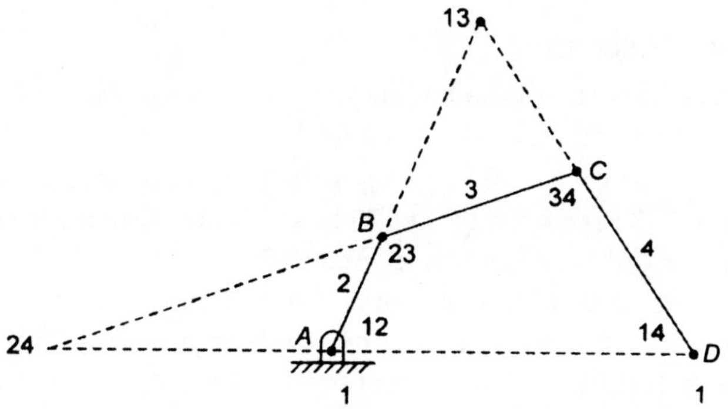

Fig. 2.34 Instantaneous centres in a four-bar mechanism.

2.4.5 Location of instantaneous Centres

The following observations are quite helpful in locating the instantaneous centres :

1. For a pivoted or pin joint, instantaneous centre for the links lies on the centre of the pin [see Fig. 2.35 (a)].
2. In a pure rolling contact of the two links, the instantaneous centre lies at their point of contant [see Fig. 2.35(b)]. This is because the relative velocity between the two links at the point of contact is zero.
3. In a sliding motion, the instantaneous centre lies at infinity in a direction perpendicular to the path of motion of the slider. This is because the sliding motion is equivalent to a rotary motion of the links with radius of curvature equal to infinity [see Fig. 2.35 (c)].
4. If the slider (link 2)moves on a curve surface (link 1), then the instantaneous centre lies at the centre of curvature of curved surface (see Fig. 2.35 (d) and (e)].

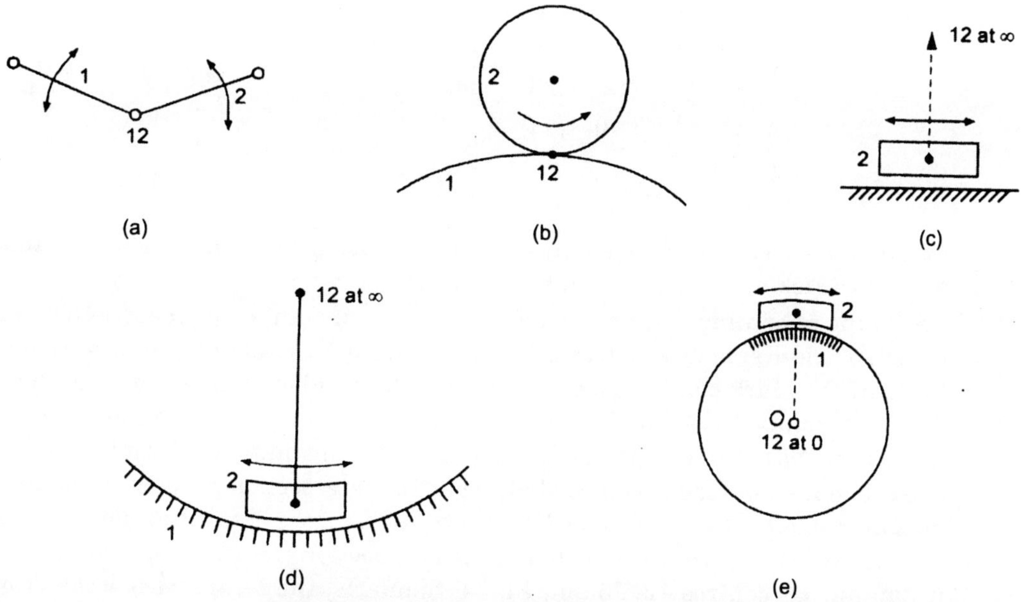

Fig. 2.35 Location of instantaneous centres.

2.4.6 Arnold-Kennedy's Theorem

This theorem states that if three plane bodies have relative motion among themselves, their three instantaneous centres must lie on a straight line.

Consider three rigid links 1, 2, 3; 1 being a fixed link. 12 and 13 are the instantaneous centres of links 1, 2, and 1, 3 respectively. Let 23 be the instantaneous centre of links 2, 3, lying outside the line joining 12 and 13, as shown in Fig. 2.36. Links 2 and 3 are moving relative to link 1. Therefore, the motion of their instantaneous centre 23 is to be the same whether it is considered in body 2 or 3. If the point 23 is considered on link, 2, then its velocity v_2 is perpendicular to the line joining 12 and 23. If the point 23 lies on link 3, then its velocity v_3 must be perpendicular to the line joining 13 and 23. The velocities v_2 and v_3 of the point 23 are in different directions, which is not possible. The velocities v_2 and $v_3$of instantaneous centre 23 will be equal only if it lies on the line joining 12 and 13. Hence all the three instantaneous centres 12,13 and 23 must lie on a straight line.

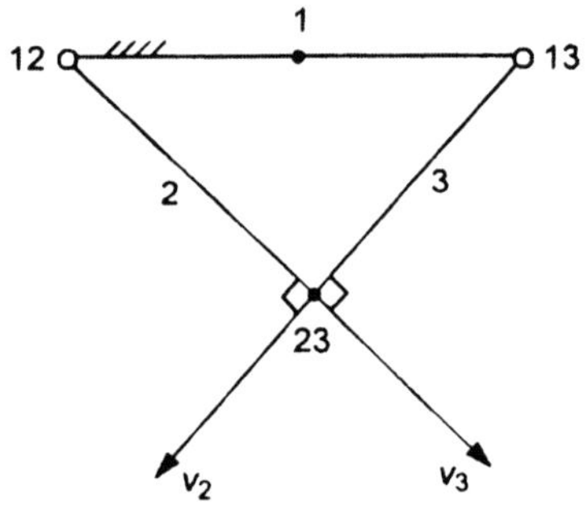

Fig. 2.36 Arnold-Kennedy's theorem.

2.4.7 Method of Locating Instantaneous Centres

The following procedure may be adopted to locate the instantaneous centres :

1. Determine the number of instantaneous centres from, $N = n\ (n - 1)/2$.
2. Make a list of all the instantaneous centres by writing the link numbers in the first row and instantaneous centres in ascending order in columns. For example, for a four-bar chain shown in Fig. 2.37 (a), we have

1	2	3	4
✓	✓	✓	
12	23	34	
(13)	(24)		
✓			
14			

3. Locate the fixed and permanent instantaneous centres by inspection, as explained in art. 2.4.5. Tick mark (✔) these instantaneous centres.
4. Locate the remaining neither fixed nor permanent instantaneous centres (circled) by using Arnold-Kennedy's theorem. This is done by a circle diagram, as shown in Fig. 2.37 (b). Mark points on a circle equal to the number of links in the mechanism. Join the points by solid lines for which instantaneous centres are known by inspection. Now join the points forming the other instantaneous centres by dotted lines. The instantaneous centre shall lie at the intersection of the lines joining the instantaneous centres of the two adjacent triangles of the dotted line. For example, in Fig. 2.37 (b), the centre 13 is located by the intersection of lines (produced) joining the instantaneous centres 12, 23 and 14, 34. Similarly, the centre 24 is located by the intersection of lines (produced) joining the centres 23, 34 and 12, 14.

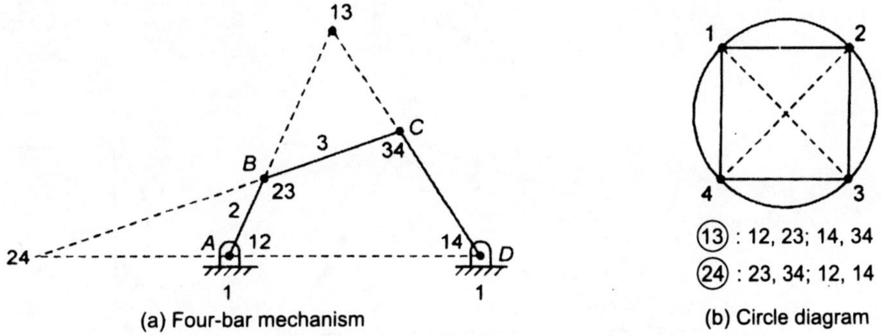

Fig. 2.37 Locating instantaneous centres of a four-bar mechanism.

2.4.8 Angular Velocity Ratio Theorem

According to this theorem, angular velocites of two links vary invrsely as the distances from their common instantaneous centre to this respective centres of rotation relative to the frame. For example, for the four-bar mechanism shown in Fig. 2.37 (a), if ω_2 is the angular velocity of link 2, then angular velocity ω_4 of link 4 will be given by the following relationship :

$$\frac{\omega_4}{\omega_2} = \frac{(24 - 12)}{(24 - 14)}$$

where 24 is the common instantaneous centre of links 2 and 4. 12 and 14 are instantaneous centers of links 2 and 4 w.r.t. the fixed link 1 (*i.e.*, frame) respectively.

If the respective centers of rotation are on the same side of the common instantaneous centre then the direction of angular velocities will be same. However, if the respective centres of rotation are on the opposite sides then the direction of angular velocities wil be opposite.

Similarly,
$$\frac{\omega_3}{\omega_2} = \frac{(23 - 12)}{(23 - 13)}$$

■ **Example 2.22** In the four-bar mechanism shown in Fig. 2.38 (a), the link 2 is rotating at angular velocity of 15 rad/s cw. Locate all the instantaneous centres of the mechanism and find (a) angular speeds of links 3 and 4, (b) linear velocities of links 3 and 4, and (c) linear velocities of points E and F.

$AB = 200\,\text{mm}, BC = 250\,\text{mm}, CD = 300\,\text{mm}, AD = 500\,\text{mm}, BAD = 60°, BE = FD = 150\,\text{mm}.$

Solution :

Given $n = 4, \omega = 15$ rad/s

Number of instantaneous centres, $N = \dfrac{4\,(4-1)}{2} = 6$

The instantaneous centres are :

1	2	3	4
✓	✓	✓	
12	23	34	
(13)	(24)		
✓			
14			

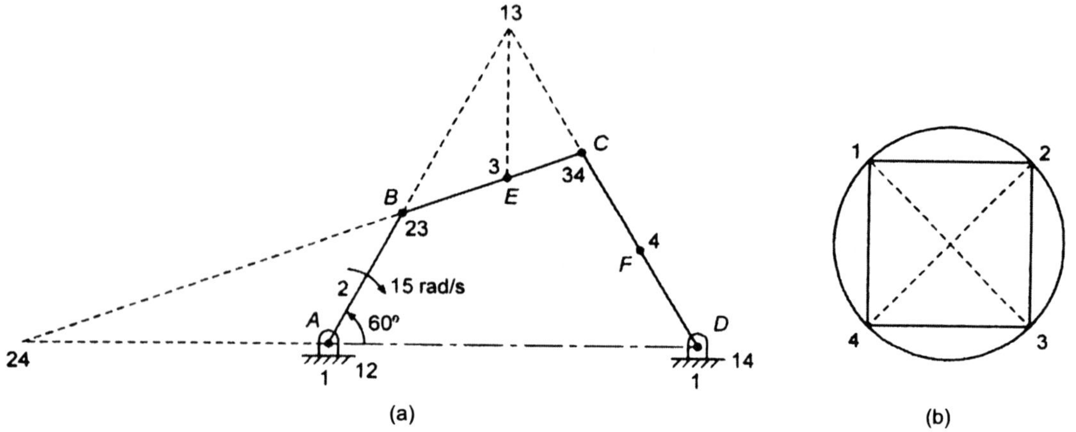

Fig. 2.38

1. Draw the configuration diagram of the four-bar mechanism to a convenient scale, as shown in Fig. 2.38 (a).
2. Locate the fixed and permanent instantaneous centres 12, 14, 23 and 34.
3. Locate the centres 13 and 24 by using Arnold-Kennedy's theorem of three centres from Fig 2.38 (b).

(13) : 12, 23; 14, 34

(24) : 12, 14; 23, 34

4. By measurement, we have

$$23 - 13 = 280 \text{ mm}$$

$$24 = 12 = 420 \text{ mm}$$

$$24 - 14 = 920 \text{ mm}$$

$$13 - 34 = 190 \text{ mm}$$

$$13 - E = 200 \text{ mm}$$

(a)
$$\frac{\omega_3}{\omega_2} = \frac{23 - 12}{23 - 13} = \frac{200}{280}$$

or
$$\omega_3 = \left(\frac{200}{280}\right) \times 15 = 10.71 \text{ rad/s}$$

$$\frac{\omega_4}{\omega_2} = \frac{24 - 12}{24 - 14} = \frac{420}{920}$$

or
$$\omega_4 = 420 \times \frac{15}{920} = 6.848 \text{ rad/s}$$

(b) $v_b = \omega_2 \,.\, AB = 15 \times 0.2 = 3 \text{ m/s}$

$$v_c = v_b \frac{13 - 23}{13 - 23}$$

$$= 3 \times \frac{190}{280} = 2.04 \text{ m/s}$$

$$v_{bc} = v_b \left[\frac{BC}{13-23}\right]$$

$$= 3 \times \frac{250}{280} = 2.678 \text{ m/s}$$

(c) $v_e = \omega_3 \,.\, (13 - E)$

$= 10.71 \times 0.2 = 2.142$ m/s

$$v_f = v_c \left(\frac{FD}{CD}\right)$$

$$= 2.04 \times \frac{150}{300} = 1.02 \text{ m/s.}$$

■ **Example 2.23** Locate the instantaneous centre of the slider crank mechanism shown in Fig. 2.39 (a). Find the velocity of slider. $OA = 160$ mm, $AB = 470$ mm, and $OB = 600$ mm, $\omega_2 = 12$ rad/s cw.

Solution :

Given $n = 4$, $\omega_2 = 2$ rad/s cw, lengths of links.

Number of instantaneous centres, $N = \dfrac{4(4-1)}{2} = 6$

1	2	3	4
✓	✓	✓	
(12)	(23)	34	
13	24		
✓			
14			

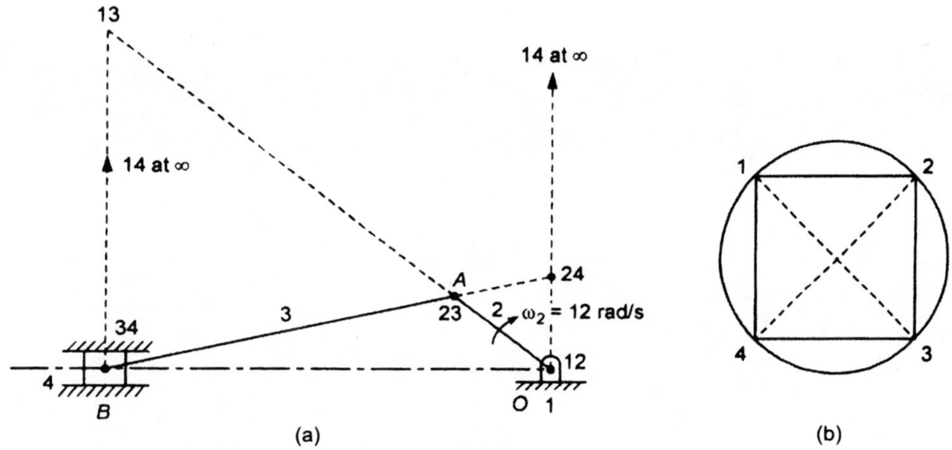

Fig. 2.39

1. Draw the configuration diagram to a convenient scale as shown in Fig. 2.39 (a).
2. Locate fixed centres 12 and 34.
3. Locate permanent centre 23. Instantaneous centre 14 is infinity.

4. Locate neither fixed nor permanent centres 13 and 24 by using Arnold-Kennedy's three centres theorem.

   As shown in Fig. 2.39 (b), we have

   (13) : 12, 23; 14, 34

   (24) : 12, 14; 23, 34

5. By measurements, we have

$$13 - 23 = 550 \text{ mm}$$
$$13 - 34 = 390 \text{ mm}$$
$$v_a = \omega_2 \,.\, OA = 12 \times 0.16 = 1.92 \text{ m/s}$$
$$= \omega_3 \,.\, (23 - 13)$$

Hence
$$\omega_3 = \omega_2 \left[\frac{(23-12)}{(23-13)}\right]$$
$$= 12 \times \frac{160}{550} = 3.491 \text{ m/s}$$

where ω_3 is the angular velocity of link 3 about 13.

Velocity of slider

$$v_b = \omega_3 \,(24 - 13)$$
$$= \omega_2 \left[\frac{(23-12)}{(23-13)}\right](34 - 13)$$
$$= \omega_2 \,.\, OA \left[\frac{(23-12)}{(23-13)}\right]$$
$$= 12 \times 0.16 \times \frac{390}{550} = 1.36 \text{ m/s.}$$

■ **Example 2.24** Locate the instantaneous centres of the mechanism shown in Fig. 2.40 (a).

Solution :

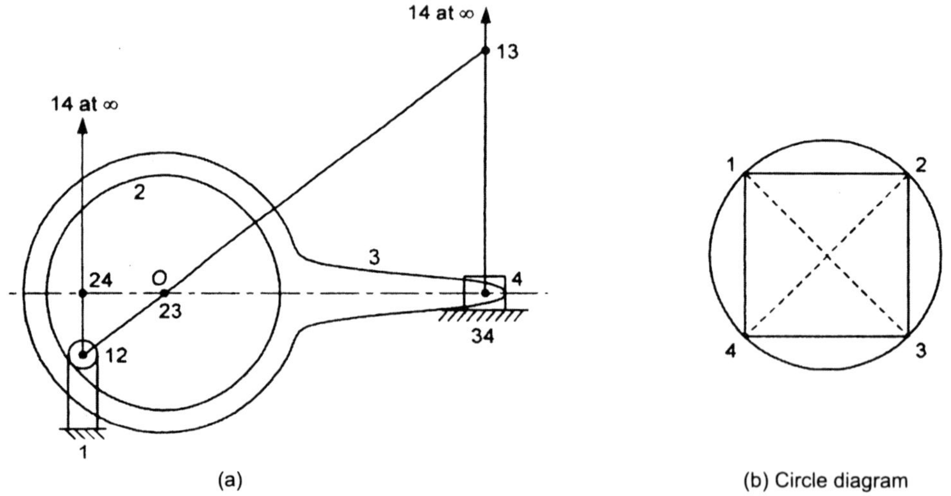

Fig. 2.40 Locating instantaneous centres of mechanisms.

Here $n = 4$

Number of instantaneous centres $= 4 \times \dfrac{3}{2} = 6$

1	2	3	4
✓	✓	✓	✓
12	23	34	
(13)	(24)		
✓			
14			

Instantaneous centres 12, 14, 23, 34 are located by inspection and centres 13, 24 are located by using Arnold-Kennedy's theorem as follows :

13 : 12, 23; 14, 34

24 : 23, 34; 12, 14

■ **Example 2.25** Using relative velocity method find the absolute velocity of the slider E in the mechanism shown in Fig. 2.41 (a). The crank O_1A rotates at 60 rpm. $O_1A = 50$ mm, $O_2B = 120$ mm, $AB = 270$ mm, $O_2D = 90$ mm and $DE = 180$ mm.

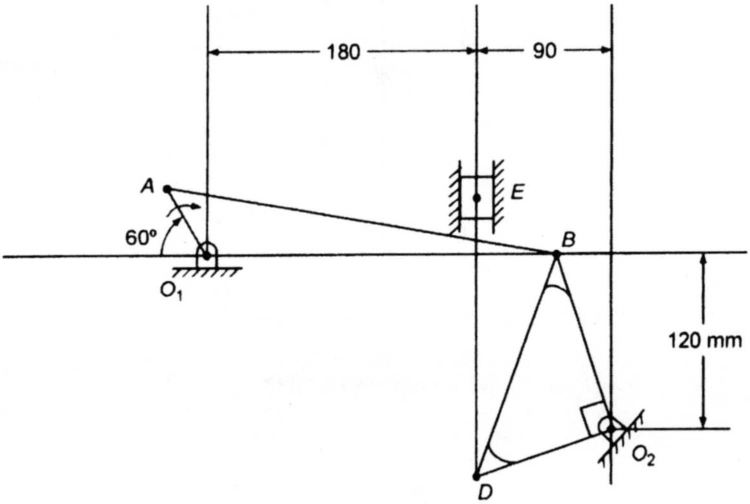

Scale: 1 cm = 50 mm

(a) Configuration diagram

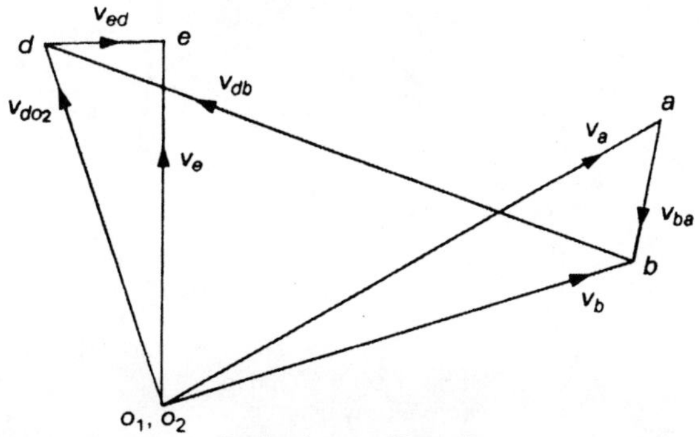

Scale: 1 cm = 50 mm/s

(b) Velocity diagram

Fig. 2.41

Solution :

$$\omega = \frac{2\pi N}{60} = \frac{2\pi \times 60}{60} = 6.28 \text{ rad/s}$$

$$v_a = \omega \times O_1A = 6.28 \times 50 = 314.16 \text{ mm/s}$$

1. Draw $v_a = o_1a \perp O_1A$, $o_1a = 6.28$ cm.
2. Draw $ab \perp AB$ and $o_2b \perp O_2B$ to meet at point b.
3. Draw $bd \perp BD$ and $o_2d \perp O_2D$ to meet at point d.
4. Draw $de \perp DE$ and $o_1e \perp O_1B$ to meet at point e.

Absolute velocity of the slider $v_e = o_1e = 4 \text{ cm} = 200 \text{ mm/s}$.

■ **Example 2.26** Figure 2.42 (a) shows a swivelling joint mechanism in which AB is the driving crank which rotates at 240 rpm clockwise.

$AB = 100$ mm, $BC = 750$ mm, $CD = 200$ mm, $AD = 600$ mm, $BE = EC$, $EF = 340$ mm, $FG = 400$ mm.

Determine (a) the velocity of slider G, (b) angular velocity of link EF and (c) the velocity of link EF in the swivel block.

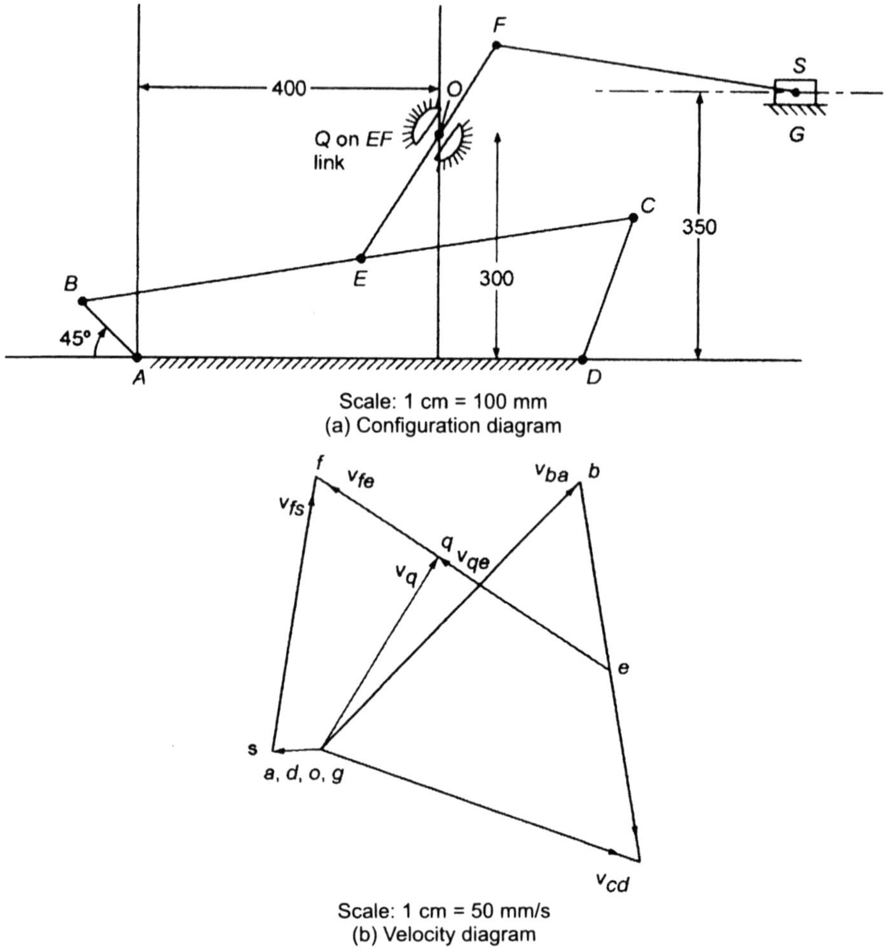

Fig. 2.42

Solution :

$$\omega = \frac{2\pi \times 240}{60} = 25.13 \text{ rad/s}$$

$$v_b = \omega \times AB = 25.13 \times 100 = 2513 \text{ mm/s}$$

1. Draw $v_{ba} = ab \perp AB$, $ab = 5.02$ cm.
2. Draw $bc \perp BC$ and $cd \perp CD$ to meet at c.
3. Since $BE = EC$, therefore $ce = eb$. Locate point e.
4. Let Q be a point on link EF at joint O. Draw $eq \perp EF$ and $oq \parallel EF$ to locate point q. Measure $eq = 2.8$ cm.
5. Now $\frac{ef}{eq} = \frac{EF}{EQ}$, $ef = 2.8 \times \frac{340}{200} = 4.76$ cm. Extend eq to f so that $ef = 4.76$ cm.

   Draw $fs \perp FS$ and $gs \parallel$ line of stroke of slider to met at points s.

   (a) Velocity of slider $s = gs = 0.6$ cm $= 300$ mm/s.

   (b) Angular velocity of link $EF = \frac{v_{fe}}{EF} = \frac{4.76 \times 500}{340} = 7$ rad/s, ccw.

   (c) Velocity of link EF in swivel block $= oq = 3$ cm 1500 mm/s.

SOLVED QUESTIONS FROM UNIVERSITY EXAMINATIONS

■ **Example 2.27** Describe relative velocity method to determine velocity of links of mechanisms by taking a suitable example. [KU, Dec., 2009]

Solution : Refer to act 2.3.1 and act 2.3.7.

■ **Example 2.28** Sketch an elliptical trammel and locate on it all instantaneous centres of rotation. [KU., Dec. 2006]

Solution :

Number of links, $n = 4$

Number of instantaneous centres,

1	2	3	4
✓	✓	✓	
12	23	34	
	✓		
(13)	24		
✓			
14			

(13) : 14, 34; 12, 23

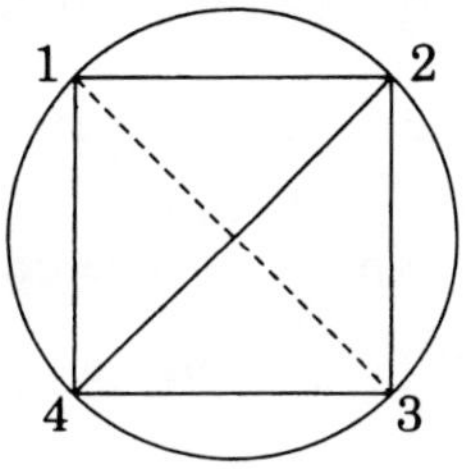

$$N = \frac{n(n-1)}{2} = \frac{4 \times 3}{2} = 6$$

The instantaneous centres are shown in Fig. 2.43.

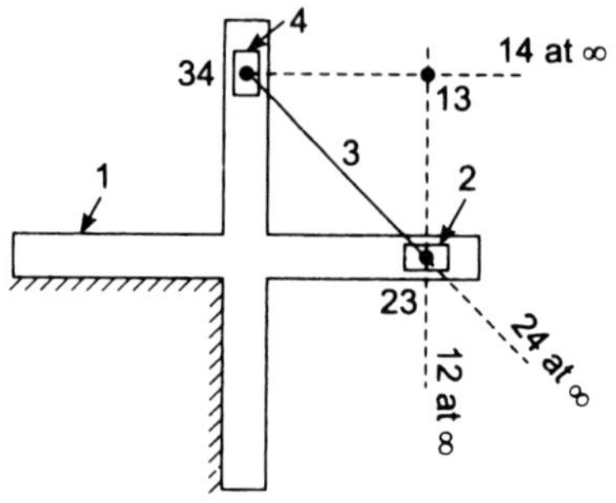

Fig. 2.43

■ **Example 2.29** In a four-bar mechanism *ABCD*, points *A* and *C* are fixed points 30 cm apart and *AB*, *CD* are bars 60 cm and 70 cm long respectively, which are connected by a bar *BD* which is 50 cm long. If *AB* rotates with uniform speed of 60 rpm, determine:

(i) Velocity of *D* when *AB* is perpendicular to *AC* and also when it makes 10° on either side of perpendicular, and

(ii) The instantaneous centre of the bar *BD* and its angular velocities in the three positions. [MDU, May 2009]

Solution : $n = 4$

Number of instantaneous centres,

$$N = \frac{n\,(n-1)}{2} = \frac{4\,(4-1)}{2} = 6$$

1	2	3	4
✓	✓	✓	
12	23	34	
(13)	(24)		
✓			
14			

(13) : 12, 23; 14, 34

(24) : 12, 14; 23, 34

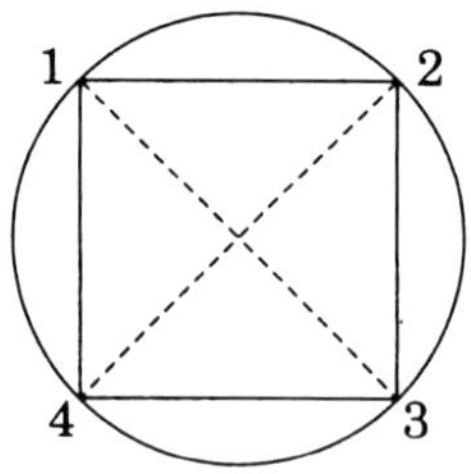

Instantaneous centres 12, 23, 34, 14 are located by inspection and 13, 24 by using Kennedy's three moment theorem.

The instantaneous centres have been located in Fig. 2.44.

$$N = 60\,\text{rpm},\ \omega_2 = \frac{2\pi N}{60} = \frac{2\pi \times 60}{60} = 2\pi\ \text{rad/s}$$

$$v_b = \omega_2 \times AB = 2\pi \times 0.6 = 3.77\ \text{m/s}$$

(i) Velocities of *D*

By measurement, we have

$$23 - 13 = 8.8\,\text{cm} = 176\,\text{cm}$$

$$24 - 12 = 18.1\,\text{cm} = 362\,\text{cm}$$

$$24 - 14 = 392\,\text{cm}$$

$$13 - 34 = 6\,\text{cm} = 190\,\text{cm}$$

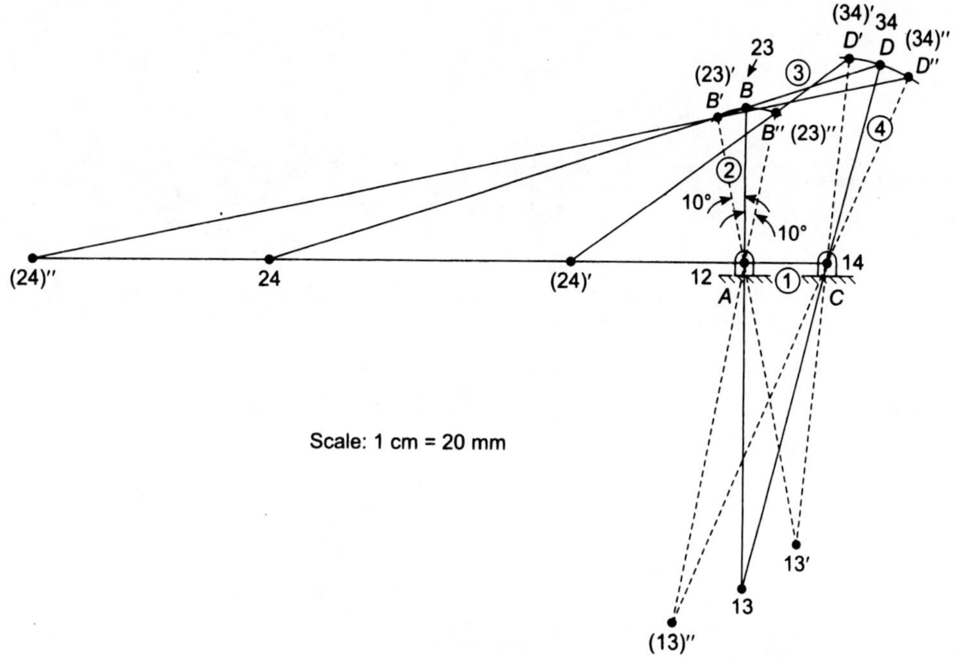

Fig. 2.44

$$\frac{\omega_3}{\omega_2} = \frac{23-12}{23-13} = \frac{60}{176}, \omega_3 = 2\pi \times \frac{60}{176} = 2.14 \text{ rad/s}$$

$$v_d = v_b \left(\frac{13-34}{13-23}\right) = 3.77 \times \frac{190}{176} = 4.07 \text{ m/s}$$

Angular velocity of link $BD = \omega_3 = 2.14$ rad/s

Velocity of D, $v_d = 4.07$ m/s

For the 10° positions on either side of AB, calculations can be made similarly.

■ **Example 2.30** In a 4-bar mechanism *ABCD*, *A* and *D* are fixed joints 100 mm apart. *AB*, *BC*, *CD* are bars of 50, 56 and 66 mm respectively. At an instant when $\angle DAB = 60°$, link *AB* has an angular velocity of 10.5 rad/s in counter clockwise direction.

Determine : (a) Velocity of point *C*, (b) Velocity of point *E* on link *BC*, when $BE = 40$ mm (c) Angular velocities of links *BC* and *CD*, (d) Velocity of an offset point *F* on link *BC*, if $BF = 45$ mm and $CF = 30$ mm and *BCF* is read clockwise. Draw neat sketches. [MDU, Dec. 2009]

Solution :

Given : $AD = 100$ mm, $AB = 50$ mm, $BC = 56$ mm, $CD = 66$ mm, $\angle DAB = 60°$, $\omega_{AB} = 10.5$ rad/s (ccw), $BE = 40$ mm, $BF = 45$ mm, $CF = 30$ mm

Let $$CG = DG = 40 \text{ mm}$$

$$v_b = \omega_{AB} \times AB = 10.5 \times 0.05 = 0.525 \text{ m/s}$$

1. Draw the configuration diagram as shown in Fig. 2.45 (a).
2. Draw the velocity diagram as shown in Fig. 2.45 (b).

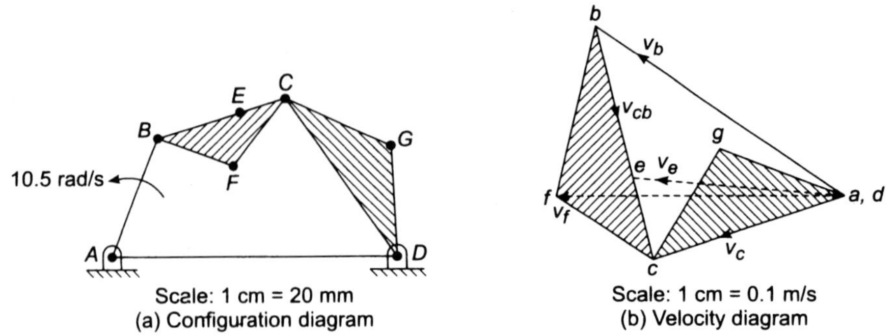

Fig. 2.45

(i) Draw $ab = v_b = 0.525$ m/s perpendicular to AB to a scale of 1 cm = 0.1 m/s

(ii) Draw $bc \perp BC$ and $dc \perp CD$ to meet at c. Then

$$bc = v_{cb} = 4 \text{ cm} = 0.4 \text{ m/s}, \; \omega_{bc} = \frac{v_{bc}}{BC} = \frac{0.4}{0.056} = 7.14 \text{ rad/s (ccw)}$$

$$dc = v_c = 3.7 \text{ cm} = 0.37 \text{ m/s}, \; \omega_{cd} = \frac{v_c}{CD} = \frac{0.37}{0.066} = 5.6 \text{ rad/s (ccw)}$$

(iii) $\frac{be}{bc} = \frac{BE}{BC}, \; be = 4 \times \frac{40}{56} = 2.86 \text{ cm}$

Locate point e in bc and join ae. Then

$$ae = v_e = 3.7 \text{ cm} = 0.37 \text{ m/s}$$

(iv) Draw $bf \perp BF$ and $cf \perp CF$ to meet at f. Join af. Then

$$af = v_f = 4.4 \text{ cm} = 0.44 \text{ m/s}$$

(v) Draw $eg \perp CG$ and $dg \perp DG$ to meet at g.

Then bcf is the velocity image of BCF and cdg is the velocity image of CDG.

■ **Example 2.31** In a 4-bar mechanism $ABCD$, points A and D are fixed points 40 mm apart. AB, BC and CD are the bars of lengths 60, 70 and 50 mm. If AB rotates with a uniform speed of 68 rpm. Determine :

(i) Velocity of C when AB is perpendicular to AD and also when its makes 30° on either side of perpendicular, and.

(ii) The instantaneous-centre of the bar BC and its angular velocities in three positions.

[MDU, Dec. 2010]

Solution :

$$\omega_2 = \frac{2\pi \times 68}{60} = 7.12 \text{ rad/s}$$

$$v_b = \omega_2 \times AB = 7.12 \times 0.06 = 0.427 \text{ m/s}$$

(i) $AB \perp AD$:

$$23 - 13 = 6 \text{ cm} = 120 \text{ mm}$$

$$24 - 12 = 11 \text{ cm} = 220 \text{ mm}$$

$$24 - 14 = 180 \text{ mm}$$

$$13 - 34 = 6 \text{ cm} = 120 \text{ mm}$$

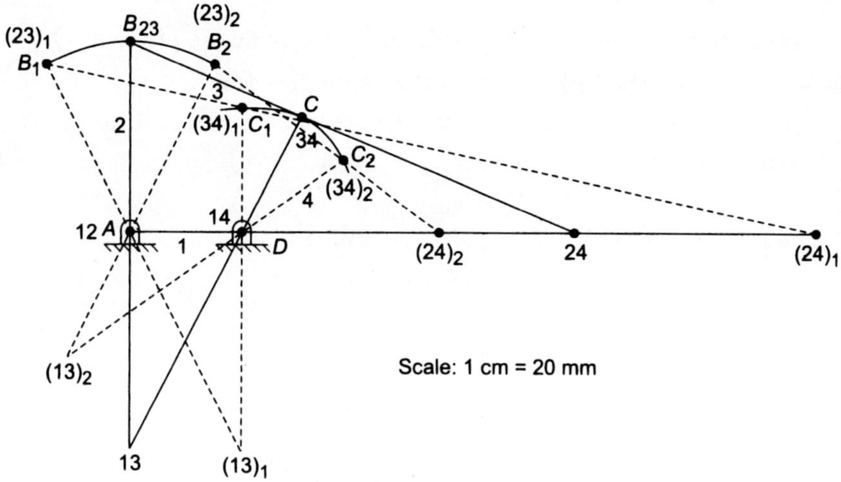

Fig. 2.46

$$\frac{\omega_3}{\omega_2} = \frac{23-12}{23-13} = \frac{60}{120}, \omega_3 = 7.12 \times \frac{1}{2} = 3.56 \text{ rad/s}$$

$$v_c = v_b\left(\frac{13-34}{13-23}\right) = 0.427\left(\frac{120}{120}\right) = 0.427 \text{ m/s}$$

Angular velocity of link $BC = \omega_3 = 3.56$ rad/s

Velocity of C, $v_c = 0.427$ m/s

For the other positions of bar AB, calculations can be made similarly.

SUMMARY

1. If ω is the angular speed of a link rotating about a point, then the linear velocity v of a point on the link at a distance r from the point of rotation is given by : $v = r\omega$
2. The relative velocity of a body A with respect to a body B is obtained by adding to the velocity of A is the reversed velocity of B. $v_{ab} = v_a - v_b$.
3. The velocity of any point on the kinematic link is given by the vector sum of the velocity of some other point in the link and the velocity of the first point relative to the other.
4. The magnitude of the velocity of any point on the kinematic link relative to the other point in the kinematic link is the product of the angular velocity of the link and distance between the two points under consideration.
5. The direction of the velocity of any point on a link relative to any other point on the link is perpendicular to the line joining the two points.
6. If C is any point on a link AB, then the corresponding point c in the velocity diagram will divide vector ab in the same ratio as the point C divides the link AB, *i.e.*, $\frac{ac}{ab} = \frac{AC}{AB}$.
7. Relative angular velocity of link 1 w.r.t. link 2,

$$\omega_{12} = \omega_1 - \omega_2$$

and relative angular velocity of 2. w.r.t. 1,

$$\omega_{21} = \omega_2 - \omega_1 = -\omega_{12}$$

8. If r = radius of the pin at joint O, then rubbing velocity at the pin joint O

$= (\omega_1 - \omega_2)\,r$, when the links move in the same direction

$= (\omega_1 + \omega_2)\,r$, when the links move in the opposite direction.

9. Forces in a mechanism

$$F_a \times v_a = F_b \times v_b$$

Considering the effect of friction, the efficiency of transmission,

$$\eta = \frac{\text{output}}{\text{input}} = \frac{F_b v_b}{F_a v_a}$$

$$F_b = \frac{\eta F_a v_a}{v_b}$$

10. Mechanical advantage,

$$MA = \frac{F_b}{F_a}$$

For a mechanism,

$$MA = \frac{\text{Output torque}}{\text{Input torque}} = \frac{T_b}{T_a} = \frac{\omega_a}{\omega_b}$$

Considering the effect of friction, $MA = \eta\omega_a/\omega_b$.

11. The number of instantaneous centres in a mechanism,

$$N = \frac{n\,(n-1)}{2}$$

where n = number of links.

12. The instantaneous centre is a point about which a link as a whole may be considred to be rotating at a given instant. The velocity of the link is zero at the instantaneous centre.

13. The angular velocities of two links vary lindearly as the distances from their common instantaneous center to their respective centres of rotation relative to the frame.

14. Arnold-Kennedy's theorem states that if three plane bodies have relative motion among themselves, their three centres must line on a straight line.

SHORT ANSWER TYPE QUESTIONS

1. What is a configuration diagram?
2. What is a velocity diagram? What are its uses?
3. What is a velocity image? Given its uses.
4. What is the velocity of rubbing? How this is determined?
5. Define instantaneous centre.
6. What are the various types of instantaneous centres?
7. How do you determine number of instantaneous centres?
8. State Arnold-Kennedy's three centres theorem.

9. Define centrode of a body.
10. State angular velocity theorem.
11. State and prove the angular velocity ratio theorem for direct contact of links.
12. State and prove Kennedy's theorem of three instantaneous centres.
13. Define Instantaneous Centres of Rotation. Give the types of instantaneous centres. With the help of neat sketch of a four-bar mechanism, show all the types of instantaneous centres.

EXERCISES

1. The dimensions of a four-bar chain shown in Fig. 2.47 are : $AD = BC = 120$ mm, $AB = 30$ mm and $CD = 60$ mm. The crank AB rotates at 100 rpm, Determine the angular speed of link CD. [**Ans.** 3.9 rad/s, cw]

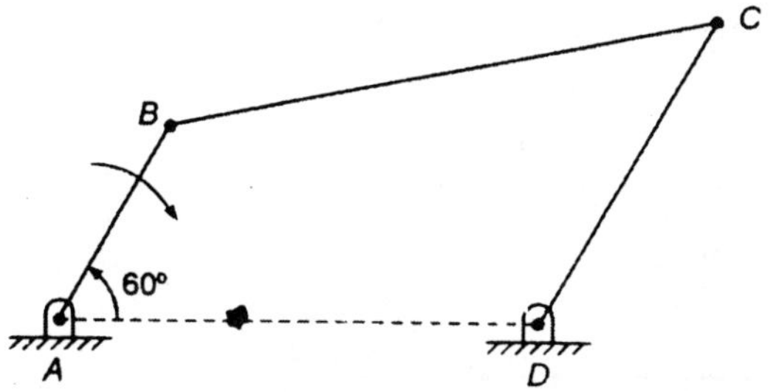

Fig. 2.47

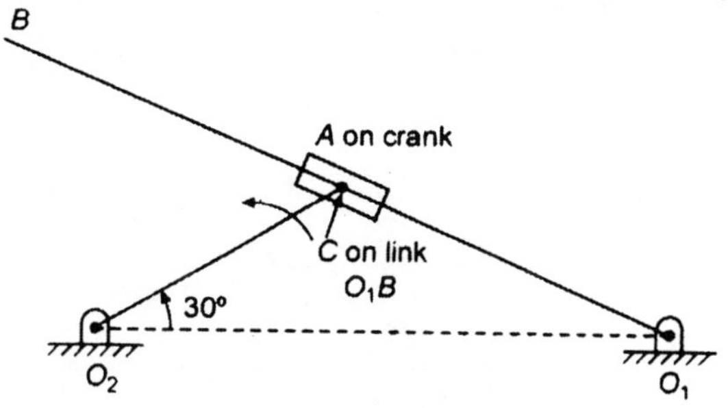

Fig. 2.48

2. In the mechanism shown in Fig. 2.48, $O_1O_2 = 210$ mm, $O_1B = 300$ mm and $O_2A = 60$ mm. The crank O_2A rotates at 300 rpm in the ccw direction. Find (a) angular speed of link O_1A, and (b) velocity of slider. [**Ans.** 8.9 rad/s ccw 1.425 m/s]
3. The dimensions of the various links of the mechanism shown in Fig. 2.49 are : $AD = DE = 150$ mm, $BC = CD = 450$ mm, $EF = 375$ mm.

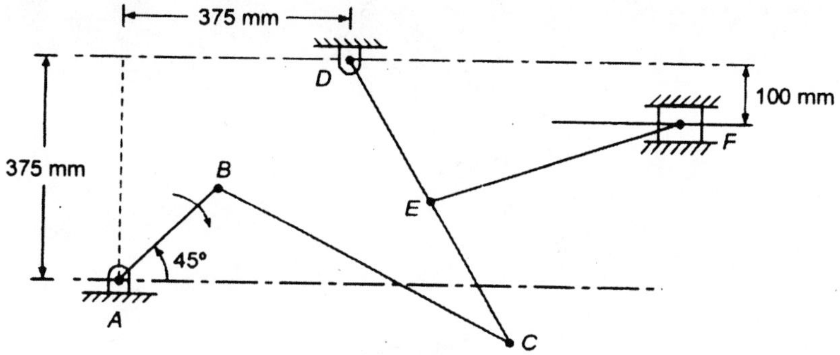

Fig. 2.49

The crank AB rotates at 120 rpm. The lever DC oscillates about the fixed point D. Determine (a) velocity of slider F, and (b) angular speed of CD. [**Ans.** 0.7 m/s 4.9 rad/s ccw]

4. In the toggle mechanism shown in Fig. 2.50, the crank OA rotates at 180 rpm and the slider is constrained to move on a horizontal path. $OA = 180$ mm, $BC = 240$ mm, $AB = 360$ mm, and $BD = 540$ mm.

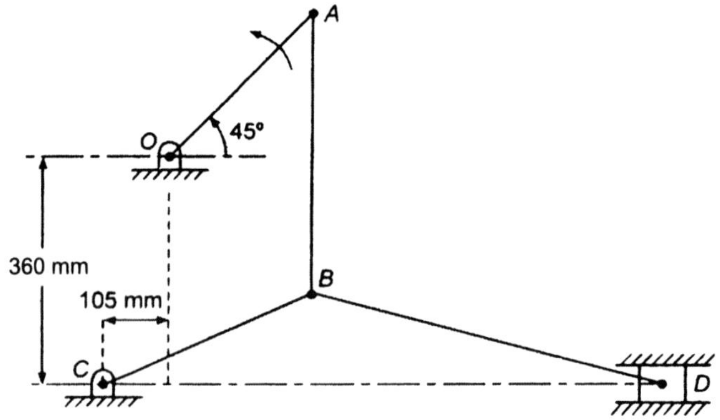

Fig. 2.50

Find (a) velocity of slider D, (b) angular speed of links AB, BC and BD, (c) velocity of rubbing on the pins of diameter 30 mm at A and D, and (d) torque applied to crank OA for a force of 2 kN at D.
[**Ans.** 1.75 m/s 2.33 rad/s ccw, 11.04 rad/s ccw, 4.53 rad/s cw; 0.1836 m/s 0.0679 m/s; N.m]

5. The crank OA of the mechanism shown in Fig. 2.51 rotates at 100 rpm clockwise. Using instantaneous centre method determine the linear velocities of points B, C and D, and angular speeds of links AB, BC and CD. [**Ans.** 3.0 m/s, 1.5 m/s, 0.98 m/s; 2.69 rad/s, 7.50 rad/s, 2.08 rad/s]

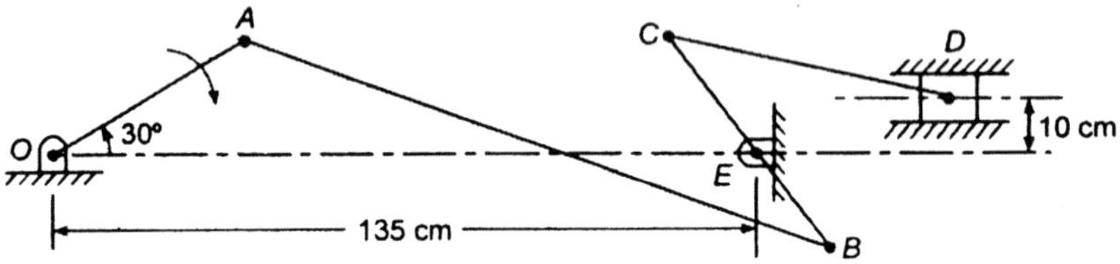

Fig. 2.51

6. Find the velocity of point C in the mechanism shown in Fig. 2.52 by using relative velocity method. Crank O_2A rotates at 20 rad/s clockwise.

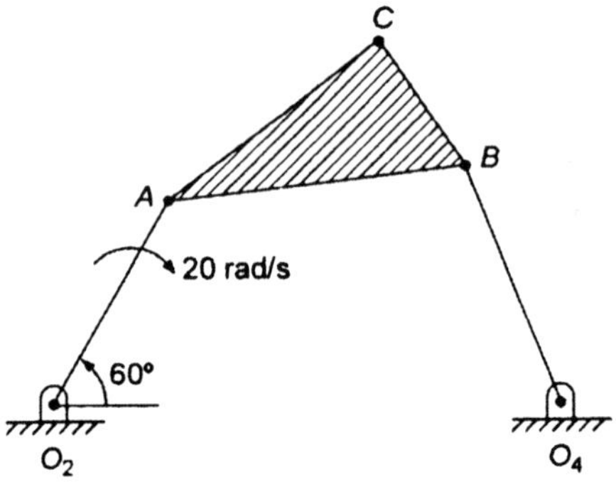

Fig. 2.52

$O_2O_4 = 350$ mm, $O_2A = 140$ mm, $AB = 250$ mm, $O_4B = 210$ mm, $BC = 100$ mm and $AC = 160$ mm
[**Ans.** 1.70 m/s]

7. The wheels 2 and 4 roll on a fixed link 1, as shown in Fig. 2.53. The angular speed of wheel 2 is 10 rad/s. Determine the angular speed of links 3 and 4 and the relative velocity of point D with respect to point E. [**Ans.** 2.5 rad/s ccw, 13 rad/s ccw; 1.5 cm/s]

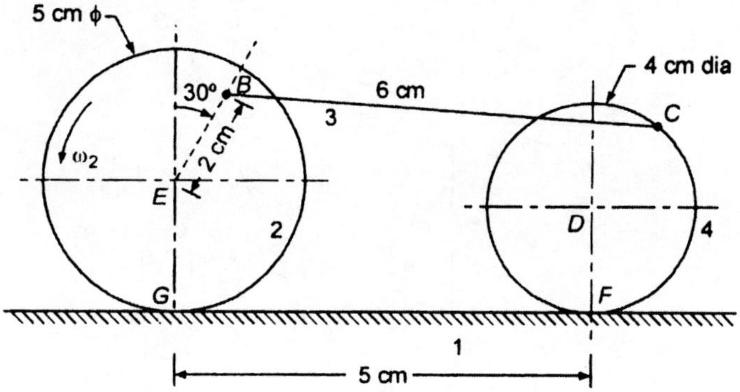

Fig. 2.53

8. Using instantaneous centre method, find the angular velocity of BC, and velocity of C and E in the mechanism shown in Fig. 2.54. Link AB renolves at 30 rpm clockwise. $AB = 0.5$ m, $BC = 1.75$ m, $CD = 1.5$ m, $AD = 2$ m, and $BE = 0.75$ m.

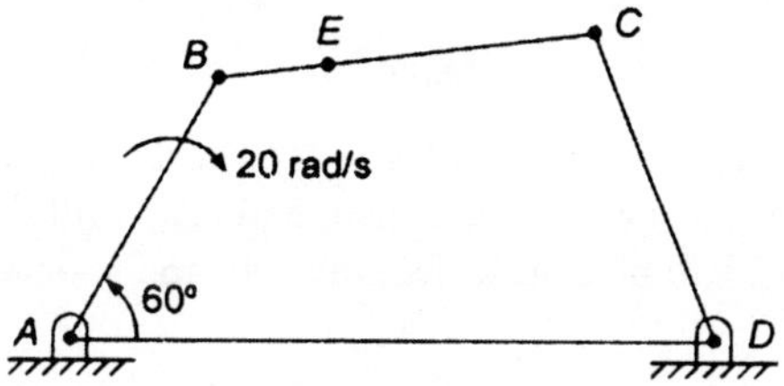

Fig. 2.54

9. Find 2.55 shows an oscillating link OAB pivoted at O which is revolving at 90 rpm ccw. $OA = 150$ mm, $AB = 75$ mm, and $AC = 250$ mm. Calculate (a) velocity of slider C, (b) angular speed of link AC, and (c) the rubbing velocity at pins O, A and C, each of diameter 20 mm.

[**Ans.** 1.32 m/s; 1.524 rad/s, cw; 30 π mm/s, 15 mm/s, 53 mm/s]

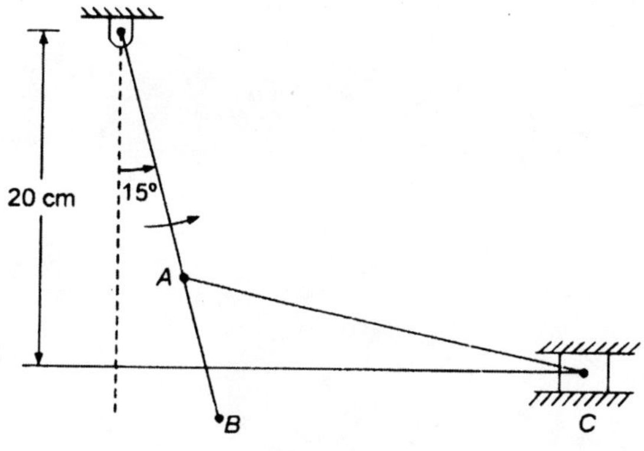

Fig. 2.55

10. The driving crank O_2A of a quick-return motion mechanism shown in Fig. 2.56 is rotating at 200 rpm cw. Find the velocity of tool post C and angular speed of link O_1B. $O_2A = 75$ mm, $O_1, B = 375$ mm, $O_1O_2 = 200$ mm, $BC = 500$ mm.

[**Ans.** 1.6 m/s, 4.53 rad/s]

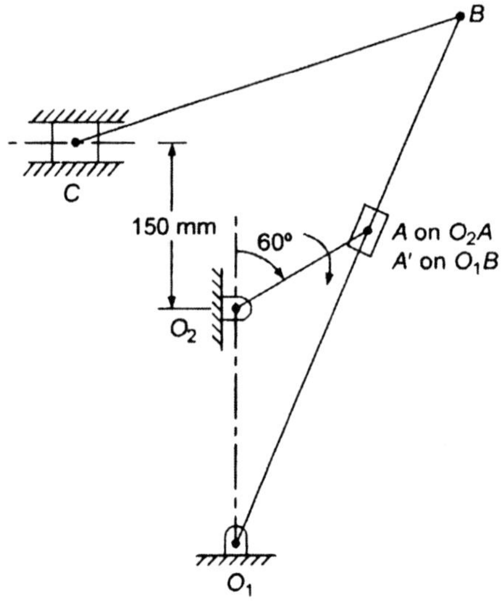

Fig. 2.56

11. In a four-bar chain *ABCD*, *AD* is the fixed link and *AB* the driving link, which rotates at 120 rpm cw under a constant driving torque of 50 N.m. At a particular instant the angular velocity of driven link *CD* is 2 rad/s. Calculate the ideal mechanical advantage and the resisting torque if the efficiency of four-bar chain is 75%. [**Ans.** 6.28, 235.5 N.m]

❑❑❑

Chapter

3

Acceleration in Mechanisms

Inside This Chapter

3.1 INTRODUCTION

In the second chapter, we have defined the concept of velocity. In this chapter, we shall study the concept of acceleration. The acceleration may be defined as the rate of change of velocity of a body with respect to time. The acceleration can be linear or angular. Linear acceleration is the rate of change of linear velocity of a body with respect to time. Angular acceleration is the rate of change of angular velocity of a body with respect to time.

Acceleration diagram : It is the graphical representation of the acceleration of the various links of a mechanism drawn on a suitable scale. It helps as to determine the acceleration of various links of the mechanism.

The determination of acceleration of various links is important from the point of view of calculating the forces and torques in the various links to carry out the dynamic analysis.

3.2 ACCELERATION OF A BODY MOVING IN A CIRCULAR PATH

Consider a body moving in a circular path of radius r with angular speed ω, as shown in Fig. 3.1(a). The body is initially at point A and in time δt moves to point B. Let the velocity change from v to $v + \delta v$ at B and the angle covered be $\delta\theta$.

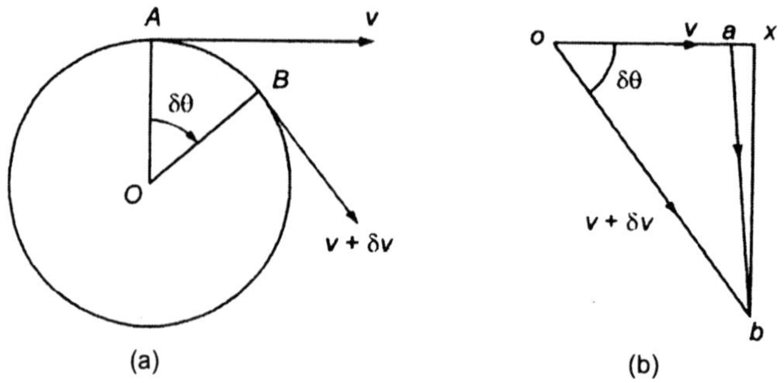

Fig. 3.1 Acceleration for a link.

The change in velocity can be determined by drawing the velocity diagram as shown in Fig. 3.1 (b). In this diagram, $oa = v$, $ob = v + \delta v$, and ab = change in velocity during time δt. The vector ab is resolved into two components ax and xb, parallel and perpendicular to oa respectively.

Now $$ax = ox - oa = ob \cos \delta\theta - oa = (v + \delta v) \cos \theta - v$$

$$xb = ob \sin \delta\theta = (v + \delta v) \sin \delta\theta$$

The rate of change of velocity is defined as the acceleration. It has two components : tangential and normal. The tangential component of acceleration f^t is the acceleration in the tangential direction. It is defined by,

$$f^t = \text{change of velocity in the tangential direction per unit time}$$

$$= \frac{ax}{\delta t} = \frac{(v + \delta v) \cos \theta - v}{\delta t}$$

$$\approx = \frac{(v + \delta v) - v}{\delta t} \quad \text{[Because for small angle } \cos \delta\theta \approx 1]$$

$$= \frac{\delta v}{\delta t} = \frac{\delta v}{dt} = \frac{d(\omega r)}{dt} = r\left(\frac{d\omega}{dt}\right) = r\alpha \quad \text{...(3.1)}$$

where α = angular acceleration

The normal component of acceleration f^n is the acceleration in the direction normal to the tangent at that instant. This component is directed towards the centre of the circular path. It is also called the radial or centripetal acceleration. It is defined by,

$$f^n = \text{change of velocity component in a direction normal to tangent per unit time}$$

$$= \frac{xb}{\delta t}$$

$$= \frac{(v + \delta v) \sin \theta}{\delta t}$$

Acceleration in Mechanisms

$$\approx \frac{(v + \delta v)\delta\theta}{\delta t} \quad \text{[Because for small angle } \sin \delta\theta \approx \delta\theta]$$

$$= \frac{v\, \delta\theta}{\delta t} + \frac{\delta v\, \delta\theta}{\delta t} \approx \frac{v\, \delta\theta}{\delta t} \quad \text{[Neglecting second term, being small]}$$

$$= \frac{v\, d\theta}{dt} = v\omega = r\omega^2 = \frac{v^2}{r} \quad \text{...(3.2)}$$

Two cases arise regarding the motion of the body.

(i) When the body is rotating with uniform angular velocity, then $d\omega/dt = 0$, so that $f^t = 0$. The body will have only normal acceleration, $f^n = r\omega^2$.

(ii) When the body is moving on a straight path, r will be infinitely large and $\frac{v^2}{r}$ will tend to zero, so that $f^n = 0$. The body will have only tangential acceleration, $f^t = r\omega$

3.3 ACCELERATION DIAGRAMS

3.3.1 Total Acceleration of a Link

Consider two points A and B on a rigid link, as shown in Fig. 3.2 (a), such that the point B moves relative to point A with an angular velocity ω and angular acceleration α.

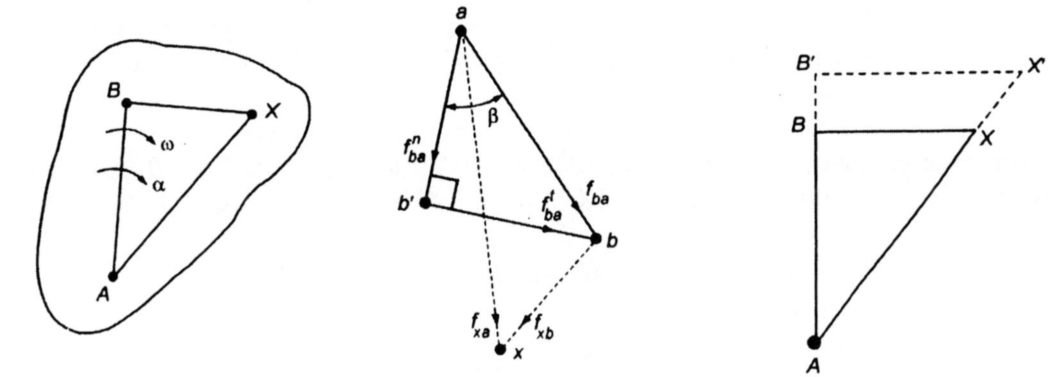

(a) Link subjected to acceleration (b) Acceleration diagram (c) Acceleration image

Fig. 3.2

Centripetal (or normal or radial) acceleration of point B with respect to point A is,

$$f^t_{ba} = \omega^2 \cdot AB = \frac{v^2_{ba}}{AB} \qquad \text{...(3.3)}$$

Tangential acceleration of point B with respect to point A,

$$f^t_{ba} = \alpha \cdot AB \qquad \text{...(3.4)}$$

Total acceleration of B w.r.t. A,

$$f_{ba} = f^n_{ba} + f^t_{ba} \qquad \text{(vector sum)}$$

$$= \omega^2 \times AB + \alpha \times AB$$

$$= (\omega^2 + \alpha) \cdot AB \qquad \text{...(3.5)}$$

$$\tan\beta = \frac{f^t_{ba}}{f^n_{ba}} = \frac{\alpha}{\omega^2} \qquad \text{...(3.6)}$$

The acceleration diagram has been represented in Fig. 3.2 (b).

3.3.2 Acceleration of a Point on a Link

The accelerations of any point X on the rigid link w.r.t. A as shown in Fig. 3.2 (a) are :

$$f_{xa}^n = \omega^2 \cdot AX \quad ...(3.7)$$

$$f_{xa}^t = \alpha \cdot AX \quad ...(3.8)$$

Total acceleration $$f_{xa} = f_{xa}^n + f_{xa}^t \quad \text{(vector sum)} ...(3.9)$$

Therefore, f_{xa} denoted by ax in the acceleration diagram shown in Fig. 3.2 (b) is inclined to XA at the same angle β. Triangles abx and ABX are similar. Thus, point x can be fixed on the acceleration image, corresponding to point X on the link. Total acceleration of X relative to A,

$$f_{xa} + ax$$

Total acceleration of X relative to B,

$$f_{xb} = bx$$

Acceleration Image

The concept of velocity image was explained in art 2.3.7 (b) in Chapter 2. It was stated that the velocity images are useful in finding velocities of effect points of links. In the same way, acceleration images are also helpful to find the accelerations of offset points of the links. The acceleration image is obtained in the same manner as a velocity image.

An easier method of making Δabx similar to ΔABX is by making AB' on AB equal to ab and drawing a line parallel to BX, meeting AX in X'. $AB'X'$ is the exact size of the triangle to be made on ab.

Take $$ax = AX' \text{ and } bx = B'X'$$

Thus the point x is located.

The method is illustrated in Fig. 3.2(c).

3.3.3 Absolute Acceleration for a Link

Consider the rigid link AB such that point B is rotating about A with angular velocity ω and angular acceleration α, as shown in Fig. 3.3 (a). The point A itself has acceleration f_a. The acceleration diagram is shown in Fig. 3.3 (b). Absolute acceleration of B,

$$f_b = f_a + f_{xa}^n + f_{xa}^t \quad \text{(Vector Sum)}$$

$$= f_a + \omega^2 \cdot AB + \alpha \cdot AB \quad ...(3.10)$$

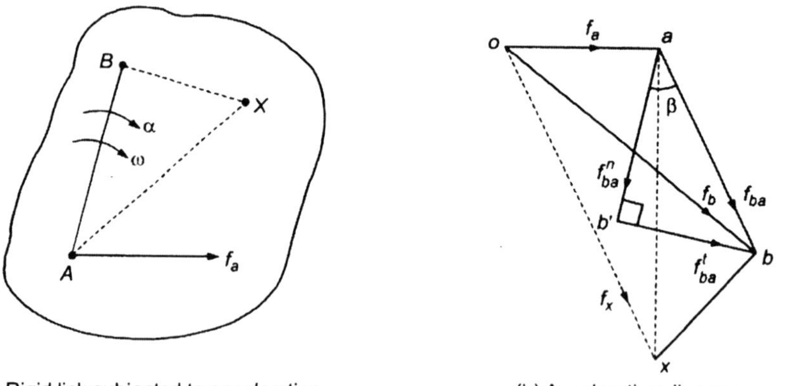

(a) Rigid link subjected to acceleration (b) Acceleration diagram

Fig. 3.3 Absolute acceleration of a rigid link.

Similarly for any other point X

$$f_x = f_a + f_{xa}^n + f^t xa$$
$$= f_a + \omega^2 \cdot AX + \alpha \cdot AX \quad ...(3.11)$$

3.3.4 Acceleration Centre

Consider the rigid link AB whose ends A and B have accelerations f_a and f_b respectively, as shown in Fig. 3.4 (a). The acceleration diagram is shown in Fig. 3.4 (b). If we select a point O on the link such that triangles aob and AOB are similar, then the acceleration of point O relative to a fixed link or fixed point O is zero. The point O is called the instantaneous centre of acceleration of link AB or **acceleration centre.**

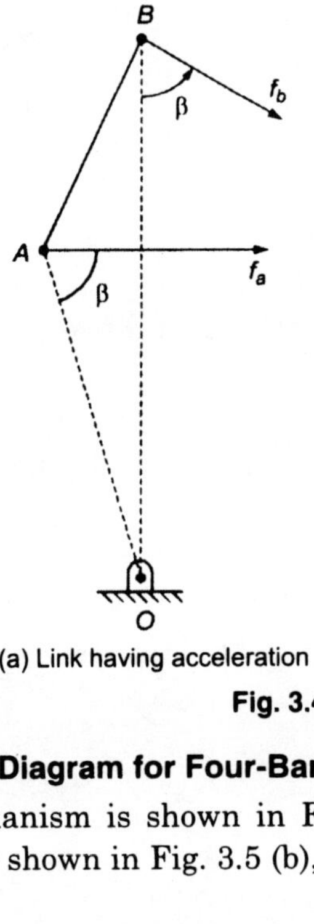

(a) Link having acceleration

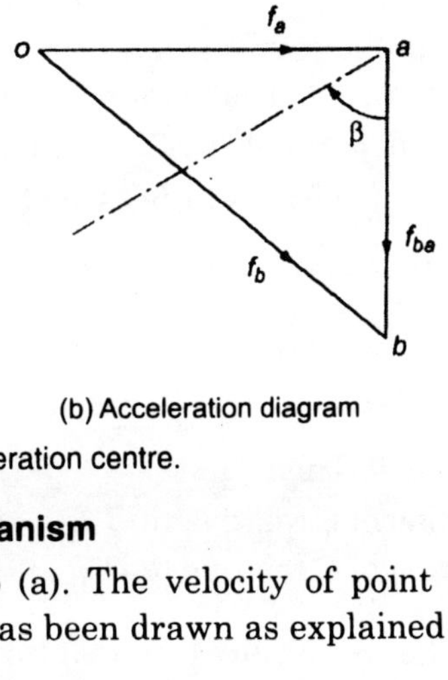

(b) Acceleration diagram

Fig. 3.4 Acceleration centre.

3.3.5 Acceleration Diagram for Four-Bar Mechanism

The four-bar mechanism is shown in Fig. 3.5 (a). The velocity of point B, $v_b = \omega \cdot AB$. The velocity diagram is shown in Fig. 3.5 (b), and has been drawn as explained in art. 2.3.7 (i). In this diagram,

$$v_b = ab \perp AB$$
$$dc \perp DC$$

and
$$bc \perp BC$$

Then Velocity of C relative to D, $v_{cd} = v_c = dc$

Velocity of C relative to B, $v_{cb} = bc$

Now are calculate the acceleration of various points and links.

$$f_{ba}^n = f_b^n = \frac{v_2^2}{AB} = ab$$

$$f_{cd}^n = \frac{v_{cb}^2}{CB} = bc'$$

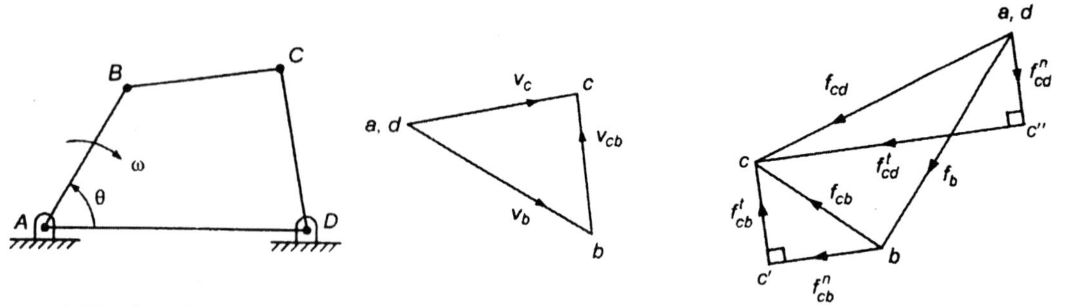

(a) Configuration diagram (b) Velocity diagram (c) Acceleration diagram

Fig. 3.5 Acceleration diagram for four-bar mechanism.

$$f_{cd}^n = \frac{v_c^2}{CB} = dc''$$

$$f_{cb} = f_{cb}^n + f_{cb}^t$$

$$bc = bc' + c'c$$

$$f_{cd} = f_{cd}^n + f_{cd}^t$$

$$dc = dc'' + c''c$$

The acceleration diagram is shown in Fig. 3.5 (c) to a suitable scale. To construct the acceleration diagram, proceed as follows :

1. Draw $ab = f_b$ parallel to AB, which is known in magnitude and direction.
2. Draw $bc' = f_{cb}^n$ parallel to BC, which is known in magnitude and direction.
3. Draw cc', representing f_{cb}^n perpendicular to bc', which is known in direction only.
4. Now draw $dc'' = f_{cd}^n$ parallel to CD, which is known in magnitude and direction.
5. Draw $c''c$, representing f_{cd}^t, perpendicular to dc'' to intersect $c'c$ at point c.
6. Join dc and bc. Then
   Acceleration of C relative to B, $f_{cb} = bc$
   Acceleration of C relative to D, $f_{cd} = f_c = dc$

3.3.6 Four-Bar Mechanisms with Ternary Link

The four-bar mechanism with a ternary link is shown in Fig. 3.6 (a), in which the driving crank has angular velocity ω and angular acceleration $\alpha \cdot v_a = \omega \times O_1A$. The velocity diagram is shown in Fig. 3.6 (b), and has been drawn as explained in art. 2.3.7 (ii). In this diagram,

(i) Velocity Diagram

$$v_a = o_1a \perp O_1A$$

$$ab \perp AB$$

$$o_2b \perp O_2B$$

Then velocity of B relative to A, $v_{ba} = ab$

Velocity of B relative to O_2, $v_{bo2} = v_b = o_2b$

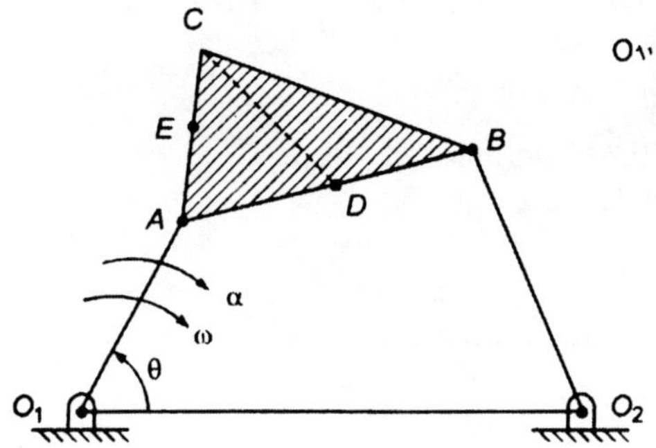

(a) Configuration diagram

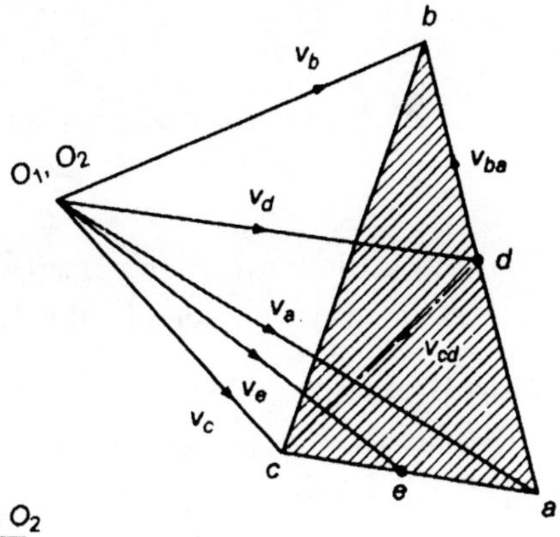

(b) Velocity diagram

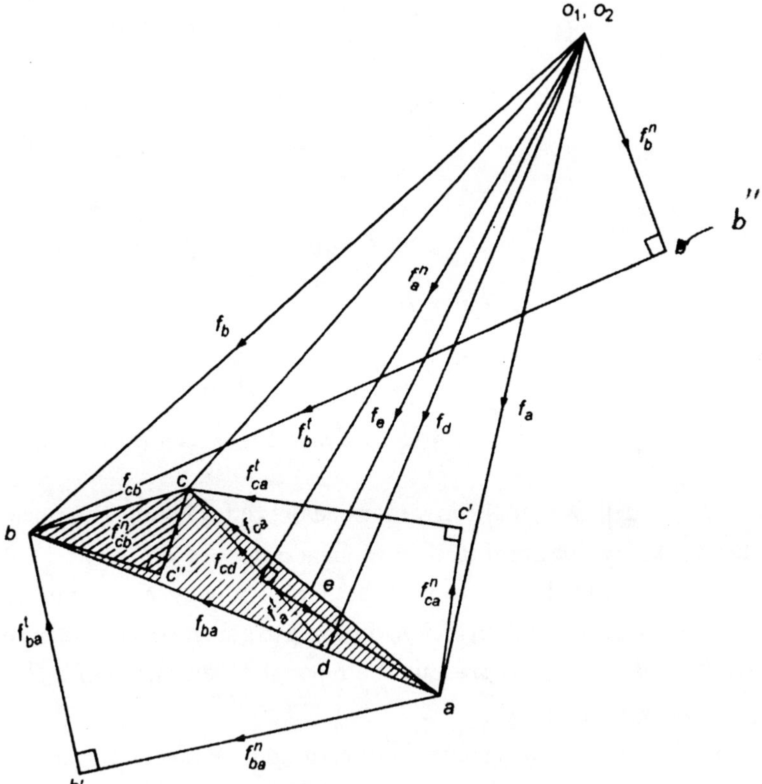

(c) Acceleration diagram

Fig. 3.6 Acceleration diagram for four bar claim with termary link.

Now $\qquad bc \perp BC$

and $\qquad ac \perp AC$

which locates point *c*.

Velocity of *C* relative to O_1, $v_c = o_1c$

The Δabc is the velocity image of ternary link *ABC*.

Now $$\frac{ad}{ab} = \frac{AD}{AB}$$

Locate point d in ab. Join o_1d and cd. Then $v_d = o_1d$ and $v_{cd} = dc$

Now $$\frac{ae}{ac} = \frac{AE}{AC}.$$

Locate point e in ac. Join o_1e, then velocity of point E, $v_e = o_1e$

This completes the construction of velocity diagram.

(ii) Acceleration Diagram

Now we calculate the accelerations of various points and links.

$$f_a^n = f_{ao1}^n = \frac{v_a^2}{O_1A} = o_1a'$$

$$f_a^t = f_{ao1}^t = \alpha \times O_1A = a'a$$

Total acceleration of A, $f_a = f_{ao1} = f_{ao1}^n + f_{ao1}^t = o_1a$ (vector sum)

$$f_b^n = f_{bo2}^n = \frac{v_b^2}{O_2B} = o_2b''$$

$$f_{ba}^n = \frac{v_{ba}^2}{AB} = ab'$$

$$b'b' = f_{ba}^t \perp ab'$$

$$b''b' = f_{bo2}^t = f_b^t \perp o_2b''$$

$$f_b = f_{bo2} = f_{ao1} + f_{ba} \qquad \text{(vector sum)}$$

$$= f_{ao1}^n + f_{ao1}^t + f_{ba}^n + f_{ba}^t \qquad \text{(vector sum)}$$

The following steps may be adopted to draw the acceleration diagram as shown in Fig. 3.6(c).

1. Draw $f_a^n = o_1a' \parallel O_1A$, to a convenient scale, to represent the normal acceleration of A, which is known in magnitude and direction.
2. Draw $f_a^t = a'a \perp O_1A$ at a' to represent the tangential acceleration of A, which is known in magnitude and direction. Join o_1a to get the total acceleration of A, $f_a = o_1\alpha$.
3. Draw $f_{ba}^n = ab' \parallel AB$ to represent the normal acceleration of AB, which is known in magnitude and direction.
4. Draw $f_{ba}^1 \perp AB$ at b' to represent the tangential acceleration of AB, which is not known in magnitude.
5. Draw $f_b^a = o_2b'' \parallel O_2B$ to represent the normal acceleration of B, which is known in magnitude and direction.
6. Draw $f_b^t \perp O_2B$ at b'' to represent the tangential acceleration of B, which is not known in magnitude.

7. The lines of f_{ba}^t and f_b^t meet at point b. Join ab and o_2b. Then

   Total acceleration of B relative to A, $f_{ba} = ab$

   Total acceleration of B relative to O_2, $f_b = o_2b$

(iii) Acceleration of Intermediate Points

8. To calculate the acceleration of point D in AB, we have

$$\frac{ad}{ab} = \frac{AD}{AB}$$

   This locate point d in ab. Join o_1d. Then acceleration of point D, $f_d = o_1d$

   Similarly acceleration of any point in link O_1A and O_2B can be determined.

(iv) Acceleration of Offset Points

9. Draw a line ac' parallel to AC to represent f_{ac}^n which is known in magnitude and direction. Draw another line perpendicular to AC at c' to represent f_{ac}^t, which is unknown in magnitude.
10. Draw a line bc'' parallel to BC to represent f_{cb}^n which is known in magnitude and direction. Draw another line perpendicular to BC at c'' to represent f_{cb}^t, which is unknown in magnitude. These two lines meet at c. Join ac and bc.

    The triangle abc and ABC are similar. Then Δabc is the **acceleration image** of ternary link ABC. To find the acceleration of offset point C, join o_1c. Then

    Total acceleration of C, $f_c = o_1c$
11. To find the acceleration of point E in AC, we have

$$\frac{ae}{ac} = \frac{AE}{AC}$$

    Thus locate point e in ac. Join o_1e. Then

    Total acceleration of E, $f_e = o_1e$

(v) Angular Acceleration of Links

It may be observed that the tangential component of acceleration occurs due to the angular acceleration of a link. Thus, angular acceleration can be determined if the tangential acceleration is known.

In Fig. 3.6 (c), tangential acceleration of B relative to A, $f_{ba}^t = b'b$

i.e., tangential acceleration B relative to A is in a direction b' to b or in a counter-clockwise direction about A.

Now
$$f_{ba}^t = \alpha_{ba} \times AB$$

$$\therefore \quad \alpha_{ba} = \frac{f_{ba}^t}{AB}$$

Tangential acceleration of A relative to B, $f_{ab}^t = bb'$

i.e., tangential acceleration A relative to B is in a direction b to b' or in a counter-clockwise direction about B with magnitude $\alpha_{ab} = \dfrac{f_{ab}^t}{AB}$, which is the same as α_{ba}.

Thus, angular acceleration of a link about one end of the link is the same in magnitude and direction as the angular acceleration about the other end

Tangential acceleration of B relative to O_2, $f_b^t = f_{bo2}^t = b''b$

i.e., B relative to O_2 moves in a direction from b'' to b or B moves in the counter-clockwise direction about O_2.

$$\alpha_b = \alpha_{bo2} = \frac{f_b^t}{O_2B} \text{ ccw}$$

3.3.7 Acceleration Diagram for Slider-Crank Mechanism

(i) Velocity Diagram

For the slider crank mechanism is shown in Fig. 3.7 (a). $v_a = \omega \cdot OA$ The velocity diagram is shown in Fig. 3.7 (b) and has been drawn as explained in art. 2.3.8. In this diagram,

$$v_a = oa \perp OA$$
$$ab \perp AB$$
$$ob \parallel OB$$

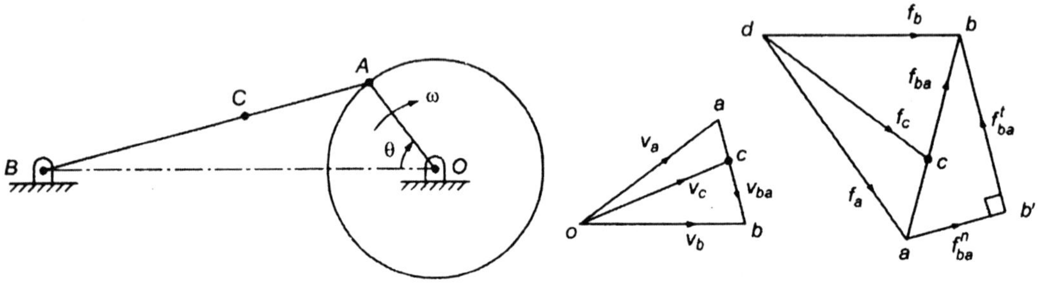

(a) Configuration diagram (b) Velocity diagram (c) Acceleration diagram

Fig. 3.7 Acceleration diagram of slider-crank mechanism.

Then

Velocity of slider, $v_b = ob$

Velocity of connecting rod, $v_{ba} = ab$

(ii) Acceleration Diagram

The acceleration diagram is shown in Fig. 3.7 (c), in which

$$f_a = \omega^2 \times OA = \frac{v_a^2}{OA} = oa$$

$$f_{ba}^n = \frac{v_{ba}^2}{AB} = ab'$$

The following steps may be adopted to draw the acceleration diagram :

1. Draw $oa = f_a$ parallel to OA on a convenient scale to represent the normal acceleration of the crank OA.
2. Draw $ab' = f_{ba}^n$ parallel to AB to represent the normal acceleration of the connecting rod, which is known in magnitude and direction.

3. Draw a line perpendicular to *AB* at *b′* to represent the tangential acceleration f^t_{ba} of *AB*, which is unknown in magnitude.
4. Draw a line parallel to *OB* at *o* to represent the acceleration f_b of the slider *B*. Let these two lines intersect at *b*. Join *ab*. Then
   Linear acceleration of slider *B*, $f_b = ob$
5. To find the acceleration of any point *C* in *AB*, we have

$$\frac{ac}{ab} = \frac{AC}{AB}$$

   Locate point *c* in *ab*. join *ac*. Then
   Acceleration of point *C*, $f_c = oc$

■ **Example 3.1** In the four-bar mechanism shown in Fig. 3.8, the lengths of the various links are : $AB = 190$ mm, $BC = CD = 280$ mm, $AD = 500$ mm, $\angle BAD = 55°$. The crank *AB* rotates at 10 rad/s in the clockwise direction. Determine (a) the acceleration of the links *BC* and *CD*, and (b) angular accelerations of *BC* and *CD*.

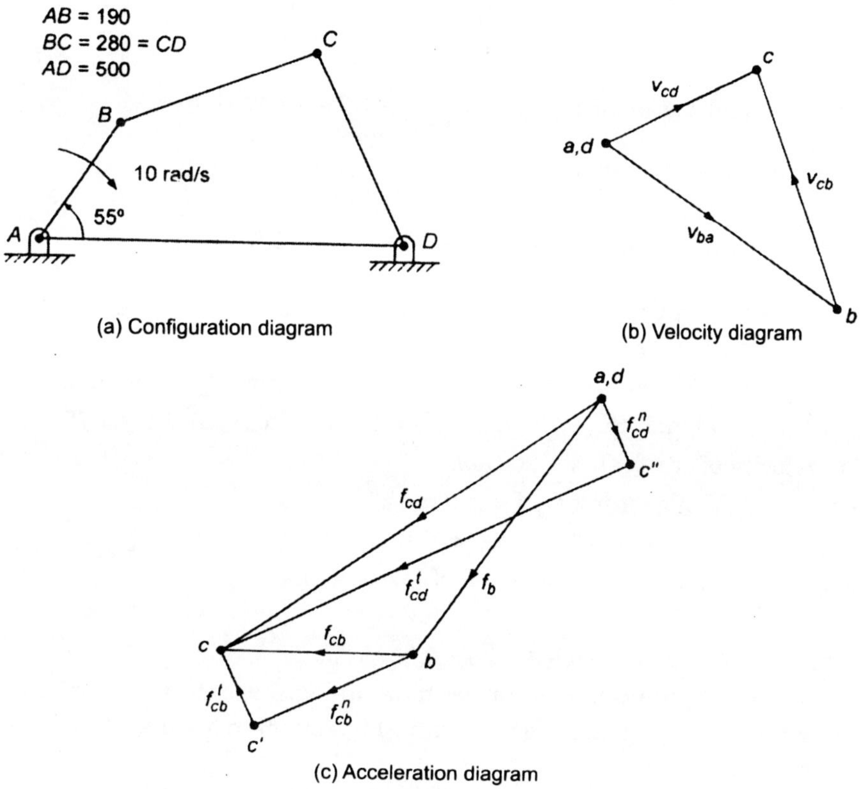

Fig. 3.8

Solution :

Linear velocity of point $B, v_b = v_{ba} = AB \times \omega = 0.19 \times 10 = 1.9$ m/s

1. Draw the configuration diagram to a scale of 1 cm = 100 mm, as shown in Fig. 3.8 (a).
2. Draw the velocity digram shown in Fig. 3.8 (b), as explained in art 3.3.5, to a scale of 1 cm = 0.5 m/s.

3. Measure the velocity of C relative to $B, v_{cb} = bc = 3.3 \text{cm} = 1.65$ m/s, and the velocity of C relative to $D, v_{cd} = dc = 2.2$ cm $= 1.1$ m/s.
4. Calculate the normal accelerations :

$$f_b^n = f_{ba}^n = \frac{v_{ba}^2}{AB} = \frac{(1.9)^2}{0.9} = 19 \text{ m/s}^2$$

$$f_{cb}^n = \frac{v_{cb}^2}{BC} = \frac{(1.65)^2}{0.28} = 9.72 \text{ m/s}^2$$

$$f_{cd}^n = \frac{v_{cd}^2}{CD} = \frac{(1.1)^2}{0.28} = 4.32 \text{ m/s}^2$$

5. Draw the acceleration diagram to a scale of 1 cm $= 5 \text{m/s}^2$, as shown in Fig. 3.8 (c) and as explained in art. 3.3.5.
6. Measure the acceleration of C relative to $B, f_{cb} = bc = 2.1$ cm $= 10.5 \text{ m/s}^2$.

   Acceleration of C relative to $D, f_{cd} = dc = 5.1$ cm $= 25.5 \text{m/s}^2$
7. Tangential acceleration of BC, $f_{cb}^t = cc' = 0.9$ cm $= 4.5 \text{ m/s}^2$

   Angular acceleration of BC, $\alpha_{bc} = \dfrac{f_{bc}^t}{BC} = \dfrac{4.5}{0.28} = 16.07 \text{ rad/s}^2$ cw

   Tangential acceleration of CD, $f_{cd}^t = cc'' = 5.2$ cm $= 26 \text{ m/s}^2$

   Angular acceleration of CD, $\alpha_{cd} = \dfrac{f_{cd}^t}{CD} = \dfrac{26}{0.28} = 92.857 \text{ rad/s}^2$ ccw

■ **Example 3.2** A four-bar mechanism with ternary link is shown in Fig. 3.9 (a). The lengths of various links is given as below :

$$O_1O_2 = 600 \text{ mm}, O_1A = 300 \text{ mm}, AB = 400 \text{ mm}, O_2B = 50 \text{ mm},$$
$$AC = 300 \text{ mm}, BC = 250 \text{ mm}, AD = 100 \text{ mm, and } \angle AO_1O_2 = 75°.$$

Angular velocity of crank $O_1A = 20$ rad/s

Angular acceleration of crank $O_1A = 100 \text{ rad/s}^2$

Determine (a) acceleration of coupler. AB, (b) acceleration of lever O_2B, (c) acceleration of points C and D, and (d) angular acceleration of ternary link.

Solution :

Linear velocity of $A, v_a = \omega \times O_1A = 20 \times 0.3 = 6$ m/s

1. Draw the configuration diagram as shown in Fig. 3.9 (a) to a scale of 1 cm = 100 mm.
2. Draw the velocity diagram as shown in Fig. 3.9 (b) to a scale of 1 cm = 1 m/s, and as explained in art. 3.3.6.
3. Measure $ab = v_{ba} = 3.5$ cm $= 3.5$ m/s; $o_2b = v_b = 4.7$ cm $= 4.7$ m/s; $o_1c = v_c = 3.5$ cm $= 3.5$ m/s $v_{ca} = ac = 2.7$ cm $= 2.7$ m/s; $v_{cb} = 2.1$ cm $= 2.1$ m/s
4. Now $\dfrac{ad}{ab} = \dfrac{AD}{AB}$, and $\dfrac{3.5 \times 100}{4000} = 0.875$ cm. Locate point d in ab and join o_1d. Then $v_d = o_1d = 5.5$ cm $= 5.5$ m/s.

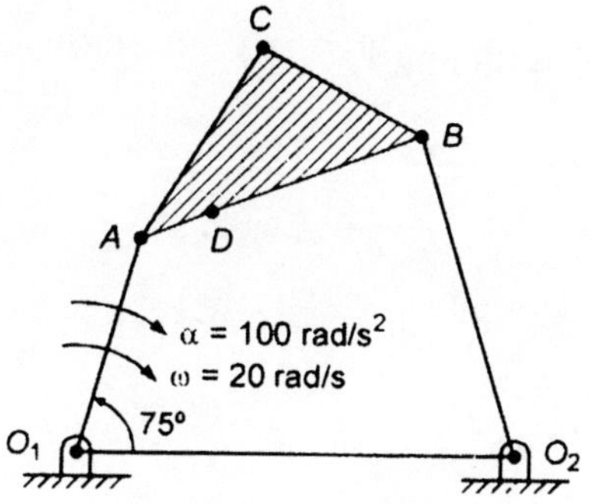

Scale: 1 cm = 100 mm
(a) Configuration diagram

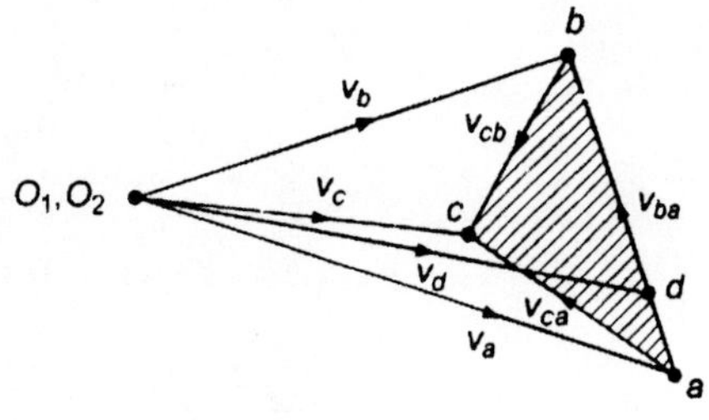

Scale: 1 cm = 1 m/s
(b) Velocity diagram

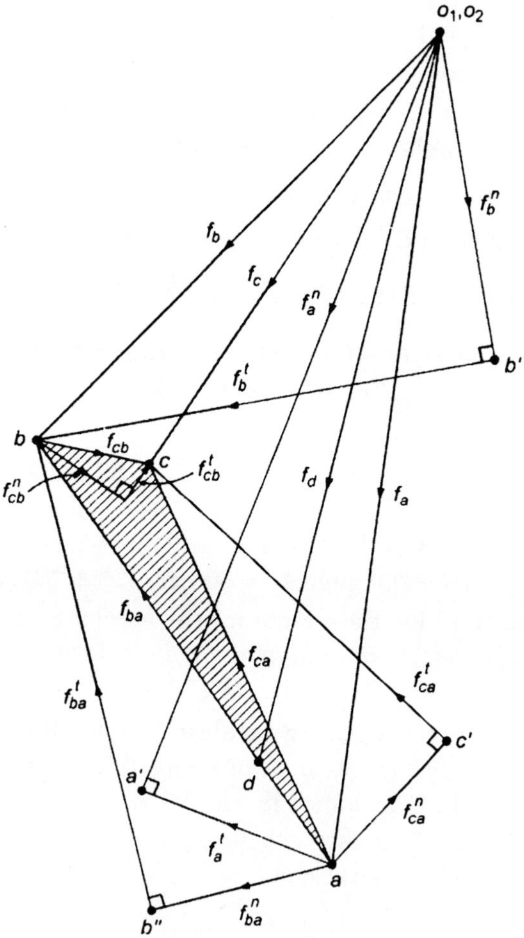

Scale: 1 cm = 10 m/s²
(c) Acceleration diagram

Fig. 3.9

5. Calculate the acceleration of various points as follows :

$$f_a^n = \frac{v_a^2}{O_1A} = \frac{6^2}{0.3} = 120 \text{ m/s}^2$$

$$f_a^t = \alpha \times O_1A = 100 \times 0.3 = 30 \text{ m/s}^2$$

$$f_{ba}^n = \frac{v_{ba}^2}{AB} = \frac{(3.5)^2}{0.4} = 30.625 \text{ m/s}^2$$

$$f_b^n = \frac{v_b^2}{O_2B} = \frac{(4.7)^2}{0.45} = 49.1 \text{ m/s}^2$$

$$f_{ac}^n = \frac{v_{ac}^2}{AC} = \frac{(2.7)^2}{0.3} = 24.3 \text{ m/s}^2$$

$$f_{cb}^n = \frac{v_{cb}^2}{BC} = \frac{(2.1)^2}{0.25} = 17.64 \text{ m/s}^2$$

6. Draw the acceleration diagram on a scale of 1 cm = 10 m/s^2, as shown in Fig. 3.9 (c), and as explained in art. 3.3.6.
7. Measure the various required lengths on the acceleration diagram.
   (a) Acceleration of coupler AB, $f_{ba} = ab = 7.8 \text{ cm} = 78 \text{ m/s}^2$
   (b) Acceleration of lever O_2B, $f_b = o_2b = 8.6 \text{ cm} = 86 \text{ m/s}^2$
   (c) Acceleration of point C, $f_c = o_1c = 7.9 \text{ cm} = 79 \text{ m/s}^2$
8. Locate point d in ab from the relation, $ad = \frac{ab \times AD}{AB} = 1.95$ cm. Then

   Acceleration of point D, $f_d = o_1d = 11 \text{ cm} = 110 \text{ m/s}^2$.
9. Tangential acceleration of AB, $f_{ba}^t = b'' b = 7.2 \text{ cm} = 72 \text{ m/s}^2$

   Angular acceleration of AB, $\alpha_{ab} = \frac{f_{ba}^t}{AB} = \frac{72}{0.4} = 180 \text{ rad/s}^2$ cw.

■ **Example 3.3** In the slider-crank shown in Fig. 3.10 (a), the lengths of the various links are :

$OA = AC = 200$ mm, $AB = 600$ mm, $\angle AOB = 30°$.

The crank rotates at 10 rad/s. Determine (a) the acceleration of the connecting rod AB, (b) acceleration of slider B, and (c) acceleration of a point C in AB.

Solution : Linear velocity of point A, $v_a = \omega \times OA = 10 \times 0.2 = 2$ m/s

1. Draw the configuration diagram to a scale of 1 cm = 100 mm, as shown in Fig. 3.10 (a).
2. Draw the velocity diagram to a scale of 1 cm = 0.5 m/s, as shown in Fig. 3.10 (b), and following the procedure explained in art. 3.3.7.
3. Measure the velocity of B relative to A, $v_{ba} = ab = 3.5 \text{ cm} = 1.75$ m/s.
4. Calculate the normal accelerations :

$$f_a^n = \frac{v_a^2}{OA} = \frac{2^2}{0.2} = 20 \text{ m/s}^2$$

$$f_{ba}^n = \frac{v_{ba}^2}{AB} = \left(\frac{1.75}{0.6}\right)^2 = 5.1 \text{ m/s}^2$$

5. Draw the acceleration diagram as shown in Fig. 3.10(c) to a scale of 1 cm = 4 m/s^2, and following the procedure as explained in art. 3.3.7.

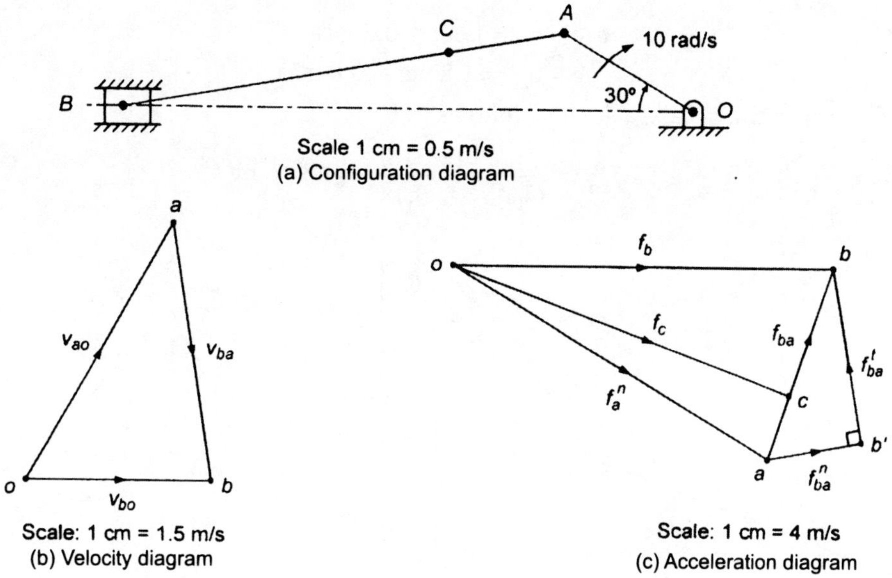

Fig. 3.10

6. Measure the acceleration of connecting rod, $f_{ba} = ab = 2.7 \text{ cm} = 10.8 \text{ m/s}^2$

   Acceleration of slider B, $f_b = ob = 5.2 \text{ cm} = 20.8 \text{ m/s}^2$

7. Now $\dfrac{a}{ab} = \dfrac{AC}{AB}$

$$ac = \frac{2.7 \times 200}{600} = 0.9 \text{ cm}$$

8. Locate point c in ab and join oc. Then

   Acceleration of point C, $f_c = oc = 4.9 \text{ cm} = 19.6 \text{ m/s}^2$.

■ **Example 3.4** In the mechanism shown in Fig. 3.11(a), determine the acceleration of the slider C. $O_1A = 100$ mm, $AB = 120$ mm, $O_2B = 150$ mm, and $BC = 350$ mm. The crank O_1A rotates at 240 rpm.

Solution :

Angular velocity of crank O_1A, $\omega = \dfrac{2\pi \times 240}{60} = 25.13$ rad/s

Linear velocity of point A, $v_a = \omega\, O_1A = 25.13 \times 0.1 = 2.513$ m/s

1. Draw the configuration diagram to a scale of 1 cm = 50 mm, as shown in Fig. 3.11(a).
2. Draw the velocity diagram as shown in Fig. 3.11(b) to a scale of 1 cm = 0.5 m/s.
3. Draw $v_a = o_1a \perp OA = 5.5$ cm.
4. Draw a line perpendicular to AB at 'a' and another line perpendicular to O_2B at o_2 to meet at b.
5. Draw aline perpendicular to BC at b and another line parallel to the line of stroke of the slider C at o_1. Then

$$o_2b = v_b = 1.3 \text{ cm} = 0.65 \text{ m/s}$$
$$ab = v_{ba} = 4.8 \text{ cm} = 2.4 \text{ m/s}$$

$$o_1c = v_c = 1 \text{ cm} = 0.5 \text{ m/s}$$

$$bc = v_{cb} = 1.2 \text{ cm} = 0.6 \text{ m/s}$$

6. Calculate the accelerations as follows :

$$f_{ao1}^n = \frac{v_a^2}{O_1A} = \frac{(2.513)^2}{0.1} = 63.15 \text{ m/s}^2$$

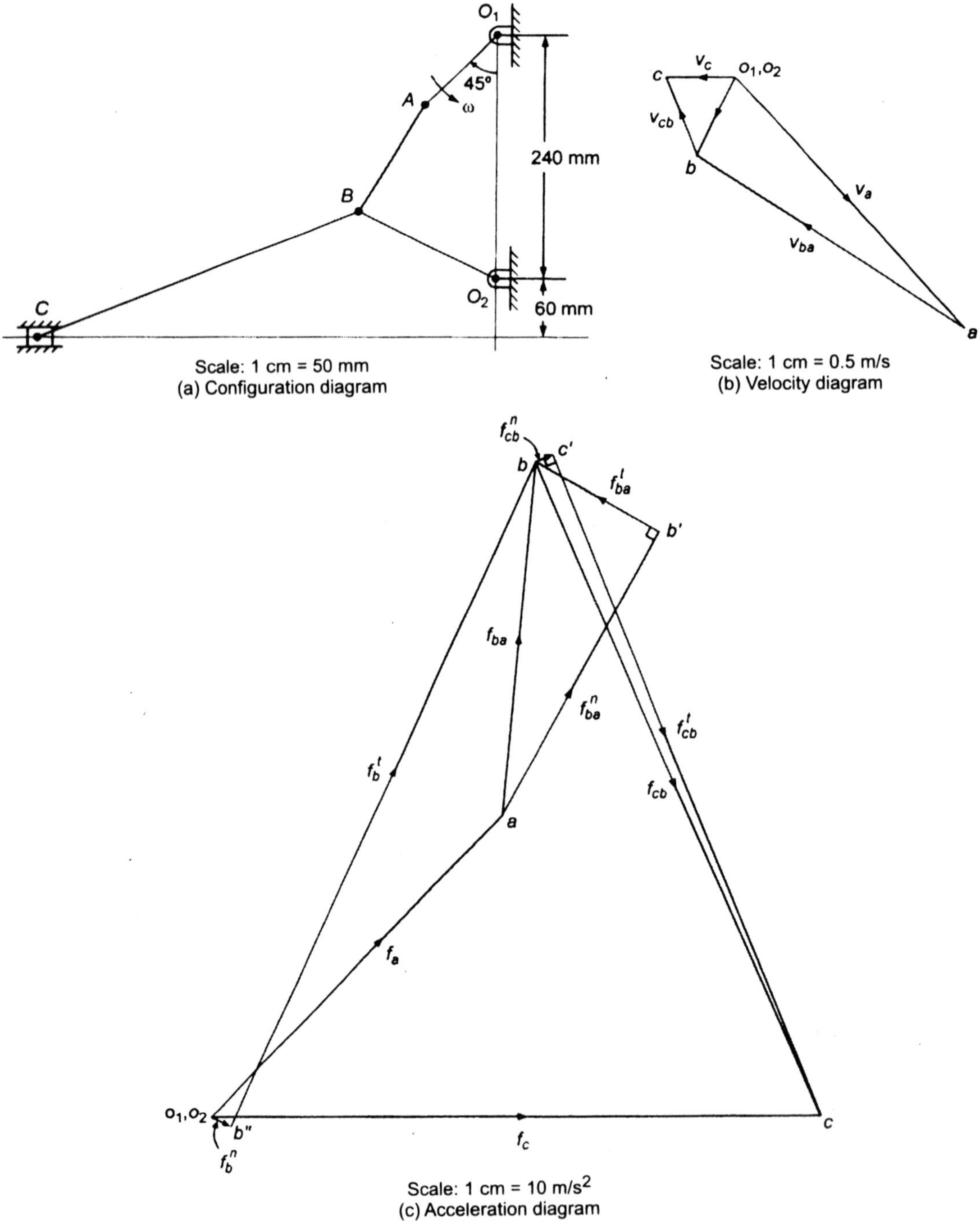

Fig. 3.11

$$f_{ba}^n = \frac{f_{ba}^2}{AB} = \frac{(2.4)^2}{0.12} = 48 \text{ m/s}^2$$

$$f_{bo2}^n = \frac{v_b^2}{O_2B} = \frac{(0.65)^2}{0.15} = 2.81 \text{ m/s}^2$$

$$f_{cb}^n = \frac{v_{cb}^2}{BC} = \frac{(0.6)^2}{0.35} = 1.03 \text{ m/s}^2$$

7. Draw acceleration diagram as shown in Fig. 3.11 (c) to a scale of a cm = 10m/s^2.
8. Draw $f_a^n = o_1a = 6.3$ cm parallel to O_1A.
9. Draw $ab' = f_{ba}^n = 4.8$ cm at 'a' parallel to AB.
10. At b' draw a line perpendicular to AB representing f_{ba}^t whose magnitude is unknown.
11. Draw $o_2b'' = f_b^n = 0.28$ cm parallel to O_2B.
12. Draw a line perpendicular to O_2B at b'' representing f_b^t whose magnitude is unknown.
    This line intersects the line b' at b. Join ab to get $f_{ba} = ab$.
13. Draw a line $bc' = 0.1$ cm at b parallel to BC representing f_{cb}^n. Draw a line at c' perpendicular to BC to represent f_{cb}^t and another line at o_1 parallel to the line of stroke of the slider at C meeting the first line at c. Then
    Acceleration of the slider at $C, f_c = o_1c = 9.1$ cm $= 91 \text{ m/s}^2$

3.4 CORIOLIS ACCELERATION

It has been observed in art. 3.2 that the total acceleration of a point with respect to another point in a rigid link is the vector sum of the its normal and tangential components. This holds true when the distance between the two points is fixed and the relative acceleration of the two points on a moving rigid link has been considered. If the distance between the two points varies, *i.e.*, the second point which was considered stationary, nor slides, the total acceleration will contain one additional component, called **Coriolis** component of acceleration.

Consider a slider B on a link OA such that when the link OA is rotating clockwise with angular velocity ω the slider B moves outward with linear velocity v, as shown in Fig. 3.12. Let in time $\delta\theta$ the angle turned through by link OA be $\delta\theta$ to occupy the new position OA' and the slider moves to position E. The slider can be considered to move from B to E as follows :

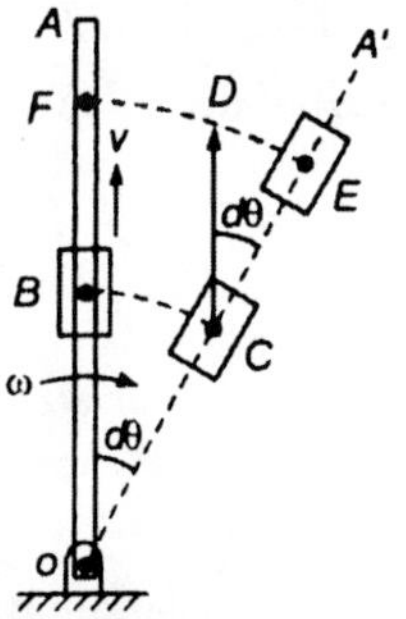

Fig. 3.12

1. From B to C due to angular velocity ω of link OA.
2. C to D due to outward velocity v of the slider.
3. D to E due to acceleration perpendicular to the rod, *i.e.*, due to coriolis acceleration.

   Now and DE = arc EF − arc FD

   = arc EF − arc BC

$$= OF \cdot d\theta - OB \cdot d\theta$$
$$= (OF - OB)d\theta$$
$$= BF \cdot d\theta$$
$$= CD \cdot d\theta$$

Now $CD = v \cdot dt$

and $d\theta = \omega \cdot dt$

Hence $\text{arc } DE = (v \cdot dt) \cdot (\omega \cdot dt)$

Now arc DE = chord DE, as $d\theta$ is very small.

Therefore, $$DE = v \cdot \omega \, (dt)^2 \quad \text{...(1)}$$

But $$DE = \frac{1}{2} \cdot f^{cr} \, (dt)^2 \quad \text{...(2)}$$

where f^{cr} is the constant coriolis acceleration of the particle. Hence from (1) and (2), we get

$$f^{cr} = 2v\omega \quad \text{...(3)}$$

The direction of the coriolis acceleration component is such to as to rotate the sliding velocity vector in the same sense as the angular velocity of OB. This is achieved by turning the sliding velocity vector through 90° in a manner that the velocity of this vector is the same as that of angular velocity of OA. The method of finding the direction of coriolis acceleration is illustrated in Fig. 3.13.

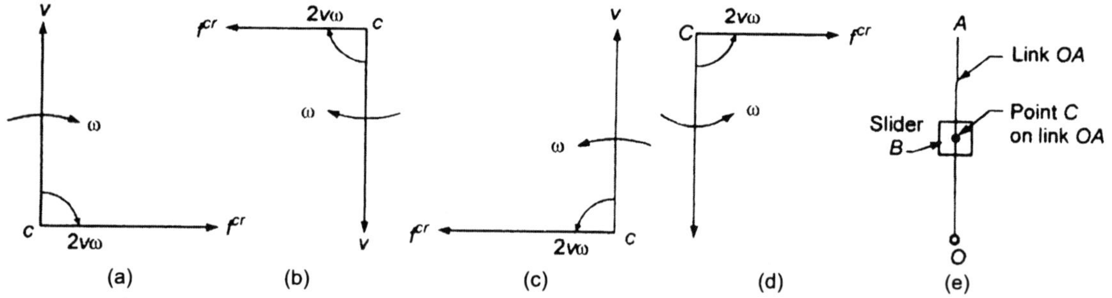

Fig. 3.13 Finding direction of coriolis acceleration.

■ **Example 3.5** In the crank and slotted lever type quick return motion mechanism shown in Fig. 3.14 (a), the crank AB rotates at 120 rpm. Determine (a) velocity of ram at D, (b) magnitude of coriolis acceleration component, and (c) acceleration of ram at $D \cdot AB = 200$ mm, $OC = 800$ mm, $CD = 600$ mm, $OA = 300$ mm.

Solution : Velocity diagram :

$$\omega = 2\pi \times \frac{120}{60} = 12.57 \text{ rad/s}$$

$$v_b = v_{ba} = \omega \cdot AB = 12.57 \times 0.2 \times 2.51 \text{ m/s}$$

1. Draw the configuration diagram to a scale of 1 cm = 100 mm, as shown in Fig. 3.14 (a). By measurement, $OP = 450$ mm.
2. Draw the velocity diagram as shown in Fig. 3.14 (b) to a scale of 1 cm = 0.5 m/s, by adopting the following steps.
3. Draw $ba \perp AB = 5.02$ cm to represent the linear velocity of point B on link AB.

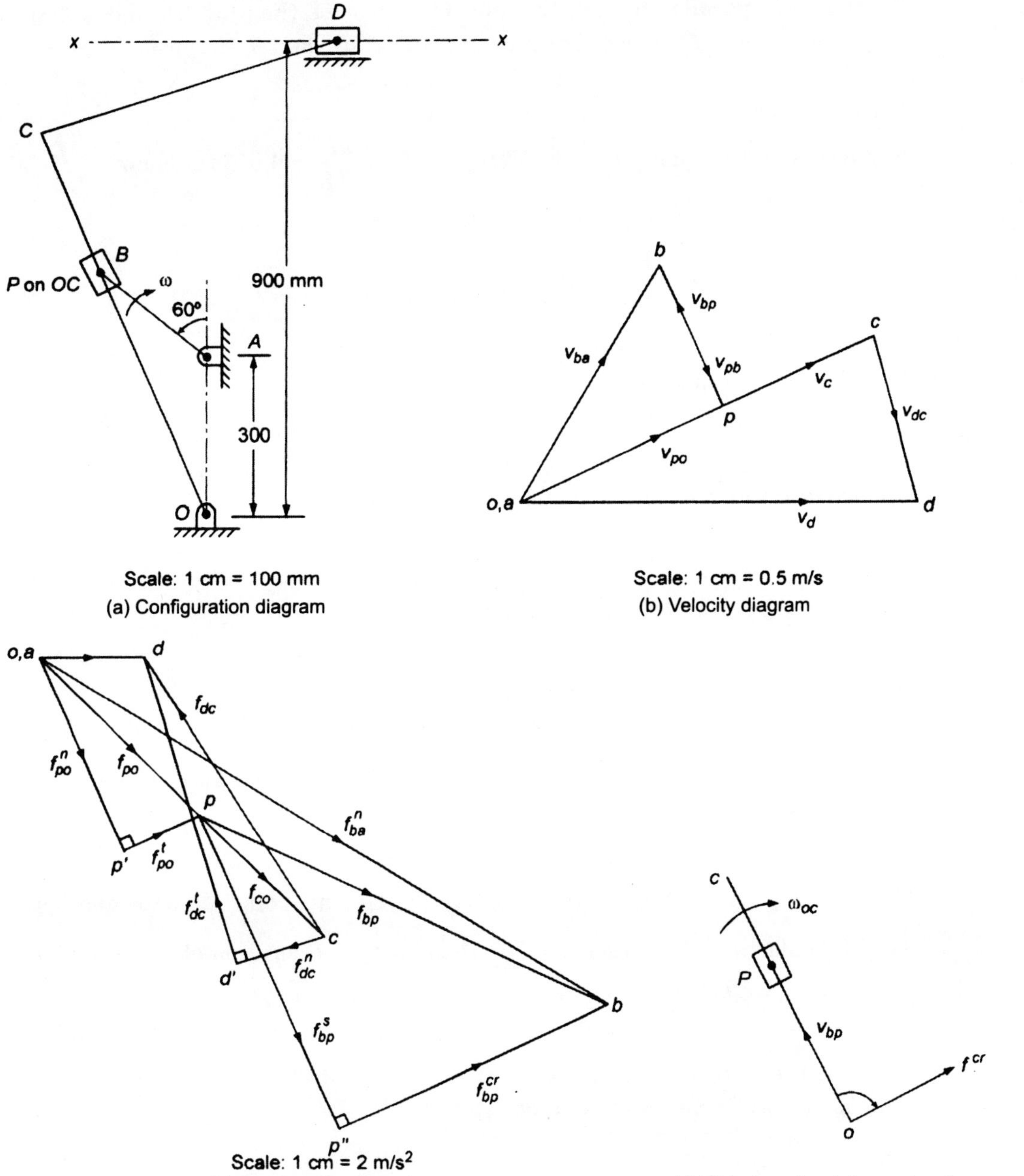

Scale: 1 cm = 100 mm
(a) Configuration diagram

Scale: 1 cm = 0.5 m/s
(b) Velocity diagram

Scale: 1 cm = 2 m/s²
(c) Acceleration diagram

(d) Direction of coriolis acceleration

Fig. 3.14

4. Let the coincident point of B on the slotted lever OC be P. Draw a line at a perpendicular to OC and another line at b parallel to OC to meet at point p.
5. By measurement, $op = 4.2$ cm. Then $ac = \dfrac{op \times OC}{OP} = \dfrac{4.2 \times 800}{600} = 7.47$ cm
6. Produce op to oc.

7. Draw a line parallel to XX, the line of stroke of the ram D, and a line at c perpendicular to CD to meet the previous line at d. Then
Velocity of the slider $D, v_d = od = 7.8\text{ cm} = 3.9\text{ m/s}$
Velocity of point $C, v_c = oc = 7.47\text{ cm} = 3.74\text{ m/s}$
Angular velocity of slotted lever $OC, \omega_{oc} = \dfrac{v_c}{OC} = \dfrac{3.74}{0.8} = 4.675\text{ rad/s cw}$

$$v_{po} = op = 4.2\text{ cm} = 2.1\text{ m/s}$$
$$v_{bp} = pb = 3\text{ cm} = 1.5\text{ m/s}$$
$$v_{cd} = cd = 3.2\text{ cm} = 1.6\text{ m/s}$$

Acceleration Diagram

1. Calculate the acceleration of various points.
Coriolis acceleration of B relative to $P, f_{bp}^{cr} = 2v_{bp} \times \omega_{oc} = 2 \times 1.5 \times 4.675 =$ 14.025 m/s. The direction of coriolis component of acceleration is shown in Fig. 3.14(d), which is perpendicular to OC.

$$f_{ba}^n = \frac{v_{ba}^2}{AB} - \frac{(2.51)^2}{0.2} = 31.5\text{ m/s}^2$$

$$f_{po}^n = \frac{v_{po}^2}{OP} = \frac{(2.1)^2}{0.45} = 9.8\text{ m/s}^2$$

$$f_{dc}^n = \frac{v_{dc}^2}{CD} = \frac{(1.6)^2}{1.6} = 4.27\text{ m/s}^2$$

2. Draw the acceleration diagram as shown in Fig. 3.14 (c) to a scale of 1 cm = 2 m/s^2, by adopting the following steps.
3. Draw $ab = f_{ba}^n = 15.75\text{ cm}$ parallel to AB.
4. Draw $p'' b = f_{bp}^{cr} = 7.0\text{ cm}$ perpendicular to OC at point b, and another line perpendicular to OC representing f_{bp}^s, the sliding acceleration of point B relative to point $P \cdot f_{bp} = pb$ is the total acceleration of B relative to P.
5. Draw $op' = f_{po}^n = 4.9\text{ cm}$ parallel to OP and another line $p'p$ perpendicular to it to meet the previous line at point p.
6. Join bp and op. By measurement, $op = 5.3\text{ cm}$.
7. Now $oc = \dfrac{op \times OC}{OP} = \dfrac{5.3 \times 800}{450} = 9.4\text{ cm}$
8. Extend op to point c so that $oc = 9.4\text{ cm}$.
9. Draw $cd' = f_{dc}^n - 2.13\text{ cm}$ parallel to CD and draw a line perpendicular to CD at d' to represent the tangential acceleration of CD.
10. Draw a line at o parallel to the line of action of the ram at D meeting the previous line at d. Join cd. Then
Acceleration of the ram at $D, f_d = od = 2.5\text{ cm} = 5\text{ m/s}^2$
(a) Velocity of ram = 3.9 m/s

(b) Coriolis acceleration = 14.025 m/s^2

(c) Acceleration of ram = 2.5 m/s^2

■ **Example 3.6** Draw the acceleration diagram for the Whitworth mechanism shown in Fig. 3.15 (a).

$$O_1O_2 = 300 \text{ mm}, O_1A = 200 \text{ mm},$$
$$AB = 700 \text{ mm}, BC = 800 \text{ mm},$$
$$\angle AO_1O_2 = 45°.$$

The crank O_1A rotates at 120 rpm clockwise. Determine (a) Velocity of ram C, (b) coriolis acceleration, (c) angular acceleration of AB, (d) angular acceleration of BC, and (e) acceleration of ram C.

Solution :

$$\omega = 2\pi \times \frac{120}{60} \; 12.57 \text{ rad/s}$$

$$v_a = 12.57 \times 0.2 = 2.51 \text{ m/s}$$

Velocity diagram

1. Drw the configuration diagram to a scale of 1 cm = 100 cm, as shown in Fig. 3.15 (a). By measurement, O_2P = 460 mm. Point P is the coincident point of A on the link AB, whereas point A is on the crank O_1A.
2. Draw the velocity diagram to a scale of 1 cm = 0.5 m/s, as shown in Fig. 3.15 (b), by adopting the following steps.
3. Draw $o_1a = v_a = 5.02 \text{ cm} \perp O_1A$.
4. Draw a line at 'a' parallel to PB and another line at o_2 perpendicular to PB to meet at point $p \cdot v_{ap} = pa$ represents the velocity of sliding between A and P.
5. By measurement, o_2p = 4.5 cm.

   Now $$bp = o_2p \times \frac{PB}{O_2P} = \frac{4.5 \times 700}{460} = 6.85 \text{ cm}$$

6. Extend line po_2 to b such that pb = 6.85 cm.
7. Draw a line at b perpendicular to BC and another line parallel to the path of the ram at C to meet at point c. Then

$$v_{ap} = pa = 2.3 \text{ cm} = 1.15 \text{ m/s}$$
$$v_{po2} = o_2p = 4.5 \text{ cm} = 2.25 \text{ m/s}$$
$$v_{cb} = bc = 0.8 \text{ cm} = 0.4 \text{ m/s}$$

Velocity of ram $C, v_c = o_1c = 1.9 \text{ cm} = 0.85 \text{ m/s}$

Angular velocity of link O_2P, $\omega_{o2p} = \frac{v_{po2}}{O_2P} = \frac{2.25}{0.46} = 4.89$ rad/s cw (see figure)

Angular velocity of BC, $\omega_{bc} = \frac{v_{bc}}{BC} = \frac{0.4}{0.8} = 0.5$ rad/s

Acceleration Diagram

Coriolis acceleration, $f_{ap}^{cr} = 2v_{ap}\,\omega_{o2p} = 2 \times 1.15 \times 4.89 = 11.247 \text{ m/s}^2$

Refer to figure for direction of coriolis acceleration.

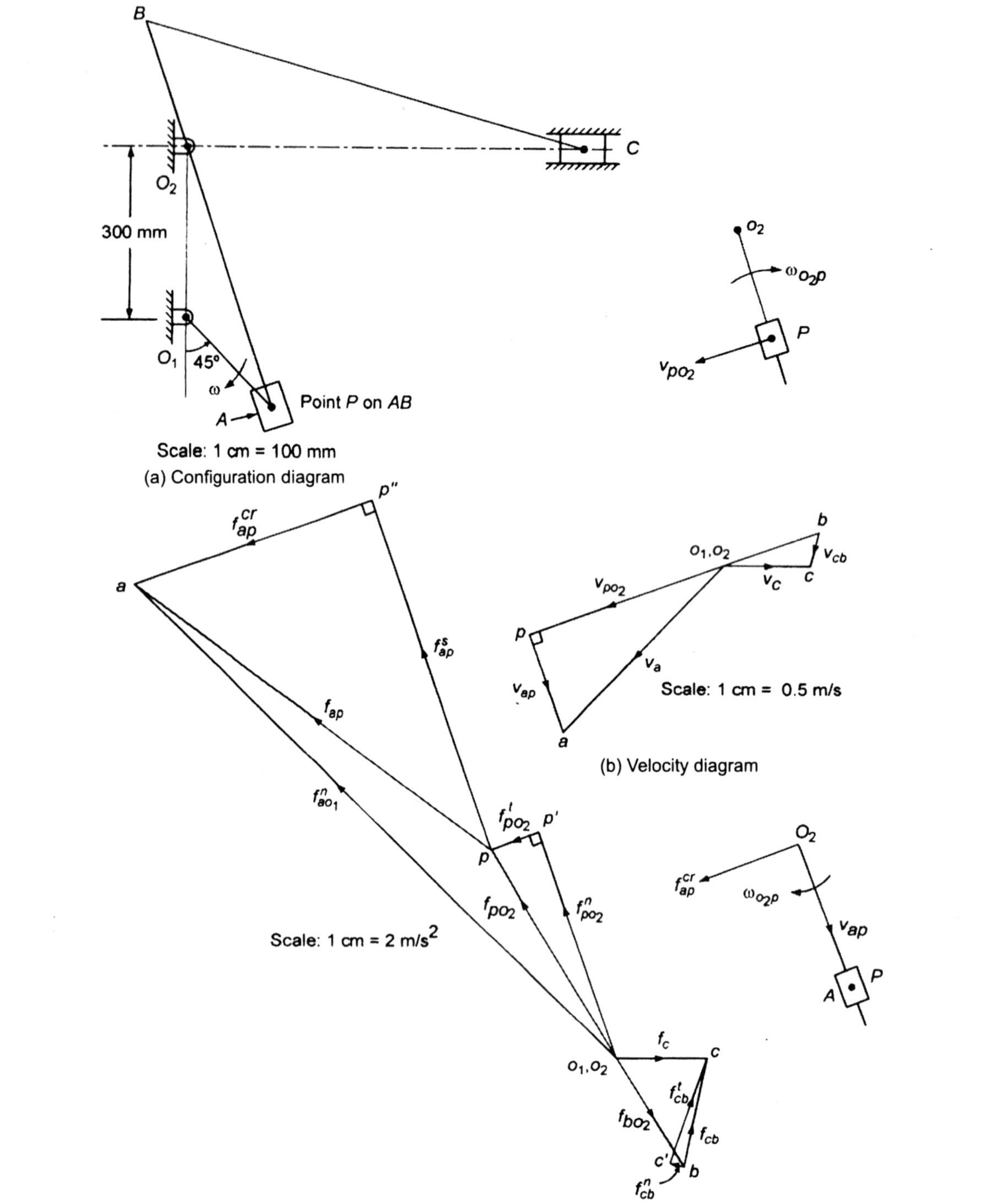

(a) Configuration diagram

(b) Velocity diagram

(c)

Fig. 3.15

$$f_a^n = \frac{v_a^2}{O_1A} = \frac{(2.51)^2}{0.2} = 31.5 \text{ m/s}^2$$

$$f^n_{po2} = \frac{v^2_{po2}}{O_2P} = \frac{(2.25)^2}{0.46} = 11 \text{ m/s}^2$$

$$f^n_{cb} = \frac{v^2_{cb}}{BC} = \frac{(0.4)^2}{0.8} = 0.2 \text{ m/s}^2$$

1. Draw the acceleration diagram as shown in Fig. 3.15 (c) to a scale of 1 cm = 2 m/s^2, by adopting the following procedure.
2. Draw $o, a = f^n_a$ = 15.75 cm parallel to O_1A.
3. At point 'a' draw a line $ap'' = f^{cr}_{ap}$ = 5.6 cm perpendicular to PB. At point p' draw another line perpendicular to PB representing the acceleration of sliding f^s_{ap} between A and $P \cdot f_{ap} = pa$ is the total acceleration of A relative to P.
4. Draw $o_2p' = f^n_{po2}$ = 5.5 cm parallel to O_2P. At p' draw another line perpendicular to O_2P representing f^t_{o2o} to meet the line f^s_{ap} at point p. Join ap and o_2p. Then

$$f_{ap} = pa \text{ and } f_{po2} = o_2p$$

5. By measurement po_2 = 5.7 cm = 11.4 m/s^2
6. Now $pb = \dfrac{po_2 \times O_2B}{O_2P} = \dfrac{5.7 \times 700}{460} = 8.67$ cm.
7. Produce po_2 to b so that pb = 8.67 cm.
8. At point b draw $bc' = f^n_{cb}$ = 0.1 cm parallel to BC and draw a line at c' perpendicular to BC representing f^t_{cb}.
9. Draw a line at o_2 parallel to the path of ram at C meeting the line drawn at c' at c. Then acceleration of ram C, $f_c = o_2c$ = 2.2 cm = 4.4 m/s^2
10. Calculate the angular acceleration as follows :
Tangential acceleration of O_2P, $f^t_{po2} = p'p$ = 1.2 cm = 2.4 m/s^2

$$\text{Angular acceleration of } O_2P = \frac{f^t_{po2}}{O_2P} = \frac{2.4}{0.46} = 5.21 \text{ rad/s}^2$$

Tangential acceleration of BC, $f_{cb} = c'c$ = 2.7 cm = 5.4 m/s^2

$$\text{Angular acceleration of } BC = \frac{f^t_{cb}}{BC} = \frac{5.4}{0.8} = 6.75 \text{ rad/s}^2$$

(a) 0.85 m/s (b) 11.247 m/s^2 (c) 5.21 rad/s^2
(d) 6.75 rad/s^2 (e) 4.4 m/s^2

3.5 LINK SLIDING IN A SWIVELLING PIN

Consider a link AB sliding through a swivelling pin O, as shown in Fig. 3.16 (a). C is a point on link AB. The point A moves up with velocity v_a and acceleration f_a.

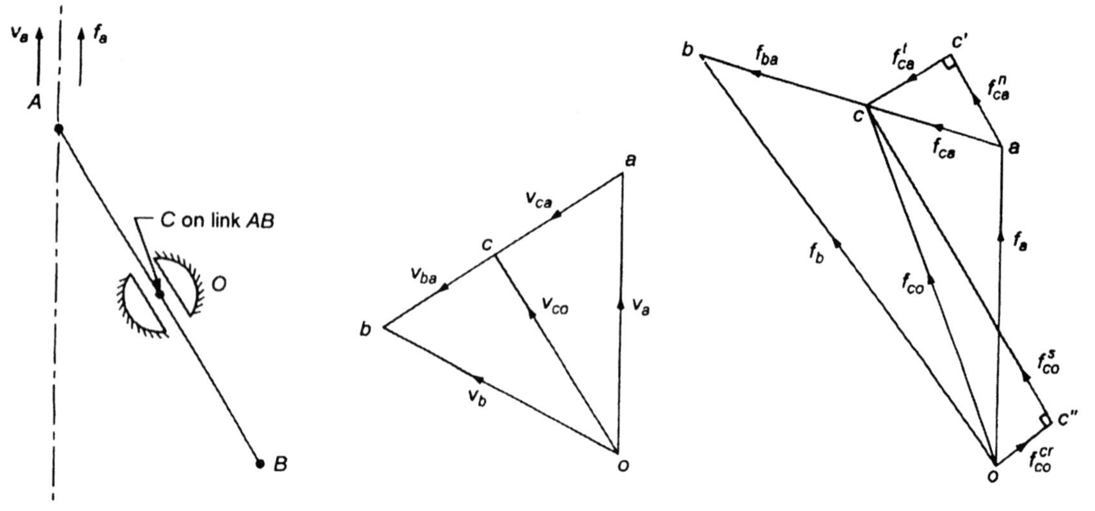

(a) Configuration diagram (b) Velocity diagram (c) Acceleration diagram

Fig. 3.16 Link sliding in a swivelling pin.

Velocity Diagram

1. Draw $oa = v_a$ parallel to the path of motion of point A, as shown in Fig. 3.16 (b).
2. From 'a' draw a line perpendicular to AB to represent the velocity of A relative to B, v_{ba}.
3. From o draw a line parallel to AB to represent the sliding velocity between C and O, v_{co}, to meet the above line at c.
4. Produce ac to b such that $\frac{ac}{ab} = \frac{AC}{AB}$. Join o to b. Then

$$v_b = ob$$
$$v_{co} = oc$$
$$v_{ca} = ac$$
$$v_{bc} = cb$$

Acceleration Diagram

$$f_{co}^{cr} = 2\, vco \times \omega_{ab}$$
$$f_{ca}^{n} = \frac{v_{ca}^2}{AC}$$

1. Draw $oa = f_a$ parallel to the path of A, as shown in Fig. 3.16 (c).
2. Draw $ac' = f_{ca}^n = \frac{v_{ca}^2}{CA} \cdot = \frac{ac^2}{CA}$ parallel to AB and a line perpendicular to AB at c' to represent the tangential acceleration f_{ca}^t.
3. Draw the coriolis acceleration of CO, $f_{co}^{cr} = oc''$ perpendicular to AB at o.
4. Draw the sliding acceleration f_{co}^s between C and O parallel AB to c'' to meet the f_{ca}^t line at c.
5. Join ac and oc.

6. Measure ac. Then $\frac{ab}{ac} = \frac{AB}{AC}$.
7. Produce ac to b.
8. Join ab. Then $f_b = ob$.

■ **Example 3.7** In the swivelling joint mechanism shown in Fig. 3.17 (a), $AB = 300$ mm, $BC = 800$ mm, $CD = 400$ mm, $AD = 500$ mm, $BE = 400$ mm, $EF = 50$ mm, $ES = 250$ mm, and $FR = 600$ mm. The Crank ABrotates at 20 rad/s. Determine (a) the coriolis acceleration and (b) the sliding acceleration of link EF in the trunnion.

Solution : Given : $\omega = 20$ rad/s, $\alpha = 200$ rad/s^2

$$v_{ba} = 20 \times 0.3 = 6 \text{ m/s}$$

Velocity Diagram

1. Draw the configuration diagram to a scale of 1 cm = 100mm, as shown in Fig. 3.17 (a).
2. Draw the velocity diagram to a scale of 1 cm = 1 m/s, as shown in Fig. 3.17 (b).
3. Draw $ab = v_{ba} = 6$ cm.
4. Draw bc perpendicular to BC and dc perpendicular to CD to meet at c. Then $v_{cb} = bc$ and $v_{cd} = dc$.
5. By measurement, $bc = 5.3$ cm. Then $\frac{be}{bc} = \frac{BE}{BC}$, so that $be = \frac{5.3 \times 400}{259} = 2.65$ cm.

   Locate point e in bc.
6. Draw a line perpendicular to EF at e and a line parallel to EF at g to meet at s.
7. By measurement, $es = 4.2$ cm. Then $\frac{ef}{es} = \frac{EF}{ES}$, so that $ef = \frac{4.2 \times 500}{259} = 8.4$ cm.
8. Produce es to f.
9. Draw a line at f perpendicular to FR and another line at 'a' parallel to the path of ram R to meet at r.
10. By measurements, we have

$$v_{se} = es = 4.2 \text{ cm} = 4.2 \text{ m/s}$$
$$v_{sg} = gs = 3.6 \text{ cm} = 3.6 \text{ m/s}$$
$$v_{fe} = ef = 8.4 \text{ cm} = 8.4 \text{ m/s}$$
$$v_{rf} = fr = 5.8 \text{ cm} = 5.8 \text{ m/s}$$
$$v_r = ar = 3.4 \text{ cm} = 3.4 \text{ m/s}$$
$$\omega_{ef} = \frac{v_{fe}}{EF} = \frac{8.4}{0.5} = 16.8 \text{ rad/s cw}$$

Acceleration Diagram

Coriolis acceleration,

$$f_{sg}^{cr} = 2\, v_{sg}\, \omega_{ef} = 2 \times 3.6 \times 16.8 = 120.96 \text{ m/s}^2$$

$$f_{ba}^{n} = \frac{v_{ba}^2}{AB} = \frac{t^2}{0.3} = 120 \text{ m/s}^2$$

$$f_{ba}^{t} = \alpha \times AB = 200 \times 0.3 = 60 \text{ m/s}^2$$

$$f_{cb}^n = \frac{v_{cb}^2}{BC} = \frac{(5.3)^2}{0.8} = 35.1 \text{ m/s}^2$$

$$f_{cd}^n = \frac{v_{cd}^2}{CD} = \frac{(6.3)^2}{0.4} = 99.2 \text{ m/s}^2$$

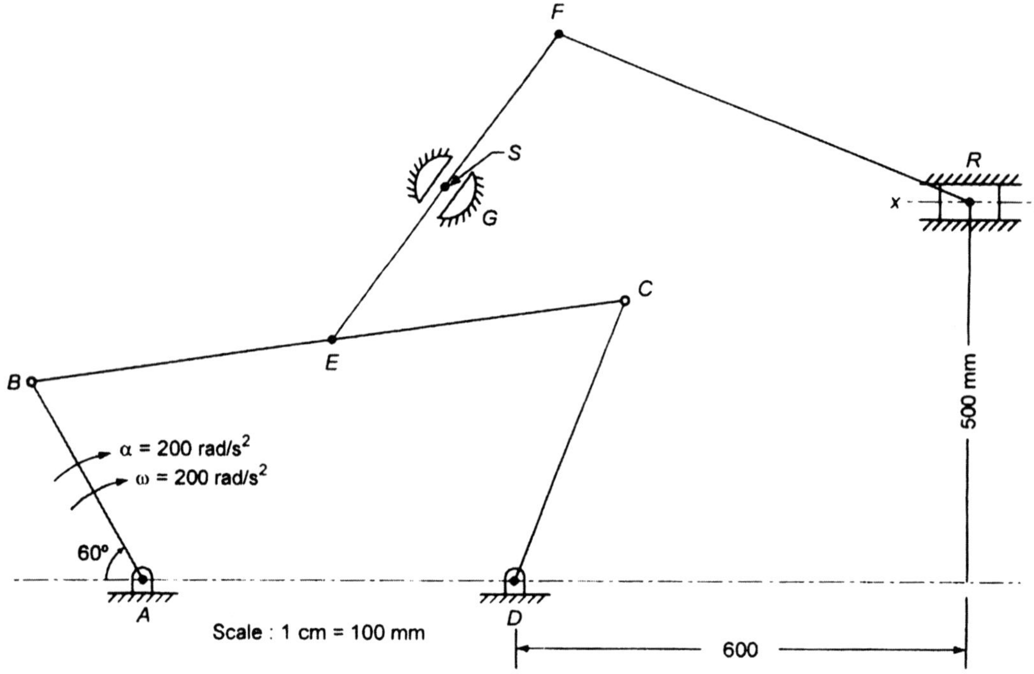

(a) Configuration diagram

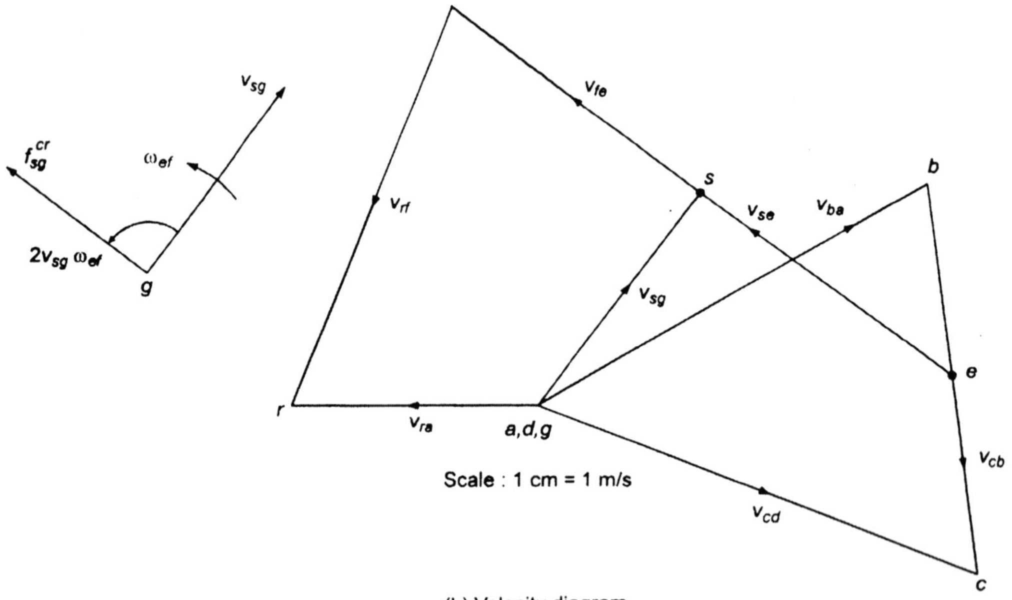

(b) Velocity diagram

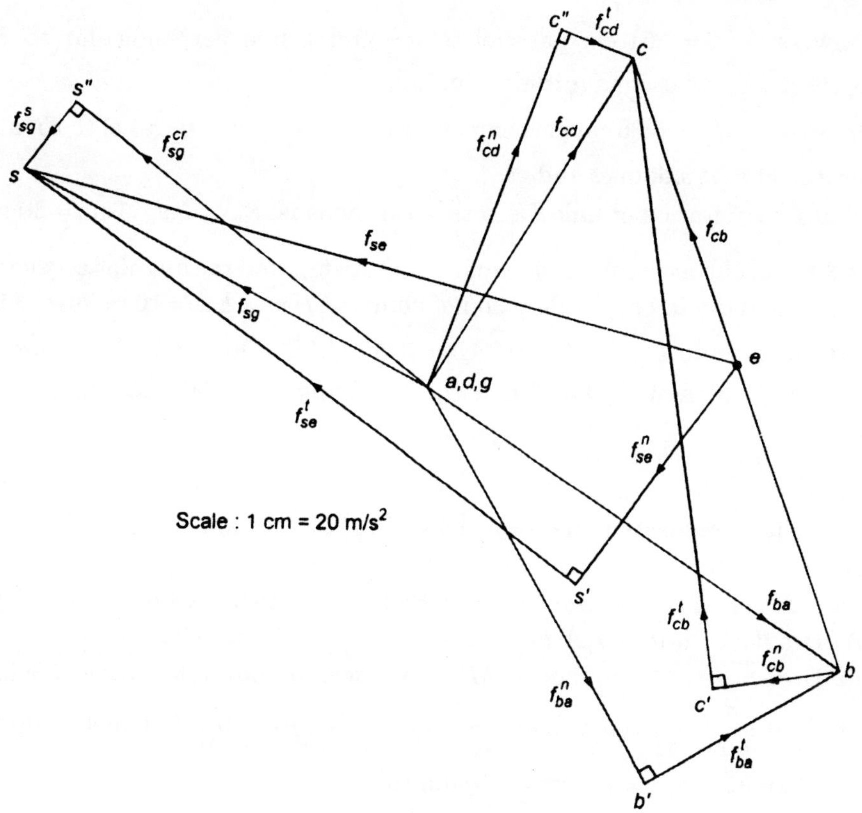

(c) Acceleration diagram

Fig. 3.17

$$f_{fe}^n = \frac{v_{fe}^2}{EF} = \frac{(8.4)^2}{0.5} = 141.1 \text{ m/s}^2$$

$$f_{rf}^n = \frac{v_{rf}^2}{FR} = \frac{(5.8)^2}{0.6} = 56.0 \text{ m/s}^2$$

$$f_{se}^n = \frac{v_{se}^2}{ES} = \frac{(4.2)^2}{0.25} = 70.56 \text{ m/s}^2$$

1. Draw the acceleration diagram to a scale of 1 cm = 20 m/s^2, as shown in Fig. 3.17 (c).
2. Draw $ab' = f_{ba}^n = 6$ cm parallel to AB and $b'b = f_{ba}^t = 3$ cm perpendicular to AB to meet at b. Join ab to given f_{ba}.
3. Draw $bc' = f_{cb}^n = 1.75$ cm parallel to BC and a line perpendicular to BC at c' representing f_{bc}^t, whose magnitude is unknown.
4. Draw $dc = f_{cb}^n = 4.96$ cm parallel to CD and a line perpendicular to CD at c'' to meet the line at c' at c. Join bc and dc.
5. By measurement, $bc = 8.5$ cm. Now $\frac{be}{bc} = \frac{BE}{BC}$, so that be $= \frac{8.5 \times 400}{800} = 4.25$ cm.

   Locate point e in bc.

6. Draw $es' = f_{se}^{n} = 3.51$ cm parallel to ES and a line perpendicular to ES at s' to represent f_{se}^{t}, whose magnitude is unknown.
7. Draw $as'' = f_{sg}^{cr} = 6.05$ cm perpendicular to EF and a line parallel to EF at s'' to meet the line at s' at s. Join es and gs.
8. Sliding acceleration of link EF in the trunnions is, $f_{sg}^{s} = s''s = 1\text{ cm} = 20\text{ m/s}^2$.

■ **Example 3.8** For the mechanism shown in Fig. 3.18 (a), find the angular acceleration of the links AB, BO_2 and the linear acceleration of points C, D and E. $\omega = 10$ rad/s, $\alpha = 200$ rad/s^2

$$O_1O_2 = 100\text{ mm}, O_1A = 50\text{ mm}, AB = 45\text{ mm}, AD = 30\text{ mm}, AC = 45\text{ mm},$$
$$BC = 30\text{ mm}, O_2B = 55\text{ mm}, O_2E = 45\text{ mm, and } BE = 25\text{ mm}.$$

Solution :

Velocity Diagram

1. Draw the configuration diagram shown in Fig. 3.18 (a) to a scale of 1 cm = 10 mm.
2. Given : $\omega = 10$ rad/s, $v_a = \omega \times O_1A = 10 \times 15$ mm/s
3. Draw the velocity diagram shown in Fig. 3.18 (b) to a scale of 1 cm = 100 mm/s.
   (i) Draw $v_a = o_1a \perp O_1A$, $o_1a = 5$ cm.
   (ii) Draw $o_2b \perp O_2B$ and $ab \perp AB$ to intersect at point b. Measure $ab = 5.4$ cm.
   (iii) Now $\frac{ad}{ab} = \frac{AD}{AB}$, $ad = 5.4 \times \frac{30}{45} = 3.6$ cm. Locate point d in ab. Join o_1d.
   Then $v_d = o_1d = 4.4\text{ cm} = 440\text{ mm/s}$

   $$v_b = o_2b = 5\text{ cm} = 500\text{ mm/s}$$
   $$v_{ba} = 5.4\text{ cm} = 540\text{ mm/s}$$

   (iv) Draw $be \perp BE$ and $o_2e \perp O_2E$ to meet at e, then $v_e = 4.2\text{ cm} = 420\text{ mm/s}$
   (v) Draw $bc \perp BC$ and $ac \perp AC$ to meet at c. Δabc is the velocity image of ABC and o_2be that of O_2BE.

Acceleration Diagram

$$f_{ao1}^{n} = \frac{v_a^2}{O_1A} = \frac{500^2}{50} = \text{mm/s}^2$$

$$f_{ao1}^{t} = \alpha \times O_1A = 200 \times 50 = 10{,}000\text{ mm/s}^2$$

$$f_{bo2}^{n} = \frac{v_b^2}{O_2B} = \frac{500^2}{55} = 4545.5\text{ mm/s}^2$$

$$f_{ba}^{n} = \frac{v_{ba}^2}{AB} = \frac{540^2}{45} = 6480\text{ mm/s}^2$$

$$f_{eo2}^{n} = \frac{v_e^2}{O_2E} = \frac{420^2}{45} = 3920\text{ mm/s}^2$$

1. Draw acceleration diagram as shown in Fig. 3.18 (c) to a scale of 1 cm = 100 mm/s^2
2. Draw $f_{ao1}^{n} = o_1a' \parallel O_1A$, $o_1a' = 5$ cm.

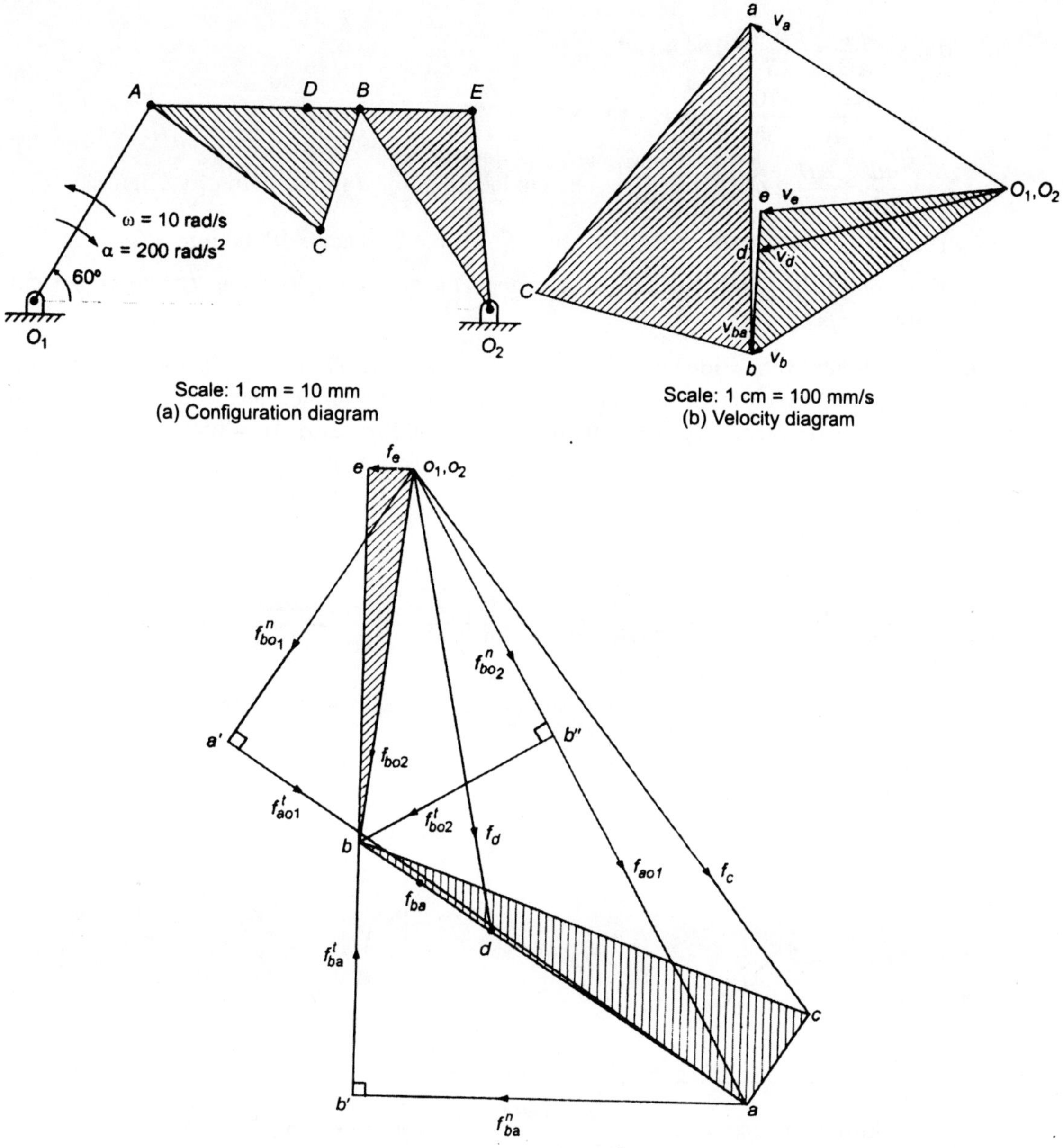

Scale: 1 cm = 10 mm
(a) Configuration diagram

Scale: 1 cm = 100 mm/s
(b) Velocity diagram

Scale: 1 cm = 1000 mm/s^2
(c) Acceleration diagram

Fig. 3.18

3. Draw $f^t_{ao1} = a'a \perp O_1A$, $aa' = 10$ cm. Join $o_1a . o_1a = f_{ao1}$
4. Draw $f^n_{ba} = ab' \parallel AB$ and $f^n_{bo2} = o_2b'' \parallel O_2B$, $o_2b^2 = 4.54$ cm.
5. Draw $f^t_{ba} = b'b \perp ab'$ and $f^t_{bo2} = b''\, b \perp$ to intersect at point b. Join ab and o_2b.
6. Measure $f_{ba} = ab = 7.4$ cm $= 7400$ mm/s^2 and $f_{bo2} = o_2b = 5.7$ cm $= 5700$ mm/s^2.

7. $\alpha_{ba} = \dfrac{f_{ba}}{AB} = \dfrac{7400}{45} = 164.4 \text{ rad/s}^2$

$\alpha_{bo2} = \dfrac{f_{bo2}}{O_2B} = \dfrac{5700}{55} = 10.364 \text{ rad/s}^2$

8. Now $\dfrac{ad}{ab} = \dfrac{AD}{AB}$, $ad = 7.4 \times \dfrac{30}{45} = 4.86$ cm locate point d in ab. Join o_1d. Then $f_d = o_1d$.

9. Draw $ac \perp AC$ and $be \perp BC$. Join $o_{1c} \cdot f_e = o_1c = 10.6 \text{ cm} = 10600 \text{ mm/s}^2$.

10. Draw $be \perp BC$ and $o_2e \perp O_2E$ to meet at e. Join o_2e. Then $f_e = o_2e = 0.4$ cm $= 400 \text{ mm/s}^2$.

■ **Example 3.9** For the slider-crank mechanism shown in Fig. 3.19 (a), determined (a) acceleration of slider B, (b) acceleration of point C, and (c) acceleration of link AB. The crank OA rotates at 180 rpm. $OA = 500$ mm, $AB = 1500$ mm, and $AC = 250$ mm.

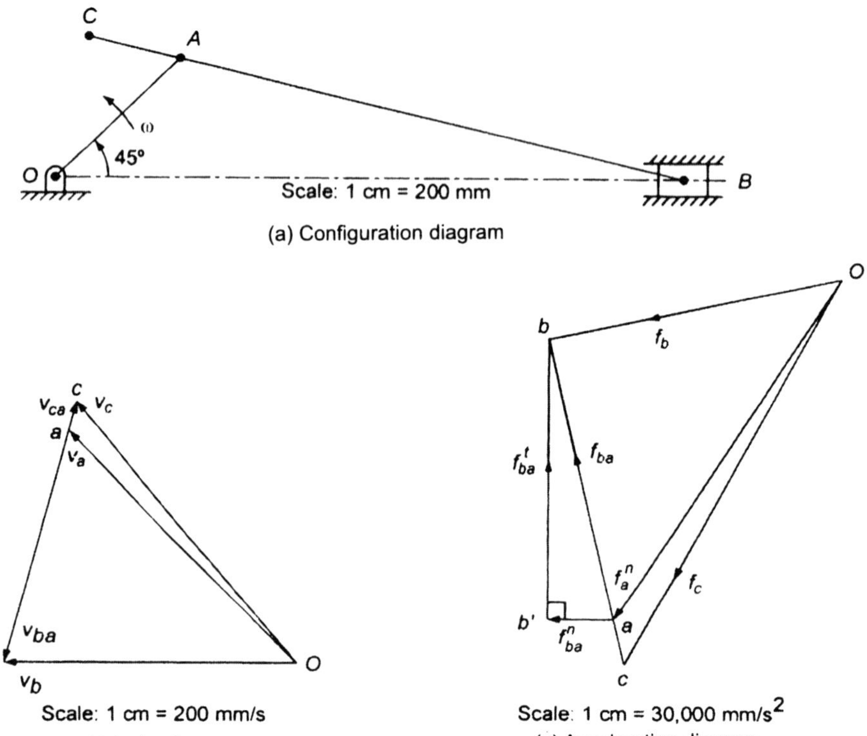

(a) Configuration diagram

(b) Velocity diagram

(c) Acceleration diagram

Fig. 3.19

Solution :

1. Draw the configuration diagram shown in Fig. 3.19 (a) to a scale of 1 cm = 200 mm.

2. Angular speed of crank OA, $\omega = 2\pi \times \dfrac{180}{60} = 18.85$ rad/s

$$v_a = \omega \times OA = 18.85 \times 500 = 4424.8 \text{ mm/s}$$

3. Draw the velocity diagram shown in Fig. 3.19 (b) to a scale of 1 cm 200 mm/s.

4. Draw $va = oa \perp OA$, $oa = 4.71$ cm.

5. Draw $ab \perp AB$ and $ob \parallel OB$ to meet at point b. $v_b = ab = 4.5$ cm = 8400 mm/s.

$$v_{ba} = ab = 3.5 \text{ cm} = 7000 \text{ mm/s}.$$

6. Now $\frac{ac}{ab} = \frac{AC}{AB}$, $ac = 3.5 \times \frac{250}{1500} = 0.538$ cm. Produce ba to c so that $ac = 0.583$ cm.

Join oc, then $v_c = oc = 5$ cm = 10,000 mm/s .

Acceleration Diagram

$$f_{oa}^n = \frac{v_a^2}{OA} = \frac{(9424.8)^2}{500} = 1{,}77{,}654 \text{ mm/s}^2$$

$$f_{ba}^n = \frac{v_{ba}^2}{AB} = \frac{7000^2}{1500} = 32667 \text{ mm/s}^2$$

1. Draw the acceleration diagram shown in Fig. 3.19 (c) to a scale of 1 cm = 30,000 mm/s^2.
2. Draw $f_{oa}^n = oa \parallel OA$, $oa = 5.92$ cm.
3. Draw $f_{ba}^n = ab' \parallel AB$, $ab' = 1.08$ cm. $f_{ba}^t = b'b \perp AB$ and $f_b = ob \parallel OB$ to at point b. Join ab.
4. Measure $f_b = ob = 4.3$ cm = 1,29,000 mm/s^2

$$f_{ba} = ab = 4.2 \text{ cm} = 1{,}26{,}000 \text{ mm/s}^2$$

5. Now $ac = ab \times \frac{AC}{AB} = 4.2 \times \frac{250}{1500} = 0.7$ cm.
6. Produce ba to c so that $ac = 0.7$ cm. Join ac. Then $f_c = oc$ 6.4 cm = 1,92,000 mm/s^2

■ **Example 3.10** Draw the acceleration diagram for the shaper mechanism shown in Fig. 3.20(a). $OB = 150$ mm, $CB = 225$ mm, $OC = 150$ mm. Find the coriolis acceleration of slider B.

Solution : Velocity Diagram

1. Draw the configuration diagram shown in Fig. 3.20 (a) to a scale of 1 cm = 50 mm.
2. $\omega = 20$ rad/s (given), $v_b = \omega \times OB = 20 \times 150 = 3000$ rad/s.
3. Draw the velocity diagram shown in Fig. 3.20 (b) to a scale of 1 cm = 1000 mm/s^2.
4. Draw $v_b = ob \perp OB$. $ob = 3$ cm.
5. Draw $cd \perp AC$ and $db \parallel AC$ to meet at d. Then, velocity of slider B along ABC, $v_{bd} = db = 2$ cm = 2000 mm/s^2.

$$v_{dc} = cd = 1.5 \text{ cm} = 1.5 \text{ m/s}$$

6. Angular speed of ABC link, $\omega_{AC} = \frac{v_{dc}}{CD} = \frac{1.5 \times 100}{225} = 6.67$ rad/s.

Acceleration Diagram

Coriolis acceleration of slider B, $f_{bd}^{cr} = 2v_{bd}\,\omega_{AC} = 2 \times 2000 \times 6.67 = 26680$ mm/s^2 and is perpendicular to CB.

$$f_{bo}^n = \frac{v_b^2}{OB} = \frac{3000^2}{150} = 60{,}000 \text{ mm/s}^2 \text{ along } BO.$$

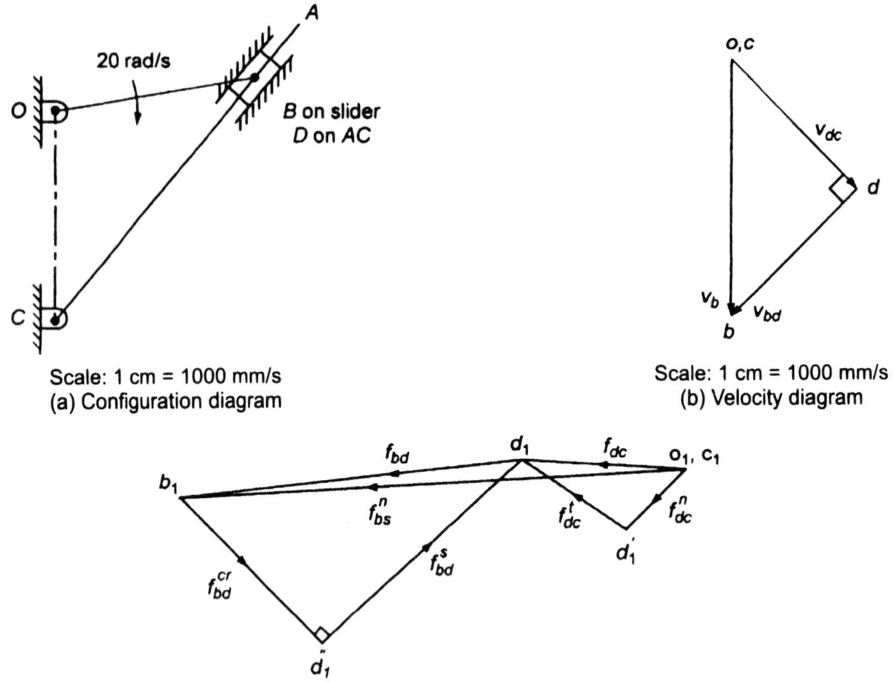

Scale: 1 cm = 1000 mm/s
(a) Configuration diagram

Scale: 1 cm = 1000 mm/s
(b) Velocity diagram

(c) Acceleration diagram

Fig. 3.20

$$f^n_{dc} = \frac{v^2_{dc}}{CD} = \frac{1500^2}{225} = 10{,}000 \text{ mm/s}^2 \text{ along } CD.$$

1. Draw acceleration diagram shown in Fig. 3.20 (c) to a scale of 1 cm = 10,000 mm/s^2.
2. Draw $f^n_{bn} = o_1b_1 \parallel OB, o_1b_1 = 6$ cm.
3. Draw $f^{cr}_{bd} = b_1d''_1 \perp CD, b_1d''_1 = 2.67$ cm
4. Draw $f^s_{bd} \perp b''_1$.
5. Draw $f^n_{dc} = c_1d'_1 \parallel CD, c_1d'_1 = 1$ cm
6. Draw $f^t_{bc} \perp c_1d'_1$ to intersect f^s_{bd} at d_1.
7. Join b_1d_1 and c_1d_1.

3.6 KLEIN'S CONSTRUCTION

Klein's construction is used to draw the acceleration diagram for a single slider-crank mechanism.

The following steps may be adopted to draw the acceleration diagram, as shown in Fig. 3.21 :

1. Draw the configuration diagram *OAB* of the slider-crank mechanism to a convenient scale.
2. Draw $OI \perp OB$ and produce *BA* to meet *OI* at *C*.
3. With *AC* as the radius and *A* as the centre, draw a circle.

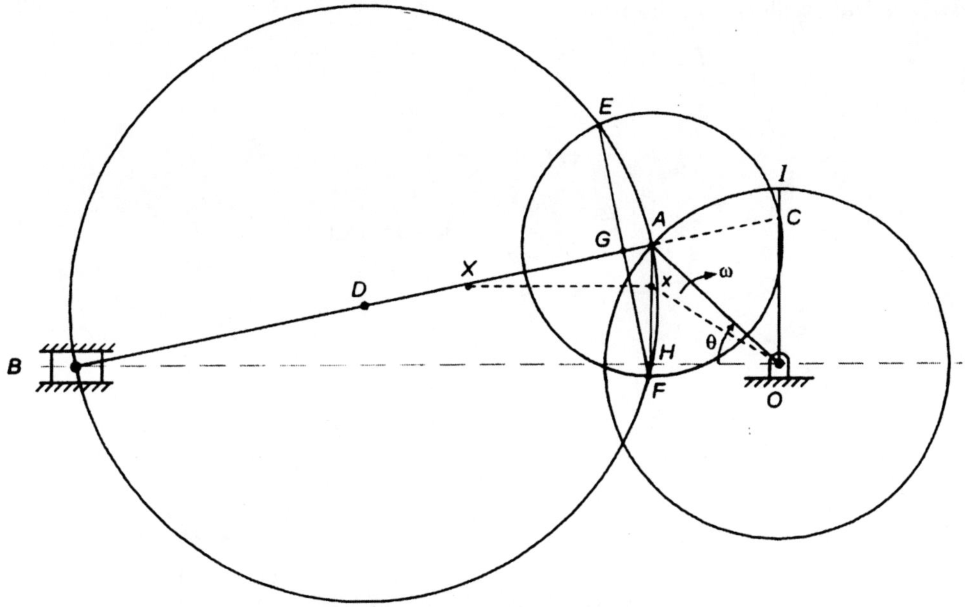

Fig. 3.21 Klein's construction.

4. Locate the mid-point D of AB. With D as centre and DA as radius, draw the circle to intersect the previously drawn circle at E and F. Join EF intersecting AB at G.
5. EF meets OB at H. If it does not meet OB then produce EF to meet OB at H. Join AH. Then $OAGH$ is the acceleration diagram.

$$OA = f_{ao}^{n}$$

$$AG = f_{ba}^{n}$$

Acceleration of piston (slider B), $f_b = \omega^2 \cdot OH$

$$f_{ba}^{n} = \omega^2 \cdot AG$$

$$f_{ba}^{t} = \omega^2 \cdot GH$$

6. To find the acceleration fo any point X in AB, draw a line $XX \parallel OB$ to intersect HA at X. Join OX. Then

$$f_x = \omega^2 \cdot OX$$

$$\alpha_{AB} = \frac{f_{ba}^{t}}{AB} = \frac{\omega^2 \cdot GH}{AB} \text{ (ccw)}$$

■ **Example 3.11** The crank of an engine 300 mm long rotates at a uniform speed of 300 rpm. The ratio of connecting rod length to crank radius is 4. Determine (a) acceleration of the piston, (b) angular acceleration of the rod, and (c) acceleration of a point X on the connecting rod at 400 mm from crank pin. The crank position is 60° from inner dead centre.

Solution : The acceleration diagram using Klein's construction has been drawn in Fig. 3.22, following the steps as explained in art. 3.6.

$OAGH$ is the acceleration diagram.

$$f_{ba}^{n} = \omega^2 \cdot AG$$

$$f_{ba}^{t} = \omega^2 \,.\, GH$$

$$f_{ba} = \omega^2 \cdot AH$$

$$\omega^2 = \left(\frac{2\pi \times 300}{60}\right)^2 = 986.96 \text{ (rad/s)}^2$$

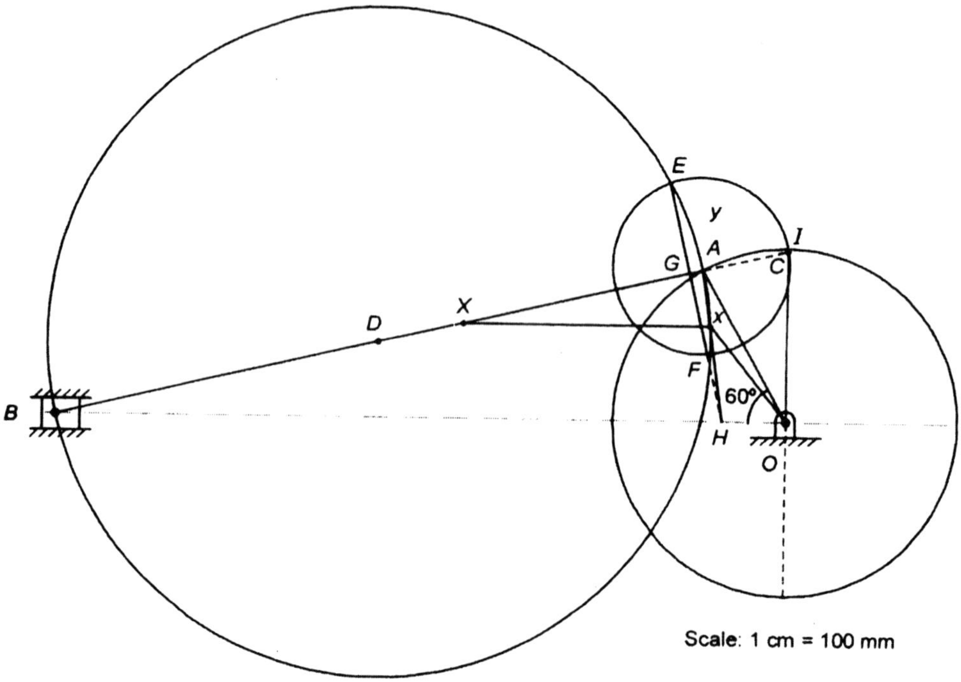

Fig. 3.22

Locate point X in AB, such that AX = 400 mm. Draw Xx parallel to BO to meet AH at x. Join Ox. Then

$$Ox \propto f_x$$

$$f_b = \omega^2 OH$$

(a) Acceleration of piston,

$$f_b = \omega^2 \times OH = \frac{986.96 \times 1.1 \times 100}{1000} = 108.56 \text{ m/s}^2$$

(b) Tangential acceleration of rod,

$$f_{ba}^{t} = \omega^2 \times GH = \frac{986.96 \times 2.6 \times 100}{1000} = 265.61 \text{ m/s}^2$$

Angular acceleration of rod,

$$\alpha_{AB} = \frac{f_{ba}^{t}}{AB} = \frac{256.61}{1.2} = 213.84 \text{ rad/s}^2$$

(c) Acceleration of point X on rod

$$= \omega^2 \times Ox = \frac{986.96 \times 2.2 \times 100}{1000} = 217.13 \text{ m/s}^2$$

3.7 ANALYTICAL ANALYSIS OF SLIDER-CRANK MECHANISM

Consider the slider-crank mechanism shown in Fig. 3.23. Let θ be the angle turned through by the crank $OA = r$ when the slider B has moved by an amount x to the right, and ϕ the angle which the connecting rod $AB = l$ makes with the line of stroke.

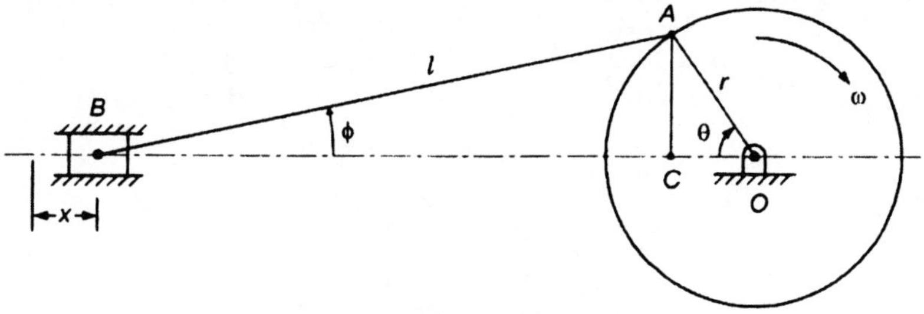

Fig. 3.23 Slider crank mechanism.

$$x = r + l - (OC + BC)$$
$$= r + l - (r \cos\theta + l \cos\phi)$$
$$= r(1 - \cos\theta) + l(1 - \cos\phi)$$

Now $$AC = r \sin\theta = l \sin\phi$$

$$\sin\phi = \frac{r}{l} \sin\theta$$

Let $$n = \frac{l}{r}$$

Then $$\cos\phi = (1 - \sin^2\phi)^{0.5}$$

$$= \left[\frac{1 - \sin^2\theta}{n^2}\right]^{0.5}$$

$$x = r(1 - \cos\theta) + l\left\{1 - \frac{\sin^2\theta}{n^2}\right\}^{0.5}$$

$$= r\left[(1 - \cos\theta) + n\left\{1 - \left(1 - \frac{\sin^2\theta}{n^2}\right)^{0.5}\right\}\right]$$

$$= r\left[(1 - \cos\theta) + n\left\{1 - \left(1 - \frac{\sin^2\theta}{2n^2} - \ldots\right)\right\}\right]$$

$$\approx r\left[(1-\cos\theta)+\frac{\sin^2\theta}{2n}\right]$$

$$\approx r\left[(1-\cos\theta)+\frac{(1-\cos 2\theta)}{4n}\right] \qquad \ldots(3.13)$$

Velocity of slider, $\quad v_b = \frac{dx}{dt} = \frac{dx}{d\theta}\cdot\frac{d\theta}{dt} = \frac{dx}{d\theta}\cdot\omega$

$$= \omega r\left[\sin\theta + \frac{\sin 2\theta}{2n}\right] \qquad \ldots(3.14)$$

Acceleration of slider,

$$f_b = \frac{d^2x}{dt^2} = \frac{dv_b}{dt} = \frac{dv_b}{d\theta}\cdot\frac{d\theta}{dt} = \omega\cdot\frac{dv_b}{d\theta}$$

$$= \omega^2 r\left[\cos\theta + \frac{\cos 2\theta}{n}\right]$$

Now $\quad \sin\phi = \frac{\sin\theta}{n}$

$$\cos\phi\cdot\frac{d\phi}{dt} = \frac{\cos\theta}{n}\times\frac{d\theta}{dt}$$

$$= \frac{\omega\cos\theta}{n}$$

Angular velocity of connecting rod,

$$\omega_{ba} = \frac{d\phi}{dt} = \frac{\omega}{n}\cdot\frac{\cos\theta}{\cos\phi}$$

$$= \frac{\omega\cos\theta}{(n^2-\sin^2\theta)^{0.5}}$$

$$\approx\left(\frac{\omega}{n}\right)\cos\theta \qquad \ldots(3.15)$$

Angular acceleration of connecting rod,

$$\alpha_{ba} = \frac{d\omega_{ba}}{dt} = -\omega^2\sin\theta\,\frac{(n^2-1)}{(n^2-\sin^2\theta)^{3/2}}$$

$$\approx -\frac{\omega^2}{n}\cdot\sin\theta \qquad \ldots(3.16)$$

■ **Example 3.12** In example 3.14, calculate analytically, the acceleration of the piston and angular acceleration of the rod.

Solution : Given : $n = \frac{l}{r} = 4, \theta = 60°, r = 300$ mm, $N = 300$ rpm

$$\omega^2 = \left(\frac{2\pi\times 300}{60}\right)^2 = 986.96 \text{ rad/s}^2$$

Acceleration of piston,

$$f_b = \omega^{2r}\left[\cos\theta + \frac{\cos 2\theta}{n}\right]$$

$$= 986.96 \times 0.3\left[\cos 60° + \frac{\cos 120°}{4}\right]$$

$$= 111.0 \text{ m/s}^2$$

Angular acceleration of rod $\cong \dfrac{\omega^2}{n}\sin\theta$

$$= \frac{986.96 \times 60°}{4}$$

$$= 213.68 \text{ rad/s}^2$$

■ **Example 3.13** The crank of an engine is 20 cm long and the connecting rod is 4 cm long. Determine the acceleration of the piston and angular acceleration of the connecting rod when the crank has turned through 45° from the inner dead centre. The crank rotates at a uniform speed of 240 rpm.

Solution : Given $r = 20$ cm, $l = 40$ cm, $\theta = 45°$, $N = 240$ rpm.

Angular speed of crank, $\omega = \dfrac{2\pi N}{60} = \dfrac{2\pi \times 240}{60} = 25.13$ rad/s

$$n = \frac{l}{r} = \frac{40}{20} = 2$$

Acceleration of piston $= \omega^2 r\left[\cos\theta = \dfrac{\cos 2\theta}{n}\right]$

$$= (25.13)^2 \times 0.2\left[\cos 45° + \frac{\cos 90°}{2}\right]$$

$$= 89.33 \text{ m/s}^2$$

Angular acceleration of connecting rod $= \dfrac{-\omega^2(n^2 - 1)\sin\theta}{(n^2 - \sin^2\theta)^{3/2}}$

$$= \frac{(25.13)^2 \times (4-1)\sin 45°}{(4 - \sin^2 45°)^{3/2}} = -204.6 \text{ rad/s}^2$$

■ **Example 3.14** An engine mechanism shown in Fig. 3.24 has a crank length of 0.1 m and connecting rod 0.3 m long with centre of gravity G 0.1 m from end B. The crankshaft has a speed of 75 rad/sec and an angular acceleration of 1200 rad/sec^2. Find acceleration of point G and angular acceleration of connecting rod.

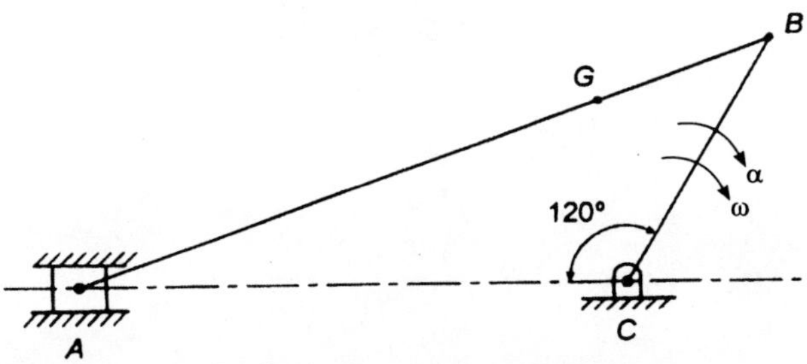

Fig. 3.24

Solution : Given $r = CB = 0.1$ m, $l = AB = 0.3$ m, $BG = 0.1$ m, $\omega = 75$ rad/s, $\alpha = 1200$ rad/s^2
$v_{bc} = \omega \times BC = 75 \times 0.1 = 7.5$ m/s, $f_{bc}^t = r\alpha = 0.1 \times 1200 = 120$ m/s^2

Velocity Diagram

1. Draw configuration diagram to a scale of 1 cm = 0.05 m as shown in Fig. 3.25(a).
2. Draw velocity diagram as shown in Fig. 3.25 (b) to a scale of 1 cm = 1 m/s.
3. Draw $v_{bc} = cb \perp CB$, $cb = 7.5$ cm
4. Draw $ba \perp AB$ and $c\alpha \parallel AC$ to meet at α.
5. By measurement : $v_{ab} = ba = 4$ cm = 4 m/s, $v_{ac} = ca = 5.3$ cm 5.3 m/s.
6. Now $\dfrac{bg}{ba} = \dfrac{BG}{BA}$, $bg = 4 \times \dfrac{0.1}{0.3}$ 1.33 cm.

   Locate point g in ba. join cg. Then $v_{gc} = cg = 6.6$ cm = 6.6 m/s.

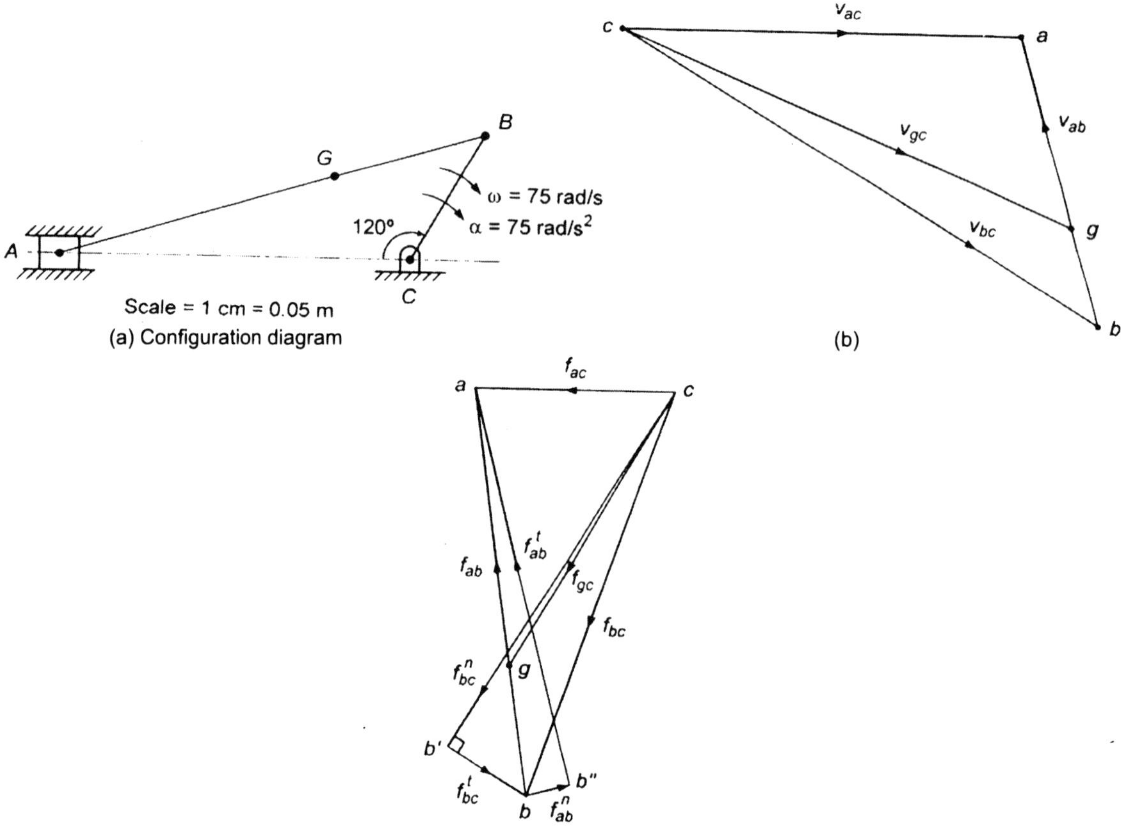

Scale = 1 cm = 0.05 m
(a) Configuration diagram

(b)

(c)

Fig. 3.25

Acceleration Diagram

$$f_{bc}^n = \frac{v_{bc}^2}{BC} = \frac{(7.5)^2}{0.1} = 562.5 \text{ m/s}^2$$

$$f_{bc}^t = 120 \text{ m/s}^2, \; f_{ab}^n = \frac{v_{ab}^2}{AB} = \frac{4^2}{0.3} = 53.33 \text{ m/s}^2$$

1. Draw acceleration diagram to a scale of 1 cm = 100 m/s^2 as shown in Fig. 3.25 (c).
2. Draw $f^n_{bc} = cb' \parallel BC$, $cb' = 5.625$ cm.
3. Draw $f^t_{bc} = b'b \perp BC$, $b'b = 1.2$ cm. Join cb. Then $cb = f_{bc}$.
4. Draw $f^n_{ab} = bb'' \parallel AB$, $bb'' = 0.53$ cm.
5. Draw $b''a \perp AB$ representing f^t_{ab} and $ca \parallel AC$ representing f_{ac} to meet at 'a'.
6. Join ab. Then $f_{ab} = ba = 5.5$ cm $= 550$ m/s^2
7. Now $\dfrac{bg}{ba} = \dfrac{BG}{BA}$, $bg = 5.5 \times \dfrac{0.1}{0.3} = 1.83$ cm.

   Locate g in ab. Join cg. Then $cg = f_{gc} = 4.2$ cm $= 420$ m/s^2.

   Angular acceleration of connecting rod $= \dfrac{f^t_{ab}}{AB}$

$$f^t_{ab} = b''a = 5.45 = 545 \text{ m/s}^2$$

$$\alpha_{AB} = \frac{545}{0.3} = 1816.67 \text{ rad/s}^2 \text{ ccw (Fig. 3.26)}$$

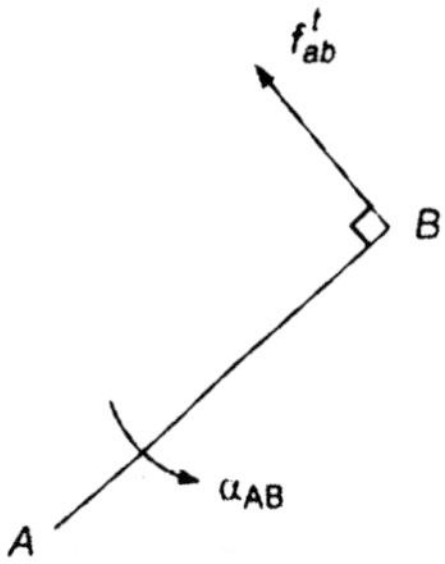

Fig. 3.26

SOLVED QUESTIONS FROM UNIVERSITY EXAMINATIONS

■ **Example 3.15** What is meant by Corioli's components of acceleration? Determine magnitude and direction. [KU, Dec. 2006]

Solution : Refer to art. 3.4

■ **Example 3.16** The crank of a single cylinder diesel engine is 50 mm long and its connecting rod 220 mm. If the engine rotates 900 rpm at, determine by analytical method the acceleration of the piston and angular acceleration of the connecting rod. Angular displacement of crank is 45° from IDC. [KU, Dec. 2006]

Solution : Given $r = 50$ mm, $l = 220$ mm, $N = 900$ rpm, $\theta = 45°$

$$n = \frac{l}{r} = \frac{220}{50} = 4.4$$

$$\omega = \frac{2\pi N}{60} = \frac{2\pi \times 900}{60} = 94.248 \text{ rad/s}$$

$$\text{Displacement of piston } \alpha = r\left[(1 - \cos\theta) + \frac{\sin^2\theta}{2n}\right]$$

$$= 50\left[(1 - \cos 45°) + \frac{\sin^2 45°}{2 \times 4.4}\right]$$

$$= 17.485 \text{ mm}$$

Acceleration of piston, $f_p = \omega^2 r\left[\cos\theta + \dfrac{\cos 2\theta}{n}\right]$

$$= (94.248)^2 \times 0.05\left[\cos 45° + \frac{\cos 90°}{4.4}\right]$$

$$= 314.05 \text{ m/s}^2$$

Angular acceleration of connecting rod,

$$\alpha_c = -\frac{\omega^2(n^2 - 1)\sin\theta}{(n^2 - \sin^2\theta)^{3/2}}$$

$$= -\frac{(94.248)^2\,[(4.4)^2 - 1]\sin 45°}{[(4.4)^2 - \sin^2 45°]^{3/2}}$$

$$= -1407.96 \text{ rad/s}^2$$

■ **Example 3.17** The crank OA of an oscillating cylinder engine is 6 cm long while the piston rod AB is 18 cm long. The trunion axis of the cylinder is 30 cm from the crank shaft axis, the crank OA rotates uniformly about O at 240 r.p.m. When the crank is inclinded at 45° to the IDC, determine :

(a) Veclocity of the piston relative to cylinder walls.

(b) Angular velocity of piston rod AB.

(c) Coriolis-components of acceleration. [KU, Dec. 2007]

Solution : Given $OA = 6$ cm, $AB = 18$ cm, $OC = 30$ cm, $N = 240$ rpm, $\theta = 45°$

$$\omega = \frac{2\pi N}{60} = \frac{2\pi \times 240}{60} = 8\pi \text{ rad/s}$$

Velocity of A, $v_a = \omega \times OA = 8\pi \times 6 = 150.8$ cm/s or 1.508 m/s.

1. Draw the configuration diagram as shown in Fig. 3.27 (a) to a scale of 1 cm = 5 cm.
2. Draw the velocity diagram as shown in Fig. 3.27 (b) to a scale of 1 cm = 0.25 m/s.
   (i) Draw $oa \perp OA$ such that $oa = 1.508$ m/s

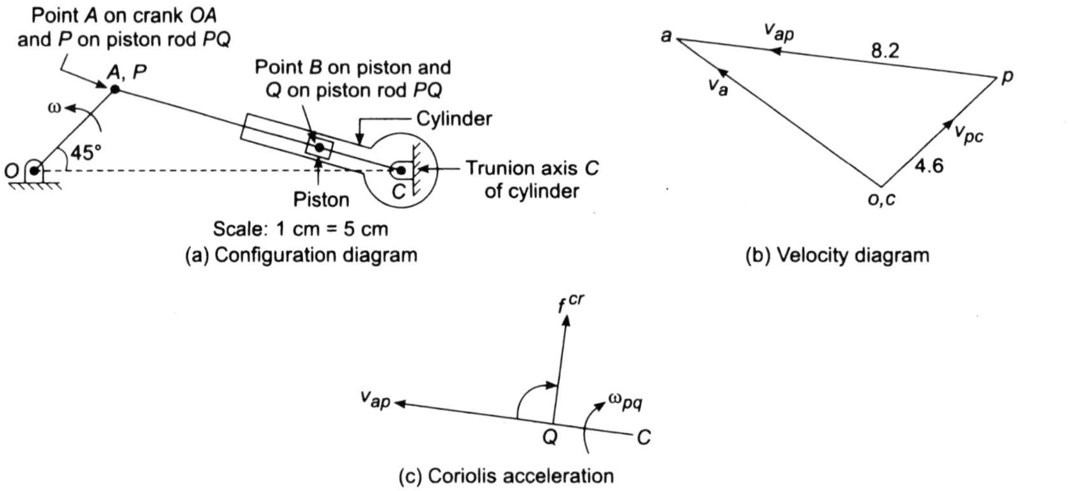

Fig. 3.27

(ii) Draw a line from '*a*' parallel to *PC* and another line from *c* perpendicular to *PC* to meet at *p*.

(iii) By measurement, $v_{ap} = pa = 8.2 \text{ cm} = 2.05 \text{ m/s}$
and $v_{pc} = cp = 4.6 \text{ cm} = 1.15 \text{ m/s}$

$$pc = 5.2 \text{ cm} = 26 \text{ cm}$$

(iv) Velocity of piston w.r.t. cylinder wall, *i.e.*, $v_{bq} = v_{ap} = 2.05$ m/s

(v) Angular velocity of piston rod , $\omega, pq = \frac{v_{pc}}{pc} = \frac{1.15 \times 100}{26} = 4.42$ rod/s (cw)

(vi) Coriolis component of acceleration is obtained by rotating the velocity vector v_{ap} through 90° in clockwise direction as shown in Fig. 3.27(c).

$$f = 2v_{pc}\omega_{pq} = 2 \times 2.05 \times 4.42 = 18.12 \text{ m/s}^2$$

■ **Example 3.18** In a slider crank mecahnism the crank *OA* rotates with a uniform velocity of 900 rpm. Lenth of connecting rod link *AB* is 32 cm and that of crank *OA* = 8 cm. For a position when crank *OA* makes as angle of 45° with the line joining *O* and slider *B*. Find :

(a) Linear acceleration of slider *B*.

(b) Acceleration of point *C* located at a distance of 10 cm from point *A* on *AB*.

[KU. Dec., 2007]

Solution : Given *N* = 900 rpm, *AB* = 32 cm, *OA* = 8 cm, θ = 45°, *AC* = 10 cm

$$\omega = \frac{2\pi N}{60} = \frac{2\pi \times 900}{60} = 30\,\pi \text{ rad/s}$$

1. Draw the configuration diagram as shown in Fig. 3.28 (a) to a scale of 1 cm = 4 cm.

$$\text{OB} = 9.3 \text{ cm} = 37.2 \text{ cm}$$

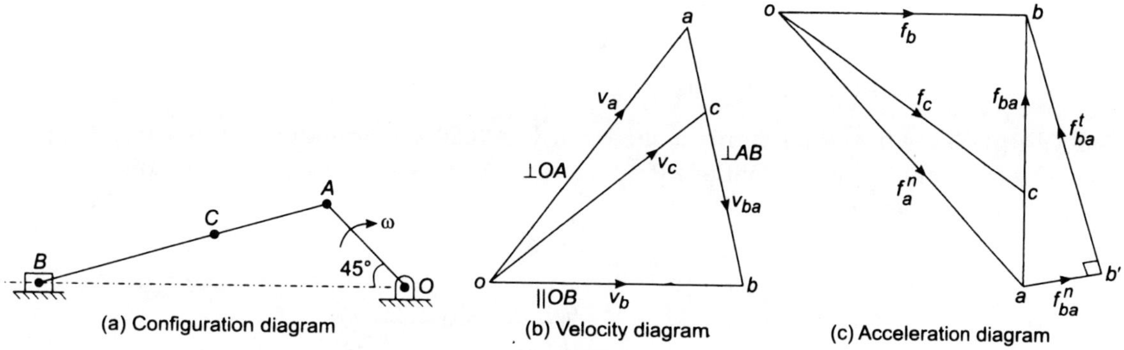

Fig. 3.28

2. Draw the velocity diagram as shown in Fig. 3.28 (b) to a scale of 1 cm = 1 m/s

$$v_a = \omega \times OA = 30\pi \times 0.08 = 7.54 \text{ m/s}$$

(i) Draw $oa \perp OA$ such that $oa = 7.54$ m/s

(ii) From point '*a*' draw a line perpendicular to *AB* and another line at *o* parallel to *OB* to intersect at point *b*.

(iii) By measurement $v_{ba} = ab = 5.4 \text{ cm} = 5.4 \text{ m/s}$

$$v_b = ob = 6 \text{ cm} = 6 \text{ m/s}$$

(iv) $\frac{ac}{ab} = \frac{AC}{AB}$

$$ac = 5.4 \times \frac{10}{32} = 1.68 \text{ cm}$$

(v) Locate point C on ab and join oc then

$$v_c = oc = 6.7 \text{ cm} = 6.7 \text{ m/s}$$

3. Calculate acceleration of various points.
Noraml acceleration of A,

$$f_a^n = \frac{v_a^2}{OA} = \frac{(7.54)^2}{0.08} = 710.6 \text{ m/s}^2$$

Normal acceleration of AB, $f_{ba}^n = \frac{v_{ba}^2}{AB} = \frac{(5.4)^2}{0.32} = 91.1 \text{ m/s}^2$

4. Draw the acceleration diagram as shown in Fig. 3.28(c) to a scale of 1 cm = 100 m/s^2
   (i) Draw oa parallel to OA such that $oa = 710.6$ m/s^2.
   (ii) Draw $ab' = f_{ba}^n = 91.1$ m/s^2 parallel to AB.
   (iii) At point b' draw a line perpendicular to AB representing f_{ba}^t.
   (iv) At o draw a line parallel to OB representing f_b to meet the above line at b.
   (v) Join above measure $ab = f_{ba} = 5$ cm $= 5$ m/s^2.
   (vi) $\frac{ac}{ab} = \frac{AC}{AB}$

$$ac = 5 \times \frac{10}{32} = 1.56 \text{ cm}$$

Locate point c in ab. Join oc. Then $f_c = oc = 6.2$ cm $= 620$ m/s^2

5. Linear acceleration of slider B, $f_b = ob = 5.1$ cm $= 510$ m/s^2

■ **Example 3.19** In the mechanism shown in Fig. 3.29 the crank OA rotates at 210 r.p.m. clockwise. For the given configuration, determine the acceleration of the slider D and angular acceleration of link CD. [KU, Dec., 2008]

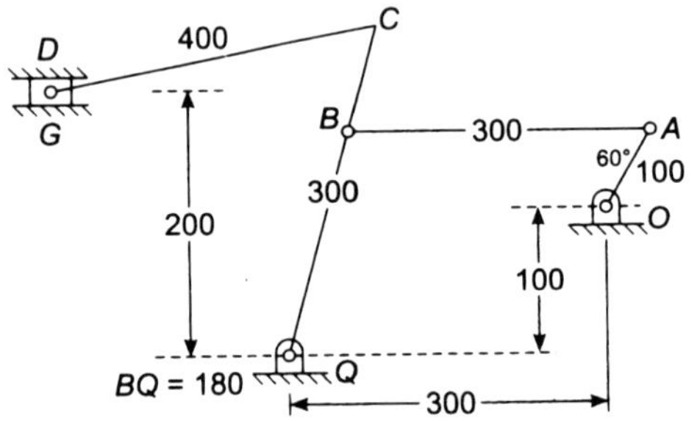

Fig. 3.29

Solution : $\omega = \frac{2\pi N}{60} = \frac{2\pi \times 210}{60} = 22$ rad/s

$$v_a = \omega \times OA = 22 \times 0.1 = 2.2 \text{ m/s}$$

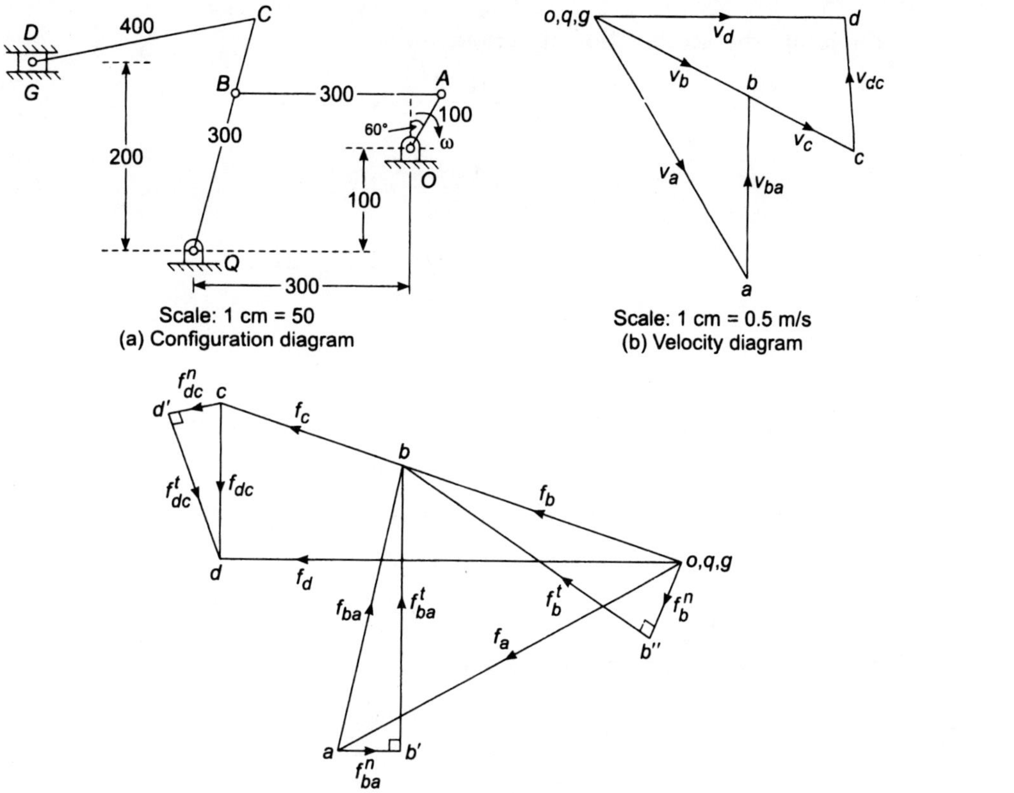

Scale: 1 cm = 50
(a) Configuration diagram

Scale: 1 cm = 0.5 m/s
(b) Velocity diagram

Scale: 1 cm = 5 m/s²
(c) Acceleration diagram

Fig. 3.30

1. Draw configuration diagram as shown in Fig. 3.30(a) to a scale of 1 cm = 50 mm. By measurement, $BC = 125$ mm and $QB = 175$ mm.
2. Draw the velocity diagram as shown in Fig. 3.30(b) to scale of 1 cm = 0.5 m/s.
   (i) Draw oa perpendicular to OA such that $oa = 2.2$ m/s.
   (ii) From point 'a' draw $ab \perp AB$ and from point q draw $qb \perp QB$ to meet at point b.
   (iii) Measure $qb = 2.4$ cm.

$$\frac{bc}{qb} = \frac{BC}{QB}$$

$$bc = 2.4\,\frac{125}{175} = 1.71 \text{ cm}$$

   (iv) Locate point c on qb produced.
   (v) Draw $cd \perp CD$ and $gd \| GD$ to meet at point d.
   (vi) Measure $v_{ba} = ab = 2.8 \text{ cm} = 1.4 \text{ m/s}$

$$v_b = qb = 2.4 \text{ cm} = 1.2 \text{ m/s}$$

$$v_c = qc = 4.2 \text{ cm} = 2.05 \text{ m/s}$$

$$v_{dc} = cd = 2 \text{ cm} = 1.0 \text{ m/s}$$

$$v_d = gd = 3.3 \text{ cm} = 1.65 \text{ m/s}$$

3. Calculate the acceleration of various links.

$$f_a^n = \frac{v_a^2}{OA} = \frac{(2.2)^2}{0.1} = 48.4 \text{ m/s}^2$$

$$f_{ba}^n = \frac{v_{ba}^2}{AB} = \frac{(1.4)^2}{0.3} = 6.53 \text{ m/s}^2$$

$$f_b^n = \frac{v_b^2}{QB} = \frac{(1.2)^2}{0.175} = 8.2 \text{ m/s}^2$$

$$f_{dc}^n = \frac{v_{dc}^2}{CD} = \frac{1}{0.4} = 2.5 \text{ m/s}^2$$

4. Draw the acceleration diagram as shown in Fig. 3.30(c) to a scale of 1 cm = 5 m/s^2

(i) Draw $oa = f_a^n = f_a = 48.4 \text{ m/s}^2$ parallel to OA.

(ii) At point 'a' draw $ab' = f_{ba}^n = 6.53 \text{ m/s}^2$ parallel to AB.

(iii) At b' draw a line perpendicular to AB representing f_{ba}^t.

(iv) At point q draw $f_b^n = ob'' = 8.23 \text{ m/s}^2$ parallel to QB.

(v) At b'' draw a line perpendicular to QB representing f_b^t to meet the f_{ba}^t line at b.

(vi) Join qb and ab. Then $ab = f_{ba}$ and $qb = f_b$.

By measurement, $qb = 7.3$ cm

$$\frac{qc}{qb} = \frac{QC}{QB}$$

$$qc = 7.3 \times \frac{300}{175} = 12.5 \text{ cm}$$

Extend line qb to qc.

(vii) At point c draw $cd' = f_{dc}^n = 2.5 \text{ m/s}^2$.

(viii) At d' draw a line perpendicular to CD representing f_{dc}^t.

(ix) At g draw a line parallel to the line of strope of slider D to meet the above line at d. Then

Acceleration of slider D, $f_d = gd = 12 \text{ cm} = 60 \text{ m/s}^2$

Tangential acceleration of CD, $f_{dc}^t = d'\,d = 3.4 \text{ cm} = 17 \text{ m/s}^2$

Angular acceleration of

$$CD = f_{dc}^t / CD = \frac{17}{0.4} = 42.5 \text{ rad / s}^2 \text{ (ccw)}$$

■ **Example 3.20** What is coriolis component of acceleration? Describe an expression for evaluating it and explaing how the direction is fixed. [KU, Dec., 2009]

Solution : Refer to art. 3.4.

■ **Example 3.21** How will you determine acceleration in case of slider crank mechanism by graphical method using velocity polygon? [KU, Dec., 2009]

Solution : Refer to art. 3.7.

■ **Example 3.22** Fig. 3.31 represents a pump mechanism. The dimensions are :

$$OA = 320 \text{ mm}, AC = 680 \text{ mm}, OQ = 650 \text{ mm}.$$

For the configuration determine :

(a) Angular velocity of the cylinder,

(b) Sliding velocity of planger,

(c) Absolute velocity of the plunger,

(d) Linear acceleration of the plunger relative to cylinder,

(e) Angular acceleration of the plunger.

The crank OA rotates at 20 rad/s clockwise. [MDU, Dec., 2007]

Solution :

Acceleration Diagram

Produce CA to R. Line AC is integraded with the plunger and thus A can be considered to be a point on it. Assume the cylinder to be of such a length that a point B is located on it just beneath the point at Fig. 3.31(a).

$$v_a = \omega \times OA = 20 \times 0.320 = 6.4 \text{ m/s}$$

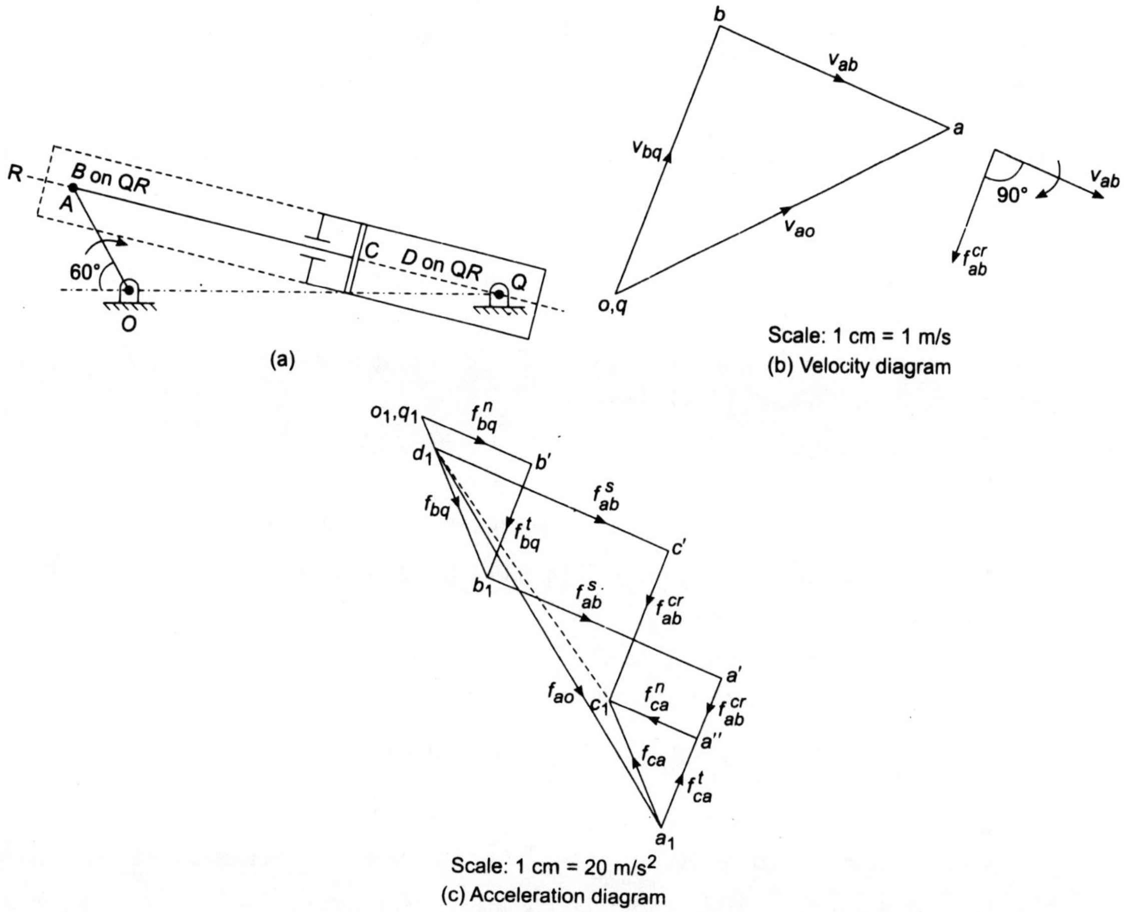

Fig. 3.31

Draw the velocity diagram as shown in Fig. 3.43(b).

(i) Draw $oa = v_a = 6.4$ m/s perpendicular to OA.

(ii) Draw qb perpendicular to QB and another line at 'a' parallel to BQ to meet at b.

(iii) By measureement , we have

$$qb = v_{bq} = 4.8 \text{ cm} = 4.8 \text{ m/s}$$

$$ba = v_{ab} = 4.3 \text{ cm} = 4.3 \text{ m/s}$$

$$BQ = 860 \text{ mm}$$

Angular velocity of cylinder = angular velocity of BQ

$$\omega_{bq} = \frac{v_{bq}}{BQ} = \frac{4.8}{0.86} = 5.58 \text{ rad/s (cw)}$$

Vector	Magnitude, m/s²	Direction	Sense
f_{ao} or o_1a_1	$\frac{(oa)^2}{OA} = \frac{(6.4)^2}{0.32} = 128$	$\parallel OA$	$\to O$
f_{bq}^n or q_1b'	$\frac{(bq)^2}{BQ} = \frac{(4.8)^2}{0.86} = 26.79$	$\parallel QB$	$\to Q$
f_{bq}^t or $b'b_1$	—	$\perp QB$	—
f_{ab}^s or b_1a'	—	$\parallel QB$	—
f_{ab}^{cr} or $a'a_1$	48	$\perp QB$	—

$$f_{ab}^{ca} = 2\omega_{bq}v_{ab} = 2\frac{v_{bq}}{BQ} \times v_{ab} = 2 \times 5.58 \times 4.3 = 48 \text{ m/s}^2$$

The direction of coriolis acceleration f_{ab}^{cr} is obtained by relating v_{ba} through 90° in the direction of ω_{bq} *i.e.*, clockwise.

Draw the acceleration diagram as shown in Fig. 3.31(c) to a scale of 1 cm = 20 m/s^2.

(i) Draw $o_1a_1 = 128$ m/s^2 parallel to OA.

(ii) From q_1 draw q_1b' parallel to BQ such that $q_1b' = 26.79$ m/s^2.

(iii) At point b' draw a line perpendicular to QB.

(iv) At point a_1 draw $a_1a' = f_{ab}^{cr} = 48$ m/s^2 perpendicular to QB.

(v) At a' draw a line parallel to QB to meet the line at b' at b_1. Join q_1b_1. Then $q_1b_1 = f_{bq} = 2.3$ cm = 46 m/s^2

Let B be a point on QB beneath point C.

Accleration of C relative to cylinder walls = $f_{cd}^s = f_{ab}^s = b_1a' = 3.6$ cm = 72 m/s^2

Now $\frac{q_1d_1}{q_1b_1} = \frac{QD}{QB}$, $q_1d_1 = 2.3 \times \left(\frac{860 - 680}{860}\right) = 0.48$ cm

Locate point d_1 on q_1b_1 such that $q_1d_1 = 0.48$ cm.

To the vector f_{dq} add the vector $f_{cd}^s = f_{ab}^s$ and then $f_{cd}^{cr} = f_{ab}^{cr}$ to locate point c_1. Join a_1 to c_1. Vector a_1c_1 represents total acceleration of C relative to A. Its two components are :

$f_{ca}^t \perp CA = a_1a''$ and $f_{ca}^n || CA = a''c_1$. $f_{ca}^t = a_1a'' = 1.4$ cm = 28 m/s^2.

Angular acceleration of piston rod, $\alpha_{ca} = \frac{f_{ca}^t}{CA} = \frac{28}{0.680} = 41.17$ rad/s^2 (ccw)

Velocity Diagram

Produce AC to R. AC passes through the point Q. Let B be a point on AR beneath Q as shown in Fig. 3.32(a). Draw the velocity diagram as shown in Fig. 3.32(b) to a scale to 1 cm = 1 m/s.

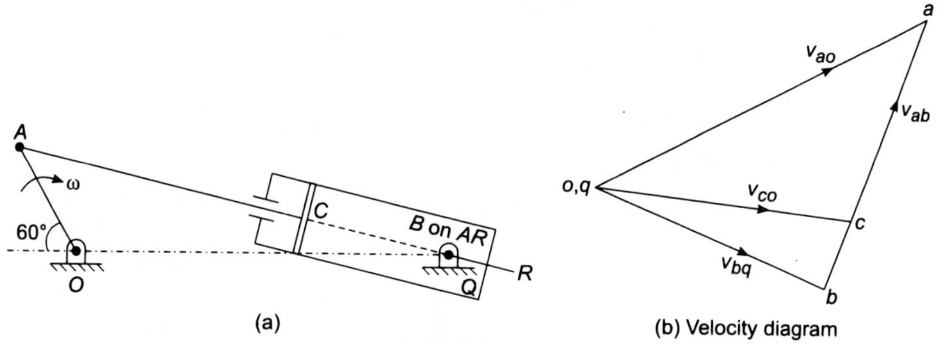

Fig. 3.32

(i) Draw $oa \perp OA$ such that $oa = 6.4$ m/s.

(ii) Draw $ab || AB$ and $ab \perp AB$ to meet at b.

(iii) By measurement, $v_{ab} = ba = 4.8$ cm

(iv) Now $\frac{ac}{ab} = \frac{AC}{AB}$, $ac = 4.8 \times \frac{680}{860} = 3.79$ cm

(v) Locate point c in ab. Join oc.

(a) Angular velocity of cylinder = angular velocity of AB

$$= \frac{v_{ab}}{AB} = \frac{4.8}{0.86} = 5.58 \text{ rad/s (cw)}$$

(b) Sliding velocity of plunger = velocity of B relative to Q.

$$= qb = 4.3 \text{ cm} = 4.3 \text{ m/s}$$

(c) Absolute velocity of plunger = oc or qc = 4.4 cm = 4.4 m/s

■ **Example 3.23** The crank of a slider-crank mechanism is 150 mm and connecting rod is 750 mm. The crank rotates at a constant speed of 300 rpm clockwise. Calculate the velocity and acceleration of the slider when crank has turned through 30° from the inner dead centre position. [MDU, May, 2008]

Solution : Given $r = 150$ mm, $l = 750$ mm, $N = 300$ rpm, $\theta = 30°$

$$n = \frac{l}{r} = \frac{750}{150} = 5$$

$$\omega = \frac{2\pi N}{60} = \frac{2\pi \times 300}{60} = 10\,\pi \text{ rad/s}$$

Velocity of slider,
$$v_s = \omega r\left(\sin\theta + \frac{\sin 2\theta}{2n}\right)$$

$$= 10\pi \times 0.15\left(\sin 30^\circ + \frac{\sin 60^\circ}{2\times 5}\right)$$

$$= 2.764 \text{ m/s}$$

Acceleration of slider, $f_s = \omega^2 r\left(\cos\theta + \frac{\cos 2\theta}{n}\right)$

$$= (10\pi)^2 \times 0.15\left(\cos 30^\circ + \frac{\cos 60^\circ}{5}\right)$$

$$= 143.014 \text{ m/s}^2$$

■ **Example 3.24** For the configuration of a slider-crank mechanism shown in Fig. 3.45(a), determine :

(a) The acceleration of slider, (b) The acceleration of point E.

(c) The angular acceleration of link AB.

The crank OA rotates at 200 rpm clockwise. OA = 450 mm, AB = 1200 mm, AE = 500 mm.

[MDU, Dec., 2008]

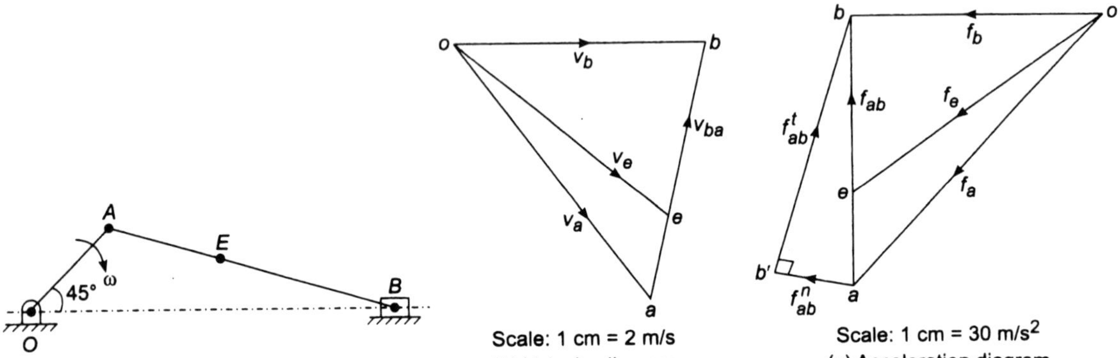

(a) Configuration diagram

Scale: 1 cm = 2 m/s
(b) Velocity diagram

Scale: 1 cm = 30 m/s²
(c) Acceleration diagram

Fig. 3.33

Solution : Given N = 200 rpm, θ = 45°, OA = 450 mm, AB = 1200 mm, AE = 500 mm.

$$\omega = \frac{2\pi N}{60} = \frac{2\pi \times 200}{60} = 20.944 \text{ rad/s}$$

$$v_a = \omega \times OA = 20.944 \times 0.45 = 9.425 \text{ m/s}$$

Draw velocity diagram as shown in Fig. 3.33(b) to a scale of 1 cm = 2 m/s.

(i) Draw $oa \perp OA$ such that oa = 9.425 m/s.

(ii) At point 'a' draw a line perpendicular to AB and another line at o parallel to OB to meet at b.

(iii) By measurement $v_b = ob$ = 4.2 cm = 8.4 m/s

$$v_{ba} = ab = 3.4 \text{ cm} = 6.8 \text{ m/s}$$

(iv) Now $\frac{ae}{ab} = \frac{AE}{AB}$, $ae = \frac{3.4 \times 500}{1200} = 1.42$ cm.

Locate point e in ab. Join oe. Then $v_e = oe$ = acceleration of A,

$$f_{ao} = \frac{v_a^2}{OA} = \frac{(9.425)^2}{0.45} = 197.4 \text{ m/s}^2$$

$$f_{ab}^n = \frac{v_{ab}^2}{AB} = \frac{(6.8)^2}{1.2} = 38.5 \text{ m/s}^2$$

Draw the acceleration diagram as shown in Fig. 3.33(c) to a scale of 1 cm = 30 m/s^2.

(i) Draw $f_a = oa = 197.4$ m/s^2 parallel to OA.

(ii) At point 'a' draw $ab' = f_{ab}^n = 38.5$ m/s^2 parallel to AB.

(iii) At b' draw a line perpendicular to AB and another line at o parallel to OB to meet at b. Join ab. Then $b'b = f_{ab}^t = 4.6$ cm $= 138$ m/s^2

$$ob = 4.5 \text{ cm} = 135 \text{ m/s}^2,\ ab = f_{ab} = 4.7 \text{ cm} = 141 \text{ m/s}^2$$

(iv) Now $\frac{ac}{ab} = \frac{AE}{AB}$, $ac = 4.7 \times \frac{500}{1200} = 1.96$ cm

Locate point c in ab. Join oe. Then $f_e = oe = 5.3$ cm $= 159$ m/s^2

(a) Acceleration of slider, $f_b = 135$ m/s^2

(b) Acceleration of point E, $f_e = 159$ m/s^2

(c) Angular acceleration of link

$$AB,\ \frac{\alpha}{ab} = \frac{f_{ab}^t}{AB} = \frac{138}{1.2} = 115 \text{ rad/s}^2$$

SUMMARY

1. A particle moving in a circular path has two components of accelerations, normal and tangential, which are perpendicular to each other.

   Normal (or radial or centripetal) component of acceleration, $f^n = \omega^2 r = \frac{v^2}{r}$.

   Tangential component of acceleration, $f^t = \alpha r$.

   where ω = angular velocity, α = angular acceleration, v = linear velocity and r = radius.

2. Total acceleration of a point B w.r.t point A on the same link (vector sum),

$$f_{ba} = f_{ba}^n + f_{ba}^t = (\omega^2 + \alpha)\, AB$$

$$\tan\beta = \frac{\alpha}{\omega^2}$$

3. Acceleration centre is a point in a link about which the acceleration is zero.

4. Coriolis component of acceleration acts when the second point which was considered stationary, now slides, *e.g.*, a slider sliding along a rotating link.

   Coriolis acceleration, $f^{cr} = 2v\omega$

   where v = linear velocity of the slider along the link

   ω = angular speed of the rotating link.

5. The direction of the coriolis component of acceleration is such as to rotate the sliding velocity vector through 90° in the same sense as the angular velocity of the rotating link.

6. Klein's construction is used to draw the acceleration diagram for a single slider-crank mechanism.

7. In a slider-crank mechanism,

Velocity of slider $= \omega r\left[\sin\theta + \dfrac{\sin 2\theta}{2n}\right]$.

Acceleration of slider $= \omega^2 r\left[\cos\theta + \dfrac{\cos 2\theta}{n}\right]$

Angular velocity of connecting rod $= \dfrac{\omega\cos\theta}{(n^2 - \sin^2\theta)^{0.5}}$

$\approx \left(\dfrac{\omega}{n}\right)\cos\theta$

Angular acceleration of connecting rod $= -\dfrac{\omega^2 \sin\theta(n^2 - 1)}{(n^2 - \sin^2\theta)^{3/2}}$

$\approx -\left(\dfrac{\omega^2}{n}\right).\sin\theta$

SHORT ANSWER TYPE QUESTIONS

1. Define normal and tangential components of acceleration.
2. How do you determine normal and tangential components of acceleration?
3. What is acceleration image? How it is helpful in determining the acceleration of offset points in a mechanism?
4. Define Coriolis component of acceleration. When it occurs?
5. How do you determine Coriolis components of acceleration?
6. What is the use of Klein's construction?
7. Find an expression for tangential acceleration and for radial acceleration when a body is moving in a circular path. What will happen to these accelerations if :
   (i) body is rotating with uniform angular velocity, and
   (ii) body is moving on a straight path?
8. Explain Klein's construction for finding acceleration of links in slider-crank mechanism.
9. What do you mean by Coriolis components of acceleration? When will it exist? Prove that this component of acceleration is equal to 2 vw. How will you find the direction of Coriolis component of acceleration?

EXERCISES

1. A link *AB* of a four-bar mechanism *ABCD* revolves uniformly at 120 rpm in a clockwise direction. Find the angular acceleration of links *BC* and *CD* and acceleration of point *E* in link *BC*. Given : *AB* = 75 mm, *BC* = 175 mm, *EC* = 50 mm, *CD* = 150 mm, *AD* = 100 mm, and ∠*BAD* = 90°

   [**Ans.** 12.57 rad/s^2, 12 rad/s^2, 10.8 cm/s^2]

2. In the steam engine mechanism show in Fig. 3.34, the crank AB rotates at 200 rpm clockwise. Find the velocities of C, D, E, F, and G, and the acceleration of slider at C. $AB = 120$ mm, $BC = 480$ mm, $CD = 180$ mm, $DE = 360$ mm, $EF = 120$ mm, and $FG = 360$ mm.

[**Ans.** 2.55 m/s, 2.5 m/s, 1.5 m/s, 1.65 m/s, 0.7 m/s, 0.85 m/s^2]

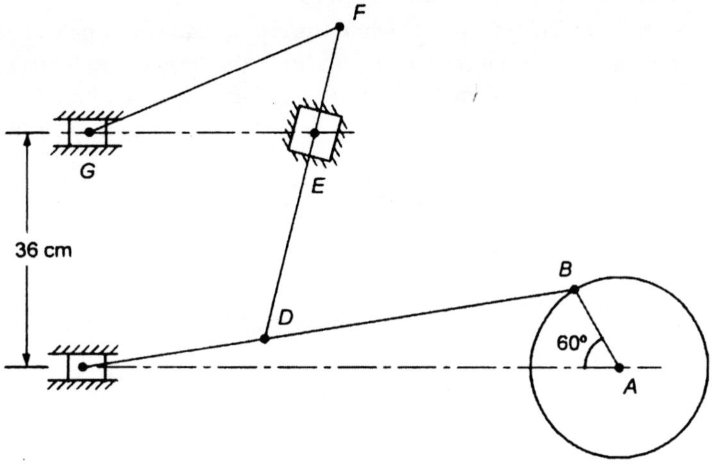

Fig. 3.34

3. In the Fig. 3.35, the washer is sliding outward on the rod with a velocity of 1.2 m/s when distance from point O is 0.6 m. Its velocity along the rod is increasing at the rate of 0.9 m/s^2. The angular velocity of rod is 5 rad/s counter-clockwise and its angular acceleration is 10 rad/s^2 clockwise. Determine the absoulute acceleration of a point on the washer. [**Ans.** 15.3 m/s^2]

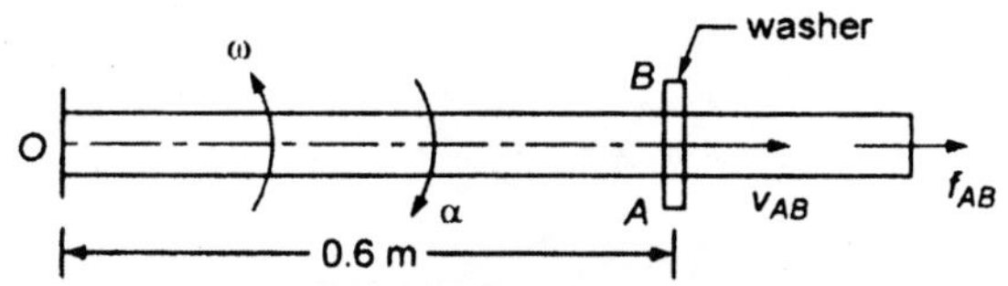

Fig. 3.35

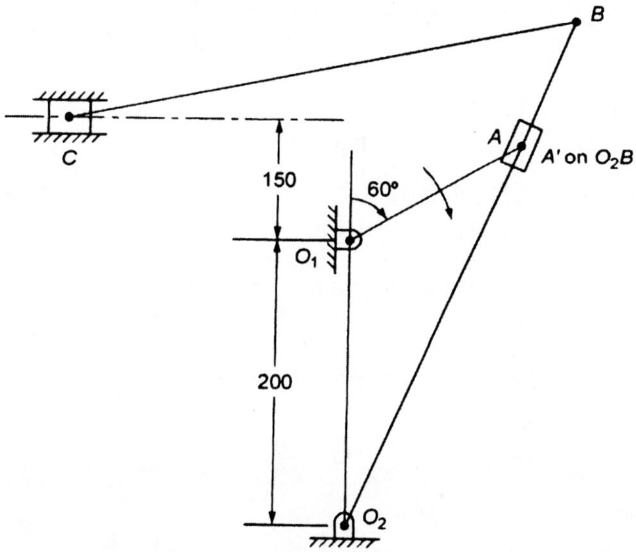

Fig. 3.36

4. The driving crank O_1A of the quick-return mechanism shown in Fig. 3.36 revolves at a uniform speed of 200 rpm clockwise. Find the velocity and acceleration of tool post C when the crank makes an angle of 60° with the vertical line of centres O_1O_2. What is the acceleration of sliding of block A along the slotted lever O_2B? [**Ans.** 1.6 m/s, 20.5 m/s^2, 22 m/s^2]

5. The crank AB of a four-bar mechanism shown in Fig. 3.37 rotates with an angular speed of 100 rad/s and an angular acceleration of 4400 rad/s^2 when the crank mades an angle of 53° to the horizontal. Determine (a) the angular acceleration of BC, and (b) linear acceleration of point E, take $AB = 75$ mm, $BC = 80$ mm, $CD = 37$ mm, $AD = 125$ mm, and $BE = 28$ mm. [**Ans.** 10,750 rad/s^2, 1160 m/s^2]

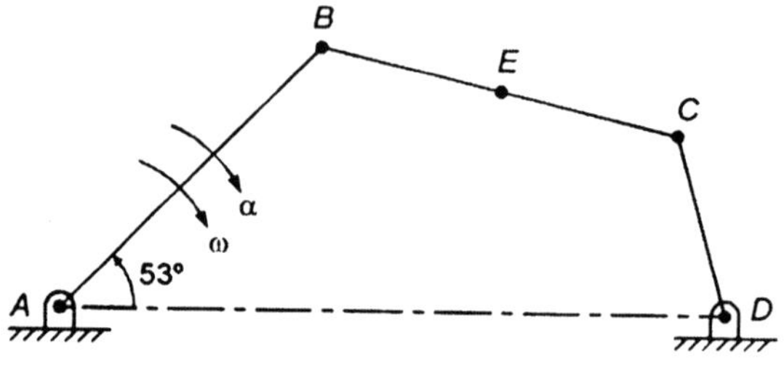

Fig. 3.37

6. Fig. 3.38 shows a mechanism in which the hydraulic actuator O_2A is expanding at a constant rate of 10 cm/s. Determine the directions and magnitudes of the angular velocity and acceleration of link O_4A. [**Ans.** 0.566 rad/s]

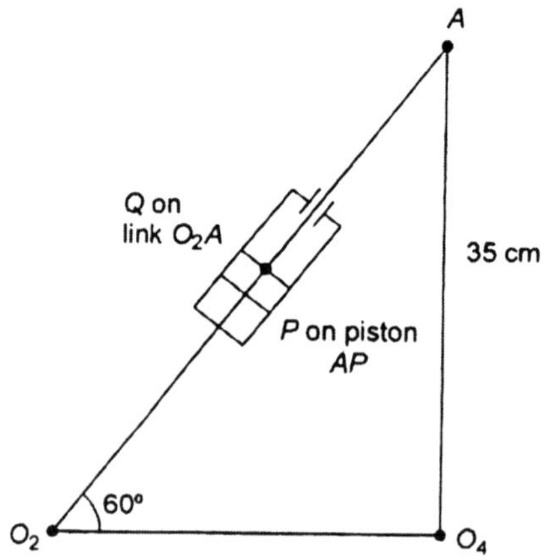

Fig. 3.38

7. For the mechanism shown in Fig. 3.39, find the acceleration of point B in link 3. Given : $\omega_2 = 30$ rad/s, $O_2A = 200$ mm, $AB = 460$ mm, $BC = 175$ mm, $AC = 600$ mm. [**Ans.** 196 m/s^2]

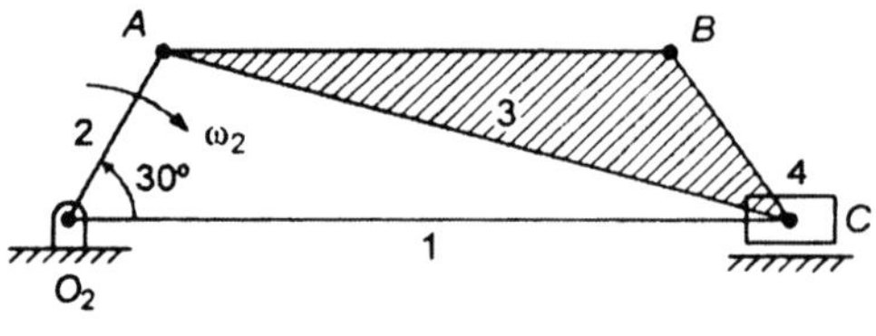

Fig. 3.39

8. Determine the angular acceleration of links 3 and 4 and the absolute acceleration of point C on link 3 in the mechanism shown in Fig. 3.40. $O_2A = 45$ mm, $AB = 130$ mm, $O_2O_4 = 90$ mm, $O_4B = 60$ mm, $AC = 55$ mm, $BC = 100$ mm. [**Ans.** 492.3 rad/s^2, 2600 rad/s, 160 m/s^2]

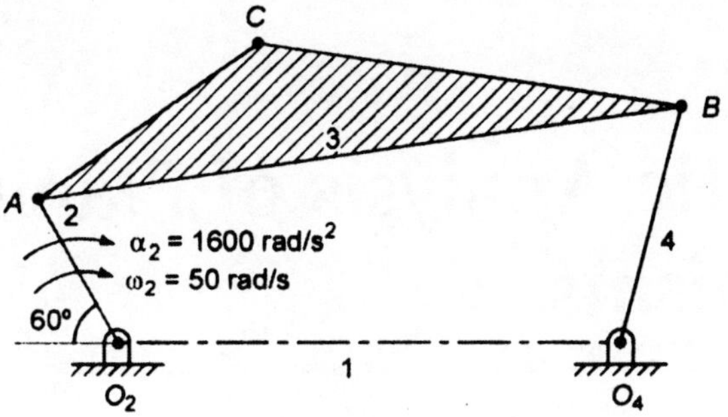

Fig. 3.40

❑❑❑

Chapter 4

Kinematic Analysis of Plane Mechanisms

Inside This Chapter

4.1 INTRODUCTION

In order to determine the displacement, velocity and acceleration of a mechanism at various portions, graphical methods are not convenient. In that case, analytical methods are preferable. In this chapter, we derive expressions for displacement, velocity and acceleration in terms of general parameters. These expressions can also be used to determine the maximum values of velocity and acceleration.

4.2 FOUR-BAR MECHANISM

A four-bar mechanism in equilibrium is shown in Fig. 4.1. $O_2O_4 = l_1$ is the fixed link, $O_2A = l_2$ is the input link, $AB = l_3$ is the coupler, and $O_4B = l_4$ is the output link. x-axis is taken along O_2O_4 and y-axis perpendicular to it. Let $\theta_1, \theta_2, \theta_3$ and θ_4 be the angles made by the links 1, 2, 3 and 4 with the x-axis in the anti-clockwise direction. The links of the mechanism can be considered as vectors, as in any configuration of the mechanism, the figure must close.

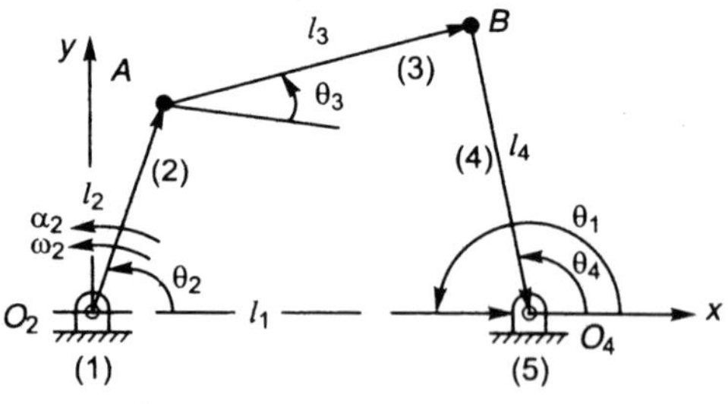

Fig. 4.1 Four-bar mechanism.

4.2.1 Displacement Analysis

The displacement along x-axis gives,

$$l_2 \cos \theta_2 + l_3 \cos \theta_3 = l_1 + l_4 \cos \theta_4 \qquad ...(4.1)$$

or

$$l_3 \cos \theta_3 = l_4 \cos \theta_4 - l_2 \cos \theta_2 + l_1$$

$$(l_3 \cos \theta_3)^2 = (l_4 \cos \theta_4 - l_2 \cos \theta_2 + l_1)^2$$

$$= l_4^2 \cos^2 \theta_4 + l_2^2 \cos^2 \theta_2 + l_1^2 - 2l_2 l_4 \cos \theta_2 \cos \theta_4 - 2l_1 l_2 \cos \theta_2 + 2l_1 l_4 \cos \theta_4 \qquad ...(4.2)$$

The displacement along y-axis gives,

$$l_2 \sin \theta_2 + l_3 \sin \theta_3 = l_4 \sin \theta_4 \qquad ...(4.3)$$

or

$$l_3 \sin \theta_3 = l_4 \sin \theta_4 - l_2 \sin \theta_2$$

$$(l_3 \sin \theta_3)^2 = (l_4 \sin \theta_4 - l_2 \sin \theta_2)^2$$

$$= l_4^2 \sin^2 \theta_4 + l_2^2 \sin^2 \theta_2 - 2l_2 l_4 \sin \theta_2 \sin \theta_4 \qquad ...(4.4)$$

Adding Eqs. (4.2) and (4.4), we get

$$l_3^2 = l_4^2 + l_2^2 + l_1^2 - 2l_2 l_4 (\cos \theta_2 \cos \theta_4 + \sin \theta_2 \sin \theta_4) - 2l_1 l_2 \cos \theta_2 + 2l_1 l_4 \cos \theta_4 \qquad ...(4.5)$$

Let

$$l_2^2 - l_3^2 + l_4^2 + l_1^2 = 2k$$

Then

$$2l_1 l_4 \cos \theta_4 - 2l_1 l_2 \cos \theta_2 - 2l_2 l_4 (\cos \theta_2 \cos \theta_4 + \sin \theta_2 \sin \theta_4) + 2k = 0$$

or

$$l_1 l_4 \cos \theta_4 - l_1 l_2 \cos \theta_2 - l_2 l_4 \cos \theta_2 \cos \theta_4 - l_2 l_4 \sin \theta_2 \sin \theta_4 + k = 0 \qquad ...(4.6)$$

From trigonometric identities, we have

$$\sin \theta_4 = \frac{2 \tan\left(\frac{\theta_4}{2}\right)}{1 + \tan^2\left(\frac{\theta_4}{2}\right)}$$

and

$$\cos \theta_4 = \frac{1 - \tan^2\left(\frac{\theta_4}{2}\right)}{1 + \tan^2\left(\frac{\theta_4}{2}\right)}$$

$$\therefore \; l_1 l_4 \left[\frac{1 - \tan^2\left(\frac{\theta_4}{2}\right)}{1 + \tan^2\left(\frac{\theta_4}{2}\right)}\right] - l_2 l_4 \cos \theta_2 \left[\frac{1 - \tan^2\left(\frac{\theta_4}{2}\right)}{1 + \tan^2\left(\frac{\theta_4}{2}\right)}\right] - l_2 l_4 \sin \theta_2 \left[\frac{2 \tan\left(\frac{\theta_4}{2}\right)}{1 + \tan^2\left(\frac{\theta_4}{2}\right)}\right] - l_1 l_2 \cos \theta_2 + k = 0$$

Multiplying throughout by $\left[1 + \tan^2\left(\frac{\theta_4}{2}\right)\right]$, we get

$$l_1 l_4 - l_1 l_4 \tan^2\left(\frac{\theta_4}{2}\right) - l_2 l_4 \cos \theta_2 + l_2 l_4 \cos \theta_2 \tan^2\left(\frac{\theta_4}{2}\right)$$

$$- 2l_2l_4 \sin\theta_2 \tan\left(\frac{\theta_4}{2}\right)$$

$$- l_1l_2 \cos\theta_2 - l_1l_2 \cos\theta_2 \tan^2\left(\frac{\theta_4}{2}\right)$$

$$+ k + k\tan^2\left(\frac{\theta_4}{2}\right) = 0$$

or $$[k - l_2(l_1 - l_4)\cos\theta_2 - l_1l_4]\tan^2\left(\frac{\theta_4}{2}\right) + (l_2l_4 \sin\theta_2)\tan\left(\frac{\theta_4}{2}\right)$$

$$+ [k - l_2(l_1 + l_4)\cos\theta_2 + l_1l_4] = 0$$

or $$A\tan^2\left(\frac{\theta_4}{2}\right) + B\tan\left(\frac{\theta_4}{2}\right) + C = 0 \qquad \text{...(4.7)}$$

where $A = k - l_2(l_1 - l_4)\cos\theta_2 - l_1l_4$

$B = - 2l_2l_4 \sin\theta_2$

$C = k - l_2(l_1 + l_4)\cos\theta_2 + l_1l_4$

Eq. (4.7) is a quadratic is $\tan\left(\frac{\theta_4}{2}\right)$. Its two roots are,

$$\tan\left(\frac{\theta_4}{2}\right) = \frac{-B \pm \sqrt{B^2 - 4AC}}{2A}$$

$$\theta_4 = 2\tan^{-1}\left[\frac{-B \pm \sqrt{B^2 - 4AC}}{2A}\right] \qquad \text{...(4.8)}$$

From Eqs. (4.1) and (4.3), we have

$$l_4 \cos\theta_4 = l_2\cos\theta_2 + l_3\cos\theta_3 - l_1 \qquad \text{...(4.9)}$$

and $$l_4 \sin\theta_4 = l_2\sin\theta_2 + l_3\sin\theta_3 \qquad \text{...(4.10)}$$

Squaring and adding the two equations, we get

$$l_4^2 = l_2^2 + l_3^2 + l_1^2 + 2l_2l_3\cos\theta_2\cos\theta_3 - 2l_1l_3\cos\theta_3 - 2l_1l_2\cos\theta_2$$

$$+ 2l_2l_3\sin\theta_2\sin\theta_3$$

Put $$l_2^2 + l_3^2 + l_1^2 - l_4^2 = 2k_1$$

$$-2l_1l_3\cos\theta_3 + 2l_2l_3\cos\theta_2\cos\theta_3 + 2l_2l_3\sin\theta_2\sin\theta_3 - 2l_1l_2\cos\theta_2 + 2k_1 = 0$$

or $$-l_1l_3\cos\theta_3 + l_2l_3\cos\theta_2\cos\theta_3 + l_2l_3\sin\theta_2\sin\theta_3 - l_1l_2\cos\theta_2 + k_1 = 0 \qquad \text{...(4.11)}$$

The Eq. (4.11) is identical to Eq. (4.6) and can be obtained from the same by substituting θ_3 for θ_4, $-l_3$ for l_4 and k_1 for k.

Then the solution of Eq. (4.11) will be given by,

$$\theta_3 = 2\tan_{-1}\left[\frac{-E \pm \sqrt{E^2 - 4DF}}{2D}\right] \qquad \text{...(4.12)}$$

where $D = k_1 - l_2(l_1 + l_3)\cos\theta_2 + l_1l_3$

$E = 2l_1l_3\sin\theta_2$

$F = k_1 - l_2(l_1 - l_3)\cos\theta_2 - l_1l_3$

Also from Eq. (4.3), we have

$$\theta_3 = \sin^{-1}\left[\frac{l_4 \sin\theta_4 - l_2 \sin\theta_2}{l_3}\right] \qquad ...(4.13)$$

4.2.2 Velocity Analysis

Let ω_2 = angular velocity given to input link 2.

Rewriting Eq. (4.1), we have

$$l_2 \cos\theta_2 + l_3 \cos\theta_3 - l_4 \cos\theta_4 - l_1 = 0 \qquad ...(4.14)$$

Differentiating Eq. (4.14) with respect to time, we have

$$\frac{d}{d\theta_2}\cdot\frac{d\theta_2}{dt}(l_2\cos\theta_2) + \frac{d}{d\theta_3}\cdot\frac{d\theta_3}{dt}(l_3\cos\theta_3) - \frac{d}{d\theta_4}\cdot\frac{d\theta_4}{dt}(l_4\cos\theta_4) - \frac{d}{dt}(l_1) = 0$$

$$-\omega_2 l_2 \sin\theta_2 - \omega_3 l_3 \sin\theta_3 + \omega_4 l_4 \sin\theta_4 = 0 \qquad ...(4.15)$$

where ω_3, ω_4 = angular velocities of links 3 and 4 respectively.

Rewriting Eq. (4.3), we have

$$l_2 \sin\theta_2 + l_3 \sin\theta_3 - l_4 \sin\theta_4 = 0 \qquad ...(4.16)$$

Differentiating w.r.t. time, we have

$$\omega_2 l_2 \cos\theta_2 + \omega_3 l_3 \cos\theta_3 - \omega_4 l_4 \cos\theta_4 = 0 \qquad ...(4.17)$$

Multiply Eq. (4.15) by $\cos\theta_3$ and Eq. (4.17) by $\sin\theta_3$ and add,

$$\omega_2 l_2(\sin\theta_3 \cos\theta_2 - \cos\theta_3 \sin\theta_2) - \omega_4 l_4(\sin\theta_3 \cos\theta_4 - \cos\theta_3 \sin\theta_4) = 0$$

or

$$\omega_2 l_2 \sin(\theta_3 - \theta_2) - \omega_4 l_4 \sin(\theta_3 - \theta_4) = 0$$

or

$$\omega_4 = \frac{\omega_2 l_2 \sin(\theta_3 - \theta_2)}{l_4 \sin(\theta_3 - \theta_4)} \qquad ...(4.18)$$

Multiply Eq. (4.15) by $\cos\theta_4$ and Eq. (4.17) by $\sin\theta_4$ and add,

$$\omega_2 l_2(\sin\theta_4 \cos\theta_2 - \cos\theta_4 \sin\theta_2) + \omega_3 l_3(\sin\theta_4 \sin\theta_3 - \cos\theta_4 \sin\theta_3) = 0$$

or

$$\omega_2 l_2 \sin(\theta_4 - \theta_2) + \omega_3 l_3 \sin(\theta_4 - \theta_3) = 0$$

$$\omega_3 = -\frac{\omega_2 l_2 \sin(\theta_4 - \theta_2)}{l_3 \sin(\theta_4 - \theta_3)} \qquad ...(4.19)$$

ω_4 and ω_3 can be calculated from Eqs. (4.18) and (4.19) as all other variables are known.

4.2.3 Acceleration Analysis

Let α_2 = angular acceleration given to link 2.

Rewriting Eqs. (4.15) and (4.17) in the following forms,

$$-\omega_2 l_2 \sin\omega_2 t - \omega_3 l_3 \sin\omega_3 t + \omega_4 l_4 \sin\omega_4 t = 0 \qquad ...(4.20)$$

$$\omega_2 l_2 \cos\omega_2 t + \omega_3 l_3 \cos\omega_3 t - \omega_4 l_4 \cos\omega_4 t = 0 \qquad ...(4.21)$$

Differentiating Eqs. (4.20) and (4.21) w.r.t. time, we get

$$(-\alpha_2 l_2 \sin\theta_2 - \omega_2^2 l_2 \cos\theta_2) - (\alpha_3 l_3 \sin\theta_3 + \omega_3^2 l_3 \cos\theta_3)$$
$$+ (\alpha_4 l_4 \sin\theta_4 + \omega_4^2 l_4 \cos\theta_4) = 0 \qquad ...(4.22)$$

$$(\alpha_2 l_2 \cos\theta_2 - \omega_2^2 l_2 \sin\theta_2) + (\alpha_3 l_3 \cos\theta_3 - \omega_3^2 l_3 \sin\theta_3)$$
$$- (\alpha_4 l_4 \cos\theta_4 - \omega_4^2 l_4 \sin\theta_4) = 0 \qquad ...(4.23)$$

where $\alpha_2 = \frac{d\omega_2}{dt}$, $\alpha_3 = \frac{d\omega_3}{dt}$ and $\alpha_4 = \frac{d\omega_4}{dt}$

Multiply Eq. (4.22) by $\cos\theta_4$ and Eq. (4.23) by $\sin\theta_4$ and add,

$$\alpha_2 l_2(\sin\theta_4\cos\theta_2 - \cos\theta_4\sin\theta_2) - \omega_2^2 l_2(\cos\theta_2\cos\theta_4 + \sin\theta_2\sin\theta_4)$$
$$- \alpha_3 l_3(\sin\theta_3\cos\theta_4 - \cos\theta_3\sin\theta_4)$$
$$- \omega_3^2 l_3(\cos\theta_3\cos\theta_4 + \sin\theta_3\sin\theta_4) + \omega_4^2 l_4 = 0$$

or
$$\alpha_2 l_2\sin(\theta_4 - \theta_2) - \omega_2^2 l_2\cos(\theta_4 - \theta_2)$$
$$- \alpha_3 l_3\sin(\theta_3 - \theta_4) - \omega_3^2 l_3\cos(\theta_4 - \theta_3) + \omega_4^2 l_4 = 0$$

$$\alpha_3 = \frac{\alpha_2 l_2\sin(\theta_4 - \theta_2) - \omega_2^2 l_2\cos(\theta_4 - \theta_2) - \omega_3^2 l_3\cos(\theta_4 - \theta_3) + \omega_4^2 l_4}{l_3\sin(\theta_3 - \theta_4)} \quad ...(4.24)$$

Multiply Eq. (4.22) by $\cos\theta_3$ and Eq. (4.23) by $\sin\theta_3$ and add,

$$\alpha_2 l_2(\sin\theta_3\cos\theta_2 - \cos\theta_3\sin\theta_2) - \omega_2^2 l_2(\cos\theta_3\cos\theta_2 + \sin\theta_3\sin\theta_2) - l_3\omega_3^2$$
$$+ \alpha_4 l_4(\sin\theta_4\cos\theta_3 - \cos\theta_4\sin\theta_3)$$
$$+ \omega_4^2 l_4(\cos\theta_3\cos\theta_4 + \sin\theta_3\sin\theta_4) = 0$$

or
$$\alpha_2 l_2\sin(\theta_3 - \theta_2) - \omega_2^2 l_2\cos(\theta_3 - \theta_2) - l_3\omega_3^2 - \alpha_4 l_4\sin(\theta_3 - \theta_4)$$
$$+ \omega_4^2 l_4\cos(\theta_3 - \theta_4) = 0$$

$$\alpha_4 = \frac{\alpha_2 l_2\sin(\theta_3 - \theta_2) - \omega_2^2 l_2\cos(\theta_3 - \theta_2) - \omega_3^2 l_3 + \omega_4^2 l_4\cos(\theta_3 - \theta_4)}{l_4\sin(\theta_3 - \theta_4)} \quad ...(4.25)$$

α_3 and α_4 can be evaluated from Eqs. (4.24) and (4.25) respectively.

4.3 COMPLEX ALGEBRA METHOD

For a four-bar mechanism, we can write

$$\bar{l}_2 + \bar{l}_3 - \bar{l}_4 - \bar{l}_1 = 0 \quad ...(4.26)$$

Transforming these vectors into complex polar form, we have

$$l_2 e^{i\theta_2} + l_3 e^{i\theta_3} - l_4 e^{i\theta_4} - l_1 = 0 \quad (\because \theta_1 = 0) ...(4.27)$$

where $e^{i\theta_2} = \cos\theta_2 + i\sin\theta_2$, etc.

Transforming Eq. (4.27) into complex rectangular form and separating the real and imaginary terms, we have

$$l_2\cos\theta_2 + l_3\cos\theta_3 = l_1 + l_4\cos\theta_4 \quad ...(4.28)$$

and
$$l_2\sin\theta_2 + l_3\sin\theta_3 = l_4\sin\theta_4 \quad ...(4.29)$$

Differentiating Eq. (4.27) w.r.t. time, we get

$$il_2\dot{\theta}_2 e^{i\theta_2} + il_3\dot{\theta}_3 e^{i\theta_3} - il_4\dot{\theta}_4 e^{i\theta_4} = 0 \quad ...(4.30)$$

or
$$l_2\omega_2 e^{i\theta_2} + l_3\omega_3 e^{i\theta_3} - l_4\omega_4 e^{i\theta_4} = 0 \quad ...(4.31)$$

Again transforming Eq. (4.31) into complex rectangular form and separating the real and imaginary terms, we get

$$l_2\omega_2 \cos\theta_2 + l_3\omega_3 \cos\theta_3 - l_4\omega_4 \cos\theta_4 = 0 \quad ...(4.32)$$

$$l_2\omega_2 \sin\theta_2 + l_3\omega_3 \sin\theta_3 - l_4\omega_4 \sin\theta_4 = 0 \quad ...(4.33)$$

Differentiating Eq. (4.31) w.r.t. time, we get

$$l_2(\ddot{\theta}_2 e^{i\theta_2} + i\dot{\theta}_2^2 e^{i\theta_2}) + l_3(\ddot{\theta}_3 e^{i\theta_3} + i\dot{\theta}_3^2 e^{i\theta_3}) - l_4(\ddot{\theta}_4 e^{i\theta_4} + i\dot{\theta}_4^2 e^{i\theta_4}) = 0 \quad ...(4.34)$$

or

$$l_2(\alpha_2 e^{i\theta_2} + i\omega_2^2 e^{i\theta_2}) + l_3(\alpha_3 e^{i\theta_3} + i\omega_3^2 e^{i\theta_3}) - l_4(\alpha_4 e^{i\theta_4} + i\omega_4^2 e^{i\theta_4}) = 0. \quad ..(4.35)$$

Transforming Eq. (4.35) into complex rectangular form and separating the real and imaginary terms, we get

$$l_2\alpha_2 \cos\theta_2 + l_3\alpha_3 \cos\theta_3 - l_4\alpha_4 \cos\theta_4 - l_2\omega_2^2 \sin\theta_2 - l_3\omega_3^2 \sin\theta_3 + l_4\omega_4^2 \sin\theta_4 = 0 \quad ...(4.36)$$

$$l_2\alpha_2 \sin\theta_2 + l_2\omega_2^2 \cos\theta_2 + l_3\alpha_3 \sin\theta_3 + l_3\omega_3^2 \cos\theta_3 - l_4\alpha_4 \sin\theta_4 - l_4\omega_4^2 \cos\theta_4 = 0 \quad ...(4.37)$$

Eqs. (4.36) and (4.37) and similar to Eqs. (4.22) and (4.23) and can be solved as before.

4.4 THE VECTOR METHOD

We have $\bar{l}_2 + \bar{l}_3 - \bar{l}_4 - \bar{l}_1 = 0$

Assume that the angles θ_3 and θ_4 are known. Differentiate the above equation w.r.t. time.

$$\omega_2 \times \bar{l}_2 + \omega_3 \times \bar{l}_3 - \omega_4 \times \bar{l}_4 = 0 \quad ...(4.38)$$

Let $\hat{i}_2, \hat{i}_3$ and $\hat{i}_4$ be the unit vectors along $\bar{l}_2, \bar{l}_3$ and $\bar{l}_4$ vectors. In plane motion mechanism, all the angular velocities are in the $\hat{k}$ direction. Therefore,

$$l_2\omega_2(\hat{k} \times \hat{i}_2) + l_3\omega_3(\hat{k} \times \hat{i}_3) - l_4\omega_4(\hat{k} \times \hat{i}_4) = 0 \quad ...(4.39)$$

Take the dot product with $\hat{i}_3$.

$$l_2\omega_2(\hat{k} \times \hat{i}_2)\cdot\hat{i}_3 + l_3\omega_3(\hat{k} \times \hat{i}_3)\cdot\hat{i}_3 - l_4\omega_4(\hat{k} \times \hat{i}_4)\cdot\hat{i}_3 = 0$$

or

$$l_2\omega_2(\hat{k} \times \hat{i}_2)\cdot\hat{i}_3 + 0 - l_4\omega_4(\hat{k} \times \hat{i}_4)\cdot\hat{i}_3 = 0$$

$$\omega_4 = \frac{l_2\omega_2(\hat{k} \times \hat{i}_2)\cdot\hat{i}_3}{l_4(\hat{k} \times \hat{i}_4)\cdot\hat{i}_3} \quad ...(4.40)$$

Taking the dot product with $\hat{i}_4$,

$$l_2\omega_2(\hat{k} \times \hat{i}_2)\cdot\hat{i}_4 + l_3\omega_3(\hat{k} \times \hat{i}_3)\cdot\hat{i}_4 - l_4\omega_4(\hat{k} \times \hat{i}_4)\cdot\hat{i}_4 = 0$$

or

$$l_2\omega_2(\hat{k} \times \hat{i}_2)\cdot\hat{i}_4 + l_3\omega_3(\hat{k} \times \hat{i}_3)\cdot\hat{i}_4 - 0 = 0$$

$$\omega_3 = -\frac{l_2\omega_2(\hat{k} \times \hat{i}_2)\cdot\hat{i}_4}{l_3(\hat{k} \times \hat{i}_3)\cdot\hat{i}_4} \quad ...(4.41)$$

Now

$$(\hat{k} \times \hat{i}_2)\cdot\hat{i}_3 = \begin{vmatrix} \hat{i} & \hat{j} & \hat{k} \\ 0 & 0 & 1 \\ \cos\theta_2 & \sin\theta_2 & 0 \end{vmatrix} \cdot\hat{i}_3 = (-\sin\theta_2\cdot\hat{i} + \cos\theta_2\cdot\hat{j})\cdot\hat{i}_3$$

$$= (-\sin\theta_2\cdot\hat{i} + \cos\theta_2\cdot\hat{j})\cdot(\cos\theta_3\hat{i} + \sin\theta_3\hat{j})$$

$$= -\sin\theta_2\cos\theta_3 + \cos\theta_2\sin\theta_3$$

$$= \sin(\theta_3 - \theta_2)$$

Similarly $\quad (\hat{k}\times\hat{i}_4)\cdot\hat{i}_3 = \sin(\theta_3 - \theta_4)$

$\therefore$
$$\omega_4 = \frac{l_2\omega_2(\hat{k}\times\hat{i}_2)\cdot\hat{i}_3}{l_4(\hat{k}\times\hat{i}_4)\cdot\hat{i}_3}$$

$$= \frac{l_2\omega_2\sin(\theta_3 - \theta_2)}{l_4\sin(\theta_3 - \theta_4)} \qquad \text{...(4.42)}$$

and
$$\omega_3 = -\frac{l_2\omega_2(\hat{k}\times\hat{i}_2)\cdot\hat{i}_4}{l_3(\hat{k}\times\hat{i}_3)\cdot\hat{i}_4}$$

$$= -\frac{l_2\omega_2\sin(\theta_4 - \theta_2)}{l_3\sin(\theta_4 - \theta_3)} \qquad \text{...(4.43)}$$

Differentiating Eq. (4.38) w.r.t. time to get the acceleration,

$$\omega_2\times l_2 + \omega_2\times(\omega_2\times l_2) + \omega_3\times l_3 + \omega_3\times(\omega_3\times l_3) - \omega_4\times l_4 - \omega_4\times(\omega_4\times l_4) = 0$$

$$\alpha_2\times l_2 + \omega_2\times(\omega_2\times l_2) + \alpha_3\times l_3 + \omega_3\times(\omega_3\times l_3) - \alpha_4\times l_4 - \omega_4\times(\omega_4\times l_4) = 0$$

or
$$l_2\alpha_2(\hat{k}\times\hat{i}_2) - l_2\omega_2^2\hat{i}_2 + l_3\alpha_3(\hat{k}\times\hat{i}_3) - l_3\omega_3^2\hat{i}_3 - l_4\alpha_4(\hat{k}+\hat{i}_4) + l_4\omega_4^2\hat{i}_4 = 0 \qquad \text{...(4.44)}$$

Taking the dot product of Eq. (4.44) with $\hat{i}_3$,

$$l_2\alpha_2(\hat{k}\times\hat{i}_2)\cdot\hat{i}_3 - l_2\omega_2^2\hat{i}_2\cdot\hat{i}_3 + 0 - l_3\omega_3^2\hat{i}_3\cdot\hat{i}_2 - l_4\alpha_4(\hat{k}\times\hat{i}_4)\cdot\hat{i}_3 + l_4\omega_4^2\hat{i}_4\cdot\hat{i}_3 = 0$$

$$\alpha_4 = \frac{l_2\alpha_2(\hat{k}\times\hat{i}_2)\cdot\hat{i}_3 - l_2\omega_2^2\hat{i}_2\cdot\hat{i}_3 - l_3\omega_3^2 + l_4\omega_4^2\hat{i}_4\cdot\hat{i}_2)}{l_4(\hat{k}\times\hat{i}_4)\cdot\hat{i}_3} \qquad \text{...(4.45)}$$

Since
$$(\hat{k}\times\hat{i}_2)\cdot\hat{i}_3 = \sin(\theta_3 - \theta_2),$$
$$\hat{i}_2\cdot\hat{i}_3 = \cos(\theta_3 - \theta_2),$$
$$\hat{i}_3\cdot\hat{i}_3 = 1,\ \hat{i}_4\cdot\hat{i}_3 = \cos(\theta_3 - \theta_4)$$

and
$$(\hat{k}\times\hat{i}_4)\cdot\hat{i}_3 = \sin(\theta_3 - \theta_4)$$

The Eq. (4.45) reduces to

$$\alpha_4 = \frac{l_2\alpha_2\sin(\theta_3 - \theta_2) - l_2\omega_2^2\cos(\theta_3 - \theta_2) - l_3\omega_3^2 + l_4\omega_4^2\cos(\theta_3 - \theta_4)}{l_4\sin(\theta_3 - \theta_4)} \qquad \text{...(4.46)}$$

Taking the dot product of Eq. (4.44) with $\hat{i}_4$, we have

$$l_2\alpha_2(\hat{k}\times\hat{i}_2)\cdot\hat{i}_4 - l_2\omega_2^2\hat{i}_2\cdot\hat{i}_4 + l_3\alpha_3(\hat{k}\times\hat{i}_3)\cdot\hat{i}_4 - l_3\omega_3^2\hat{i}_3\cdot\hat{i}_4 + l_4\omega_4^2\hat{i}_4\cdot\hat{i}_4 = 0$$

or
$$\alpha_3 = \frac{l_2\alpha_2(\hat{k}\times\hat{i}_2)\cdot\hat{i}_4 - l_2\omega_2^2\hat{i}_2\cdot\hat{i}_4 - l_3\omega_3^2\hat{i}_3\cdot\hat{i}_4 + l_4\omega_4^2}{b(\hat{k}\times\hat{i}_3)\cdot\hat{i}_4} \qquad \text{...(4.47)}$$

$$= \frac{l_2\alpha_2\sin(\theta_4 - \theta_2) - l_2\omega_2^2\cos(\theta_4 - \theta_2) - l_3\omega_3^2\cos(\theta_4 - \theta_2) + l_4\omega_4^2}{l_3\sin(\theta_3 - \theta_4)} \qquad \text{...(4.48)}$$

■ **Example 4.1** In a four-bar mechanism, the dimensions of the various links are :

$$l_1 = 120\text{ mm},\ l_2 = 50\text{ mm},\ l_3 = 70\text{ mm},\ l_4 = 60\text{ mm}$$

$$\theta_2 = 60°,\ \omega_2 = 10\text{ rad/s(ccw)},\ \alpha_2 = -25\text{ rad/s}^2$$

Determine analytically the angular displacement, angular velocity and angular accelerations of coupler and output link.

Solution : From Eq. (4.5), we have

$$2k = l_2^2 - l_3^2 + l_4^2 + l_1^2$$

$$= 50^2 - 70^2 + 60^2 + 120^2 = 15600$$

$$k = 7800 \text{ mm}^2$$

$$A = k - l_2(l_1 - l_4)\cos\theta_2 - l_1 l_4$$

$$= 7800 - 50\,(120 - 60)\cos 60° - 120 \times 60$$

$$= -900$$

$$B = -2l_2 l_4 \sin\theta_2$$

$$= -2 \times 50 \times 60 \sin 60° = -5196.15$$

$$C = k - l_2(l_1 + l_4)\cos\theta_2 + l_1 l_4$$

$$= 7800 - 50\,(120 + 60)\cos 60° + 120 \times 60 = 10500$$

$$\theta_4 = 2\tan^{-1}\left[\frac{-B \pm \sqrt{B^2 - 4AC}}{2A}\right]$$

$$= 2\tan^{-1}\left[\frac{+5196.15 \pm \sqrt{(-5196.15)^2 - 4(-900) \times 10500}}{-2 \times 900}\right]$$

$$= 2\tan^{-1}\left[\frac{+5196.15 \pm 8049.84}{-1800}\right]$$

$$= 2\tan^{-1}[1.5854 \text{ or } -7.3589]$$

$$= 115.5° \text{ or } -164.5°$$

Taking the first value, from Eq. (4.10), we have

$$l_4 \sin\theta_4 = l_2 \sin\theta_2 + l_3 \sin\theta_3$$

$$60 \sin 115.5° = 50 \sin 60° + 70 \sin\theta_3$$

$$\sin\theta_3 = 0.15505$$

$$\theta_3 = 8.92°$$

$$\omega_3 = -\frac{\omega_2 l_2 \sin(\theta_4 - \theta_2)}{l_3 \sin(\theta_4 - \theta_3)}$$

$$= -\frac{10 \times 50 \sin(115.5° - 60°)}{70 \sin(115.5° - 8.92)} = -6.14 \text{ rad/s}$$

$$\omega_4 = \frac{\omega_2 l_2 \sin(\theta_3 - \theta_2)}{l_4 \sin(\theta_3 - \theta_4)}$$

$$= \frac{10 \times 50 \sin(8.92° - 60°)}{60 \sin(8.92° - 115.5°)} = 6.76 \text{ rad/s}$$

$$\alpha_3 = \frac{\alpha_2 l_2 \sin(\theta_4 - \theta_2) - \omega_2^2 l_2 \cos(\theta_4 - \theta_2) - \omega_3^2 l_3 \cos(\theta_4 - \theta_3) + \omega_4^2 l_4}{l_3 \sin(\theta_3 - \theta_4)}$$

$$= \frac{-25 \times 50 \sin(115.5° - 60°) - 100 \times 50 \cos(115.5° - 60°) - (-6.14)^2 \times 70 \cos(115.5° - 8.92°) + (6.76)^2 \times 60}{70 \sin(8.92° - 115.5°)}$$

$$\frac{-367.292}{-6.709} = 54.75 \text{ rad/s}^2$$

$$\alpha_4 = \frac{\alpha_2 l_2 \sin(\theta_3 - \theta_2) - \omega_2^2 l_2 \cos(\theta_3 - \theta_2) - \omega_3^2 l_3 + \omega_4^2 l_4 \cos(\theta_3 - \theta_4)}{l_4 \sin(\theta_3 - \theta_4)}$$

$$= \frac{-25 \times 50 \sin(8.92° - 60°) - 100 \times 50 \cos(8.92 - 60°) - (-6.14)^2 \times 70 + (6.76)^2 \times 60 \cos(8.92° - 115.5°)}{60 \sin(8.92° - 115.5°)}$$

$$= \frac{-5590.0}{-57.5} = 97.2 \text{ rad/s}^2$$

4.5 SLIDER-CRANK MECHANISM

Consider a slider-crank mechanism in which the slider is offset by an eccentricity e, as shown in Fig. 4.2. Let $O_2C = l_1, O_2A = l_2, AB = l_3, BC = e$.

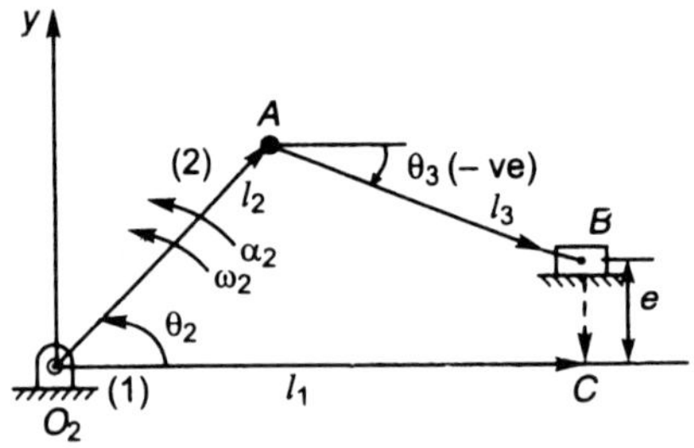

Fig. 4.2 Slider crank mechanism.

The angular velocity and acceleration of the crank are ω_2 and α_2 (ccw). Angle θ_3 measured clockwise from x-axis of link AB is –ve.

Displacement along x-axis,

$$l_2 \cos\theta_2 + l_3 \cos(-\theta_3) = l_1$$

or
$$l_3 \cos\theta_3 = l_1 - l_2 \cos\theta_2 \quad \text{...(4.49)}$$

Displacement along y-axis,

$$l_2 \sin\theta_2 - l_3 \sin(-\theta_3) = e$$

or
$$l_2 \sin\theta_2 + l_3 \sin\theta_3 = e$$

$$l_3 \sin\theta_3 = e - l_2 \sin\theta_2 \quad \text{...(4.50)}$$

Squaring Eqs. (4.49) and (4.50) and adding, we get

$$l_3^2 = [(l_1^2 + l_2^2 \cos^2\theta_2 - 2l_1 l_2 \cos\theta_2) + (l_2^2 \sin^2\theta + e^2 - 2l_2 e \sin\theta_2)]$$

$$= l_1^2 + e^2 + l_2^2 - 2l_1 l_2 \cos\theta_2 - 2l_2 e \sin\theta_2$$

or $$l_1^2 - (2l_2 \cos\theta_2)\, l_1 + (l_2^2 - l_3^2 + e^2 - 2el_2 \sin\theta_2) = 0$$

or $$l_1^2 + Al_1 + B = 0 \qquad ...(4.51)$$

where $A = -(2l_2 \cos\theta_2)$

$$B = l_2^2 - l_3^2 + e^2 - 2el_2 \sin\theta_2$$

Eq. (4.51) is a quadratic equation in l_1. Its two roots are,

$$l_1 = \frac{-A \pm \sqrt{A^2 - 4B}}{2} \qquad ...(4.52)$$

From Eq. (4.50), we have

$$\sin\theta_3 = \frac{l_2 \sin\theta_2 - e}{l_3}$$

or $$\theta_3 = \sin^{-1}\left[\frac{l_2 \sin\theta_2 - e}{l_3}\right] \qquad ...(4.53)$$

4.5.1 Velocity Analysis

Differentiating Eqs. (4.49) and (4.50) w.r.t. time, we get

$$-l_3\omega_3 \sin\theta_3 - l_2\omega_2 \sin\theta_2 - \dot{l}_1 = 0 \qquad ...(4.54)$$

and $$l_3\omega_3 \cos\theta_3 + l_2\omega_2 \cos\theta_2 = 0 \qquad ...(4.55)$$

Multiply Eq. (4.54) by $\cos\theta_3$ and Eq. (4.55) by $\sin\theta_3$ and add,

$$l_2\omega_2 (-\sin\theta_2 \cos\theta_3 + \cos\theta_2 \sin\theta_3) - \dot{l}_1 \cos\theta_3 = 0$$

$$l_2\omega_2 \sin(\theta_3 - \theta_2) - \dot{l}_1 \cos\theta_3 = 0$$

$$\dot{l}_1 = \frac{l_2\omega_2 \sin(\theta_3 - \theta_2)}{\cos\theta_3} \qquad ...(4.56)$$

From Eq. (4.55), we have

$$\omega_3 = -\frac{l_2\omega_2 \cos\theta_2}{l_3 \cos\theta_3} \qquad ...(4.57)$$

where $\dot{l}_1$ = linear velocity of slider

ω_3 = angular velocity of connecting rod or coupler link.

4.5.2 Acceleration Analysis

Differentiating Eqs. (4.54) and (4.55) w.r.t. time, we get

$$-[l_3\, \alpha_3 \sin\theta_3 + l_3\, \omega_3^2 \cos\theta_3] - [l_2\, \alpha_2 \sin\theta_2 + l_2\, \omega_2^2 \cos\theta_2] - \ddot{l}_1 = 0 \qquad ...(4.58)$$

$$l_3\alpha_3 \cos\theta_3 - l_3\, \omega_3^2 \sin\theta_3 + l_2\alpha_2 \cos\theta_2 - l_2\, \omega_2^2 \sin\theta_2 = 0 \qquad ...(4.59)$$

Multiply Eq. (4.58) by $\cos\theta_3$ and Eq. (4.59) by $\sin\theta_3$ and adding, we get

$$-l_3\, \omega_3^2 + l_2\, \alpha_2 (\sin\theta_3 \cos\theta_2 - \cos\theta_3 \sin\theta_2)$$
$$- l_2\, \omega_2^2 (\sin\theta_3 \sin\theta_2 + \cos\theta_3 \cos\theta_2) - \ddot{l}_1 \cos\theta_3 = 0$$

$$-l_3\, \omega_3^2 + l_2\, \alpha_2 \sin(\theta_3 - \theta_2) - l_2\, \omega_2^2 (\cos\theta_3 - \theta_2) - \ddot{l}_1 \cos\theta_3 = 0$$

$$\ddot{l}_1 = \frac{l_2\ \alpha_2 \sin(\theta_3 - \theta_2) - l_2\ \omega_2^2 \cos(\theta_3 - \theta_2) - l_3\ \omega_3^2}{\cos\theta_3} \quad \text{...(4.60)}$$

From Eq. (4.59), we get

$$\alpha_3 = \frac{l_3\ \omega_3^2 \sin\theta_3 - l_2\ \alpha_3 \cos\theta_2 + l_2\ \omega_2^2 \sin\theta_2}{l_3 \cos\theta_3} \quad \text{...(4.61)}$$

■ **Example 4.2** In a slider-crank mechanism, the lengths of the crank and connecting rod are 500 mm and 2000 mm respectively. It has an eccentricity of 100 mm. Assuming that the crank is rotating with an angular velocity of 25 rad/s. Calculate the following when the crank has turned through 30° in the ccw direction from IDC :

(a) The velocity and acceleration of the slider, and

(b) The angular velocity and angular acceleration of the connecting rod.

Solution. Given : $l_2 = 0.5$ m, $l_3 = 2$ m, $\theta_2 = 30°, \omega_2 = 25$ rad/s, $e = 0.1$ m

$$A = -2l_2 \cos\theta_2 = -2 \times 0.5 \cos 30° = -0.866$$

$$B = l_2^2 - l_3^2 + e^2 - 2el_2 \sin\theta_2$$

$$= (0.5)^2 - 4 + (0.1)^2 - 2 \times 0.1 \times 0.5 \sin 30° = -3.79$$

$$l_1 = \frac{-A \pm \sqrt{A^2 - 4B}}{2}$$

$$= \frac{0.866 \pm \sqrt{(-0.866)^2 - 4(-3.79)}}{2}$$

$$= 0.433 \pm 1.994 = 2.427 \text{ m}, -1.561 \text{ m}$$

We take $l_1 = 2.427$ m

$$\theta_3 = \sin^{-1}\left[\frac{l_2 \sin\theta_2 - e}{l_3}\right]$$

$$= \sin^{-1}\left[\frac{0.5 \sin 30° - 01.}{2}\right] = \sin^{-1} 0.075 = 4.3°$$

Velocity of slider,

$$\dot{l}_1 = \frac{l_2\ \omega_2 \sin(\theta_3 - \theta_2)}{\cos\theta_3}$$

$$= \frac{0.5 \times 25 \sin(4.3° - 30°)}{\cos 4.3°} = 5.436 \text{ m/s}$$

(b) Angular velocity of connecting rod,

$$\omega_3 = -\frac{l_2\omega_2 \cos\theta_2}{l_3 \cos\theta_3}$$

$$= -\frac{0.5 \times 25 \cos 30°}{2 \cos 4.3°} = -5.428 \text{ rad/s}$$

Acceleraton of slider,

$$\ddot{l}_1 = \frac{-l_2\ \omega_2 \cos(\theta_3 - \theta_2) - l_3\ \omega_3^2}{\cos\theta_3} \quad (\because \alpha_2 = 0)$$

$$= \frac{-0.5 \times (25)^2 \cos(4.3° - 30°) - 2(-5.428)^2}{\cos 4.3}$$

$$= -341.5 \text{ m/s}^2$$

Angular acceleration of connecting rod,

$$\alpha_3 = \frac{l_3 \omega_3^2 \sin\theta_3 + l_2 \omega_2^2 \sin\theta_2}{l_3 \cos\theta_3} \qquad (\because \alpha_2 = 0)$$

$$= \frac{2(-5.428)^2 \sin 4.3° + 0.5(25)^2 \sin 30°}{2\cos 4.3°}$$

$$= 80.56 \text{ rad/s}^2$$

SUMMARY

1. In analytical methods, the links of the mechanisms are considered as vectors.
2. In a four-bar mechanism :

(i) $$\theta_4 = 2\tan^{-1}\left[\frac{-B \pm \sqrt{B^2 - 4AC}}{2A}\right]$$

where $2k = l_2^2 - l_3^2 + l_4^2 + l_1^2$

$A = k - l_2(l_1 - l_4)\cos\theta_2 - l_1 l_4$

$B = -2l_2 l_4 \sin\theta_2$

$C = k - l_2(l_1 + l_4)\cos\theta_2 + l_1 l_4$

(ii) $\theta_3 = \sin^{-1}\left[\frac{l_4 \sin\theta_4 - l_2 \sin\theta_2}{l_3}\right]$

(iii) $\omega_4 = \frac{\omega_2 l_2 \sin(\theta_3 - \theta_2)}{l_4 \sin(\theta_3 - \theta_4)}$

(iv) $\omega_3 = -\frac{\omega_2 l_2 \sin(\theta_4 - \theta_2)}{l_3 \sin(\theta_4 - \theta_3)}$

(v) $\alpha_3 = \frac{\alpha_2 l_2 \sin(\theta_4 - \theta_2) - \omega_2^2 l_2 \cos(\theta_4 - \theta_2) - \omega_3^2 l_3 \cos(\theta_4 - \theta_3) + \omega_4^2 l_4}{l_3 \sin(\theta_3 - \theta_4)}$

(vi) $\alpha_4 = \frac{\alpha_2 l_2 \sin(\theta_3 - \theta_2) - \omega_2^2 l_2 \cos(\theta_3 - \theta_2) - \omega_3^2 l_3 + \omega_4^2 l_4 \cos(\theta_3 - \theta_4)}{l_4 \sin(\theta_3 - \theta_4)}$

3. In a slider-crank mechanism with offset :

(i) $$l_1 = \frac{-A + \sqrt{A^2 - 4B}}{2}$$

where $A = -2l_2 \cos\theta_2$

$B = l_2^2 - l_3^2 + e^2 - 2el_2 \sin\theta_2$

(ii) $\theta_3 = \sin^{-1}\left[\dfrac{l_2 \sin\theta_2 - e}{l_3}\right]$

(iii) $\dot{l}_1 = \dfrac{l_2\omega_2 \sin(\theta_3 - \theta_2)}{\cos\theta_3}$

(iv) $\omega_3 = \dfrac{-l_2\omega_2 \cos\theta_2}{l_3 \cos\theta_3}$

(v) $\ddot{l}_1 = \dfrac{l_2\alpha_2 \sin(\theta_3 - \theta_2) - l_2\omega_2^2 \cos(\theta_3 - \theta_2) - l_3\omega_3^2}{\cos\theta_3}$

(vi) $\alpha_3 = \dfrac{l_3\omega_3^2 \sin\theta_3 - l_2\alpha_2 \cos\theta_2 + l_2\omega_2^2 \sin\theta_2}{l_3 \cos\theta_3}$

EXERCISES

1. Derive expressions to determine the angles of the output link and coupler of a four-bar mechanism. Also deduce the relations for the angular velocity and angular acceleration of the same links.
2. Derive expressions to determine the (a) linear velocity and linear accelerations of slider and connecting rod, (b) angular velocity and angular accelerations of connecting rod.
3. In a four-bar mechanism, the dimensions of the various links are : $AB = 40$ mm, $BC = 100$ mm, $CD = 50$ mm, and $AD = 80$ mm. AD is the fixed link and AB is the input crank. If AB rotates at 25 rad/s (cw), calculate :

   Angular velocity and angular acceleration of links BC and CD, when $\theta = 30°$.
4. In a slider-crank mechanism, the crank length is 60 mm, connecting rod length is 180 mm, and eccentricity is 20 mm. When the crank has turned through 45° from the IDC at an angular velocity of 10 rad/s and angular acceleration of 15 rad/s^2, both ccw, calculate : (a) linear velocity and acceleration of the slider, and (b) the angular velocity and angular acceleraton of connecting rod.
5. Analyze analytically to find velocity and acceleration of four link mechanism. [KU, Dec., 2009]

❑❑❑

Chapter 5

Mechanisms with Lower Pairs

Inside This Chapter

5.1 INTRODUCTION

The lower pairs are those which have surface contact between them. We shall study the various mechanisms for the generation of straight line motion, both exact and approximate. These mechanisms have a vital role to play in generating the configurations for machines. The steering gears and Hooke's joint are other lower pairs which shall be discussed.

5.2 PANTOGRAPH

It is a mechanism to produce the path traced out by a point on enlarged or reduced scale. Fig. 5.1 (a) shows the line diagram of a pantograph in which $AB = CD, BC = AD$ and $ABCD$ is always a parallelogram. OQP is a straight line. P describes a path similar to that described by Q. It is used as a copying mechanism.

Proof : To prove that the path described by P is similar to that described by Q, consider Δs OAQ and OBP which are similar, because $\angle BOP$ is common.

$\angle AQO = \angle BPO$ being corresponding angles as $AQ \parallel BP$.

Hence
$$\frac{OA}{OB} = \frac{OQ}{OP} = \frac{AQ}{BP} \qquad ...(1)$$

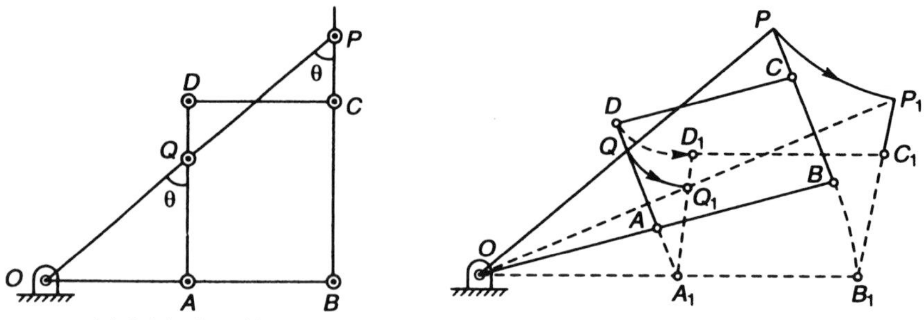

Fig. 5.1 Pantograph.

In the displaced position shown in Fig. 5.1 (b), as all inks are rigid,

$$B_1O = BO,\ D_1A_1 = DA,\ A_1O = AO$$

and
$$P_1B_1 = PB,\ B_1A_1 = BA,\ A_1Q_1 = AQ$$

Hence
$$\frac{OA_1}{OB_1} = \frac{A_1Q_1}{B_1P_1}$$

As $A_1B_1C_1D_1$ is a parallelogram, $A_1D_1 \parallel B_1C_1$, *i.e.*, $A_1Q_1 \parallel B_1P_1$.

OQ_1P_1 is again a straight line so that Δs QA_1Q_1 and OB_1P_1 are similar.

$$\frac{OA_1}{OB_1} = \frac{OQ_1}{OP_1} \qquad ...(2)$$

From Eqs. (1) and (2), we get

$$\frac{OQ}{OP} = \frac{OQ_1}{OP_1} \text{ because } OA = OA_1 \text{ and } OB = OB_1.$$

Hence QQ_1 is similar to PP_1 or they are parallel. Thus path traced by P is similar to that of Q.

The pantograph is used in geometrical instruments, manufacture of irregular objects, to guide cutting tools, and as indicator rig for cross head.

5.3 STRAIGHT LINE MOTION MECHANISMS

Now we shall study lower pairs generating straight line motion. Straight line motion can be generated by either sliding pairs or turning pairs. Sliding pairs are bulky and gets worn out rapidly. Therefore, turning pairs are preferred over sliding pairs for generating straight line motion. Straight line motion can be generated either accurately or approximately.

5.3.1 Accurate Straight Line Motion Mechanisms

Mechanisms for the generation of accurate straight line motion are :

1. Peaucellier mechanism
2. Hart mechanism
3. Scott-Russel mechanism.

1. Peaucellier Mechanism

A line diagram of the Peaucellier mechanism is shown in Fig. 5.2.

$$OR = OS \text{ and } OO_1 = O_1Q.$$

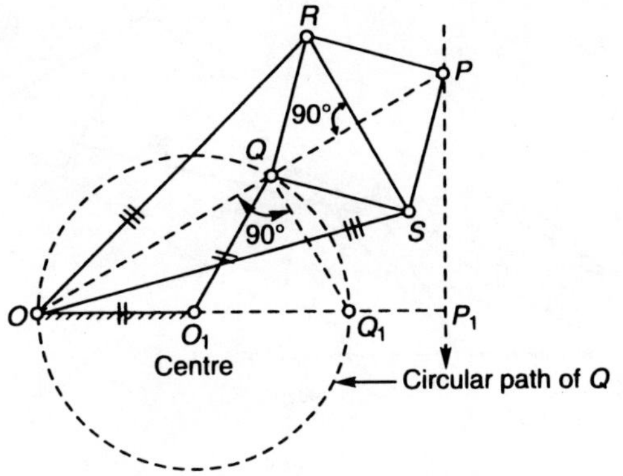

Fig. 5.2 Peaucellier straight line mechanism.

Point P describes a straight line perpendicular to OO_1 produced.

Proof : Δs ORQ and OSQ are congruent, because $OR = OS$, $OR = QS$ and OQ is common. $\angle ROQ = \angle SOQ$. Also $\angle OQR = \angle OQS$ and $\Delta s\ PRO = \Delta s\ PSQ$.

$$\angle OQR + \angle RQP = \angle OQS + \angle SQP$$

But OQ is a straight line.

$$\angle OQR + \angle RQP = \angle OQS + \angle SQP = 180°$$

Hence OQP is a straight line.

Now
$$OR^2 = OT^2 + RT^2$$
$$RP^2 = RT^2 + TP^2$$
$$OR^2 - RP^2 = OT^2 - TP^2$$
$$= (OT + TP)(OT - TP) = OP \cdot OQ$$

But OR and RP are always constant.

Hence $$OP \cdot OQ = \text{constant.}$$

Draw $PP_1 \perp OO_1$ produced and joint QQ_1. $\Delta s\ OQQ_1$ and OPP_1 are similar, because $\angle OQQ_1 = \angle OP_1P = 90°$ and $\angle QOQ_1$ is common.

Hence $$\frac{OQ}{OQ_1} = \frac{OP_1}{OP}$$

or $$OQ \cdot OP = OQ_1 \cdot OP_1 = \text{constant}$$

Now $$OQ_1 = 2OO_1 = \text{constant}$$

Hence $$OP_1 = \text{constant}$$

or point P moves in a straight line.

2. Hart Mechanism

The Hart mechanism is shown in Fig. 5.3. OO_1 is a fixed link and O_1Q is the rotating link. The point Q moves in a circle with centre O_1 and radius O_1Q. $ABCD$ is a trapezium so that $AB = CD, BC = AD$ and $BD \parallel AC$.

$$\frac{BO}{BA} = \frac{BQ}{BC} = \frac{DP}{DA}$$

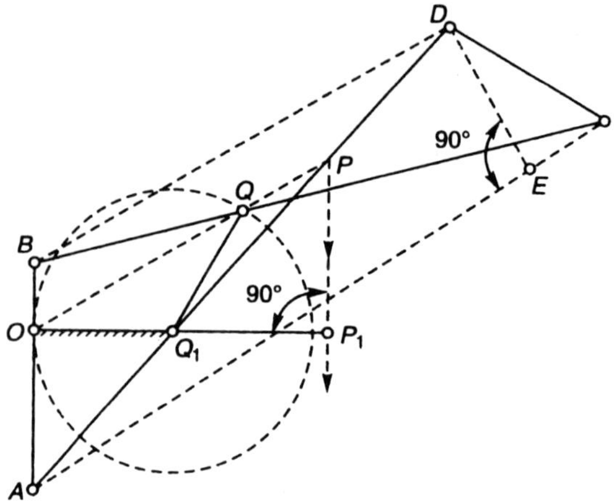

Fig. 5.3 Hart straight line mechanism.

P describes a straight line perpendicular to OO_1 produced as *Q* moves in a circle with centre O_1.

Proof : In ΔABD, $\frac{BO}{BA} = \frac{PD}{DA}$. Hence $OP \parallel BD$.

Δs *ABD* and *AOP* are similar, because $\angle AOP = \angle ABD$ being corresponding angles and $\angle DAB$ is common.

$\therefore$
$$\frac{AB}{AO} = \frac{BD}{OP}$$

or
$$OP = \frac{BD \cdot AO}{AB} \qquad ...(1)$$

Similarly as
$$\frac{BO}{AB} = \frac{BQ}{BC}$$

Hence $OQ \parallel AC$.

Now $AC \parallel BD$, $OQ \parallel BD$, $OP \parallel BD$. Hence $OP \parallel OQ$.

Since *P* is a common point, therefore *OQP* is a straight line.

Now Δs *BOQ* and *BAC* are similar.

$\therefore$
$$\frac{AC}{OQ} = \frac{AB}{OB}$$

or
$$OQ = \frac{OB \cdot AC}{AB} \qquad ...(2)$$

From Eqs. (1) and (2), we get

$$OP \cdot OQ = \left(\frac{BD \cdot AO}{AB}\right) \cdot \left(\frac{OB \cdot AC}{AB}\right)$$

$$= \left(\frac{AO \cdot OB}{AB^2}\right)(BD \cdot AC) \qquad ...(3)$$

Draw $DE \perp AC$

$$BD = AC - 2EC$$
$$AC = AE + EC$$

$$BD \cdot AC = (AC - 2EC)(AE + EC)$$
$$= (AE + EC - 2EC)(AE + EC)$$
$$= (AE - EC)(AE + EC)$$
$$= AE^2 - EC^2 \quad ...(4)$$

From ΔAED, $\quad AE^2 = AD^2 - DE^2$

From ΔGED, $\quad EC^2 = CD^2 - DE^2$

$$AE^2 - EC^2 = AD^2 - CD^2 \quad ...(5)$$

From Eqs. (4) and (5), we get

$$BD \cdot AC = AD^2 - CD^2 \quad ...(6)$$

$$OP \cdot OQ = \left(\frac{AO \cdot BO}{AB^2}\right)(AD^2 - CD^2)$$

$= $ constant, as AO, $BO \cdot AB$, AD and CD are fixed.

Hence P describes a straight line perpendicular to OO_1 produced as point Q moves in a circle with centre O_1.

3. Scott-Russel Mechanism

The Scott-Russel mechanism is shown in Fig. 5.4. It consists of a sliding pair and turning pairs. It can be used to generate approximate and accurate straight lines.

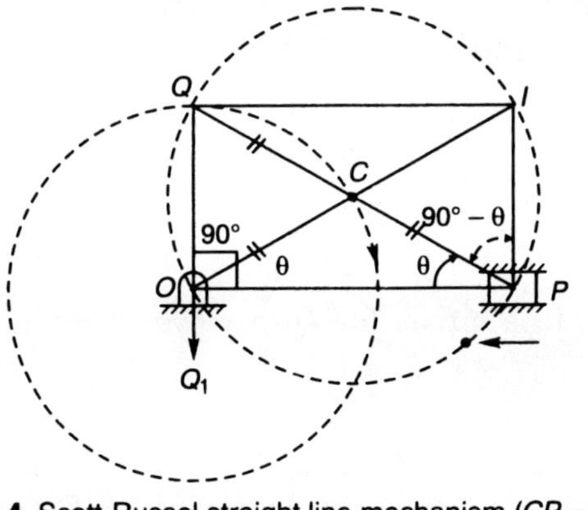

Fig. 5.4 Scott-Russel straight line mechanism ($CP = CQ$).

(a) When $OC = CP = CQ$, Q describes a straight line $QO \perp OP$ provided P moves in a straight line along OP.

(b) If $CP \neq CQ$ then Q describes an approximate straight line perpendicular to OP provided P moves along a straight line OP such that $OC = \dfrac{CP^2}{CQ}$.

Proof :

(a) As $OC = CP = CQ$, hence $\angle POQ = 90°$ and $OQ \perp OP$. If P moves along OP then Q moves along a line perpendicular to OP.

Hence $OP = OC \cos\theta + CP \cos\theta = PQ \cos\theta$

$$\angle CPI = 90° - \theta$$

Now $\angle ICP = \angle COP + \angle CPO = 2\theta$

Then $\angle CIP = 180° - \angle CPI - \angle ICP$

$$= 180° - 90° + \theta - 2\theta$$

$$= 90° - \theta$$

$$\angle CIP = \angle CPI$$

or $CI = CP$

$\therefore$ $CP = CQ = CO = CI$ and $\angle POQ = 90°$

Hence $OPIQ$ is a square. As the path of Q is $\perp QI$, I is the instantaneous centre. Q will move along a line perpendicular to OP.

(b) When $CP \neq CQ$, it will form an elliptical trammel, as shown in Fig. 5.5.

$$\frac{x^2}{CQ^2} + \frac{y^2}{CP^2} = 1$$

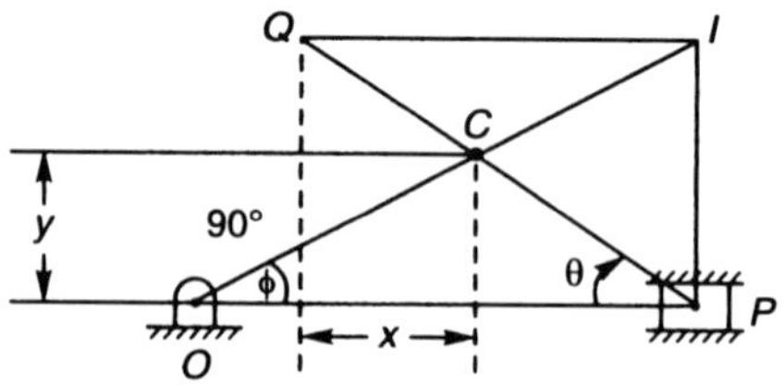

Fig. 5.5 Scott-Russel mechanism when $CP \neq CQ$.

Because $x = CQ \cos\theta$ and $y = PC \sin\theta$

$$OC = \frac{\text{semi - minor axis}^2}{\text{semi - major axis}}$$

$$= \frac{CP^2}{CQ}$$

As point C moves in a circle, for Q to move along an approximate straight line, $OC = \dfrac{CP^2}{CQ}$.

Limitations : When $OC \perp OP$, P coincides with O and $OQ = 2OC$. Here a small displacement of P shall cause a large displacement of Q, requiring a relatively small displacement of P to give displacement to Q. This requires highly accurate sliding surfaces.

5.3.2 Approximate Straight Line Motion Mechanisms

The mechanisms for the generation of approximate straight line motion are :

1. Grasshopper mechanism
2. Watt mechanism
3. Tchebicheff mechanism

1. Grasshopper Mechanism

The Grasshopper mechanism is shown in Fig. 5.6. The crank OC rotates about a fixed point $O \cdot O_1$ is a fixed pivot for link O_1P. For small angular displacements of O_1P, point Q on link PCQ will trace approximately a straight line path perpendicular to OP if

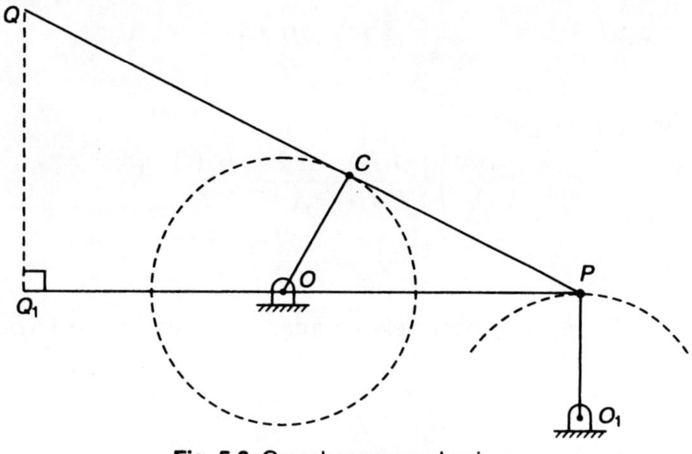

Fig. 5.6 Grasshopper mechanism.

$$OC = \frac{CP^2}{CQ}$$

2. Watt Mechanism

The Watt mechanism is shown in Fig. 5.7 (a).

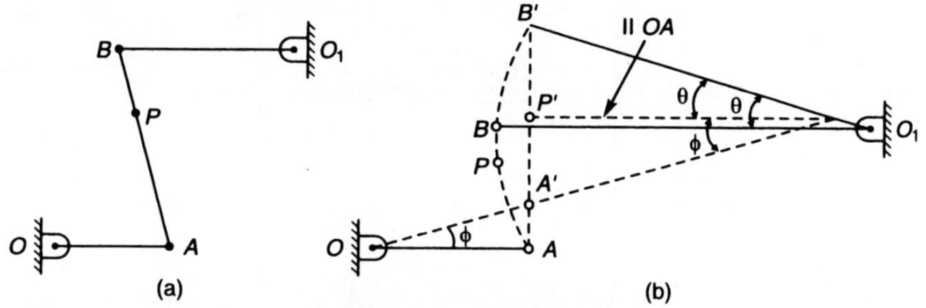

Fig. 5.7 Watt mechanism.

Links OA and O_1B oscillates about O and O_1 respectively. AB is a connecting link. P will trace an approximate straight line if $\frac{PA}{PB} \cong \frac{O_1B}{OA}$. In Fig. 5.7 (b), θ and ϕ are the amplitudes of oscillation and I is the instantaneous centre of $A'B$. P' is the point which lies on the approximate straight line described by P.

Proof :

$$\phi = \frac{AA'}{OA} \text{ and } \theta = \frac{BB}{O_1B}$$

$$\frac{\phi}{\theta} = \left(\frac{AA'}{OA}\right) \cdot \left(\frac{O_1B}{BB}\right) \qquad ...(1)$$

If $\angle B'P'I = 90°$, then

$$\sin\theta = \frac{B'P'}{B'I}$$

or

$$\theta \approx \frac{B'P'}{B'I} \text{ and } \phi \approx \frac{A'P'}{A'I}$$

$$\frac{\phi}{\theta} \approx \left(\frac{A'P'}{A'I}\right) \cdot \left(\frac{B'I}{B'P}\right) \quad ...(2)$$

From Eqs. (1) and (2), we get

$$\left(\frac{AA'}{OA}\right) \cdot \left(\frac{O_1B}{BB'}\right) \approx \frac{A'P'}{A'I} \cdot \frac{B'I}{B'P'}$$

Now
$$\frac{AA'}{BB'} = \frac{B'I}{A'I}$$

Δs OAA' and $IP'A'$ are approximately similar, as well as Δs O_1BB' and $BP'I$ are approximately similar.

$$\frac{O_1B}{OA} \approx \frac{A'P'}{B'P'} \approx \frac{AP}{BP}$$

Therefore, P divides the coupler AB in the ratio of the lengths of oscillating links. Hence P will describe an approximate straight line for a certain position of its path.

3. Tchebicheff Mechanism

The Tchebicheff mechanism is shown in Fig. 5.8. In this mechanism, $OA = O_1B$ and $AP = PB$. P is the tracing point. Let $AB = 1, OA = O_1B = x$, and $OO_1 = y$.

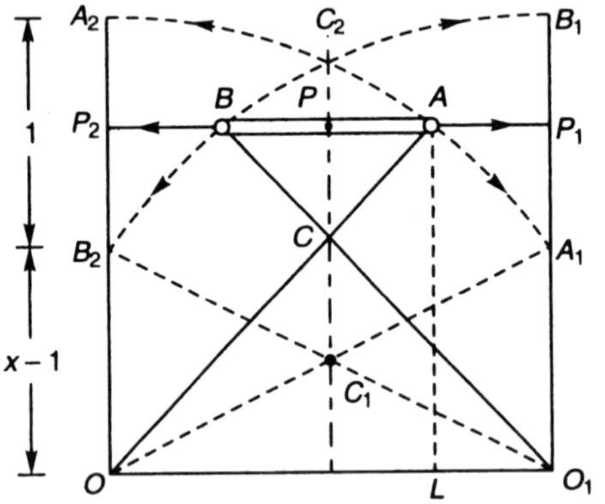

Fig. 5.8 Tchebicheff mechanism.

Now
$$OB_2^2 = B_2O_1^2 - OO_1^2$$

$\therefore$
$$(x-1)^2 = x^2 - y^2$$

or
$$x = \frac{y^2+1}{2} \quad ...(1)$$

Draw $AL \perp OO_1$, then
$$OL = OO_1 - O_1L$$

Also
$$O_1L = AP_1$$

and
$$AP_1 = \left(\frac{P_2P_1 - BA}{2}\right) = \left(\frac{y-1}{2}\right)$$

Again
$$OL = OO_1 - O_1L$$
$$= y - \frac{y-1}{2} = \frac{y+1}{2}$$

Further
$$OA^2 = AL^2 + OL^2$$
$$x^2 = \left(x - \frac{1}{2}\right)^2 + \left[\frac{y+1}{2}\right]^2$$
or
$$x = \frac{y^2}{4} + \frac{1}{2} + \frac{y}{2} \quad ...(2)$$

From Eqs. (1) and (2), we get
$$\left(\frac{y^2+1}{2}\right) = \frac{y^2}{4} + \frac{1}{2} + \frac{y}{2} \text{ or } y = 2$$

Substituting in Eq. (1), we find that, $x = 3.5$

i.e.,
$$AB : OO_1 : OA :: 1 : 2 : 3.5$$

P moves horizontally because instantaneous centre of AB will lie on point C which is the intersection of OA and BO_1. Thus C lies just below P.

■ **Example 5.1** A circle with AB as diameter has a point C on its circumference. D is a point on AC produced such that if C turns about A, the product $AC \times AD$ is constant. Prove that the point D moves in a straight line perpendicular to AB produced.

Solution : Let D_1D be perpendicular to AB produced, as shown in Fig. 5.9.

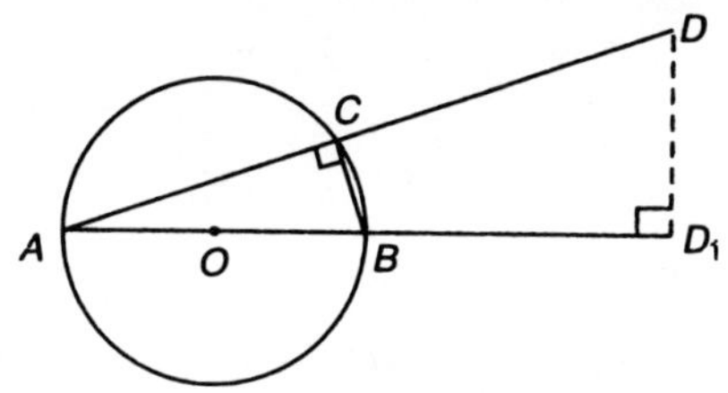

Fig. 5.9

Now $\angle ACB = 90°$, being angle in a semicircle. Also $\angle AD_1D = 90°$. Therefore, Δs ACB and AD_1D are similar, as $\angle CAB$ is common.
$$\frac{AC}{AB} = \frac{AD_1}{AD}$$
or
$$AB \times AD_1 = AC \times AD$$
or
$$AD_1 = \frac{AC \times AD}{AB} = \text{constant}$$

Because AB is fixed and $AC \times AD$ = constant.

Thus AD_1 will be constant for all positions of C. Therefore, the location of D_1, is fixed, which means that D moves in a straight line perpendicular to AB produced.

■ **Example 5.2** In a Grasshopper mechanism shown in Fig. 5.10. $OC = 100$ mm, $PC = 150$ mm, and $PQ = 375$ mm. Determine the magnitude of the vertical force at Q necessary to resist a torque of 200 N.m. applied to the link OC when it makes an angle of 30° with the horizontal.

Solution :
$$QC = PQ - PC = 375 - 150 = 225 \text{ mm}$$
$$\frac{OC}{PC} = \frac{PC}{QC}$$

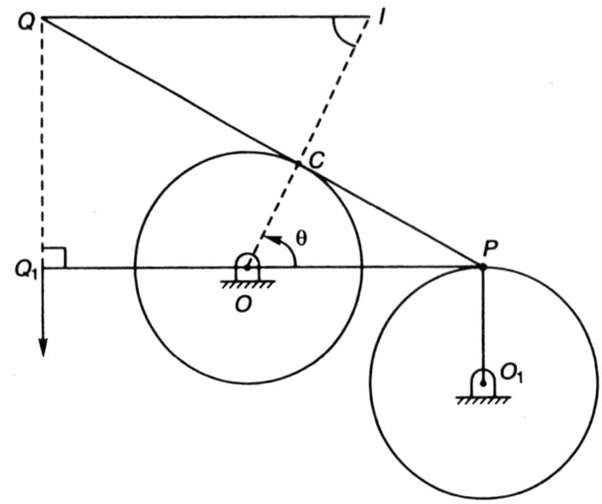

Fig. 5.10

or $$\frac{100}{150} = \frac{150}{225}$$

Thus the condition for the dimensions of a Grasshopper mechanism is satisfied, and point Q will trace an approximate straight line perpendicular to OP.

Now $$F_q \cdot v_q = T_c \cdot v_c$$

or $$F_q = \frac{T_c \cdot v_c}{v_q \cdot OC}$$

The instantaneous centre of PQ is at I. Let $\angle QIC = \theta$. Also Δs QIC and OCP are similar.

$$\frac{v_c}{v_q} = \frac{IC}{IQ} = \frac{OC}{OP}$$

Thus $$F_q = \frac{T_c}{OP}$$

When $\theta = 30°$

$$OP = 100 \cos 30° + [(150)^2 - (100 \sin 30°)^2]^{0.5} = 228 \text{ mm}$$

$$F_q = \frac{200}{0.228} = 877.2 \text{ N}$$

5.4 PARALLEL LINKAGES

These are the four-bar linkages in which the opposite links are equal in length and always form a parallelogram. There are three types of parallel linkages : parallel rules, universal drafting machine and lazy tongs. They are used for producing parallel motion.

1. **Parallel rules :** A parallel rule is shown in Fig. 5.11, in which $AB = CD = EF = GH = IJ$ and $AC = BD$, $CE = DF$, $EG = FH$, $GI = HJ$. Here AB, C, EF, GH and IJ will always be parallel to each other.

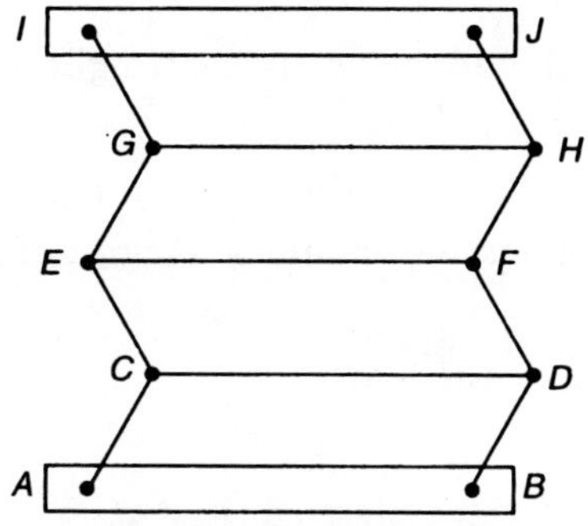

Fig. 5.11 Parallel rule.

2. **Universal drafting machine :** A universal drafting machine is shown in Fig. 5.12, in which $AB = CD$; $AC = BD$; $EF = GH$; and $EG = FH$. Position of points A and B are fixed. Similarly the positions of points E and F are fixed with respect to C and D. The positions of scales I and II are fixed with respect to points G and H. Then $ABDC$ is a parallelogram. Line CD will always be parallel to AB so that the direction of CD is fixed. Therefore, the direction of EF is fixed. Further $EFHG$ is a parallelogram, so GH is always parallel to EF such that the direction of GH is also fixed, whatever their actual position may be.

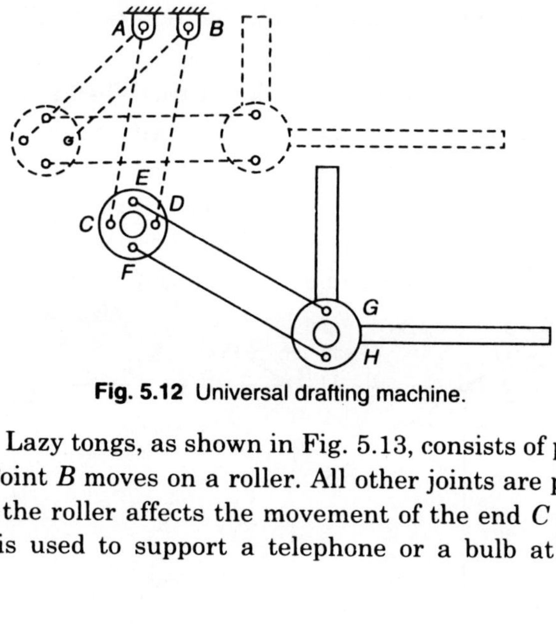

Fig. 5.12 Universal drafting machine.

3. **Lazy tongs :** Lazy tongs, as shown in Fig. 5.13, consists of pin joint A attached to a fixed point. Point B moves on a roller. All other joints are pin jointed. The vertical movement of the roller affects the movement of the end C to move in a horizontal direction. It is used to support a telephone or a bulb at point C for horizontal movements.

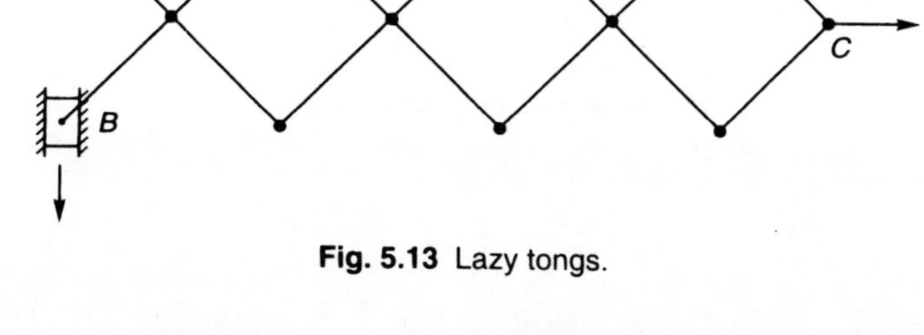

Fig. 5.13 Lazy tongs.

(i) The pencil point should move in a straight line parallel to the axis of the indicator piston.
(ii) The velocity ratio of the indicator piston and pencil should be constant.
(iii) The indicator piston motion should be magnified so as to get a good size indicator diagram.
(iv) The friction should be the least.

5.5 AUTOMOBILE STEERING GEAR MECHANISMS

The mechanisms which are used for changing the direction of motion of an automobile are called steering gears. Steering is done by changing the direction of motion of front wheels only as the gear wheels have a common axis which is fixed and moves in a straight line only.

5.5.1 Fundamental Equation for Correct Steering

When a vehicle takes a turn all the wheels should roll on the road smoothly preventing excessive tyre wear. This is achieved by mounting the two front wheels on two short axles, called stub-axles. The stub axles are pin-jointed with the main front axle which is rigidly attached to the rear axle, on which the back wheels are attached. Steering is done by front wheels only.

Figure 5.14 shows an automobile taking a right turn. When the vehicle takes a turn, the front wheels along with stub-axles turn about their respective pin-joints. The inner front wheel turns through a greater angle as compared to that of outer front wheel. In order to avoid skidding, for correct steering, the two front wheels must turn about the same instantaneous centre *I*, which lies on the axis of the gear wheels. Thus, the condition for correct steering is that all the four wheels must turn about the same instantaneous centre.

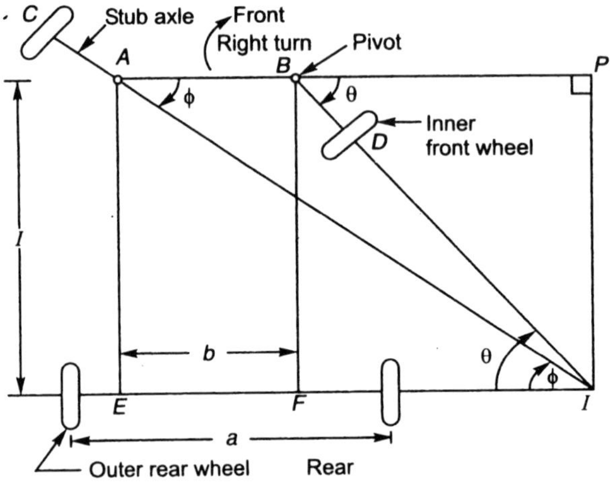

Fig. 5.14 Automobile steering gear.

Let $AE = l$ = wheel base

$CD = a$ = wheel track

$AB = b$ = distance between the pivots of front axles

I = common instantaneous centre of all wheels

θ, ϕ = angles turned through by stub-axles of the inner and outer front wheels respectively

Now
$$b = AP - BP$$
$$= l(\cot\phi - \cot\theta)$$

or
$$\cot\phi - \cot\theta = \frac{b}{l} \quad \text{...(5.1)}$$

Eq. (5.1) is called the fundamental equation for the correct steering.

5.5.2 Steering Gears

A steering gear is a mechanism for automatically adjusting values of θ, ϕ for correct steering. The following steering gears are commonly used in automobiles :

1. Davis steering gear
2. Ackermann steering gear

1. Davis Steering Gear

Davis steering gear is shown in Fig. 5.15. This type of gear has only sliding pairs. Two arms AG and BH are fixed to the stub-axles AC and BD respectively. CAG and DBH form two similar bell-crank levers pivoted at A and B respectively. KL is a cross-link which is constrained to slide parallel to AB. The ends of the cross-link KL are pin-jointed to two sliders S_1 and S_2 as shown. These sliders are free to slide on links AG and BH respectively. The whole mechanism is in front of the front wheels.

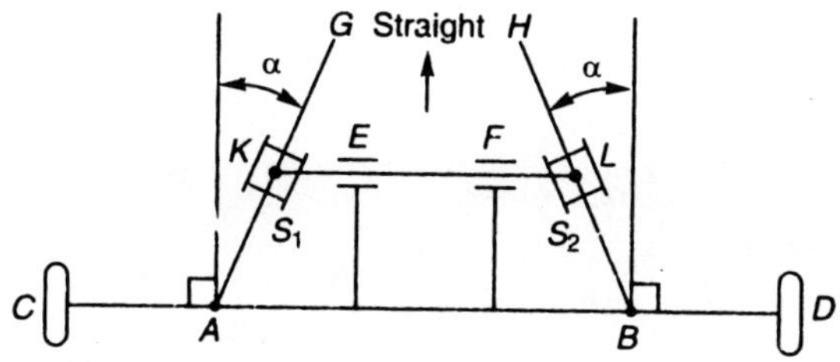

Fig. 5.15 Davis steering gear for straight drive.

During the straight motion of the vehicle, the gear is in the mid-position, with equal inclination of the arms AG and BH with the verticals at A and B. The steering is achieved by moving cross-link KL to the right or left of the mid-position. The steering gear for taking a right turn is shown in Fig. 5.16. $K'L'$ shows the position of the cross-link KL while taking a right turn. **Determination of angle** α

Let x = distance moved by KL from mid-position

$= KK' - LL'$

y = horizontal distance of points K and L from A and B respectively

h = vertical distance between AB and KL

α = angle of inclination of track arms AG and BH with the vertical in mid-position

θ, ϕ = angles turned by stub axles

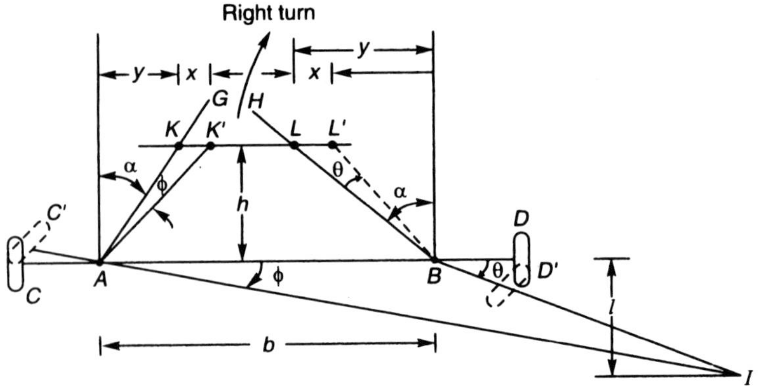

Fig. 5.16 Davis steering gear taking a right turn.

From Fig. 5.16, we have

$$y = AK \sin \alpha = BL \sin \alpha$$

$$\tan (\alpha - \theta) = \frac{y - x}{h}$$

or

$$\frac{\tan \alpha - \tan \theta}{1 + \tan \alpha \tan \theta} = \frac{y - \infty}{h}$$

Now

$$\tan \alpha = \frac{y}{h}$$

$\therefore$

$$\frac{\frac{y}{h} - \tan \theta}{1 + \frac{y}{h} \tan \theta} = \frac{y - x}{h}$$

or

$$\tan \theta = \frac{xh}{y^2 - xy + y^2} \quad \text{...(1)}$$

Similarly

$$\tan (\alpha + \phi) = \frac{y + x}{h}$$

so that

$$\tan \phi = \frac{xh}{y^2 + xy + h^2} \quad \text{...(2)}$$

From Eqs. (1) and (2), we get

$$\cot \phi - \cot \theta = \frac{2y}{h} = \frac{b}{l} \quad \text{...(5.2)}$$

or

$$\tan \alpha = \frac{y}{h} = \frac{b}{2l}$$

Generally, $\frac{b}{l} = 0.4$ to 0.5, so that $\alpha = 11.3°$ to $14.1°$. There will be friction and more wear due to sliding pairs in the Davis steering gear. Therefore, it becomes inaccurate after some use. On account of these reasons, it is not used in practice.

2. Ackermann Steering Gear

The Ackermann steering gear has only turning pairs. Fig. 5.17 shows the gear for straight drive. The turning pairs are : *AK, KL, LB* and *AB*. The two short arms *AK* and *BL* are of equal

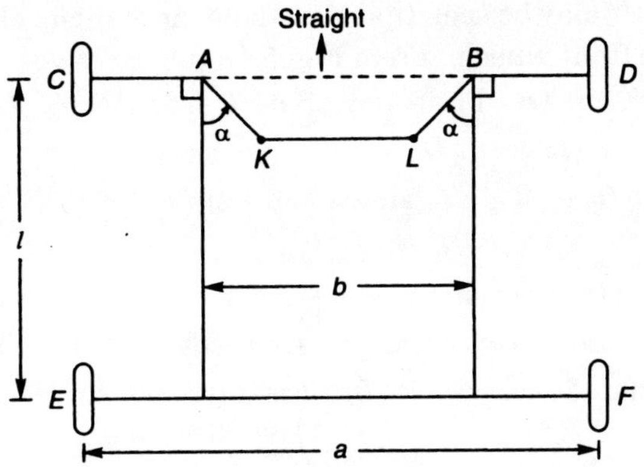

Fig. 5.17 Ackermann steering gear for straight drive.

length and are connected by pin-joints with front wheel axle *AB* at *A* and *B* respectively. *AC* and *BD* are the stub-axles so that *CAK* and *DBL* form bell-crank levers. *ABLK* form a four-bar linkage. *KL* is the track rod.

For correct steering, we have

$$\cot \phi - \cot \theta = \frac{b}{l}$$

Generally, $$\frac{b}{l} = 0.4 \text{ to } 0.5 \approx 0.455$$

Determination of angle α : Consider the Ackermann steering gear, as shown in Fig. 5.18 (a), taking a right turn. The instantaneous centre *I* lies on a line parallel to the rear axis at a distance of approximately 0.3*l*.

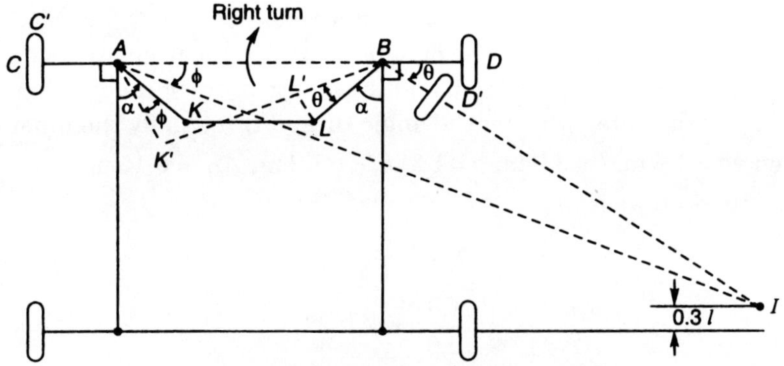

(a) Ackermann steering gear taking a right turn

(b) Determination of anlge α

Fig. 5.18

Above the rear axis. It may be seen that the whole mechanism of the Ackermann steering gear is on the back of the front wheels. From Fig. 5.18 (b), we have

Projection of arc $K'K$ on AB = Projection of arc $L'L$ on AB

or
$$K_1K'_1 = L_1L'_1$$

$$AK\ [\sin \alpha - \sin(\alpha - \phi)] = BL\ [\sin(\alpha + \theta) - \sin \alpha]$$

Now
$$AK = BL$$

$\therefore$
$$\sin \alpha - \sin(\alpha - \phi) = \sin(\alpha + \theta) - \sin \alpha$$

or
$$\sin \alpha - \sin \alpha \cos \phi + \cos \alpha \sin \phi = \sin \alpha \cos \theta + \cos \alpha \sin \theta - \sin \alpha$$

$$\sin \alpha\,(2 - \cos \phi - \cos \theta) = \cos \alpha\,(\sin \theta - \sin \phi)$$

$$\tan \alpha = \frac{\sin \theta - \sin \phi}{2 - \cos \phi - \cos \theta} \qquad ...(5.3)$$

The values of θ and ϕ are known for correct steering. Hence, the value of α can be determined from Eq. (5.3).

■ **Example 5.3** In a Davis steering gear, the distance between the pivots of the front axle is 1.2 m and the wheel base is 2.8 m. When the automobile is moving along a straight path, find the inclination of the track arms to the longitudinal axis of the automobile?

Solution : Given : $b = 1.2$ m, $l = 2.8$ m

Now
$$\tan \alpha = \frac{b}{2l}$$

$$= \frac{1.2}{5.6} = 0.2143$$

$$\alpha = 12.095°$$

■ **Example 5.4** A car with a track of 1.5 m and a wheel base of 2.9 m has a steering gear mechanism of the Ackermann type. The distance between the front stub-axle pivots is 1.3 m. The length of each track arm is 150 mm and the length of track rod is 1.2 m. Find the angle turned through by the outer wheel if the angle turned through by the inner wheel is 30°.

Solution : Given : $a = 1.5$ m, $l = 2.9$ m, $b = 1.3$ m, $c = 1.2$ m, $AK = 0.15$ m.

For correct steering, we have

$$\cot \phi - \cot \theta = \frac{b}{l}$$

$$= \frac{1.3}{2.9} = 0.44827$$

$$\cot \phi - \cot 30° = 0.44827$$

$$\cot \phi - 1.83205 = 0.44827$$

$$\cot \phi = 2.18032$$

$$\phi = 24.64°$$

5.6 HOOKE'S JOINT

It is a device to connect two shafts whose axes are neither coaxial nor parallel but intersect at a point. This is used to transmit power from the engine to the rear axle of an automobile and similar other applications. It is also called universal coupling.

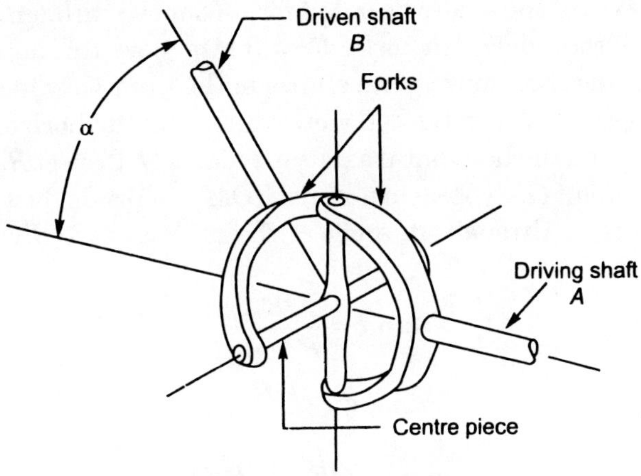

Fig. 5.19 Hooke's joint.

The Hooke's joint, as shown in Fig. 5.19, consists of two forks connected by a centre piece, having the shape of a cross or square carrying four trunnions. The ends of the two shafts to be connected together are fitted to the forks.

5.6.1 Velocities of Shafts

Let the driving shaft A and the driven shaft B be connected by a joint having two forks KL and MN. The two forks are connected by a cross $KLMN$ intersecting at O, as shown in Fig. 5.20. Let

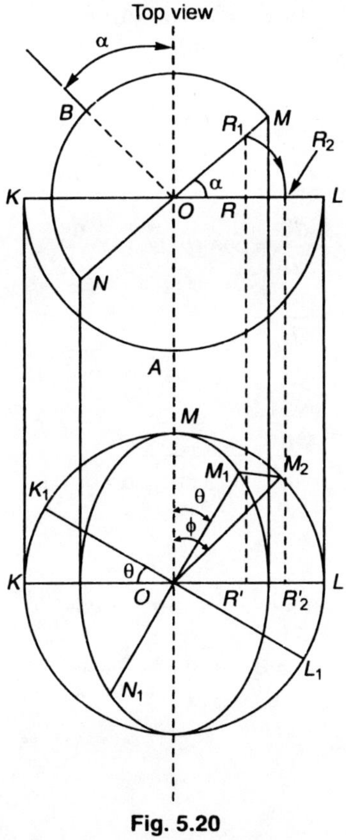

Fig. 5.20

the angle between the axes of the shafts be α. Let fork KL move through an angle θ in a circle to the position K_1L_1 in the front view. The fork MN will also move through the same angle θ. MN being not in the same plane shall move in an ellipse in the front view to its new position M_1N_1. To find the true angle, project M_1 to the top view which cuts the horizontal axis in R and fork MN in R_1. Rotate R_1 to R_2 on the horizontal axis with centre O. Project R_2 back in the front view cutting the circle in M_2. Joint OM_2 Measure angle MOM_2, which is the true angle ϕ Thus when the driving shaft A revolves through an angle θ, the driven shaft B will revolve through an angle ϕ

Now
$$\tan\phi = \frac{OR'_2}{R'_2 M_2}$$

and
$$\tan\theta = \frac{OR'}{R' M_1}$$

$$\therefore \quad \frac{\tan\phi}{\tan\theta} = \frac{OR'_2}{R'_2 M_2} \cdot \frac{R' M_1}{OR'}$$

But
$$R' M_1 = R'_2 M_2$$

$$\therefore \quad \frac{\tan\phi}{\tan\theta} = \frac{OR'_2}{OR'} = \frac{OR_1}{OR}$$

Now
$$OR = OR_1 \cos\alpha$$

$$\frac{\tan\phi}{\tan\theta} = \frac{1}{\cos\alpha}$$

or
$$\tan\theta = \cos\alpha \tan\phi \qquad ...(5.4)$$

Angular velocty of driving shaft, $\omega_A = \dfrac{d\theta}{dt}$

Angular velocity of driven shaft, $\omega_B = \dfrac{d\phi}{dt}$

Differentiating Eq. (5.4), we get

$$\sec^2\theta \cdot \frac{d\theta}{dt} = \cos\alpha \sec^2\phi \cdot \frac{d\phi}{dt}$$

$$\frac{\dfrac{d\phi}{dt}}{\dfrac{d\theta}{dt}} = \frac{\cos\alpha}{1 - \sin^2\alpha\cos^2\theta}$$

or
$$\frac{\omega_B}{\omega_A} = \frac{\cos\alpha}{1 - \sin^2\alpha\cos^2\theta} \qquad ...(5.5)$$

Case 1 : $\dfrac{\omega_B}{\omega_A} = 1$

$$\cos\alpha = 1 - \sin^2\alpha\cos^2\theta$$

or
$$\cos\theta = \pm\sqrt{\frac{1}{1 + \cos\alpha}} \qquad ...(5.6)$$

This condition occurs once in each quadrant, as shown in Fig. 5.21.

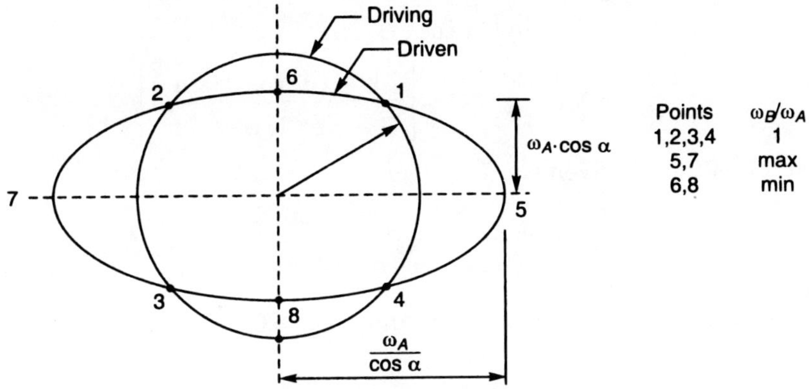

Fig. 5.21 Speed polar diagram.

Case 2 : $\frac{\omega_B}{\omega_A}$ is minimum when the denominator is maximum, *i.e.*, $\sin^2\alpha\cos^2\theta$ must be minimum, or $\cos\theta = 0°$. Thus $\frac{\omega_B}{\omega_A}$ is minimum at $\theta = 90°, 270°$. Then

$$\frac{\omega_B}{\omega_A} = \cos\alpha \qquad ...(5.7)$$

Case 3 : $\frac{\omega_B}{\omega_A}$ is maximum when denominator is minimum, *i.e.*, when $\cos^2\theta = 1$, or $\cos\theta = \pm 1$,

i.e., $\theta = 0°$ or $180°$. Then

$$\frac{\omega_B}{\omega_A} = \frac{1}{\cos\alpha} \qquad ...(5.8)$$

Case 4 : Maximum fluctuation of velocity of driven shaft

$$= \frac{(\omega_B)_{max} - (\omega_B)_{min}}{(\omega_B)_{mean}}$$

Now $\qquad (\omega_B)_{mean} = \omega_A$

$$\text{Maximum speed fluctuation} = \frac{\frac{\omega_A}{\cos\alpha} - \omega_A\cos\alpha}{\omega_A}$$

$$= \frac{1-\cos^2\alpha}{\cos\alpha}$$

$$= \tan\alpha\sin\alpha \qquad ...(5.9)$$

For α to be small, $\tan\alpha \approx \alpha$ and $\sin\alpha \approx \alpha$

$$\text{Maximum speed fluctuation} \approx \alpha^2 \qquad ...(5.10)$$

5.6.2 Angular Acceleration of Driven Shaft

$$\frac{\omega_B}{\omega_A} = \frac{\cos\alpha}{1 - \sin^2\alpha\cos^2\theta}$$

$$\frac{d\omega_B}{dt} = \omega_A \cdot \frac{d\theta}{dt} \cdot \cos\alpha \left[\frac{2\sin^2\alpha \cdot \cos\theta \cdot \sin\theta}{(1 - \sin^2\alpha\cos^2\theta)^2}\right]$$

Angular acceleration of driven shaft,

$$\alpha_B = \frac{\omega_A^2 \cos\alpha \sin^2\alpha \sin 2\theta}{(1 - \sin^2\alpha\cos^2\theta)^2} \quad ...(5.11)$$

For acceleration to be maximum or minimum $\frac{d\alpha_B}{d\theta} = 0$

$$\frac{d\alpha_B}{d\theta} = \omega_A^2 \cos\alpha \sin^2\alpha$$

$$\times \left[\frac{2\sin 2\theta\,(1 - \sin^2\alpha\cos^2\theta)\sin^2\alpha\sin 2\theta - (1 - \sin^2\alpha\cos^2\theta)^2 \cdot 2\cos 2\theta}{(1 - \sin^2\alpha\cos^2\theta)^4}\right]$$

$$= \cos^2 2\theta + \left(\frac{2}{\sin^2\alpha} - 1\right)\cos 2\theta - 2$$

$$= 0$$

or

$$\cos 2\theta \approx \frac{2\sin^2\alpha}{2 - \sin^2\alpha} \quad ...(5.12)$$

5.7 DOUBLE HOOKE'S JOINT

The double Hooke's joint, as shown in Fig. 5.22, is used to maintain the speed of driven shaft equal to the driving shaft at every instant. To achieve this, the driving and the driven shafts

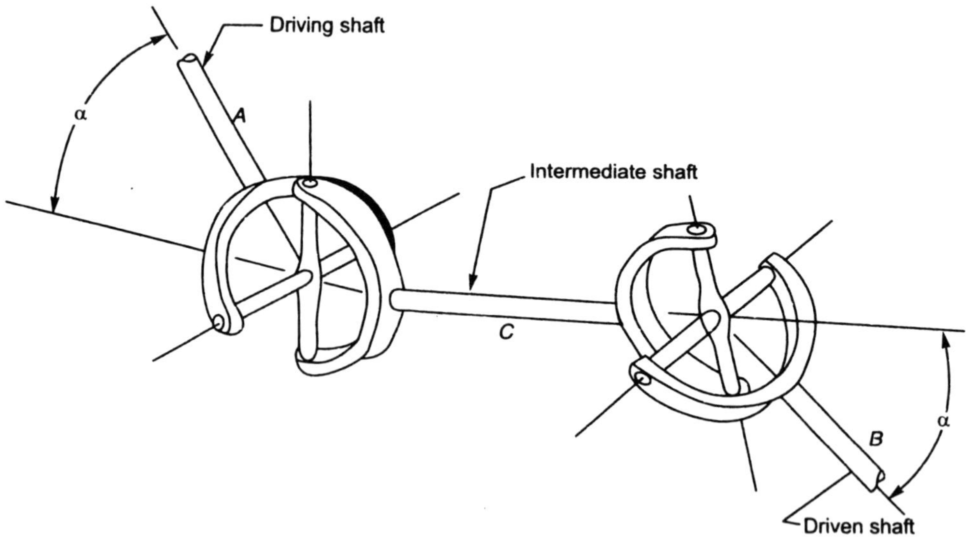

Fig. 5.22 Double Hooke's joint.

should make equal angles with the intermediate shaft and the forks of the intermediate shaft should lie in the same plane. Let γ be the angle turned by the intermediate shaft while the angle turned by the driving shaft and the driven shaft be θ and ϕ respectively.

Then $$\tan\theta = \cos\alpha \tan\gamma$$

and $$\tan\phi = \cos\alpha \tan\gamma$$

$\therefore$ $$\theta = \phi$$

However, if the forks of the intermediate shafts lie in perpendicular planes to each other, then the variation of speed of the driven shaft will be there.

$$\left(\frac{\omega_c}{\omega_A}\right)_{min} = \cos\alpha$$

$$\left(\frac{\omega_B}{\omega_C}\right)_{min} = \cos\alpha$$

$\therefore$ $$\left(\frac{\omega_B}{\omega_A}\right)_{min} = \cos^2\alpha \qquad ...(5.13)$$

Similarly, $$\left(\frac{\omega_B}{\omega_A}\right)_{max} = \frac{1}{\cos^2\alpha} \qquad ...(5.14)$$

■ **Example 5.5** The angle between the axes of two shafts connected by Hooke's joint is 25. Determine the angle turned through by the driving shaft when the velocity ratio is unity.

Solution : For velocity ratio to be unity,

$$\cos\theta = \pm\left[\frac{1}{1+\cos\alpha}\right]^{0.5}$$

$$= \pm\left[\frac{1}{1+\cos 25°}\right]^{0.5}$$

$$= \pm\, 0.72427$$

$$\theta = 43.59° \text{ or } 136.41°$$

■ **Example 5.6** Two shafts are connected by a Hooke's joint. The driving shaft revolves uniformly at 600 rpm. The total variation in the speed of the driven shaft is not to exceed ± 6% of the mean speed. Find the greatest permissible angle between the centre lines of the shafts.

Solution : Total fluctuations in speed of the driven shaft

$$= 0.12\,\omega_m$$

$$= \omega_m\left[\frac{1-\cos^2\alpha}{\cos\alpha}\right]$$

or $$\cos^2\alpha + 0.12\cos\alpha - 1 = 0$$

$$\cos\alpha = \left[-\frac{0.12 \pm \{(0.12)^2 + 4t\}^{0.5}}{2}\right]$$

$$= 0.9418$$

$$\alpha = 19.64°$$

■ **Example 5.7** Two shafts are connected by the Hooke's joint. The driving shaft rotates at a uniform speed of 1000 rpm. The angle between the shafts is 20°. Calculate the maximum and minimum speeds of the driven shaft, when the acceleration of the driven shaft is maximum?

Solution : Maximum speed,

$$N_{max} = \frac{N}{\cos\alpha} = \frac{1000}{\cos 20°} = 1064.2 \text{ rpm}$$

Minimum speed,

$$N_{min} = N\cos\alpha = 1000\cos 20° = 939.7 \text{ rpm}$$

For acceleration to be maximum,

$$\cos 2\theta \approx \frac{2\sin^2\alpha}{2-\sin^2\alpha} \approx \frac{2\sin^2 20°}{2-\sin^2 20°} = \frac{2\times 0.11698}{2-2\times 0.11698} = 0.13248$$

$$\theta = 41.19°, 138.61°, 221.19°, \text{ and } 318.81°$$

■ **Example 5.8** The driving shaft of a Hooke's joint runs at a speed of 400 rpm. The angle between the shafts is 25°. The driven shaft with attached masses has a mass of 50 kg at a radius of gyration of 200 mm. If a steady torque of 900 N.m resist rotation of the driven shaft, find the torque required at the driving shaft when angle turned through by the driving shaft is 45°.

Solution : Given : $N_A = 400 \text{ rpm}, \alpha = 25°, m = 50 \text{ kg}, K = 0.2 \text{ m}, T = 900 \text{ N.m}, \theta = 45°$

Moment of inertia of the driven shaft,

$$I = mK^2 = 50\times(0.2)^2 = 2 \text{ kg. m}^2$$

Angular acceleration of the driven shaft

$$\alpha_B = -\omega_A^2\left[\frac{\cos\alpha\sin 2\theta\sin^2\alpha}{(1-\cos^2\theta\sin^2\alpha)^2}\right]$$

$$= -\left(\frac{2\pi\times 400}{60}\right)^2\left[\frac{\cos 25°\sin 90°\sin^2 25°}{(1-\cos^2 45°\sin^2 25°)^2}\right]$$

$$= -1754.6\left[\frac{0.16187}{0.910697}\right]$$

$$= -311.868 \text{ rad/s}^2$$

Torque required to accelerate the driven shaft $= I\alpha_B$

$$= -2 \times 311.868$$

$$= -623.73 \text{ N. m}$$

Total torque required on the driven shaft,

$$T_B = 900 - 623.73$$

$$= 276.26 \text{ N. m}$$

Torque required on the driving shaft,

$$T_A = \frac{T_B \omega_B}{\omega_A}$$

$$= T_B \left[\frac{\cos \alpha}{1 - \cos^2 \theta \sin^2 \alpha} \right]$$

$$= 276.26 \left[\frac{\cos 25°}{1 - \cos^2 45° \sin^2 25°} \right]$$

$$= 274.93 \text{ N. m}$$

■ **Example 5.9** In a double Hooke's coupling, the driving and the driven shafts are parallel and the angle between each and the intermediate shaft is 30°. Find the maximum and minimum velocities of the driven shaft if the axis of the driving pin carried by the intermediate shaft has inadvertently been placed 90° in advance of the correct position. The driving shaft rotates uniformly at 350 rpm.

Solution :

$$(N_B)_{max} = \frac{N_A}{\cos^2 \alpha}$$

$$= \frac{350}{\cos^2 30°}$$

$$= 466.67 \text{ rpm}$$

$$(N_B)_{min} = N_A \cos^2 \alpha$$

$$= 350 \cos^2 305°$$

$$= 262.5 \text{ rpm}$$

■ **Example 5.10** A torque of 100 N.m is applied to the link *OC* of a Grasshopper mechanism shown in Fig. 5.23. Link *OC* makes an angle of 20° with the horizontal. Find the magnitude of the vertical force exerted at *Q* to overcome this torque. *OC* = 80 mm, *PC* = 120 mm, and *PQ* = 300 mm. Also calculate the force required if the link makes an angle of 0°.

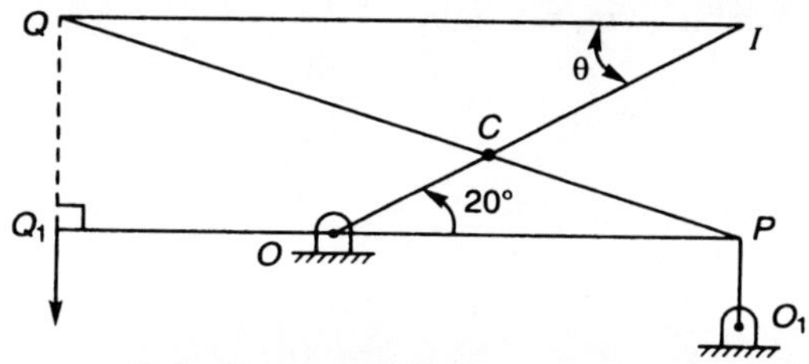

Fig. 5.23 Grasshopper mechanism.

Solution : $QC = QP - CP = 300 - 120 = 180$ mm

$$\frac{OC}{PC} = \frac{PC}{QC}$$

$$\therefore \quad \frac{80}{120} = \frac{120}{180} = \frac{2}{3}$$

Let F_q = vertical force exerted at Q

Then
$$F_q \times v_q = T_c \times v_c$$

$$F_q = \frac{T_c \times v_c}{v_q \times OC}$$

$$\frac{v_c}{v_q} = \frac{IC}{IQ} = \frac{OC}{OP} = \frac{80}{OP}$$

$$F_q = \frac{T_c \times 80}{80 \times OP} = \frac{T_c}{OP}$$

$$OP = OC \cos 20° = 80 \cos 20° = 75.17 \text{ mm}$$

$$F_q = \frac{100}{75.17 \times 10^{-3}} = 1330\ N$$

When $\theta = 0°$, then

$$OP = OC = 80 \text{ mm}$$

$$F_q = \frac{100}{80 \times 10^{-3}} = 1250 \text{ N}$$

Example 5.11 The distance between the fixed centres of a Watt's straight line mechanism shown in Fig. 5.24 is 320 mm. The lengths of links are : $OA = 300$ mm, $AB = 400$ mm and $BO_1 = 250$ mm. Locat the position of a point P on AB which will trace approximately straight line.

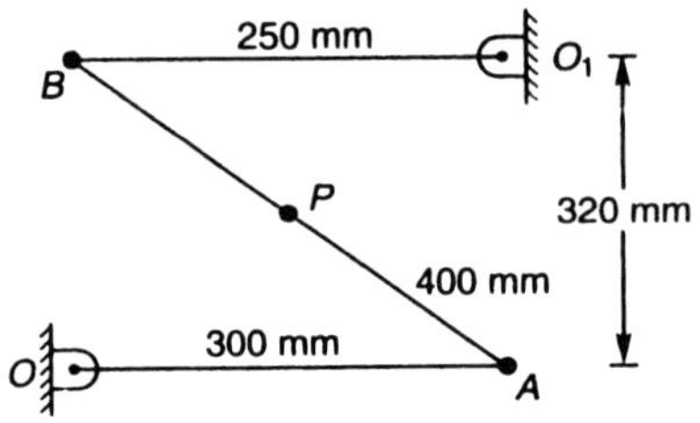

Fig. 5.24 Watt's mechanism.

Solution : For the Watt's mechanism, we have

$$\frac{AP}{BP} = \frac{O_1B}{OA} = \frac{250}{300} = \frac{5}{6}$$

$$AP = \frac{5 \times 400}{11} = 181.8 \text{ mm}$$

$$BP = 213.2 \text{ mm}$$

■ **Example 5.12** In a Robert mechanism shown in Fig. 5.25 (a), $AB = BC = CD = \frac{AD}{2}$. Locate the point P on the central vertical arm that approximately describes a straight line.

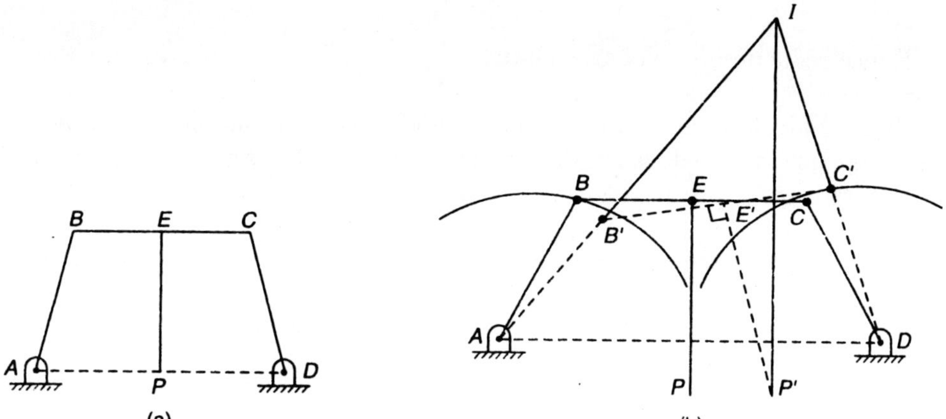

Fig. 5.25 Robert mechanism.

Solution : Draw the mechanism with $BC = 30$ mm (assumed) in its displaced position $AB\,C'\,D$, as shown in Fig. 5.25 (b). Produce AB and DC' to meet at I, the instantaneous centre. Draw a vertical line from I to intersect the perpendicular on $B\,C'$ at E' at P'. Measure $E'\,P'$. Then

$$E'\,P' = 3.6 \text{ cm} = 1.2\ BC$$

■ **Example 5.13** Fig. 5.26 shows a pantograph mechanism. Where will the pen be placed to copy a diagram to a reduced scale? What will be the reduction ratio of areas?

$$OA = AB = BC + CO = BQ = 10 \text{ cm}$$

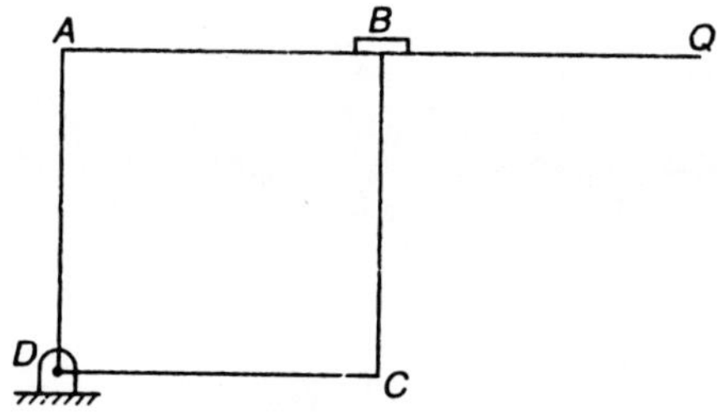

Fig. 5.26

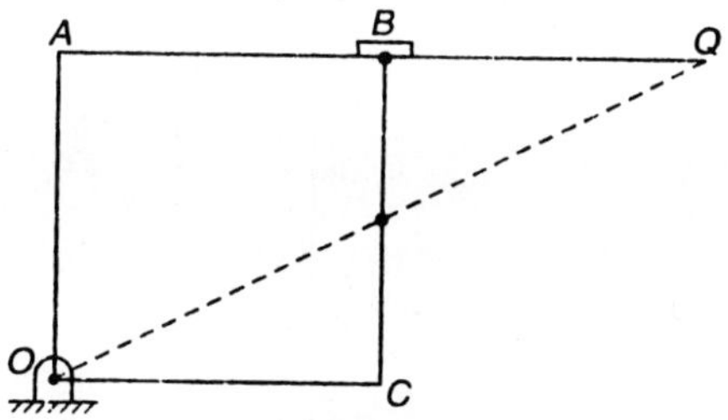

Fig. 5.27

Solution : Pantograph

1. Draw the configuration diagram to a scale of 1 cm = 4 cm, as shown in Fig. 5.27.
2. Join OQ to intersect BC at point P. Point P describes a path similar to that described by point Q.
3. To copy a diagram on a reduced scale, the pen should be placed at point P.
4. Δs PQB and OAQ are similar.

$$\frac{BQ}{AG} = \frac{PQ}{OQ} = \frac{BP}{OA} = \frac{5}{10} = \frac{1}{2}$$

$$\frac{\text{Area } \Delta\, PQB}{\text{Area } \Delta\, OAQ} = \frac{\frac{1}{2} \times 5 \times 10}{\frac{1}{2} \times 10 \times 20} = \frac{25}{100} = \frac{1}{4}$$

Hence reduction ratio of area shall be $\frac{1}{4}$.

■ **Example 5.14** Fig. 5.28 shows an arrangement of copying mechanism. A circle of diameter 9 cm is to be copied to a reduced scale. Determine the positions of scribing pens and ratio of reduction obtained.

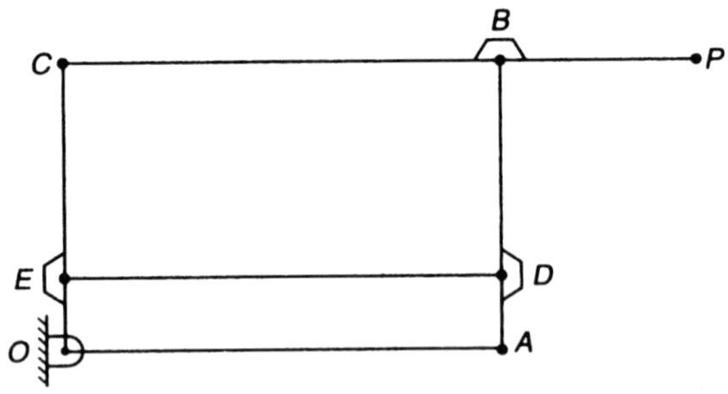

Fig. 5.28

Solution :

$$OC = OE + EC = 2 + 6 = 8 \text{ cm}$$

$$CP = CB + BP = 10 + 4 = 14 \text{ cm}$$

$$OP = \sqrt{OC^2 + CP^2}$$

$$= \sqrt{8^2 + 14^2} = \sqrt{260} = 16.124 \text{ cm}$$

For the copying mechanism shown in Fig. 5.29, we have

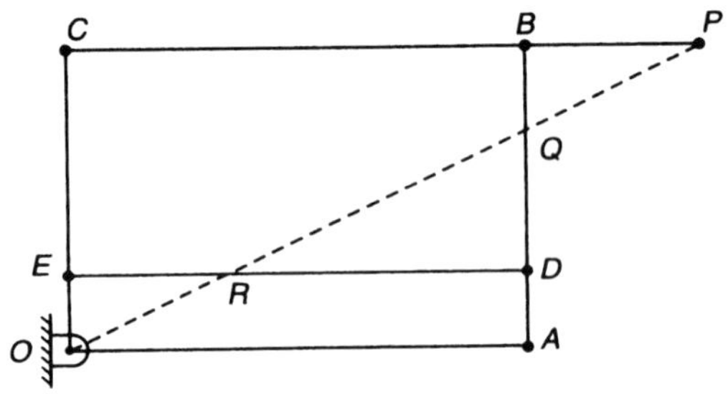

Fig. 5.29

$$\frac{OC}{OE} = \frac{OP}{OR} = \frac{CP}{ER}$$

$$\therefore \quad OR = OP \times \frac{OE}{OC} = 16.124 \times \frac{2}{8} = 4.03 \text{ cm}$$

$$ER = \sqrt{OR^2 - OE^2} = \sqrt{(4.03)^2 - 4} = 3.5 \text{ cm}$$

$$\frac{BQ}{CO} = \frac{PB}{PC}, \; BQ = 8 \times \frac{4}{14} = 2.286 \text{ cm},$$

$$QP = \sqrt{BP^2 - BQ^2} = \sqrt{16 + (2.286)^2} = 4.61 \text{ cm}$$

$$OQ = 16.124 - 4.61 = 11.518 \text{ cm}$$

The position of scribing pens is at Q and R.

$$\text{Ratio of reduction} = \frac{OR}{OP} = \frac{4.03}{16.12} = \frac{1}{4}$$

and

$$\frac{OQ}{OP} = \frac{11.517}{16.124} = \frac{1}{1.4}$$

■ **Example 5.15** Fig. 5.30 shows Peaucllier's mechanism. Describe the path of the tracing point Q.

$$OA = AP = PS = SQ = QR = RP = 15\,\text{cm}$$

$$OR = OS = 40\,\text{cm}$$

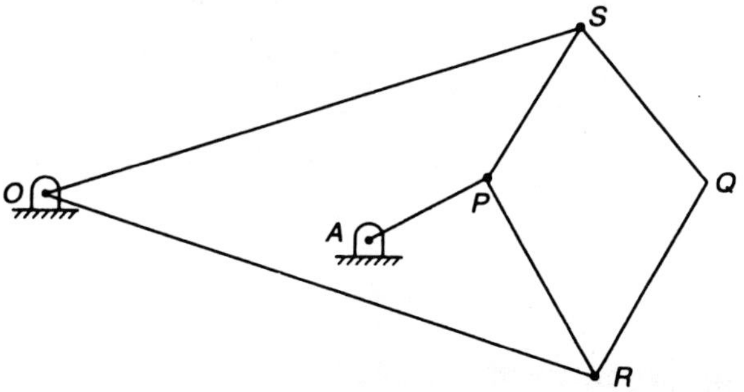

Fig. 5.30

Solution : Given : $OA = AP = PS = SQ = QR = RP = 15\,\text{cm}$

$$OR = OS = 40\,\text{cm}$$

It can be proved that as the link AP rotates about A, Q moves in a straight line perpendicular to OA produced (Fig. 5.31).

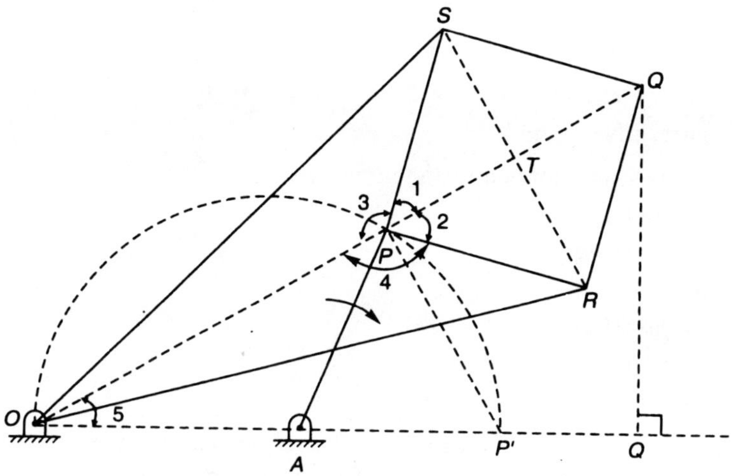

Fig. 5.31

Since $PRQS$ is a rhombus, PQ always bisects $\angle RPS$, *i.e.*, $\angle 1 = \angle 2$ in all positions. In Δs OPS and OPR, OP is common, $OS = OR$ and $PS = PR$.

$\therefore$ Δs are congruent in all positions, or 0 3 = 0 4.

$\therefore$

$$\angle 1 + \angle 3 = \angle 2 + \angle 4$$

But $\angle 1 + \angle 2 + \angle 3 + \angle 4 = 360°$

$\therefore$ $\angle 1 + \angle 3 = \angle 2 + \angle 4 = 180°$

or O, P and S lies on a straight line.

Let QQ' be the perpendicular on OA produced. Δs OPP' and OQQ' are similar because $\angle 5$ is common and $\angle OPP' = OQ'Q = 90°$

$$\therefore \quad \frac{OP}{OQ'} = \frac{OP'}{OQ}$$

$$\text{or} \quad OQ' \times OP' = OQ \times OP = (OT + TQ)(OT - TP)$$
$$= (OT - TP)(OT + TP)$$
$$= OT^2 - TP^2$$
$$= (OS^2 - ST^2) - (SP^2 - ST^2)$$
$$= OS^2 - SP^2$$

$$\text{or} \quad OQ' = \frac{OS^2 - SP^2}{OP'} = \text{const.}$$

Because OS, SP and OP' are always fixed.

This means that the projections of Q and OA produced is constant for all the configurations. Thus QQ' is always perpendicular to OA produced or Q moves in a straight line perpendicular to OA.

SOLVED QUESTIONS FROM UNIVERSITY EXAMINATIONS

■ **Example 5.16** Sketch and describe the crosby engine indicator. [KU, Dec., 2006]

Solution : Refer to art. 5.5.1.

■ **Example 5.17** Sketch and describe the HART mechanism. [KU, Dec., 2006, 2009]

Solution : Refer to art. 5.3.1.

■ **Example 5.18** Write short notes on the following :

(i) Davis steering gear.
(ii) Antifriction bearings.
(iii) Klein's construction. [KU, Dec., 2007]

Solution : (i) Refer to art. 5.6.2.
(ii) Refer to art. 6.10.
(iii) Refer to art. 3.6.

■ **Example 5.19** Under what conditions Scott Russell mechanism traces out a straight line and an ellipse? State the limitation of Scott Russell mechanism. [KU, Dec., 2007]

Solution : Refer to art. 5.3.1

■ **Example 5.20** A Hooke's joint is used to connect two shafts whose axes are inclined at 20°. The driving shaft rotates uniformly at 6000 r.p.m. What are the extreme angular velocities of the driven shaft? Find the maximum value of retardation and acceleration, and state the angle where both will occur. [KU, Dec., 2007]

Solution : Given : $\alpha = 20°$, $N_A = 6000$ rpm

$$\frac{\omega_B}{\omega_A} = \frac{\cos\alpha}{1 - \sin^2\alpha\cos^2\theta}$$

$\frac{\omega_B}{\omega_A}$ is minimum at $\theta = 90°$ and $270°$

$$(\omega_B)_{min} = \omega_A \cos\alpha = 6000\cos 20°$$
$$= 5638.16 \text{ rpm or } 590.43 \text{ rad/s}$$

$\frac{\omega_B}{\omega_A}$ is maximum at $\theta = 0°$ and $180°$

$$(\omega_B)_{max} = \frac{\omega_A}{\cos\alpha} = \frac{6000}{\cos 20°}$$
$$= 6385 \text{ rpm or } 668.63 \text{ rad/s}$$

For angular acceleration to be maximum or minimum

$$\cos 2\theta = \frac{2\sin^2\alpha}{2 - \sin^2\alpha} = \frac{2\sin^2 20°}{2 - \sin^2 20°} = 0.01242$$
$$2\theta = 82.86° \text{ or } 277.14°$$
$$\theta = 41.43° \text{ or } 138.57°$$
$$\alpha_B = \frac{\omega_A^2\cos\alpha\sin^2\alpha\sin 2\theta}{(1 - \sin^2\alpha\cos^2\theta)^2}$$

For $\theta = 41.43°$

$$\alpha_B = \frac{\left(\frac{2\pi \times 6000}{60}\right)^2 \cos 20° \sin^2 20° \sin 82.86°}{(1 - \sin^2 20° \cos^2 41.43°)^2}$$
$$= -49334 \text{ rad/s}^2$$

At $\theta = 138.57°$

$$\alpha_B = \frac{\left(\frac{2\pi \times 6000}{60}\right)^2 \cos 20° \sin^2 20° \sin 227.14°}{(1 - \sin^2 20° \cos^2 138.57°)^2}$$
$$= -49334 \text{ rad/s}^2$$

■ **Example 5.21** Sketch a Paucelier mechanism. Show that it can be used to trace a straight line. [KU, Dec., 2008]

Solution : Refer to art. 5.3.1.

■ **Example 5.22** Briefly discuss the following :

(i) Straight line mechanism

(ii) Pantograph [MDU, May, 2008]

Solution : (i) Refer to art. 5.3.

(ii) Refer to art. 5.2.

■ **Example 5.23** Describe the following :

(i) Ackermann steering gear mechanism

(ii) Scott-Russel Mechanism. [MDU, Dec., 2010]

Solution : (i) Refer to art. 5.6.2.

(ii) Refer to art. 5.3.1.

■ **Example 5.24** Two shafts are connected by a Hooke's joint. The driving shaft revolves uniformly at 500 rpm. If the total permissible variation in speed of the driven shaft is not to exceed ± 5% of the mean speed, find the greatest permissible angle between the centre line of the shafts. Calculate also the maximum the minimum speeds of the driven shaft.

[MDU, Dec., 2008]

Solution : Given : $N_A = 500$ rpm $\Delta N = \pm 5\%$

$$\text{Maximum speed fluctuation} = \frac{1-\cos^2\alpha}{\cos\alpha} = 0.1$$

$$1-\cos^2\alpha = 0.1\cos\alpha$$

$$\cos^2\alpha + 0.1\cos\alpha - 1 = 0$$

$$\cos\alpha = \frac{1}{2}[-0.1 \pm \sqrt{(0.1)^2 + 4\times 1\times 1}]$$

$$= 0.95125$$

$$\alpha = 17.96°$$

$$(N_2)_{max} = \frac{N_1}{\cos\alpha} = \frac{500}{\cos 17.96°} = 525.62 \text{ rpm}$$

$$(N_2)_{min} = N_1\cos\alpha = 500\cos 17.96° = 475.63 \text{ rpm}$$

SUMMARY

1. A pantograph is a mechanism to produce a path traced out by a point on enlarged or reduced scale.
2. Accuxate straight line mechanisms are : Peaucellier, Hart and Scott Russel.
3. Appropriate straight line mechanisms are : Grasshopper, Watt and Tchebicheff.
4. Steering gear mechanism is used for changing the direction of motion of an automobile.
5. Davis steering gear lies in front of the front wheels.
6. To avoid skidding the two front wheels must turn about the same instantaneous centre.
7. The fundamental equation for correct steering is :

$$\cot\phi - \cot\theta = \frac{b}{l}$$

where θ, ϕ = angles turned through by stub-axles of inner and outer front wheels.

b = distance between the pivots of front axles.

l = wheel base

8. For the Davis steering gear,

$$\tan \alpha = \frac{b}{2l}$$

where α = angle of inclination of teach arm with vertical in mid-position.

9. For the Ackermann steering gear,

$$\tan \alpha = \frac{\sin \theta - \sin \phi}{2 - \cos \theta - \cos \phi}$$

10. For the Davis steering gear, the instantaneous centre lies on the axis of rear wheels.

11. For Ackermann steering gear, the instantaneous centre lies at aproximately 0.3 l above the rear wheels axis.

12. For a Hooke's coupling,

$$\tan \theta = \cos \alpha \tan \phi$$

where θ, ϕ = angles turned though by driving and driven shafts respectively.
α = angle of inclination of driven shaft with driving shaft.

13. The ratio of angular speeds of driven to driving shaft is :

$$\frac{\omega_B}{\omega_A} = \frac{\cos \alpha}{1 - \sin^2 \alpha \cos^2 \theta}$$

14. For $\frac{\omega_B}{\omega_A} = 1, \cos \theta = \sqrt{\frac{1}{1 + \cos \alpha}}$

$$\left(\frac{\omega_B}{\omega_A}\right)_{max} = \frac{1}{\cos \alpha}$$

$$\left(\frac{\omega_B}{\omega_A}\right)_{min} = \cos \alpha$$

15. Maximum fluctuation of speed of driven shaft

$$\frac{(N_B)_{max} - (N_B)_{min}}{(N_B)_{mean}} = \tan \alpha \times \sin \alpha \simeq \alpha^2$$

16. Angular acceleration of driven shaft,

$$\alpha_B = \frac{-\omega_A^2 \cos \alpha \sin^2 \alpha \sin 2\theta}{(1 - \sin^2 \alpha \cos^2 \theta)^2}$$

17. For angular acceleration to be maximum or minimum,

$$\cos 2\theta = \frac{2 \sin^2 \alpha}{2 - \sin^2 \alpha}$$

18. Double Hooke's joint is used to have a constant speed ratio of driving and driven shafts.

19. For a double Hooke's joint, the speed ratios are :

$$\left(\frac{\omega_B}{\omega_A}\right)_{max} = \frac{1}{\cos^2 \alpha}$$

$$\left(\frac{\omega_B}{\omega_A}\right)_{min} = \cos^2 \alpha$$

SHORT ANSWER TYPE QUESTIONS

1. What is a pantograph? What are its uses?
2. Name the exact straight line mechanisms.
3. List the approximate straight line mechanisms.
4. Define a steering gear.
5. What is the function of a steering gear?
6. Name the steering gears used in automobiles.
7. Compare Davis and Ackermann steering gears.
8. Write the fundamental equation for correct steering.
9. What is a Hooke's joint? Where it is used?
10. Write the expression for ratio of angular velocities of shafts for a Hooke's joint.
11. Draw the polar velocity diagram of a Hooke's joint and mark its salient features.
12. Write the expressions for maximum and minimum speeds of driven shaft in a Hooke's joint.
13. What is a double Hooke's joint? What is its use?
14. Write the condition for the speeds of two shafts to be same in a Hooke's joint.

EXERCISES

1. Two shafts connected by a Hooke's joint have their axes inclined at 20°. The driving shaft rotates at 1440 rpm and the driven shaft carries a flywheel of mass 20 kg and radius of gyration of 10 cm. Find the maximum torque in the driven shaft. **[Ans.** 568.36 N.m]
2. The axes of two shafts connected by a Hooke's joint are inclined at 20°. At what positions of the driving shaft, the velocities of two shafts are equal? State whether the accelerations are positive or negative at these positions.
   If the driving shaft rotates at 1800 rpm, determine acceleration of the driven shaft at any one of the above positions. **[Ans.** 44.1°, 135.9°, 224.1°, 315.9°, +ve at 135.9°, 315.9° and –ve at 44.1°, 224.1°, –4421.23 rad/s^2 at 44.1°]
3. The driving shaft of a Hooke's joint is rotating at a uniform speed of 600 rpm. The speed of the driven shaft must be 575 and 625 rpm. Determine the maximum permissible angle between the shafts. **[Ans.** 16.43°]
4. A shaft running at 1200 rpm is connected to a second shaft by a Hooke's joint. The angle between the axes of the shafts is 15°. Determine the velocity and acceleration of the driven shaft when it has turned through an angle of 10° from the plane containing the shaft axes. **[Ans.** 1239.5 rpm –400 rad/s^2]
5. The intermediate shaft of a double Hooke's joint is inclined at 15° to each. The input and output forks on the intermediate shaft have been assembled by mistake at 90° to one another. Determine the maximum and minimum speeds of output shaft if the speed of input shaft is 600 rpm. Also calculate the coefficient of fluctuation in speed. **[Ans.** 643.1 rpm, 559.75 rpm; 13.9%]
6. Two shafts are connected by a Hooke's joint. The driving shaft rotates uniformly at 600 rpm. If the total permissible variation in speed of the driven shaft is not to exceed ± 5% of the mean speed, find the greatest permissible angle between the centre lines of the shafts. **[Ans.** 17.96°]

7. The moment of inertia of the driven shaft in a Hooke's joint is 35 kg.m^2. The driven shaft is inclined at 30° to the axis of the driving shaft. The driving shaft rotates at 2700 rpm with a steady torque of 275 N.m. Determine the maximum fluctuation of output torque. [**Ans.** 96.91% of input torque]

8. In a Davis steering gear, the distance between the pivots of the front axle is 1.2 m and the wheel base is 2.8 m. Find the inclination of each arm to the longitudinal axis of the car, when it is moving along a straight path. [**Ans.** 12.1°]

9. The distance between the pivots of the front stub axles of a car is 1.35 m. The length of track rod is 1.25 m. The wheel track is 1.5 m and the wheel base is 2.8 m. What should be the length of the track arm if the gear is to be given a correct steering, when rounding a corner of radius 5 m.

[**Ans.** 0.213 m]

10. In a Scott-Russel mechanism shown in the Fig. 5.32, $OA_1 = 30$ cm, $A_1C_1 = 40$ cm. Find the length of the extended link A_1B_1 so that point B_1 is generating an approximate straight line motion.

[**Ans.** 53.3 cm]

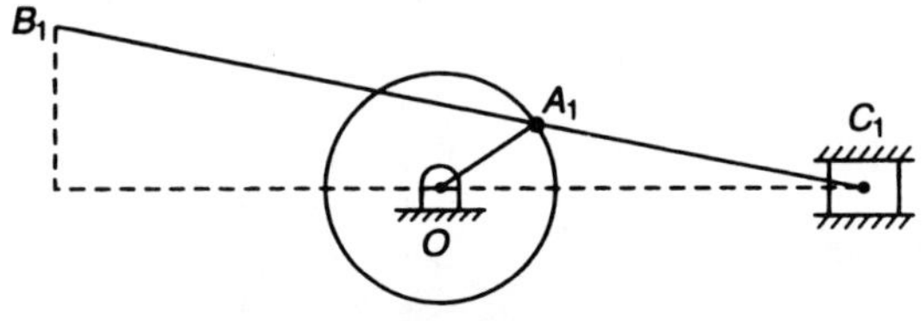

Fig. 5.32

❑❑❑

Chapter 6

Friction

Inside This Chapter

6.1 INTRODUCTION

When a body moves or tends to move on another body, the property of the two bodies by virtue of which a force is developed between the two bodies, which opposes the motion, is called *friction,* and the opposing force is called the *force of friction.* The force of friction acts opposite to the impending motion.

Friction : A Blessing or a Curse

Friction is a blessing when it is necessary to increase the force of friction between two contacting bodies when power is being transmitted through them, as in the case of a belt and pulley and friction clutches. In the case of brakes also, it is necessity to have friction between contacting surfaces to reduce the speed of the moving member or stop it altogether.

In the case of lathe slides, journal bearings, etc., it is necessary to reduce the force of friction in order to decrease the power lost due to friction. In such situations, friction is a curse. To decrease the friction, proper lubrication has to be provided.

6.2 TYPES OF FRICTION

The friction is of the following types :

- *Static friction :* It is the friction experienced by two bodies in contact, while at rest.
- *Dynamic (or kinetic) friction :* It is the friction experienced by two bodies in contact, while in motion. It is less than the static friction.
- *Sliding friction :* It is due to sliding of two bodies on each other.
- *Rolling friction :* It is due to rolling of two bodies on each other.
- *Pivot friction :* It is the friction experienced by two bodies due to the rotation of one body around the other, as in the case of a foot step bearing.
- *Dry (or solid) friction :* It is due to two dry and unlubricated surfaces in contact.
- *Boundary (or skin or greasy) friction :* It is the friction experienced by two bodies separated by a very thin layer of a lubricant.
- *Fluid (or film) friction :* It is the friction experienced by two bodies in contact when separated by a thick film of a lubricant.

6.3 LAWS OF FRICTION

(a) Static Friction

1. The force of friction always acts in a direction opposite to the impending motion.
2. The limiting force of friction is directly proportional to the normal reaction.
3. The force of friction is independent of the area of contact between the two surfaces.
4. The force of friction depends upon the roughness of the surfaces of two materials.

(b) Kinetic Friction

1. The force of friction always acts in a direction opposite to that in which the body is moving.
2. The magnitude of kinetic bears a constant ratio to the normal reaction between the two surfaces.
3. The force of friction is independent of the relative velocity between the two surfaces in contact, but it decreases slightly with increase in velocity.
4. The force of friction increases with reversal of motion.
5. The coefficient of friction changes slightly due to temperature changes.

(c) Fluid Friction

1. The force of friction is almost independent of the load.
2. The force of friction reduces with the increase of temperature of the lubricant.
3. The force of friction depends upon the type and viscosity of the lubricant.
4. The force of friction is independent of the nature of surfaces.
5. The frictional force increases with the increase in the relative velocity of the frictional surfaces.

6.4 DEFINITIONS

(a) Coefficient of Friction

Consider a block of weight W resting on a plane rough surface, as shown in Fig. 6.1. The normal reaction because of the weight of the block is R. Let a force P be applied to the block towards the right.

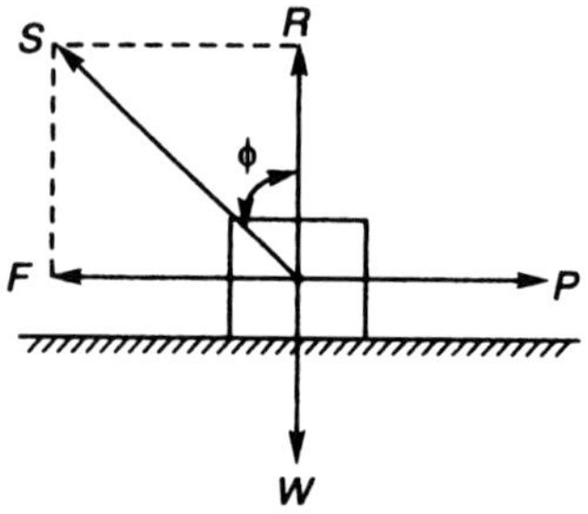

Fig. 6.1 Block resting on rough horizontal plane.

The frictional force F will set up between the block and the rough surface. The direction of F will be towards the left. The *coefficient of friction* (μ) is the ratio of the force of friction (F) to the normal reaction (R). Thus

$$\mu = \frac{F}{R} \qquad ...(6.1)$$

(b) Angle of Friction

Let S be the resultant of F and R making an angle of ϕ with R that is called the angle of friction, such that

$$\tan \phi = \frac{F}{R} \qquad ...(6.2)$$

From Eqs. (6.1) and (6.2), we find that

$$\mu = \tan \phi (6.3)$$

Therefore, in the case of limiting friction, the coefficient of friction is equal to the tangent of the angle of friction.

Now consider the block resting on rough inclined plane, as shown in Fig. 6.2. By resolving the forces along and perpendicular to the plane, we have

$$F = W \sin \theta$$

$$R = W \cos \theta$$

or

$$\frac{F}{R} = \tan \theta$$

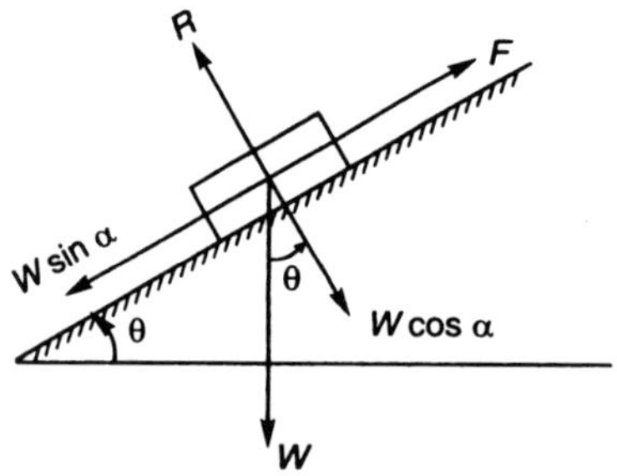

Fig. 6.2 Block resting on rough inclined plane.

Comparing with Eq. (6.2), we find that

$$\tan\theta = \tan\phi$$

or $$\theta = \phi \quad ...(6.4)$$

Thus, in the case of limiting friction, the angle of the plane is equal to the angle of friction. The angle of the plane when motion of an object on the plane is impending is called the angle of *repose*. This is the maximum angle that a heap of sand or similar materials with make with the horizontal.

(c) Cone of Friction

If the force P is made to revolve about a vertical axis, the resultant S will also remove about a vertical axis. As S revolves, it will generate a cone of vertex angle 2ϕ. This cone is called the *cone of friction*, as show in in Fig. 6.3.

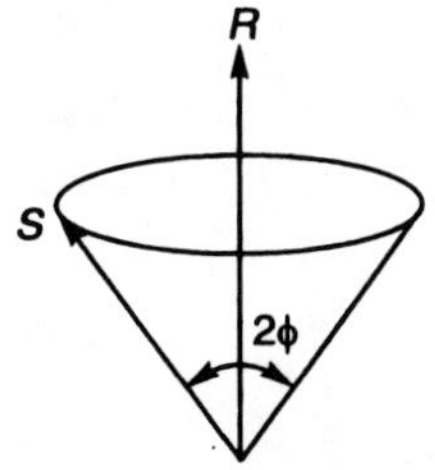

Fig. 6.3 Cone of friction.

6.5 FORCE ANALYSIS OF A SLIDING BODY

A rough body resting on another rough body may be made to slide by a force of the pull or push type. Now we study the force required to slide the body under various situations.

(a) Boyd Resting on a Horizontal Plane

Consider a body of weight W resting on a rough horizontal plane being pulled by a force P inclined at an angle θ, as shown in Fig. 6.4(a). By resolving the forces horizontally and vertically, we have

$$P\cos\theta = F$$

$$R + P\sin\theta = W$$

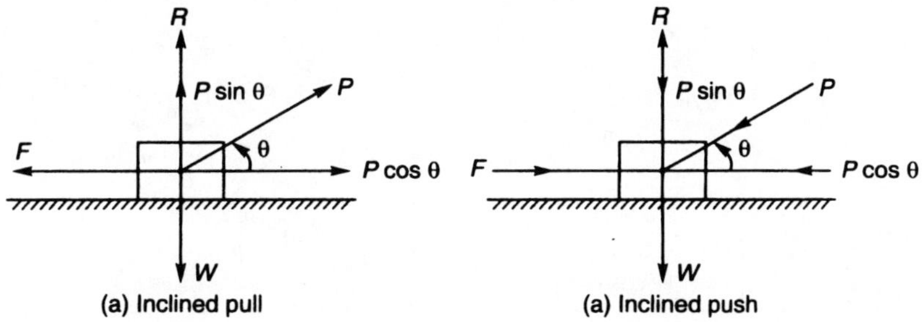

Fig. 6.4 Force analysis of a sliding body on horizontal plane.

Now, $$F = \mu R$$

Therefore, $$P\cos\theta = \mu R$$

$$\frac{P\cos\theta}{\mu} + P\sin\theta = W$$

$$P\left(\frac{\cos\theta}{\mu} + \sin\theta\right) = W$$

or
$$P(\cos\theta\cos\phi + \sin\phi\sin\theta) = W\sin\phi$$

$$P\cos(\theta - \phi) = W\sin\phi$$

$$P = \frac{W\sin\phi}{\cos(\theta - \phi)} \quad \text{...(6.5)}$$

For P to be minimum, $\cos(\theta - \phi)$ should be maximum, that is,

$$\cos(\theta - \phi) = 1$$

or
$$\theta = \phi$$

and
$$P_{min} = W\sin\theta \quad \text{...(6.6)}$$

If the force P is of the push type, as shown in Fig. 6.4 (b), then

$$P\cos\theta = F = \mu R$$

$$P\sin\theta + W = R$$

$$= P\frac{\cos\theta}{\mu}$$

$$P(\cos\theta\cos\phi - \sin\theta\sin\phi) = W\sin\phi$$

$$P\cos(\theta + \phi) = W\sin\phi$$

or
$$P = \frac{W\sin\phi}{\cos(\theta + \phi)} \quad \text{...(6.7)}$$

(b) Body Resting on an Inclined Plane with Applied Force Inclined to the Plane

1. **Body going up the plane :** Consider the body resting on the inclined plane as shown in Fig. 6.5(a). Body resolving the forces perpendicular and along the plane, we have

$$P\cos\theta = W\sin\alpha + \mu R$$

$$W\cos\alpha = P\sin\theta + R$$

or
$$P\cos\theta = W\sin\alpha + \tan\phi(W\cos\alpha - P\sin\theta)$$

or
$$P(\cos\theta + \sin\theta\tan\phi) = W(\sin\alpha + \cos\alpha\tan\phi)$$

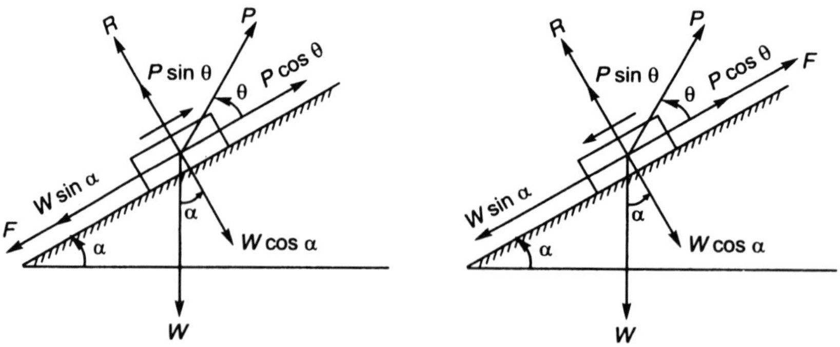

(a) Body going up the plane

(b) Body going down the plane

Fig. 6.5 Inclined pull acting on body resting on inclined plane.

or $$P(\cos\theta \cos\phi \sin\theta \sin\phi) = W(\sin\alpha \cos\phi + \cos\alpha \sin\phi))$$

$$P\cos(\theta - \phi) = W \sin(\alpha + \phi) \qquad \text{...(6.8)}$$

$$P = \frac{W \sin(\alpha + \phi)}{\cos(\theta - \phi)}$$

For P to be minimum, $\cos(\theta - \phi) = 1$, or $\theta = \phi$. In that case,

$$P_{min} = W \sin(\alpha + \theta) \qquad \text{...(6.9)}$$

2. **Body going down the plane :** When the body is going down the plane, as shown in Fig. 6.5(b), then we get

$$P = \frac{W \sin(\alpha + \phi)}{\cos(\theta + \phi)} \qquad \text{...(6.10)}$$

For P to be minimum, $\cos(\theta + \phi) = 1$, or $\phi = -\theta$. In the case

$$P_{min} = W \sin(\alpha - \theta). \qquad \text{...6.(11)}$$

(c) Body Resting on an Inclined Plane with Horizontal Applied Force

1. **Body going up the plane :** Consider the body as shown in Fig. 6.6(a). By resolving the forces perpendicular and along the plane, we have

$$P\cos\alpha = W\sin\alpha + \mu R$$

$$P\sin\alpha + W\cos\alpha = R$$

or $$P\cos\alpha = W\sin\alpha + \tan\phi(P\sin\theta + W\cos\alpha)$$

$$P(\cos\alpha\cos\phi - \sin\alpha\sin\phi) = W(\sin\alpha\cos\phi + \cos\alpha\sin\phi)$$

$$P\cos(\alpha + \phi) = W\sin(\alpha + \phi)$$

$$P = W\tan(\alpha + \phi) \qquad \text{...(6.12)}$$

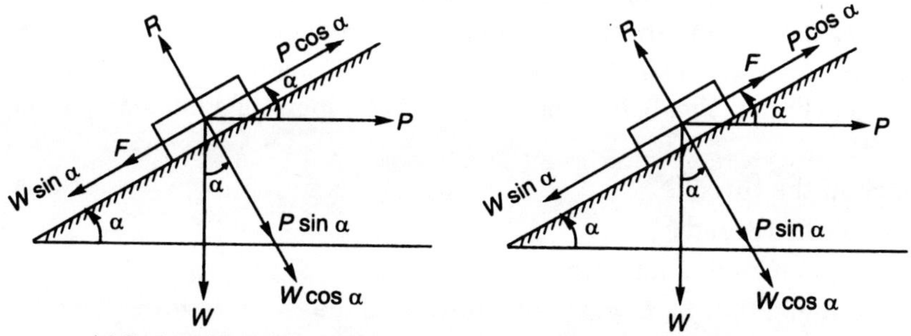

(a) Body going up the plane (b) Body going down the plane

Fig. 6.6 Horizontal pull acting on body resting on inclined plane.

2. **Body going down the plane :** For the body shown in Fig. 6.6(b), we get

$$P = W\tan(\phi - \alpha) \qquad \text{...(6.13)}$$

(d) Efficiency of the Inclined Plane

The efficiency of the inclined plane is defined as the ratio of the effort required P_0 without friction and P with friction.

$$\eta = \frac{P_0}{P}$$

1. For body given up the plane with inclined applied force,

$$\eta = \frac{\dfrac{\sin \alpha}{\cos \theta}}{\dfrac{\sin (\alpha + \phi)}{\cos (\theta - \phi)}}$$

$$= \frac{1 + \mu \tan \theta}{1 + \mu \cot \alpha} \qquad \text{...(6.14)}$$

2. For body going down the plane with inclined applied force

$$\eta = \left(- \frac{\sin \alpha}{\cos \theta} \right) \times \left[\frac{\cos (\theta + \phi}{\sin (\phi - \alpha} \right]$$

$$= \frac{1 - \mu \tan \theta}{1 - \mu \cos \alpha} \qquad \text{...(6.15)}$$

3. For body going up the plane with horizontal applied force

$$\eta = \frac{\tan \alpha}{\tan (\alpha + \phi)} \qquad \text{...(6.16)}$$

4. For body going down the plane with horizontal applied force

$$\eta = \frac{\tan (\alpha - \phi)}{\tan \alpha} \qquad \text{...(6.17)}$$

6.6 SCREW THREADS

Screws are used for fastening, load lifting, and power transmission purposes. The screws may have single start or multi-start threads. Leads is the product of pitch and number of starts. A screw thread is obtained when the hypotenuse of right-angled triangle is wrapped round the circumference of a cylinder. Fig. 6.7(a) shows the development of a helix of diameter d and lead L or pitch p.

Let p = pitch of the threads

L = lead of the threads

d_m = mean pitch circle diameter of the screw

α = helix angle, that is, angle of inclination of the threads with a line perpendicular to the axis of the screw.

Then

$$\tan \alpha = \frac{p}{\pi d_m} \text{ or } \frac{L}{\pi d_m} \qquad \text{...(6.18)}$$

(a) Square Threads

In the case of square threads as shown in Fig. 6.7(b), the faces of the threads are normal to the axis of the spindle. The force P acting horizontally required at the screw threads to slide the load W up the inclined plane is given by Eq. (6.12), that is

$$P = W \tan (\alpha + \phi)$$

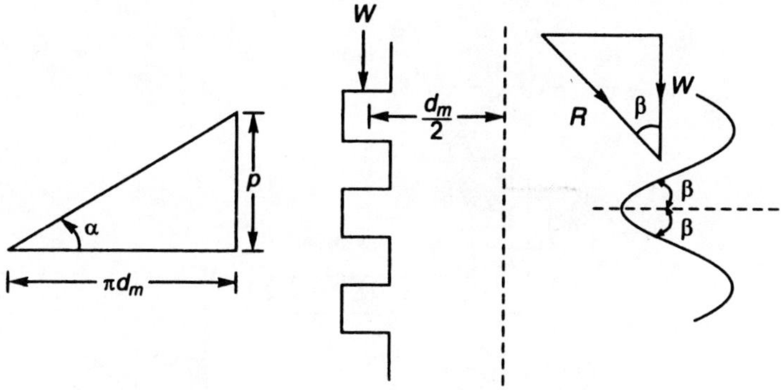

(a) Helix-angle (c) Square theads (c) V-threads

Fig. 6.7 Square and V-threads.

(b) *V*-threads

For a screw having *V*-threads, as shown in Fig. 6.7(c), let 2β be the angle of the threads. Then normal reaction on the threads is given by :

$$R = \frac{W}{\cos \beta}$$

Frictional force,

$$F = \mu R = \left(\frac{\mu}{\cos \beta}\right) \cdot W = \mu_e W$$

where $\mu_e = \mu / \cos \beta$ is called the virtual or equivalent coefficient of friction.

Therefore, for a screw having *V*-threads, the virtual coefficient of friction should be used to calculate the torque required to lift the load.

6.7 SCREW JACK

A screw jack is a device for lifting of loads. The principle of working of a screw jack is similar to that of an inclined plane. Fig. 6.8 shows the common form of the screw jack.

Let P = effort applied at the circumference of the screw to lift the load

W = load to be lifted

μ = coefficient of friction between the screw and the nut

$= \tan \phi$

1. Resting of Load

As derived in section 6.5, for raising the load, we have

$$P = W \tan (\alpha + \phi)$$

Torque required to overcome friction between the screw and nut is

$$T_1 = \frac{P \times d_m}{2} = \frac{W d_m \tan (\alpha + \phi)}{2}$$

Torque required to overcome friction at the collar,

$$T_2 = \frac{\mu_c W d_c}{2}$$

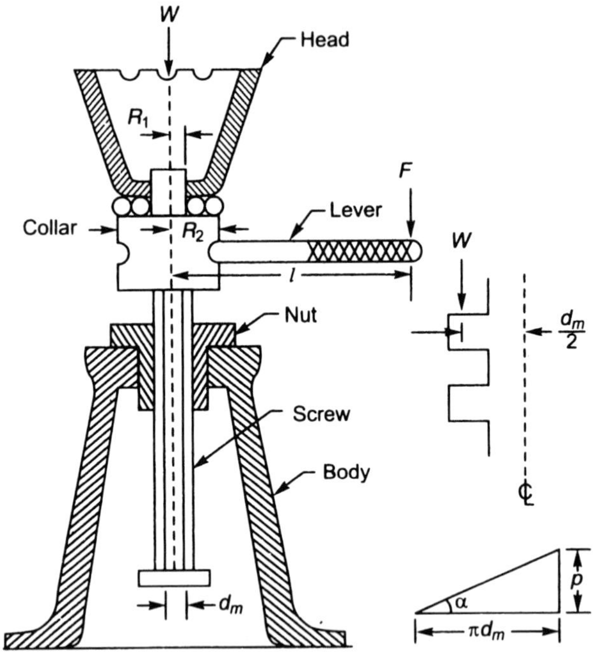

Fig. 6.8 Screw jack.

where μ_c = collar coefficient of friction

d_c = mean diameter of the collar = $R_1 + R_2$

Total torque, $$T = T_1 + T_2 = \frac{1}{2}(Pd_m + \mu_c W d_c) \quad ...(6.19)$$

Let F = effort applied at the end of a lever length l

Then $$T = Fl$$

2. Lowering of Load

For lowering of load, we have

$$P = W \tan(\phi - \alpha)$$

Torque required to overcome friction between screw and nut,

$$T_1 = \frac{Pd_m}{2} = \frac{1}{2} W d_m \tan(\phi - \alpha)$$

Torque required to overcome collar friction,

$$T_2 = \frac{1}{2} \mu_c W d_c$$

Total torque, $$T = T_1 + T_2$$

$$= \frac{1}{2}(Pd_m + \mu_c W d_c)$$

Effort F required at the end of lever of length l,

$$F = \frac{T}{l}$$

3. Efficiency of Screw Jack

Efficiency, $$\eta = \frac{\tan\alpha}{\tan\alpha - \phi}, \text{ for raising of load, and} \quad \text{...(6.20a)}$$

$$= \frac{\tan(\phi - \alpha)}{\tan\alpha}, \text{ for lowering of load} \quad \text{...(6.20b)}$$

For efficiency to be maximum, $= \frac{d\eta}{d\alpha} = 0$

Now $$\eta = \frac{\tan\alpha\,(1 - \tan\alpha \times \tan\phi)}{\tan\alpha + \tan\phi}$$

$$= \left(\frac{\sin\alpha}{\cos\alpha}\right)\frac{\left[1 - \left(\frac{\sin\alpha}{\cos\alpha}\right)\left(\frac{\sin\phi}{\cos\phi}\right)\right]}{\left(\frac{\sin\alpha}{\cos\alpha}\right) + \left(\frac{\sin\phi}{\cos\phi}\right)}$$

$$= \frac{\sin\alpha[\cos\alpha\cos\phi - \sin\alpha\sin\phi}{\cos\alpha\,[\sin\alpha\cos\phi + \sin\phi\cos\alpha]}$$

$$= \frac{\sin\alpha\cos(\alpha + \phi)}{\cos\alpha\sin(\alpha + \phi} = \frac{\sin(2\alpha + \phi) - \sin\phi}{\sin(2\alpha + \phi) + \sin\phi}$$

For efficiency to be maximum, $\sin(2\alpha + \phi) = 0$

or $$2\alpha + \phi = 90°$$

or $$\alpha = 45° - \frac{\phi}{2} \quad \text{...(6.21)}$$

$$\eta_{max} = \frac{\sin 90° - \sin\phi}{\sin 90° + \sin\phi} = \frac{1 - \sin\phi}{1 + \sin\phi}$$

4. Self-locking and Over-hauling Screw

The effort required for lowering of load is,

$$P = W\tan(\phi - \alpha)$$

Torque, $$T = \frac{Pd_m}{2} = W\left(\frac{d_m}{2}\right)\tan(\phi - \alpha)$$

Torque will remain positive if $\phi > \alpha$. Such a screw is called *self-locking screw.* If $\phi < \alpha$, then the torque will become negative. In other words, the screw will lower of its own. Such a screw is called *over-hauling screw.*

For $\alpha = \phi$, the efficiency of the screw for raising of load becomes,

$$\eta = \frac{\sin\phi\cos 2\phi}{\cos\phi\sin 2\phi} = \frac{\tan\phi}{\tan 2\phi}$$

For a self-locking screw,

$$\eta \le \frac{\tan\phi}{\tan 2\phi} \le \frac{2\,\phi(1 - \tan^2\phi}{2\tan\phi} \le \left[\frac{1}{2} - \frac{\tan^2\phi}{2}\right]$$

$$\eta \le 50\% \quad \text{...(6.23)}$$

6.8 FRICTION IN BEARINGS

When a shaft is rotating in a bearing, a pivot or a collar is provided on the shaft to take up the axial thrust. Common examples are steam turbines, hydraulic turbines, propeller shaft of a ship, etc.

6.8.1 Flat Pivot Bearing

When the axial force is taken by the end of a shaft that is inserted in a recess to take up the thrust, it is called a pivot bearing or foot step bearing.

A flat pivot bearing like the foot step bearing is shown in Fig. 6.9(a).

Let W = load on the bearing

r = radius of bearing surface

p = intensity of pressure between rubbing surfaces

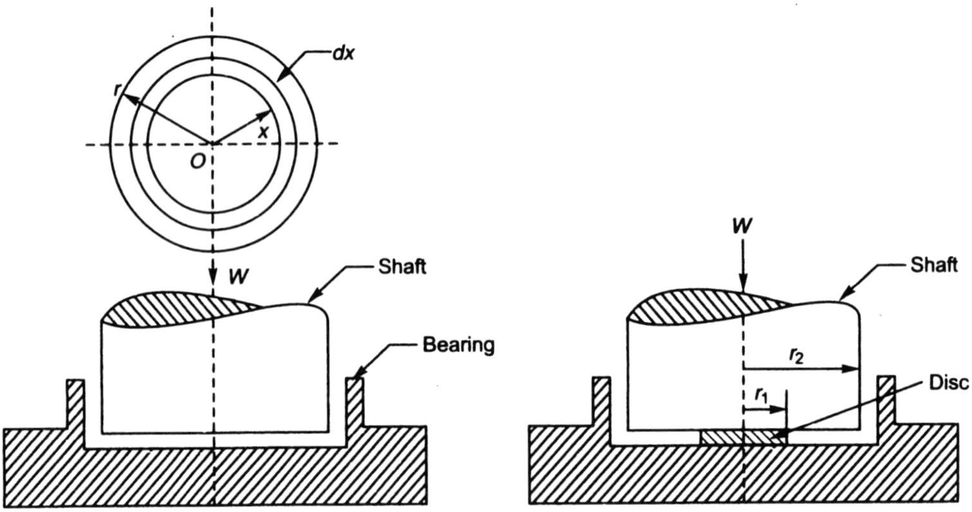

Fig. 6.9 Flat pivot bearing.

(a) Uniform Pressure

When the pressure is uniformly distributed over the bearing surface area, then

$$p = \frac{W}{\pi r^2} \quad \text{...(6.24)}$$

Consider a ring of radius x and thickness dx of the bearing area, as shown in the Fig. 6.9.

Area of bearing surface, $\delta A = 2\pi x \cdot dx$

Load transmitted to the ring, $\delta W = p \cdot \delta A$

Frictional resistance to sliding at radius r,

$$F_r = \mu \cdot \delta W = 2p\mu \cdot px\ dx$$

Frictional torque on the ring,

$$dT_f = F_f x = 2\pi\mu \cdot px^2\ dx$$

Total frictional torque,

$$T_f = \int_0^r 2\pi\mu\ px^2\ dx = 2\pi\mu p \int_0^r x^2\ dx = 2\pi\mu p \left|\frac{x^2}{3}\right|_0^r$$

$$= \frac{2\pi\mu\ pr^3}{3}$$

$$= \frac{2\mu Wr}{3} = \mu W r_m \quad \text{...(6.25)}$$

where $$n_m = \frac{2}{3} r \quad \text{...(6.26)}$$

Power lost in friction, $$P = T_f \cdot \omega \quad \text{...(6.27)}$$

Where $\omega = \dfrac{2\pi N}{60}$ rad/s and N is the speed of the shaft in rpm.

If the shaft of radius r_2 is resting on a disc of radius r_1 as shown in Fig. 6.9(b), then

$$p = \frac{W}{\pi(r_2^2 - r_1^2)} \quad \text{...(6.28)}$$

$$T_f = 2\pi\,\mu\ p \int_{r_1}^{r_2} x^2\ dx = 2\pi\,\mu p \left|\frac{x^3}{3}\right|_{r_1}^{r^2}$$

$$= 2\pi\mu p \frac{(r_2^3 - r_1^3)}{3} = 2\mu W \left(\frac{r_2^3 - r_1^3}{r_2^2 - r_1^2}\right) = \mu W r_m \quad \text{...(6.29)}$$

where $$r_m = \frac{2}{3}\left(\frac{r_2^3 - r_1^3}{r_2^2 - r_1^2}\right) = \text{mean radius} \quad \text{...(6.30)}$$

(b) Uniform Wear

The rate of wear depends upon the intensity of pressure and the rubbing velocity. The rubbing velocity increases with the increase in radius. Therefore, the wear rate is proportional to the pressure p and radius x. For uniform wear,

$$px = C$$

or $$p = \frac{C}{x}$$

where C is a constant.

Load transmitted to the ring, $$\delta W = p \cdot 2\pi x\ dx$$

$$= 2\pi C \cdot dx$$

Total load transmitted to the bearing,

$$W = 2\pi C \int_0^r dx = 2\pi C\,|x|_0^r$$

$$= 2\pi C r$$

or $$C = \frac{W}{2\pi r}$$

Frictional torque acting on the ring,

$$dT_f = 2\pi\mu p x^2 \, dx = 2\pi\mu C \, dx$$

Total frictional on the bearing,

$$T_f = 2\pi\mu C\int_0^r x \, dx = 2\pi\mu C \left|\frac{x^2}{2}\right|_0^r = \pi\mu C r^2 = \frac{\pi W r}{2} \quad \text{...(6.31)}$$

If the shaft of radius r_2 is resting on a disc of inner radius r_1, then

$$T_f = 2\pi\mu C \int_{r_1}^{r_2} x \, dx = 2\pi\mu \, C\left|\frac{x^2}{2}\right|_{r_1}^{r_2} = \pi\mu \, C(r_2^2 - r_1^2) = \frac{\mu W (r_2 + r_1)}{2}$$

$$T_f = W r_m \quad \text{...(6.32)}$$

where $r_m = \frac{1}{2}(r_1 + r_2)$ is the mean radius.

6.8.2 Conical Pivot Bearing

Consider the conical pivot bearing as shown in Fig. 6.10.

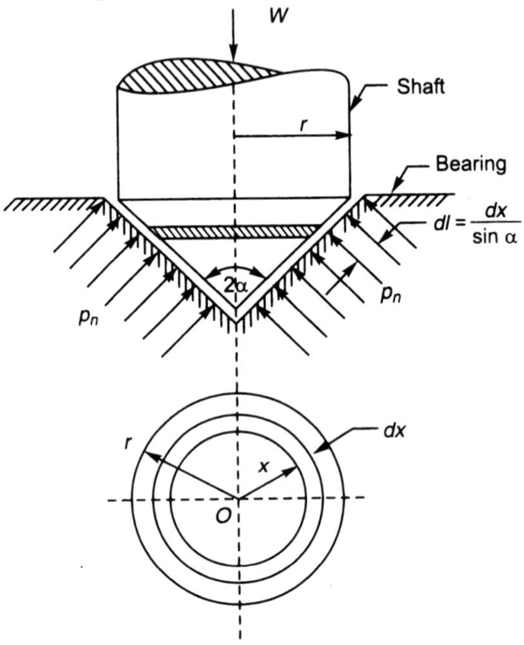

Fig. 6.10 Conical pivot bearing.

Let p_r = intensity of normal pressure on the cone

α = semicone angle

μ = coefficient of friction between the shaft and the bearing

r = radius of the shaft

Consider a small ring of radius x and thickness dx. Let dl be the length of the ring along the cone, so that

$$dl = dx \cdot \text{cosec } \alpha$$

Area of the ring, $$dA = 2\pi x \cdot dl = 2\pi x \cdot dx \text{ cosec } \alpha$$

(a) Uniform Pressure

Normal load acting on the ring, $\delta W_n = 2\pi x \times p_n \times dx \operatorname{cosec} \alpha$

Vertical load acting on the ring, $\delta W = \delta W_n \sin \alpha$

Total vertical load transmitted to the bearing,

$$W = \int_0^r \delta W = 2\pi\, p_n \int_0^r x\, dx = \pi\, p_n \,|x^2|_0^r$$

$$W = \pi r^2 p_n$$

or
$$p_n = \frac{W}{\pi r^2}$$

Frictional force acting on the ring tangentially at radius x, $F_f = \mu \times \delta W_n$

Frictional torque acting on the ring,

$$dT_f = F_f\,, x = 2\pi\,\mu \cdot p_n \cdot \operatorname{cosec} \alpha \cdot x^2 \cdot dx$$

Total frictional torque,

$$T_f = 2\pi\mu \cdot p_n \cdot \operatorname{cosec} \alpha \int_0^r x^2\, dx = \frac{2}{3}\pi\mu p_n \operatorname{cosec} \alpha \,|x^3|_0^r$$

$$= \left(\frac{3}{2}\right) \cdot \pi r^3 \cdot \mu p_n \cdot \operatorname{cosec} \alpha$$

$$T_f = \left(\frac{2}{3}\right) \cdot \mu W r \operatorname{cosec} \alpha \qquad \text{...(6.34)}$$

(b) Uniform Wear

In the case of uniform wear, the intensity of normal pressure varies inversely with the distance.

Therefore,

$$p_n \cdot x = C, \text{ where } C \text{ is a constant}$$

Load transmitted to the ring,

$$\delta W = p_n \cdot 2\pi x \cdot dx = 2\pi C \cdot dx$$

Total load transmitted to be bearing,

$$W = \int_0^r \delta\, W = 2\pi C \int_0^r dx = 2\pi C \cdot r$$

or
$$C = \frac{W}{2\pi r}$$

Frictional torque acting on the ring,

$$dT_f = 2\pi\mu \cdot p_n \cdot \operatorname{cosec} \alpha \cdot x^2 \cdot dx$$

$$= 2\pi\mu \cdot C \cdot \operatorname{cosec} \alpha \cdot x \cdot dx$$

Total frictional torque acting on the bearing,

$$T_f = 2\pi\mu \cdot C \cdot \operatorname{cosec} \alpha \int_0^r x \cdot dx$$

$$= \pi\mu \cdot C \cdot r^2 \cdot \operatorname{cosec} \alpha$$

$$= T_f = \left(\frac{1}{2}\right) \cdot \mu\, W r \operatorname{cosec} \alpha \qquad \text{...(6.35)}$$

6.8.3 Truncated Conical Pivot Bearing

Consider a truncated conical pivot bearing as shown in Fig. 6.11. Then

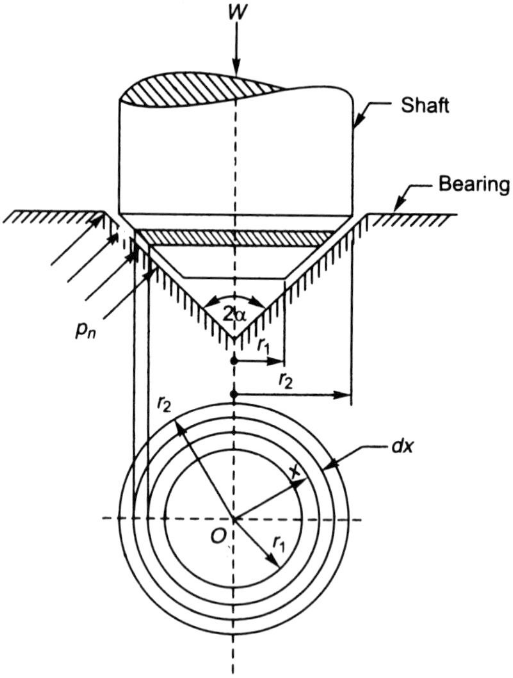

Fig. 6.11 Truncated pivot bearing.

Intensity of uniform pressure is given by:

$$p_n = \frac{W}{\pi(r_2^2 - r_1^2)}$$

(a) Uniform Pressure

Total frictional torque acting on the bearing,

$$T_f = 2\pi\mu p_n \text{ cosec } \alpha \int_{r_1}^{r_2} x^2 \, dx = 2\pi\mu p_n \text{ cosec } \alpha \frac{(r_2^3 - r_1^3)}{3}$$

$$= 2\pi\mu \left[\frac{W}{\pi(r_2^2 - r_1^2)}\right] \frac{\text{cosec } \alpha(r_2^3 - r_1^3)}{3}$$

$$T_f = \left(\frac{2}{3}\right) \cdot \mu W \cdot \text{cosec } \alpha \left(\frac{r_2^3 - r_1^3}{r_2^2 - r_1^2}\right)$$

where $= \mu W r_m \text{ cosec } \alpha$ ...(6.36)

$$r_m = \frac{2}{3} = \left[\frac{r_2^3 - r_1^3}{r_2^2 - r_1^2}\right]$$ ...(6.37)

(b) Uniform Wear

Total frictional torque acting on the bearing,

$$T_f = 2\pi\mu C \operatorname{cosec}\alpha \int_{r_1}^{r_2} x\,dx = \pi\mu C \operatorname{cosec}\alpha(r_2^2 - r_1^2)$$

$$= \pi\mu\left[\frac{W}{2\pi(r_2 - r_1)}\right]\operatorname{cosec}(r_2^2 - r_1^2) = \mu W \operatorname{cosec}\alpha\,\frac{(r_2 + r_1)}{2}$$

$$T_f = \mu W r_m \operatorname{cosec}\alpha \qquad \text{...(6.38)}$$

where $$r_m = \frac{1}{2}(r_1 + r_2) \qquad \text{...(6.39)}$$

6.8.4 Flat Collar Bearing

A single collar bearing is shown in Fig. 6.12(a) and the multiple-collar bearing in Fig. 6.12(b).

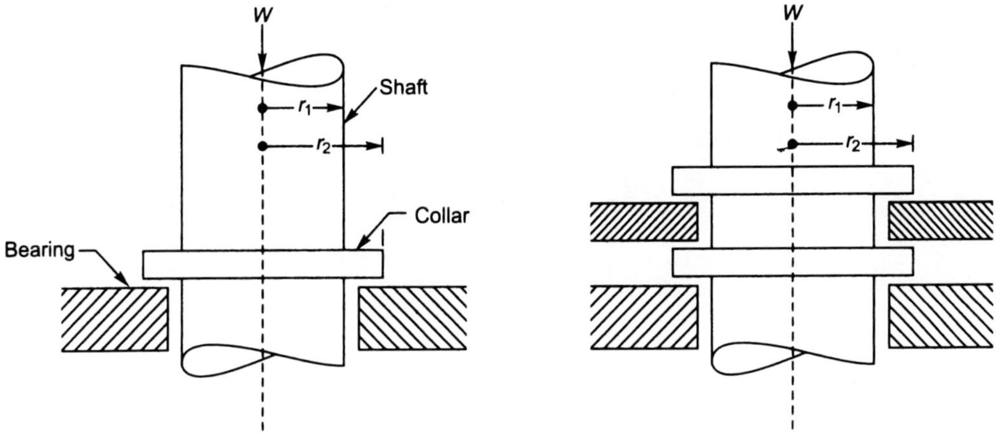

Fig. 6.12 Flat collar bearings.

Let r_1 and r_2 = inner and outer radii of the bearing, respectively.

Area of the bearing surface,

$$A = \pi(r_2^2 - r_1^2)$$

(a) Uniform Pressure

Intensity of pressure, $$p = \frac{W}{A} = \frac{W}{\pi(r_2^2 - r_1^2)}$$

Frictional torque on the ring of radius x and thickness dx,

$$dT_f = 2\pi\mu \cdot p \cdot x^2 \cdot dx$$

Total frictional torque, $$T_f = 2\pi\mu p \int_{r_1}^{r_2} x^2 \cdot dx$$

$$= \left(\frac{2}{3}\right)\cdot \mu p\,(r_2^3 - r_1^3)$$

$$= \left(\frac{2}{3}\right) \cdot \mu W \left(\frac{r_2^3 - r_1^3}{r_2^2 - r_1^3}\right) \qquad \text{...(6.40a)}$$

$$= \mu W r_m \qquad \text{...(6.40b)}$$

(b) Uniform Wear

For uniform wear, the load transmitted on the ring.

$$\delta W = p_n \cdot 2\pi x \cdot dx$$

$$= 2\pi C \cdot dx$$

Total load transmitted to the collar,

$$W = 2\pi C \int_{r_1}^{r_2} dx = 2\pi C(r_2 - r_1)$$

or

$$C = \frac{W}{2\pi(r_2 - r_1)}$$

Frictional torque on the ring,

$$dT_f = \mu \cdot \delta W \cdot r = 2\pi \cdot \mu \cdot C \cdot x \cdot dx$$

Total frictional torque on the bearing,

$$T_f = 2\pi\mu \cdot C \int_{r_1}^{r_2} x \cdot dx$$

$$= \pi\mu \cdot C(r_2^2 - r_1^2)$$

$$= \frac{\mu W(r_1 + r_2)}{2} \qquad \text{...(6.41a)}$$

$$T = \mu W r_m \qquad \text{...(6.41b)}$$

For a multi-collared bearing having n collars, multiply by n.

6.9 ROLLING FRICTION

Consider a cylinder or sphere rolling on a flat surface, as shown in Fig. 6.13(a). When there is no deformation of the surface on which rolling is taking place, then the point of contact will be a line in the case of a cylinder and a point in the case of a sphere, as shown in Fig. 6.13(a). If the surface deforms, then the shape of the surface will be as shown in Fig. 6.13(b). Let the distance between the point of contact B and the point A through which the load W passes be b and F be

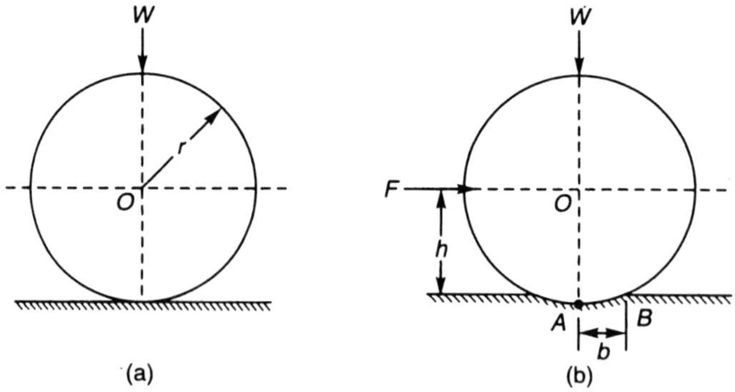

Fig. 6.13 Rolling of a cylinder or sphere.

the force required for rolling. Then rolling moment is equal to $F \cdot h$ and the resisting moment is $W \cdot b$. For the equilibrium of forces, we have

$$F \cdot h = W \cdot b$$

b is known as the coefficient of rolling friction, and has linear dimension.

Let F_r = force applied to the body for rolling

F_s = force applied to the body for sliding

Then $$F_r = \left(\frac{b}{h}\right) \cdot W$$

and $$F_s = \mu \cdot W$$

The body rolls without sliding, if $F_r < F_s$ or $\mu > b/h$. The body will slide if $F_s < F_r$ or $\mu < b/h$. The body will either roll or slide if $\mu = b/h$.

6.10 ANTI-FRICTION BEARINGS

In the case of anti-friction bearings, the point of contact between the journal and the bearing elements is either a point (as in the case of ball bearings) or a line (as in the case of roller bearings).

The ball bearing consists of a number of hardened balls mounted between two hardened races. The inner race is fitted on the shaft and the outer race is a tight fit into the bearing housing. The balls are kept at a fixed distance from one another in a brass cage. The distortion of balls is very little and the rolling friction is very low. They are mainly used to carry radial loads.

In roller, bearing, either right cylindrical or tapered rollers are present. They are used to carry heavy loads, both radial and thrust.

6.11 FRICTION CIRCLE

Consider a journal bearing, as shown in Fig. 6.14. When the journal is at stand still, then the point of contact is at A. The load W is balanced by the reaction R. When the journal starts rotating in the clockwise direction, then the point of contact shifts from A to B. The resultant of

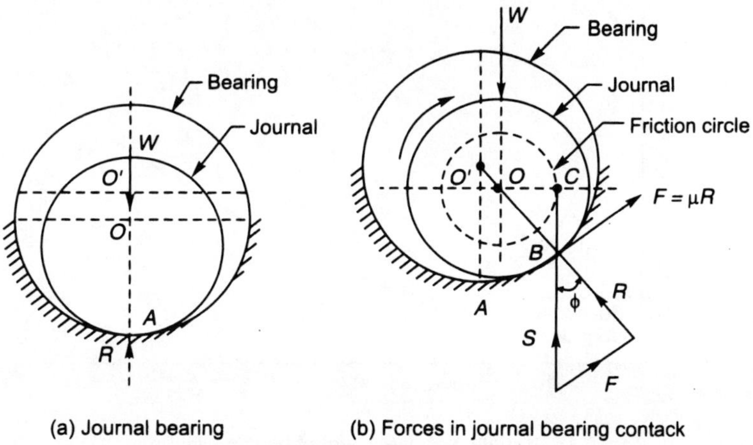

Fig. 6.14 Friction circle.

normal reaction R and force of friction $F = \mu R$ is S, as shown in Fig. 6.14(b). For the equilibrium of the journal, we have

$$W = S$$

Torque,

$$T = W \cdot OC$$
$$= W \cdot OB \sin \phi$$
$$= W \cdot r \tan \phi \quad \text{(angle } \phi \text{ being small)}$$
$$= W \cdot r\mu \qquad \text{...(6.42)}$$

A circle drawn with centre O and radius $OC = r \sin \phi \simeq r\mu$ is called the *friction circle.*

6.12 FILM FRICTION

Thin-film Lubrication

Thin-film lubrication is defined as a condition of lubrication in which the lubricant film is relatively thin and there is partial metal-to-metal contact. This made of lubrication is seen in door hinges and machine tool slides. It is also called boundary lubrication. The conditions resulting in thin-film lubrication are excessive load, insufficient surface area or oil supply, low speed, and misalignment.

Thick-film Lubrication

When the bearing and the journal are completely separated from each other by the lubricant film, then it is called thick-film lubrication. Since there is no contact between the surfaces, properties of surface, like surface finish, have little or no influence on the performance of the bearing. The resistance to relative motion between the journal and the bearing arises from the viscous resistance of the lubricant. Therefore, the performance of the bearing is affected by the viscosity of the lubricant. Thick-film lubrication could be hydrodynamic or hydrostatic.

In a journal bearing working under thick-film lubrication regime, the frictional resistance depends upon the following parameters :

1. Dynamic viscosity of the lubricant.
2. Speed of the journal.

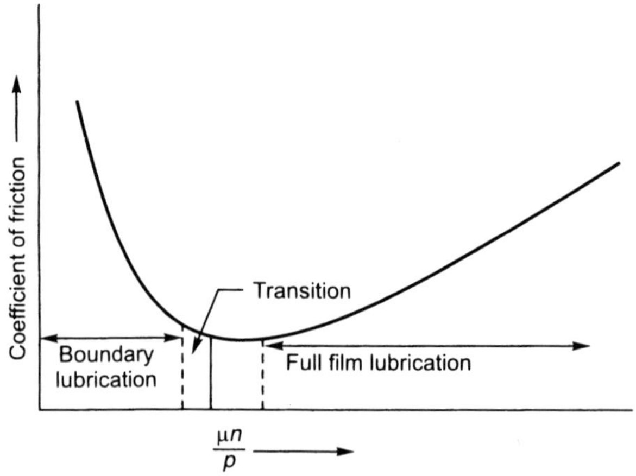

Fig. 6.15 Variation of coefficient of friction with $\mu n/p$.

3. Unit pressure of the lubricant, and
4. Radial clearance ratio.

The coefficient of friction depends upon the factor ($\mu n/p$), where μ = dynamic viscosity of lubricant, n = speed of journal, and p = unit pressure of lubricant. The value of this factor at which the coefficient of friction is minimum is called the bearing modulus, as shown in Fig. 6.15. Below this value, there exists boundary (or thin-film) lubrication and above this value there shall be full-film (or thick-film) lubrication.

■ **Example 6.1** A turnbuckle with right and left-hand single-start square threads in used to couple two railway coaches. The pitch of thread is 10 mm over a mean diameter of 30 mm. The coefficient of friction is 0.15. Find the work to be done in drawing the coaches together a distance of 300 mm against a steady load of 2.5 kN.

Solution : Given $p = 10$ mm, $d_m = 30$ mm, $W = 2.5$ kN, $\mu = 0.15$, $s = 300$ mm

$$\phi = \tan^{-1}\mu = \tan^{-1} 0.15 = 8.530°$$

$$\tan\alpha = \frac{p}{\pi d_m} = \frac{10}{\pi \times 30} = 0.10610, \alpha = 6.057°$$

$$P = W\tan(\alpha + \phi) = 2500 \tan 14.587° = 650.6 \text{ N}$$

$$\text{Torque on each rod} = \frac{Pd_m}{2} = 650.6 \times 0.015 = 9.76 \text{ N.m}$$

Total torque required on the coupling nut, $T = 2 \times 9.76 = 19.52$ N.m

For one revolution of rod, coaches are drawn nearer by a distance = $2p$ = 20 mm

$$\text{Number of turns required, } N = \frac{s}{2p} = \frac{300}{20} = 15$$

Work down by torque = $2\pi NT = 2\pi \times 15 \times 19.52 = 1839.6$ N.m

■ **Example 6.2** The spindle of a screw jack has single-start square threads with an outer diameter of 50 mm and a pitch of 10 mm. The spindle moves in a fixed nut. The load is carried on a swivel head but is not free to rotate. The bearing surface of the swivel head has a mean diameter of 60 mm. The coefficient of friction between the nut and the screw is 0.12 and that between the swivel head and the spindle is 0.10. Calculate the load that can be raised by efforts of 125 N, each applied at the end of two levers with effective length 400 mm. Also find the efficiency of the lifting system.

Solution : Given : $d_0 = 50$, $p = 10$ mm,

$$W = ?\ N, d_c = 60 \text{ mm}, l = 400 \text{ mm}, \mu = 0.12, \mu_c = 0.10, \text{and } F = 125 \text{ N}.$$

$$\phi = \tan^{-1}\mu = \tan^{-1} 0.12 = 6.843°$$

$$d_m = d_o - 0.5p = 50 - 5 = 45 \text{ mm}$$

$$\tan\alpha = \frac{p}{\pi d_m} = \frac{10}{\pi \times 45} = 0.0707, \alpha = 4.046°$$

$$P = W\tan(\alpha + \phi) = W \tan 10.889° = 0.19237\ W$$

$$2Fl = 0.5\,(Pd_m + \mu_c W d_c)$$

$$2 \times 125 \times 400 = 0.5\,(0.19237\ W \times 45 + 0.10 \times W \times 60)$$

$$100{,}000 = 7.3283\ W$$

$$W = 13645.7 \text{ N}$$

Efficiency, $$\eta = \frac{\tan\alpha}{\tan(\alpha+\phi} = \frac{\tan 4.046^\circ}{\tan 10.889^\circ} = \frac{0.0707 \times 100}{0.19237} = 36.75\%$$

■ **Example 6.3** A flat foot step bearing of 300 mm diameter supports a load of 8 kN. If the coefficient of friction is 0.10 and speed of the shaft is 80 rpm, find the power lost in friction, assuming (a) uniform pressure and (b) uniform wear.

Solution : Given : $r = 150$ mm, $W = 8$ kN, $\mu = 0.1$, $N = 80$ rpm

(a) Uniform pressure :

$$T_f = \frac{2}{3}\mu W r = \frac{2}{3} \times 0.1 \times 8000 \times 150 \times 10^{-3} = 80 \text{ N.m}$$

$$\omega = \frac{2\pi \times 80}{60} = 8.377 \text{ rad/s}$$

Power lost, $$P = T_f\ \omega = 80 \times 8.3776 = 670.2 \text{ W}$$

(b) Uniform wear :

$$T_f = 0.5\,\mu W r = 0.5 \times 0.1 \times 8000 \times 150 \times 10^{-3} = 60 \text{ N.m}$$

Power lost, $$P = T_f\ \omega = 60 \times 8.3776 = 502.6 \text{ W}$$

■ **Example 6.4** A vertical pivot bearing 200 mm diameter has a cone angle of 150°. If the shaft supports an axial load of 25 kN, and the coefficient of friction is 0.25, find the power lost in friction when the shaft rotates at 300 rpm, assuming (a) uniform pressure and (b) uniform wear.

Solution : Given : $r = 100$ mm, $W = 25$ kN, $\mu = 0.25$, $\alpha = 75^\circ$, $N = 300$ rpm

(a) Uniform pressure :

$$T_f = \frac{2}{3}\mu W r \operatorname{cosec}\alpha \left(\frac{2}{3}\right) \times 0.25 \times 25000 \times 0.1 \times \operatorname{cosec} 75^\circ = 431.365 \text{ N-m}$$

$$\omega = \frac{2\pi \times 300}{60} = 31.416 \text{ rad/s}$$

Power lost, $$P = T_f \omega = 431.365 \times 31.416 = 13.552 \text{ W}$$

(b) Uniform wear :

$$T_f = 0.5\,\mu W r \operatorname{cosec}\alpha = 0.5 \times 0.25 \times 25000 \times 0.1 \times \operatorname{cosec} 75^\circ = 323.5 \text{ N-m}$$

Power lost, $$P = T_f \omega = 323.5 \times 8.3776 = 10.164 \text{ W}$$

■ **Example 6.5** The thrust on the propeller shaft of a marine engine is taken by eight collars whose outer and inner diameters are 650 and 500 mm, respectively. The thrust pressure is 0.5 MPa and may be assumed uniform. The coefficient of friction between the shaft and collars is 0.05. If the shaft rotates at 120 rpm, find (a) total thrust on the collars and (b) power absorbed by friction at the bearing.

Solution : Given : $n = 8$, $d_o = 650$ mm, $d_i = 400$ mm, $p = 0.5$ MPa, $\mu = 0.04$, $N = 120$ rpm

(a) $$p = \frac{W}{A} = \frac{4W}{\pi(d_o^2 - d_i^2)}$$

$$W = \frac{0.5 \times \pi(650^2 - 400^2}{4} = 1{,}03{,}083 \text{ N}$$

(b) Uniform Pressure :

$$T_f = \frac{2}{3} n\mu W \left(\frac{r_0^3 - r_i^2}{r_0^2 - r_i^2} \right)$$

$$= \left(\frac{2}{3} \times 8 \times 0.04 \times 103083 \right) \left(\frac{325^3 - 200^3}{325^2 - 200^2} \right) \times 10^{-3}$$

$$= 8822.6 \text{ N-m}$$

$$\omega = \frac{2\pi \times 120}{60} = 12.566 \text{ rad/s}$$

$$P = T_f \omega = 8822.6 \times 12.566 \times 10^{-3} = 110.87 \text{ kW}$$

■ **Example 6.6** Two tie rods are connected by a turnbuckle having right and left-hand metric threads of V-type. The pitch of the threads is 5 mm on a mean diameter of 30 mm and a thread angle of 60°. Assuming coefficient of friction of 0.12, find the torque required to produce a pull of 40 kN.

Solution : Given : $p = 5\text{ mm}$, $d_m = 30\text{ mm}$, $2\beta = 60°$, $\mu = 0.12$, $W + 40\text{ kN}$

$$\mu_e = \frac{\mu}{\cos\beta} = \frac{0.12}{\cos 30°} = 0.13865$$

$$\phi = \tan^{-1} \mu_e = 7.89°$$

$$\alpha = \tan^{-1}\left(\frac{p}{\pi d_m} \right) = \frac{5}{\pi \times 30} = 3.04°$$

(a) When rods are tightened,

$$P = W \tan(\alpha + \phi)$$

$$= 40 \times 10^3 \times \tan(3.04 + 7.89)°$$

$$= 7724.5 \text{ N}$$

Torque,

$$T = \frac{P\, d_m}{2}$$

$$= 7724.5 \times 15 \times 10^{-3} = 115.86 \text{ Nm}$$

(b) When the rods are slackened

$$P = W \tan(\phi - \alpha)$$

$$= 40 \times 10^3 \times \tan(7.89 - 3.04°)$$

$$= 3394 \text{ N}$$

$$T = 3394 \times 15 \times 10^{-3}$$

$$= 50.91 \text{ N-m}$$

■ **Example 6.7** A conical pivot supports a load of 25 kN, the cone angle being 120°, and the intensity of normal pressure does not exceed 0.25 MPa. The external radius is twice the internal radius. Find the outer and inner radii of the bearing surface. If the shaft rotates at 180 rpm and the coefficient of friction is 0.15, find the power lost in friction, assuming uniform pressure.

Solution : Given : $W = 25$ kN, $2\alpha = 120°$, $p_n = 0.25$ MPa, $r_2 = 2r_1$, $N = 180$ rpm, $\mu = 0.15$

Intensity of normal pressure, $p_n = \dfrac{W}{\pi(r_2^2 - r_1^2)}$

$$0.25 = \frac{25 \times 10^3}{\pi[(2r_1)^2 - r_1^2]}$$

$$r_1 = 103 \text{ mm}$$

$$\alpha r_2 = 206 \text{ mm}$$

Frictional torque, $T_2 = \left(\dfrac{2}{3}\right) \cdot \mu W \text{ cosec } \alpha \left[\dfrac{r_2^3 - r_1^3}{r_2^2 - r_1^2}\right]$

$$= \left(\frac{2}{3}\right) \times 0.15 \times 25 \times 10^3 \times \text{cosec } 60° \left[\frac{206^3 - 103^3}{206^2 - 103^2}\right]$$

$$= 693.8 \text{ Nm}$$

Power lost in friction

$$\frac{T_f \omega}{1000} = 693.8 \times \frac{2\pi\left(\frac{180}{60}\right)}{1000} = 13.08 \text{ kW}$$

■ **Example 6.8** The mean diameter of a screw jack having pitch of 10 mm is 50 mm. A load of 25 kN is lifted through a distance of 180 mm. Find the work done in lifting the load and efficiency of the screw when

(a) The load rotates with the screw.

(b) The load rests on the loose head that does not rotate with the screw. The outer and inner diameters of the bearing surface of the loose head are 60 mm and 10 mm, respectively. The coefficient of friction for the screw as well as bearing surface may be taken as 0.10.

Solution : Given : $p = 10$ mm, $d_m = 50$ mm, $W = 25$ kN, $h = 180$ mm, $\mu = \mu_c = 0.10$, $R_2 = 30$ mm, $R_1 = 5$ mm

Helix angle of screw, $\alpha = \tan^{-1}\left(\dfrac{p}{\pi d_m}\right) = \tan^{-1}\left(\dfrac{10}{\pi \times 50}\right) = 3.6426°$

Angel of friction, $\phi = \tan^{-1} \mu = \tan^{-1} 0.10 = 5.7106°$

Force required at the circumference of screw to lift the load,

$$P = W \tan(\alpha + \phi) = 25 \tan(3.6426° + 5.7106°) = 4.118 \text{ kN}$$

Frictional torque of screw,

$$T_f = P \times \frac{d_m}{2} = 4.118 \times \frac{50}{2} = 102.94 \text{ N-m}$$

(a) Number of rotations made by screw to lift the load through 180 mm,

$$N = \frac{h}{p} = \frac{180}{10} = 18$$

Work done in lifting load

$$= 2\pi N T_f = 2\pi \times 18 \times 102.94 = 11642.6 \text{ N-m}$$

Efficiency,

$$\eta = \frac{\tan\alpha}{\tan(\alpha + \phi)} = \frac{\tan 3.6426°}{\tan(3.6426 + 5.7106°)} = 0.3865 \text{ or } 38.65\%$$

(b) Mean radius of bearing surface,

$$R_m = \frac{1}{2}(R_1 + R_2) = \frac{1}{2}(5 + 30) = 17.5 \text{ mm}$$

Total frictional torque required at the screw and collar,

$$T_f = P \times \frac{d_m}{2} + \mu_C W R_m$$

$$= 4.118 \times \frac{50}{2} + 0.01 \times 25 \times 17.5 = 146.7 \text{ N-m}$$

Work done in lifting the load $= 2\pi N T_f = 2\pi \times 18 \times 146.7 = 16591.4$ N-m

Torque required to lift the load without friction,

$$T_0 = W \tan\alpha \times \frac{d_m}{2}$$

$$= 26 \tan 3.6426° \times \frac{50}{2} = 39.79 \text{ N-m}$$

Efficiency of screw jack, $\eta = \dfrac{T_0}{T_f} = \dfrac{39.79}{146.7} = 0.2712$ or 27.12%

6.13 CLUTCHES

The friction clutches may be classified follows :

1. **Plate (or disc) clutches**
   (a) Single-plate clutches
   (b) Multiple-plate clutches
2. **Cone clutch**

6.13.1 Single-plate Clutch

Consider a single-plate clutch as shown in Fig. 6.16

Let r_1, r_2 = inner and outer radii of the plate, respectively

p = intensity of axial pressure on the plate

W = axial load on the clutch

μ = coefficient of friction between the friction surfaces and the plate

T = torque transmitted by the clutch

Consider an elementary ring of the friction surface of radius r and thickness dr.

Contact area of the friction surface, $dA = 2\pi r \cdot dr$

Arial force on the ring, $dW = p \cdot dA$

Frictional force on the ring, $dF = \mu \cdot dW$

Frictional torque on the ring, $dT_f = dF \cdot r = 2\pi\mu \cdot p \cdot r^2 \cdot dr$

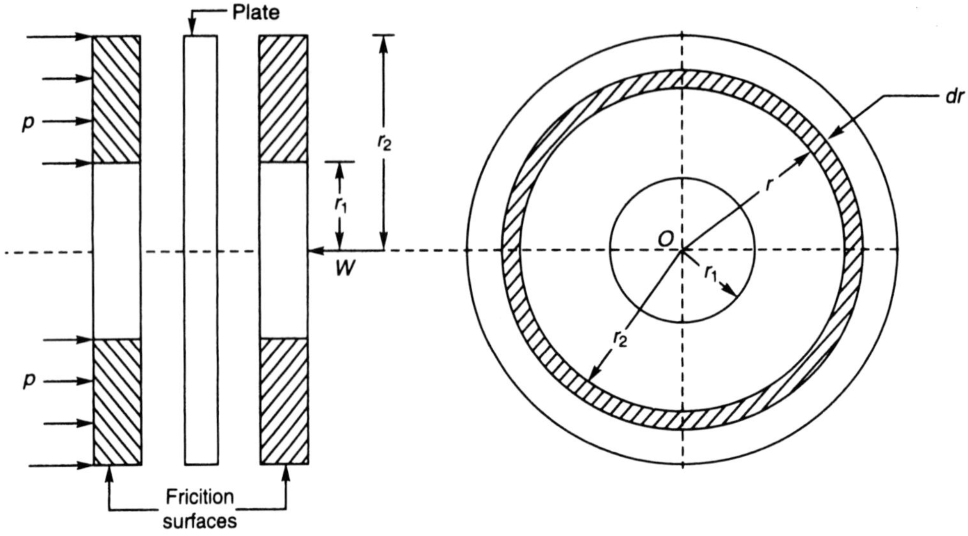

Fig. 6.16 Single-plate clutch.

(a) Uniform Pressure

When the pressure is uniform over the entire area of friction surface, then the intensity of pressure,

$$P = \frac{W}{\pi(r_2^2 - r_1^2)} \quad \text{...(16.43)}$$

Total frictional torque on the frictional surface,

$$T_f = 2\pi\mu\pi \int_{r_1}^{r_2} r^2\, dr = \frac{2}{3}\pi\mu p(r_2^3 - r_1^3)$$

$$= \left(\frac{2}{3}\right)\cdot \mu W \left[\frac{r_2^3 - r_1^3}{r_2^2 - r_1^2}\right] \quad \text{...(6.44)}$$

Mean radius of friction surface,

$$r_m = \left(\frac{2}{3}\right)\cdot\left[\frac{r_2^3 - r_1^3}{r_2^2 - r_1^2}\right] \quad \text{...(6.45)}$$

(b) Uniform Wear

For uniform wear, the intensity of pressure varies inversely with the distance, therefore

$$p \cdot r = C$$

or

$$p = \frac{C}{r}$$

At $r = r_1$, $p = p_{max}$ and $r = r_2$, $p = p_{min}$

Force acting on the ring,

$$dW = \frac{C}{r}\cdot r\pi r \cdot dr = 2\pi C \cdot dr$$

Total force acting on the friction surface,

$$W = \int_{r_1}^{r_2} 2\pi C \cdot dr = 2\pi C(r_2 - r_1)$$

or

$$C = \frac{W}{2\pi(r_2 - r_1)}$$

$$W = p_{\max} \times r_1 \times 2\pi(r_2 - r_1) \qquad \text{...(6.46)}$$

Frictional torque on the ring,

$$dT_f = 2\pi\mu \cdot C \cdot r \cdot dr.$$

Total frictional torque on the friction surface,

$$T_f = 2\pi\mu \cdot C \int_{r_1}^{r_2} r \cdot dr = \pi\mu \cdot C(r_2^2 - r_1^2)$$

$$= \mu W \frac{(r_1 + r_2)}{2} \qquad \text{...6.47)}$$

where $r_m = \dfrac{r_1 + r_2}{2}$

6.13.2 Multi-plate Clutch

A multi-plate friction clutch is shown in Fig. 6.17. Multi-plate friction clutches are used where space is a limitation, as in the case of two wheelers, scooters, etc.

Let n_1 = number of plates on the driving shaft

n_2 = number of plates on the driven shaft

Then, number of pairs of contact surfaces, $n = n_1 + n_2 - 1$ ...(6.48)

Total frictional torque transmitted, $T = n\mu W r_m$ ...(6.49)

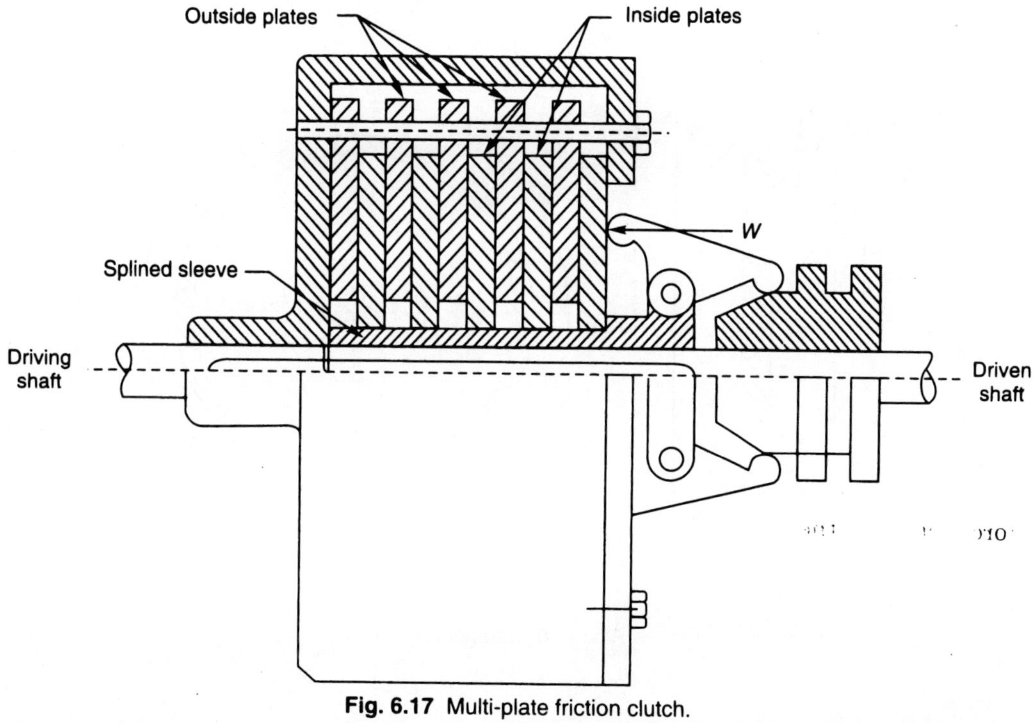

Fig. 6.17 Multi-plate friction clutch.

Dry and Wet Clutches

In case of single-plate clutches, the contacting surfaces are either one or two. Due to large surface area available, heating is not a problem and the clutch is of the dry type. In case of multi-plate clutches, the work done during engagement and disengagement is converted into heat. Heat dissipation is a serious problem in multi-clutches because of the large number of friction surfaces. Cooling oil is used to dissipate this heat. Therefore, these clutches are wet clutches.

6.13.3 Cone Clutch

Consider a cone clutch as shown in Fig. 6.18(a).

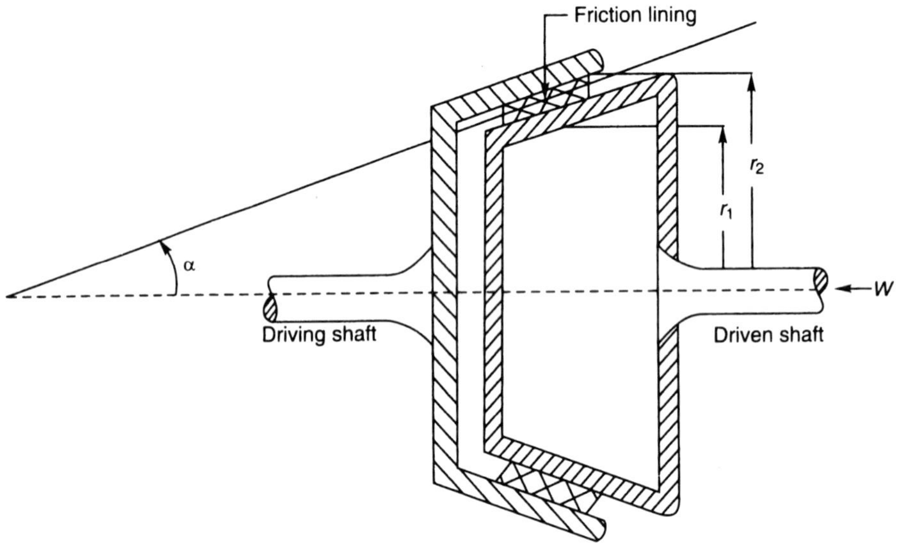

(a) Clutch geometry

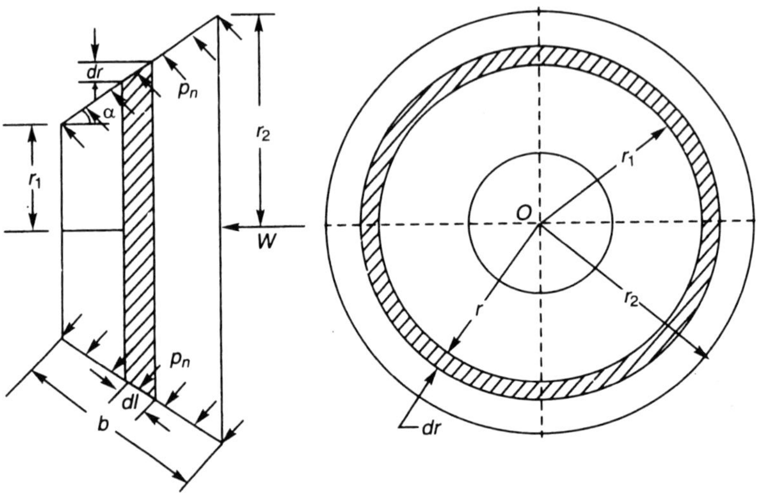

(b) Forces on cone

Fig. 6.18 Cone clutch.

Let r_1 and r_2 = inner and outer radii of the frictional conical surface, respectively

p_b = normal pressure between the contact surfaces

b = width of the conical surfaces

μ = coefficients of friction between contact surfaces

α = semi-cone angle.

Consider an elementary ring of the conical surface of radii r and $r + dr$ and of length dl, as shown in Fig. 6.18(b).

$$dl = dr \cdot \text{cosec } \alpha$$

Area of this ring, $dA = 2\pi r \cdot dl = 2\pi r \cdot dr \cdot \text{cosec } \alpha$

(a) Uniform Pressure

Normal load acting on the ring, $dW_n = p_n \cdot dA$

Axial load acting on the ring,

$$dW = dW_n \cdot \sin \alpha = 2\pi p_n \cdot r \cdot dr$$

Total axial load transmitted to the clutch.

$$W = 2\pi p_n \int_{r_1}^{r_2} r \cdot dr = \pi p_n (r_2^2 - r_1^2)$$

or

$$p_n = \frac{W}{\pi(r_2^2 - r_1^2)}$$

Frictional force on the ring, $dF = \mu \cdot dW_n$

Frictional torque, $dT_f = dF \cdot r = 2\pi\mu \cdot p_n \cdot \text{cosec } \alpha \cdot r^2\, dr$

Total frictional torque, $T_f = 2\pi\mu \cdot p_n \cdot \text{cosec } \alpha \int_{r_1}^{r_2} r^2\, dr$

$$= \left(\frac{2}{3}\right) \cdot \pi\mu \cdot p_n \text{ cosec } (r_2^3 - r_1^3)$$

$$= \left(\frac{2}{3}\right) \cdot \mu W \cdot \text{cosec } \alpha \left[\frac{r_2^3 - r_1^2}{r_2^2 - r_1^2}\right] \quad \text{...(6.50a)}$$

$$= \mu W \cdot r_m \text{ cosec } \alpha \quad \text{...(6.50b)}$$

(b) Uniform Wear

For uniform wear,

$$p_n \cdot r = C$$

or

$$p_n = \frac{C}{r}$$

Normal load on the ring, $dW_n = p_n \cdot dA$

Axial load on the ring,

$$dW = dW_n \cdot \sin \alpha = p_n \cdot 2\pi r \cdot dr = 2\pi C \cdot dr$$

Total axial load on the clutch,

$$W = 2\pi C \int_{r_1}^{r_2} dr = 2\pi C(r_2 - r_1)$$

or

$$C = \frac{W}{2\pi(r_2 - r_1)}$$

Frictional force on the ring, $dF = \mu \cdot dW_n$

Frictional torque on the ring,

$$dT_f = dF \cdot r = 2\pi\mu p_n \cdot \text{cosec}\ \alpha \cdot r_2 \cdot dr$$
$$= 2\pi\mu C \cdot \text{cosec}\ \alpha \cdot r \cdot dr$$

Frictional torque on the clutch,

$$T_f = 2\pi\mu C \cdot \text{cosec}\ \alpha \int_{r_1}^{r_2} r\ dr$$

$$= \pi\ \mu\ C\ .\ \text{cosec}\ \alpha \frac{(r_1 + r_2)}{2} \quad ...(6.51)$$

$$= \mu W r_m\ \text{cosec}\ \alpha \quad ...(6.52)$$

where $r_m = \dfrac{r_1 + r_2}{2}$ is the mean radius.

The frictional torque can also be written as,

$$T_f = 2\pi\mu \cdot p_n \cdot b \cdot r^2 m \quad ...(6.52)$$

Axial force required to engage the clutch,

$$W_e = W + \mu W_n\ \cos\alpha \quad ...(6.53)$$
$$= W_n\ \sin\alpha + \mu W_n\ \cos\alpha$$
$$= W_n (\sin\alpha + \mu\ \cos\alpha)$$

■ **Example 6.9** A single-plate clutch, with both sides effective, has inner and outer diameters of friction surface 250 mm and 350 mm, respectively. The maximum intensity of pressure is not to exceed 0.15 MPa. The coefficient of friction is 0.3. Determine the power transmitted by the clutch at a speed of 2400 rpm for (a) uniform wear and (b) uniform pressure.

Solution : (a) For uniform wear, $T_f = n \cdot \mu W \cdot r_m$

$$r_m = \frac{250 + 350}{4} = 150 \text{ mm} = 0.15 \text{ m}$$

$$n = 2$$
$$C = pr_1 = 0.15 \times 125 = 18.75 \text{ N/mm}$$
$$W = 2\pi C(r_2 - r_1) = 2\pi \times 18.75\ (175 - 125) = 5890.48 \text{ N}$$
$$T_f = 2 \times 0.3 \times 5890.48 \times 0.15 = 530.14 \text{ Nm}$$

Power transmitted,

$$P = \frac{T_f \omega}{1000} \text{ kW} = 530.14 \frac{\left(2\pi \times \dfrac{2400}{60}\right)}{1000}$$
$$= 133.24 \text{ kW}$$

(b) For uniform pressure,

$$r_m = \left(\frac{2}{3}\right) \cdot \left[\frac{r_2^3 - r_1^3}{r_2^2 - r_1^2}\right] = \left(\frac{2}{3}\right) \cdot \left[\frac{175^3 - 125^3}{175^2 - 125^2}\right]$$
$$= 151.4 \text{ mm}$$
$$W = \pi(r_2^2 - r_1^2)p$$

$$= \pi (175^2 - 125^2) \times 0.15 = 7068.58 \text{ N}$$

$$T_f = n\mu W r_m$$
$$= 2 \times 0.3 \times 7068.58 \times 151.4 \times 10^{-3}$$
$$= 642.1 \text{ Nm}$$

Power transmitted, $$P = 642.1 \times \frac{\left(2\pi \times \dfrac{2400}{60}\right)}{1000}$$
$$= 161.38 \text{ kW}$$

■ **Example 6.10** A plate clutch has three discs on the driving shaft and two discs on the driven shaft, providing four pairs of contact surfaces. The inside and outside diameters of the friction surfaces are 125 mm and 250 mm, respectively. Assuming uniform pressure and coefficient of friction equal to 0.3, find the total spring load pressing the plates together to transmit 30 kW at 1500 rpm.

Solution : Power transmitted, $$P = \frac{T_f \omega}{1000} \text{ kW}$$

$$30 = T_f \times \frac{\left(2\pi \times \dfrac{1500}{60}\right)}{1000}$$

$$T_f = \frac{(30 \times 1000 \times 60)}{(2\pi \times 1500)} = 190.986 \text{ Nm}$$

$$r_m = \left(\frac{2}{3}\right) \cdot \left[\frac{r_2^3 - r_1^3}{r_2^2 - r_1^2}\right]$$

$$= \left(\frac{2}{2}\right) \cdot \left[\frac{125^3 - 62.5^3}{125^2 - 62.5^2}\right] = 97.2 \text{ mm}$$

$$T_f = n\mu W r_m$$
$$190.986 = 4 \times 0.3 \times W \times 0.0972$$
$$W = 1637.4 \text{ N}$$

■ **Example 6.11** An engine developing 50 kW at 1200 rpm is fitted with a cone clutch. The cone is 12° and a maximum mean diameter of 500 mm. The coefficient of friction is 0.25. The normal pressure on the clutch face is not to exceed 0.1 MPa. Determine (a) the axial spring force to engage the clutch, and (b) the face width required.

Solution :

(a) $$P = \frac{T_f \omega}{1000} \text{ kW}$$

$$50 = \frac{T_f \left(2\pi \times \dfrac{1200}{60}\right)}{1000}$$

$$T_f = 397.9 \text{ Nm}$$
$$= \mu W_n r_m = 0.25 \times W_n \times 0.25$$
$$W_n = 6366.4 \text{ N}$$

Axial spring force required to engage the clutch,

$$W_e = W_n (\sin \alpha + \mu \cos \alpha)$$
$$= 6366.4 (\sin 12° + 0.25 \cos 12°)$$
$$= 2880.5 \text{ N}$$

(b) $W_n = p_n \cdot 2\pi r_m b$

$$\text{Face width, } b = \frac{2880.5}{0.1 \times 2\pi \times 250} = 18.3 \text{ mm}$$

SOLVED QUESTIONS FROM UNIVERSITY EXAMINATIONS

■ **Example 6.12** Neglecting collar friction, prove that maximum efficiency of square threaded screw moving in a nut is (1 – sin ϕ)/(1 + sin ϕ) where ϕ is friction angle. [KU, Dec., 2006]

Solution : Refer to art 6.7

■ **Example 6.13** Derive an expression for friction moment of a conical clutch, assuming uniform wear condition. [KU, Dec., 2006, 2007]

Solution : Refer to art 6.13.3

■ **Example 6.14** What is a clutch? Make a sketch of single-plate clutch and describe its working. [KU, Dec., 2008]

Solution : Refer art 6.13.1.

■ **Example 6.15** What do you understand by efficiency of the inclined plane? [KU, Dec., 2009]

Solution : Refer to art 6.5 (d).

■ **Example 6.16** In a cone clutch with semi-cone angle of 15°, maximum and minimum radii of the contact surface are 120 mm and 80 mm respectively. The speed is 800 rpm and the maximum allowable normal pressure is 150 kN/m^2. Determine the axial load and the power transmitted taking the coefficient of friction as 0.3. [KU, Dec., 2008]

Solution : Given : $\alpha = 15°$, $r_1 = 80$ mm, $r_2 = 120$ mm, $N = 800$ rpm, $p = 150$ kN/m^2, $\mu = 0.3$.

Assuming uniform wear, $p_n \; \alpha \; \frac{1}{r}$

Axial load, $\omega = 2\pi r_1 (r_2 - r_1) p_n$

$$= 2\pi \times 80 (120 - 80) \times 0.15 = 3015.93 \text{ N}$$

Frictional torque, $T_f = \mu W \frac{(r_1 + r_2)}{2} \text{ cosec } \alpha$

$$= 0.3 \times 3015.93 \left(\frac{80 + 1204}{2} \right) \times 10^{-3} \times \text{cosec } 15°$$

$$= 349.58 \text{ N-m}$$

Power transmitted,

$$P = \frac{2\pi N T_f}{60 \times 10^3} = \frac{2\pi \times 800 \times 349.58}{60 \times 10^3} = 29.286 \text{ kW}$$

■ **Example 6.17** The mean diameter of a bolt having V threads is 25 mm. The pitch of the thread is 5 mm and angle of the thread is 55°. The bolt is tightened by screwing a nut whose mean radius of bearing surface is 25 mm. The coefficient of friction for nut and bolt is 0.10 and for nut and bearing is 0.15. Find the force required at the end of lever 0.5 m long when the load on the bolt is 15 kN. [KU, Dec., 2009]

Solution : Given : $d_m = 25$ mm, $p = 5$ mm, $2\beta = 55°$, $R_m = 25$ mm, $\mu = 0.10$, $\mu_c = 0.15$,

$$l = 0.5 \text{ m}, W = 15 \text{ kN}$$

$$\alpha = \tan^{-1}\left(\frac{p}{\pi d_m}\right) = \tan^{-1}\left(\frac{5}{25\pi}\right) = 3.64°$$

$$\phi = \tan^{-1}\left(\frac{\mu}{\cos\beta}\right) = \left(\frac{0.1}{\cos 27.5°}\right) = 6.43°$$

$$P = W\tan(\alpha + \phi)$$

$$= 15 \times 10^3 \times \tan(3.64 + 6.43)° = 266.38 \text{ N}$$

$$T = \frac{Pd_m}{2} + \mu_c W R_m$$

$$= 266.3 \times 12.5 + 0.15 \times 15000 \times 25$$

$$= 33297.5 + 56250 = 89547.5 \text{ N}$$

$$Fl = T$$

$$F = \frac{89547.5}{500} = 179.1 \text{ N}$$

■ **Example 6.18** A friction clutch is required to transmit 30 kN at 1800 rpm. The clutch is to be of a single plate disc type with both sides effective. The maximum pressure that the clutch surfaces can with stand is limited to 0.08 MPa. The outer diameter of clutch plate is 275 mm. Find the required inner diameter of the plate. Also, find the total force exerted by the spings for engaging the clutch. Assume that uniform were conditions prevail and the coefficient of friction is 0.25. [MDU, Dec., 2008]

Solution : Given : $P = 30$ kW, $N = 1800$ rpm, $i = 2$, $p = 0.08$ MPa, $d_1 = 275$ mm, $\mu = 0.25$

$$P = \frac{2\pi N T_f}{60 \times 10^3}$$

$$30 = \frac{2\pi \times 1800 \times T_f}{60 \times 10^3}$$

$$T_f = 159.155 \text{ N-m}$$

$$F_a = \pi d_2 \times \frac{1}{2}(d_1 - d_2)\, p$$

$$= \pi d_2 \times \frac{1}{2}(275 - d_2) \times 0.08 = 0.1257 d_2 (275 - d_2) \text{ N}$$

$$T_f = i\,\mu F_a r_m$$

$$159.155 \times 10^3 = 2 \times 0.25 \times 0.1257 d_2 (275 - d_2) \times \frac{1}{2}(275 + d_2)$$

$$= 0.0314 d_2 [(275)^2 - d_2^2]$$

$$d_2(75625 - d_2^2) = 506830$$

$$d_2^3 - 75625d_2 + 506830 = 0$$

The value of d_2 can be estimated by hit and trial.

Let $$f(d_2) = d_2 - 75625d_2 + 506830$$

d_2, mm	200	225	230	240	235	232	231
$f(d_2)$	– 2056370	– 556370	– 158120	742630	274630	10798	– 74354

We take $$d_2 = 232 \text{ mm}$$

$$F_a = 0.1257 \times 232\,(275 - 232) = 1254 \text{ N}$$

SUMMARY

1. Friction is the property of the two bodies moving or tending to move on one another by virtue of which a force is developed between them. Force of friction is the force that opposes the motion of one body over another.
2. Friction is a blessing in some cases but a necessary evil in other cases.
3. **Static friction :** It is the friction experienced by two bodies in contact, while at rest.

   Dynamic (or kinetic) friction : It is the friction experienced by two bodies in contact, while in motion. It is less than the static friction.

   Sliding friction : It is due to sliding of two bodies on each other.

   Rolling friction : It is due to rolling of two bodies on each other.

   Pivot friction : It is the friction experienced by two bodies due to rotation of one body around other, as in the case of foot step bearing.

   Dry (or solid) friction : It is due to two dry and unlubricated surfaces in contact.

   Boundary (or skin or greasy) friction : It is the friction experienced by two bodies in contact when separated by a thin film of a lubricant.
4. Laws of static friction
   (i) The force of friction always acts in a direction opposite to the impending motion.
   (ii) The limiting force of friction is directly proportional to the normal reaction.
   (iii) The force of friction is independent of the area contact between the surfaces.
   (iv) The force of friction depends upon the roughness of the surfaces of two materials.
5. The coefficient of friction (μ) is the ratio of the force of friction (F) to the normal reaction (R), that is, $\mu = F/R$. Angle of friction is the angle that the resultant of F and R makes with R. In the case of limiting friction, the coefficient of friction is equal to the tangent of the angle of friction, that is, $\mu = \tan \phi$. In the case of limiting friction, the angle of the inclined plane is equal to the angle of friction. The angle of the plane when motion of an object on the plane is impending is called the angle of repose. Cone of friction is the cone generated by the resultant of F and R when it is rotated about the vertical axis.
6. Forces on sliding bodies :
   (a) Body resting on the horizontal plane subjected to inclined force :

   $$P = W \sin \phi / \cos (\theta - \phi) \qquad \text{for pull type force}$$
   $$= W \sin \phi / \cos (\theta + \phi) \qquad \text{for push type force}$$

   Note that pull type force is smaller than push type force.

(b) Body resting on an inclined plane with force inclined to the plane :

$$P = W \sin(\alpha + \phi)/\cos(\theta - \phi) \quad \text{for body going up the plane}$$

$$= W \sin(\phi - \alpha)/\cos(\theta + \phi) \quad \text{for body going down the plane}$$

(c) Body resting on an inclined plane with horizontal force :

$$P = W \tan(\alpha + \phi) \quad \text{for body going up the plane}$$

$$P = W \tan(\phi - \phi) \quad \text{for body going down the plane.}$$

7. Efficiency of the inclined plane :

(i) For body going up the plane with inclined force :

$$\eta = (1 + \mu \tan\theta)/(1 + \mu \cot\alpha)$$

(ii) For body going down the plane with inclined force :

$$\eta = (1 - \mu \tan\theta)/(1 - \mu \cot\theta)$$

(iii) For body going up the plane with horizontal force :

$$\eta = \tan\alpha / \tan(\alpha + \phi)$$

(iv) For body going down the plane with horizontal force :

$$\eta = \tan(\alpha - \phi)/\tan\alpha$$

8. Screw threads :

(a) Square threads :

$$P = W \tan(\alpha + \phi)$$

where $\tan F_\alpha = p/(\pi d_m)$

(b) V-threads :

Frictional force, $F = (\mu/\cos\beta)W$

9. Screw jack :

A screw jack is a device for lifting of load.

(i) Horizontal effort required :

$$= W \tan(\alpha + \phi) \text{ for raising a load}$$

$$= W \tan(\phi - \alpha) \text{ for lowering a load}$$

Torque required to overcome friction between the screw and the nut :

$$T_1 = W d_m \tan(\alpha + \phi)/2$$

Torque required to overcome friction at the collar :

$$T_2 = \mu_c W d_c / 2$$

where μ_c = collar coefficient of friction and

d_c = mean diameter of the collar

Total torque, $T = T_1 + T_2 = (Pd_m + \mu_c W d_e)/2$

F = effort applied at the end of a lever of length $l = T/l$

(ii) Efficiency :

$$\eta = \tan\alpha/\tan(\alpha + \phi), \text{ for raising of load}$$

$$= \tan(\phi - \alpha)/\tan\alpha, \text{ for lowering of load}$$

For efficiency to be maximum, $\alpha = 45° - \phi/2$

$$\eta_{max} = (1 - \sin\phi)/(1 + \sin\phi)$$

(iii) Self locking and over-hauling screw :

If $\phi > \alpha$, the the screw is called self-locking. If $\phi < \alpha$, then the screw will lower of its own. Such a screw is called over-hauling. For a self-locking screw, $\eta \le 50\%$.

10. Friction in bearings :

(i) Flat pivot bearing :

(a) Uniform pressure :

Total frictional torque,

$$T_f = 2\mu W\ R/3 = 2\pi\mu\ p(R_2 - R_1)/3$$
$$p = W/[\pi(R_2^2 - R_1^2)]$$

Power lost in friction, $P = T_f\,\omega$

(b) Uniform wear

Total frictional torque on the bearing,

$$T_f = \frac{1}{2}\mu W(R_2 + R_1) = \mu W R_m$$

(ii) Conical pivot bearing :

(a) Uniform wear

Total frictional torque, $T_f = (2/3)\mu WR \text{ cosec } \alpha$

(b) Uniform pressure

Total frictional torque on the bearing,

$$T_f = 1/2 \times \mu WR \text{ cosec } \alpha$$

(iii) Truncated conical pivot bearing :

(a) Uniform pressure :

Total frictional torque,

$$T_f = (2/3)\mu W \text{ cosec } \alpha\ [(r_2^3 - r_1^3)/(r_2^2 - r_1^2)]$$

(b) Uniform wear :

Total frictional torque, $T_f = \mu W\ r_m \text{ cosec } \alpha$

(iv) Flat collar bearing :

(a) Uniform pressure :

Intensity of pressure, $p = W/[\pi\,(r_2^2 - r_1^2)]$

Frictional torque, $T_f = (2/3)\mu W[(r_2^3 - r_1^3)/(r_2^2 - r_1^2)]$

(b) Uniform wear :

Total frictional torque on the bearing, $T_f = \mu W r_m$

11. Rolling friction :

Coefficient of rolling friction, $b = Fh/W$

Let F_r = force applied to the body for rolling = bW/h

F_s = force applied to the body for sliding = μW

The body rolls without sliding, if $F_r < F_s$ or $\mu > b/h$. The body will slide if $F_s < f_r$ or $\mu < b/h$. The body will either roll or slide if $\mu = b/h$.

12. Anti-friction bearings :

In the case of anti-friction bearings, the point of contact between the journal and the bearing elements is either a point (as in the case of ball bearings) or a line (as in the case of roller bearings).

13. Friction circle :

A circle drawn with centre O and radius $OC = r \sin \phi = r\mu$ is called the friction circle. r is the radius of the journal.

14. A clutch is a device used to transmit the rotary motion of one shaft to another when desired.

15. Friction clutches :

(i) Single-plate clutch :

(a) Uniform pressure

Intensity of pressure, $p = W/\pi(r_2^2 - r_1^2)$

Total frictional torque, $T_f = (2/3) \cdot \mu W[(r_2^3 - r_1^3)/(r_2^2 - r_1^2)]$

Mean radius of friction surface, $r_m = (2/3) \cdot [(r_2^3 - r_1^3)/(r_2^2 - r_1^2)]$

(b) Uniform wear

Total frictional torque, $T_f = \mu W(r_1 + r_2)/2$

$r_m = (r_1 + r_2)/2$

(ii) Multi-plate clutch :

n = number of plates on the driving shaft

n_2 = number of plates on the driven shaft

Number of pairs of contact surfaces, $n = n_1 + n_2 - 1$

Total frictional torque transmitted, $T = n\mu W r_m$

(iii) Cone clutch :

(a) Uniform pressure

Total frictional torque, $T_f = \mu W r_m \text{ cosec } \alpha$

(b) Uniform wear

Total frictional torque, $T_f = \mu W r_m, \text{ cosec } \alpha$

where $r_m = (r_1 + r_2)/2$

Also $T_f = 2\pi \cdot \mu p_n \cdot b r_m^2$

Axial force required to engage the clutch,

$$W_e = W_n (\sin \alpha + \mu \cos \alpha)$$

Short Answer Type Questions

1. What is friction? Is it necessary evil or blessing?
2. Describe various types of friction.
3. State the laws of friction.
4. Define coefficient of friction, angle of friction, and angle of repose.
5. When a screw jack is self-locking and self-hauling?
6. What are the limits on the efficiency of a screw jack to be of the self-locking type?
7. Explain uniform pressure and uniform wear theories for a clutch.
8. What are thin-film and thick-film lubrications?
9. What are anti-friction bearings?
10. Explain friction circle.

Exercises

1. An effort of 1 kN is required to move a certain body up an inclined plane of angle 15° when the force is acting parallel to the plane. If the angle of plane is increased to 20°, then the effort required is 1.25 kN. Find the weight of the body and coefficient of friciton.
2. A shaft has a number of collars integral with it. The outer diameter of the collars is 450 mm and the shaft diameter is 300 mm. If the uniform intensity of pressure is 0.3 MPa and the coefficient of friction is 0.05, determine (a) power absorbed in overcoming friction when the shaft is running at 120 rpm and carries a load of 150 kN, and (b) the number of collars required.
3. A thrust shaft of a ship has six collars of 600 mm outer diameter and 300 mm inner diameter. The total thrust from the propeller is 120 kN. If the coefficient of friction is 0.15 and the speed of the engine is 100 rpm, determine the power absorbed in friction when (a) pressure is uniform, and (b) wear is uniform.
4. The mean diameter of a bolt having V-threads is 25 mm. The pitch of the threads is 5 mm and the angle of threads is 55°. The bolt is tightened by screwing a nut whose mean radius of the bearing surface is 25 mm. If the coefficient of friction for nut and bolt is 0.10 and for the nut and bearing surface is 0.15, find the required at the end of a 0.5 m-long spanner when the load on the bolt is 10 kN.
5. A vertical screw with single-start square threads 50 mm mean diameter and 10 mm pitch is raised against a load of 6 kN by means of a hand wheel, the boss of which is threaded to act as a nut. The axial load is taken by a thrust collar that supports the wheel boss and has a mean diameter of 70 mm. If the coefficient of friction is 0.15 for the screw and 0.20 for the collar, find the suitable diameter of hand wheel. Assume that a person can apply a force of 150 N by each hand.
6. A single-plate clutch transmits 20 kW at 1000 rpm. The maximum pressure intensity between the plates is 0.09 MPa. The outer diameter of the plate is 350 mm and both the sides of the plate are effective. The coefficient of friction is 0.25. Determine (a) inner diameter of the plate and (b) axial force required to engage the clutch.
7. A clutch in a motor car is of single-plate type having both sides of the plate effective. It is required to transmit 35 kW at 1500 rpm. The axial thrust is 0.075 MPa. The ratio between the external and

internal radii of the plate is 1.5 and coefficient of friction is 0.25. Assuming uniform pressure, determine the dimensions of the clutch.

8. A multi-plate clutch of alternate bronze and steel plates having effective diameters of 175 mm and 72.5 mm has to transmit 25 kW at 2000 rpm. The end thrust is 1600 N and coefficient of friction is 0.1. Calculate the number of plates necessary assuming uniform pressure.

9. A cone clutch has a radii of 130 mm and 150 mm. The semi-cone angle is 20°. If coefficient of friction is 0.25 and uniform normal pressure is 0.15 MPa, find (a) necessary axial load and (b) power that can be transmitted at 1000 rpm.

10. The external and internal radii of a friction plate of a clutch are 120 mm and 60 mm respectively. The total axial thrust is 1500 N. For uniform wear, find the maximum, minimum and average pressure on the contact surfaces. [**Ans.** 0.066 MPa, 0.033 MPa, 0.044 MPa]

11. A power of 60 kW is transmitted by a multi-plate clutch at 1500 rpm. The axial intensity of pressure is not to exceed 15 MPa. The coefficient of friction for the friction surfaces is 0.15. The external radius of friction surface is 120 mm internal radius is 100 mm. Find the number of plates required to transmit the power of 60 kW.

12. A cone clutch of semi-cone angle 15° is used to transmit 30 kW at 800 rpm. The mean frictional surface radius is 150 mm normal intensity of pressure of the mean radius is not to exceed 0.15 MPa. The coefficient of friction is 0.2. Assuming uniform wear, determine : (b) Axial force required to engage the clutch. [**Ans.** 84 mm, 3089.4 N]

❑❑❑

Chapter 7

Belts, Chains and Ropes

Inside This Chapter

7.1 INTRODUCTION

Belts, chains and ropes are used for power transmission. Belts and chains are used for short centre drives whereas ropes are used for long centre distances. Belts are of two types : flat belts and *V*-belts. Chains give a more positive drive than belts. There are problems of slip and creep in belts. Belts can be either of the open type or cross type. In this chapter, we shall study belts, chains and ropes from the point of view of power transmission.

7.2 FLAT BELT DRIVE

7.2.1 Angular Velocity Ratio

Consider the open belt drive shown in Fig. 7.1. Let the smaller pulley be the driver and the bigger pulley the driven (or follower). When the driver is rotating anticlockwise, the top side of the belt will be the tight side and the bottom side the slack side.

Let d_1 = diameter of the driving pulley

d_2 = diameter of the driven pulley

n_1 = rpm of the driving pulley

n_2 = rpm of the driven pulley

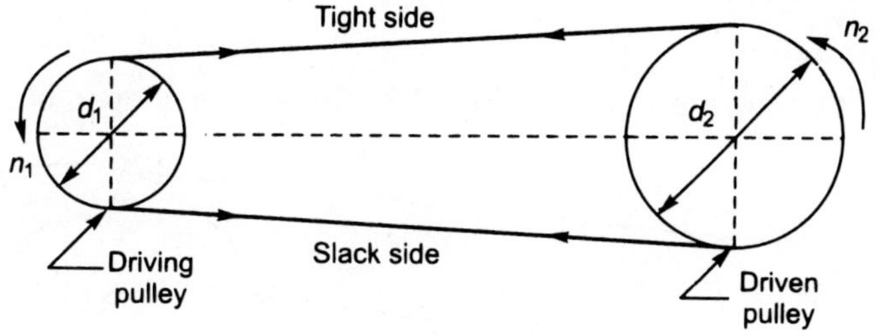

Fig. 7.1 Flat belt drive.

Then angular velocity of driver, $\omega_1 = \dfrac{2\pi n_1}{60}$ rad/s

and angular velocity of follower, $\omega_2 = \dfrac{2\pi n_2}{60}$ rad/s

Angular velocity ratio, $\dfrac{\omega_1}{\omega_2} = \dfrac{n_1}{n_2}$

Linear velocity of driver, $v_1 = \pi d_1 n_1$

Linear velocity of follower, $v_2 = \pi d_2 n_2$

Assuming there is no slip between the belt and the pulleys, $v_1 = v_2$

$$\frac{n_1}{n_2} = \frac{d_2}{d_1} \qquad ...(7.1)$$

If t is the thickness of the belt, then

$$n_1/n_2 = (d_2 + t)/(d_1 + t) \qquad ...(7.2)$$

7.6.2 Effect of Slip

Slip is the difference between the speed of the driver and the belt on the driver side and the belt and the follower on the driven side.

Let s_1 = percentage slip between the driver and the belt

s_2 = percentage slip between the belt and the follower

Linear speed of belt on driver $= v_1\left(1 - \dfrac{s_1}{100}\right)$

Linear speed of follower,
$$v_2 = v_1\left(1 - \frac{s_1}{100}\right)\left(1 - \frac{s_2}{100}\right)$$
$$= v_1\left[1 - \frac{s_1 + s_2 - 0.01\, s_1 s_2}{100}\right]$$
$$= v_1(1 - s)$$

where $s = \dfrac{(s_1 + s_2 - 0.01 s_1 s_2)}{100}$ is the total fractional slip.

Hence
$$\frac{n_2}{n_1} = \left[\frac{d_1 - t}{d_2 + t}\right](1 - s) \qquad ...(7.3)$$

7.2.3 Law of Belting

The law of belting states that the centre line of the belt as it approaches the pulley must lie in a plane perpendicular to the axis of that pulley, or must lie in the plane of the pulley, otherwise the belt will run off the pulley.

7.2.4 Length of Open Belt

Consider the open flat belt drive shown in Fig. 7.2. The length of open belt.

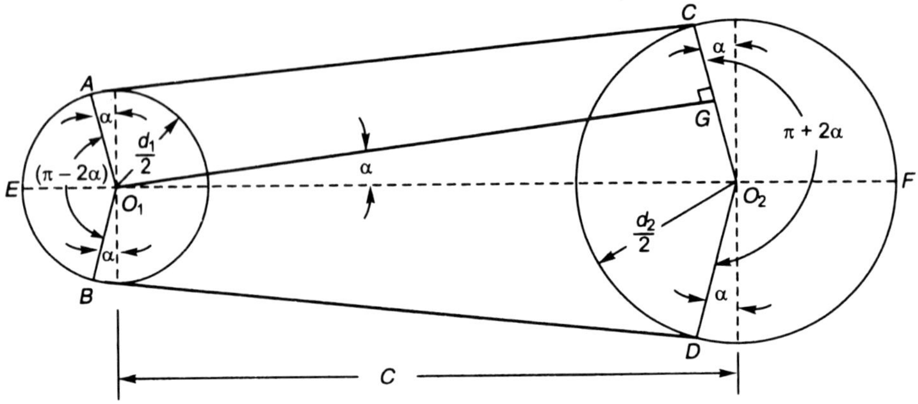

Fig. 7.2 Open flat belt drive.

$$L_o = \text{arc } AEB + \text{arc } CFD + AC + BD$$

$$\text{arc } AEB = \frac{d_1(\pi - 2\alpha)}{2}$$

$$\text{arc } CFD = \frac{d_2(\pi + 2\alpha)}{2}$$

$$AC = BD = O_1G = C\cos\alpha$$

$$L_o = \frac{d_1(\pi - 2\alpha)}{2} + \frac{d_2(\pi + 2\alpha)}{2} + 2C\cos\alpha$$

Now

$$\sin\alpha = \frac{O_2G}{O_1O_2} = \frac{(O_2C - CG)}{C} = \frac{O_2C - AO_1)}{C}$$

$$\alpha \approx \sin\alpha = \frac{d_2 - d_1}{2C}$$

$$\cos\alpha = [1 - \sin^2\alpha]^{0.5} = \left[1 - \left\{\frac{d_2 - d_1}{2C}\right\}^2\right]^{0.5}$$

$$\approx 1 - \frac{(d_2 - d_1)^2}{8C^2}$$

$$L_o = \frac{\pi(d_1 + d_2)}{2} + \alpha(d_2 - d_1) + 2C\left[1 - \frac{(d_2 - d_1)^2}{8C^2}\right]$$

$$L_o = \frac{\pi(d_1 + d_2)}{2} + \frac{(d_2 - d_1)^2}{2C} + 2C\left[1 - \frac{(d_2 - d_1)^2}{8C^2}\right]$$

$$= 2C + \frac{(d_2 - d_1)^2}{4C} + \frac{\pi(d_1 + d_2)}{2} \quad \text{...(7.4)}$$

7.2.5 Length of Cross Belt

Consider the cross belt shown in Fig. 7.3. The length of cross flat belt,

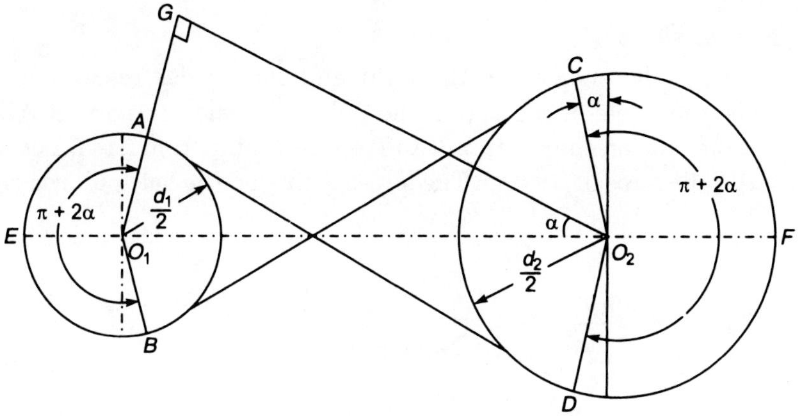

Fig. 7.3 Cross flat belt drive.

$$L_c = \text{arc } AEB + \text{arc } CEB + AD + BC$$

$$\text{arc } AEB = (\pi + 2\alpha)\frac{d_1}{2}$$

$$\text{arc } CEB = (\pi + 2\alpha)\frac{d_2}{2}$$

$$AD = BC = O_2G = C \cos \alpha$$

Now
$$\sin \alpha = \frac{O_1G}{O_1O_2} = \frac{O_1A + AG}{C} = \frac{O_1A + O_2D}{C}$$

$$= \frac{d_1 + d_2}{2C}$$

$$\cos \alpha = \left[1 - \left\{\frac{d_1 + d_2}{2C}\right\}^2\right]^{0.5}$$

$$= 1 - \frac{(d_1 + d_2)^2}{8C^2}$$

$$L_c = 2C\left[1 - \frac{(d_1 + d_2)^2}{4C}\right] + \frac{\pi(d_1 + d_2)}{2} + \frac{(d_1 + d_2)^2}{2C}$$

$$= 2C + \frac{\pi(d_1 + d_2)}{2} + \frac{(d_1 + d_2)^2}{4C} \qquad \ldots(7.5)$$

7.2.6 Angle of Arc of Contact

1. **Open belt :** The angle of arc of contact on the smaller pulley,

$$\theta = \pi - 2\alpha$$

$$= \pi - 2 \sin^{-1}\left[\frac{d_2 - d_1}{2C}\right] \qquad \ldots(7.6a)$$

The possibility of slip is more on the smaller pulley due to smaller angle of arc of contact.

2. **Cross belt :** The angle of arc of contact on either pulley,

$$\theta = \pi + 2\alpha$$

$$= \pi + 2 \sin^{-1}\left[\frac{d_1 + d_2}{2C}\right] \qquad \ldots(7.6b)$$

7.2.7 Ratio of Belt Tensions

Let T_1 and T_2 be the belt tensions on the tight and slack sides respectively, as shown in Fig. 7.4(a). Let θ be the angle of contact on the pulley. Consider a portion AB of the belt on angular arc $\delta\theta$. Let the tension change from T to $T + \delta T$ in going from A to B. Let R be the normal on the pulley and μR the force of friction. The forces acting on the belt-pulley system are shown in Fig. 7.4(b).

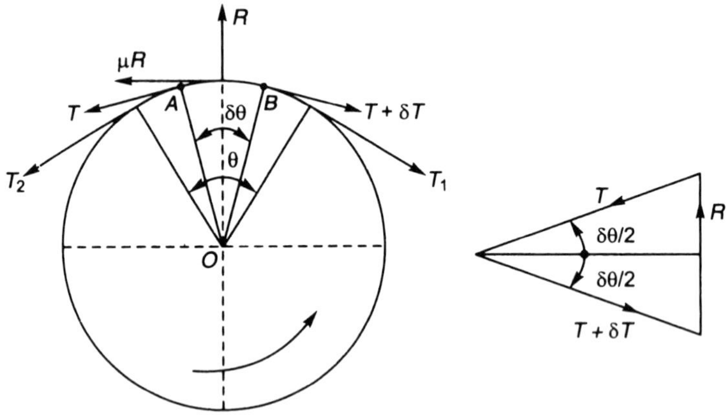

(a) Forces on the belt (b)

Fig. 7.4 Belt tensions.

Resolving the forces vertically, we have

$$R = T \sin\left(\frac{\delta\theta}{2}\right) + (T + \delta T) \sin\left(\frac{\delta\theta}{2}\right)$$

$$\approx T \frac{\delta\theta}{2} + (T + \delta T) \frac{\delta\theta}{2}$$

$$\approx T\delta\theta$$

Resolving the forces horizontally, we have

$$\mu R = (T + \delta T) \cos\left(\frac{\delta\theta}{2}\right) - T \cos\left(\frac{\delta\theta}{2}\right) = \delta T \qquad \left[\because \cos\left(\frac{\delta\theta}{2}\right) \simeq 1\right]$$

or $$\mu T\delta\theta = \delta T$$

or $$\frac{\delta T}{T} = \mu \cdot \delta\theta$$

$$\int_{T_2}^{T_1} \frac{dT}{T} = \int_0^{\theta} \mu \cdot d\theta$$

$$\ln\left(\frac{T_1}{T_2}\right) = \mu\theta$$

or $$\frac{T_1}{T_2} = \exp(\mu\theta) \qquad \ldots(7.7)$$

The ratio of belt tensions is given by Eq. (7.7).

7.2.8 Power Transmitted

Power transmitted, $$P = \frac{(T_1 - T_2)}{1000}\ \text{kW} \qquad \text{...(7.8a)}$$

$$= T_1 v \frac{[1 - \exp(\mu\theta)]}{10^3}\ \text{kW} \qquad \text{...(7.8b)}$$

7.2.9 Centrifugal Tension

The centrifugal tension is introduced in the belt due to its mass. Let T_c be the centrifugal tension in the belt. Consider the length of the belt over an angular arc $\delta\theta$, as shown in Fig. 7.5. The force acting on the belt due to belt tension,

$$= 2T_c \sin\left(\frac{\delta\theta}{2}\right)$$

$$\approx T_c\delta\theta$$

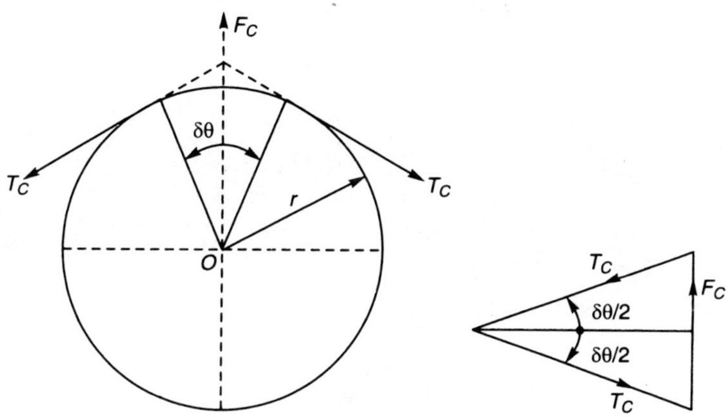

Fig. 7.5 Centrifugal tension.

Centrifugal force acting on length AB and of unit width of the belt due to angular velocity.

$$F_c = \frac{m \cdot r\delta\theta \cdot v^2}{r} = mv^2\delta\theta$$

where m = mass of the belt per unit length

r = radius of the pulley

v = speed of belt

For equilibrium of the belt,

$$T_c\delta\theta = mv^2\delta\theta$$

$$T_c = mv^2 \qquad \text{...(7.9)}$$

Centrifugal stress, $$\sigma_c = \frac{T_c}{bt} = \frac{mv^2}{bt} \qquad \text{...(7.10)}$$

where b = width of the belt and

t = thickness of the belt

Effective tension on tight side = $T_1 + T_c$

Effective tension on slack side = $T_2 + T_c$

7.2.10 Condition for Maximum Power Transmission

Power transmitted, $P = (T_1 - T_2)v$

$= T_1\,[1 - \exp(-\mu\theta)]v$

Let $1 - \exp(-\mu\theta) = k$, so that

$$P = kT_1 v$$

Now maximum belt tension, $T = T_1 + T_c$

or $T_1 = T - T_c$

$\therefore$ $P = k(T - T_c)v$

$= kTv - kT_c v$

$= kTv - kmv^3$

For P to be maximum, $\dfrac{dP}{dv} = 0$

or $kT - 3kmv^2 = 0$

or $T = 3mv^2$

or $$v = \left[\frac{T}{3m}\right]^{0.5} \qquad \text{...(7.11)}$$

$$P_{max} = kT\left(\frac{T}{3m}\right)^{0.5} - km\left(\frac{T}{3m}\right)^{1.5}$$

$$= k\left(\frac{T}{3m}\right)^{0.5}\left[T - \frac{T}{3}\right]$$

$$= \left(\frac{2}{3}\right)\frac{kT^{1.5}}{(3m)^{0.5}} \qquad \text{...(7.12)}$$

7.2.11 Initial Belt Tension

Let T_o = initial tension in belt

Resultant tension on tight side = $T_1 - T_o$

Resultant tension on slack side = $T_o - T_2$

Since the belt length remains constant, therefore

$$T_1 - T_o = T_o - T_2$$

or $$T_o = \frac{T_1 + T_2}{2} \qquad \text{...(6.13a)}$$

Considering centrifugal tension,

$$T_o = \frac{T_1 + T_2 + 2T_c}{2} \qquad \text{...(7.13b)}$$

7.2.12 Effect of Initial Tension on Power Transmission

Ratio of tensions, $\dfrac{T_1}{T_2} = \exp(\mu\theta) = c$

Initial tensions, $T_o = \dfrac{T_1 + T_2 + 2T_c}{2}$

or
$$T_1 + T_2 = 2(T_o - T_c)$$
$$T_1\left(1+\frac{1}{c}\right) = 2(T_o - T_c)$$

and
$$T_1 = \left[\frac{2c}{1+c}\right](T_o - T_c)$$
$$T_1 - T_2 = \left[\frac{2(c-1)}{c+1}\right](T_o - T_c)$$

Power transmitted,
$$P = (T_1 - T_2)\, v$$
$$= \left[\frac{2(c-1)}{c+1}\right](T_o - T_c)\, v$$
$$= \left[\frac{2(c-1)}{c+1}\right](T_o - mv^2)\, v$$

For P to be maximum, $\frac{dP}{dv} = 0$

$$\frac{2(c-1)}{(c+1)}(T_o - 3mv^2) = 0$$
$$T_o = 3mv^2 = 3T_c \quad \text{...(7.14)}$$
$$v = \left[\frac{T_0}{3m}\right]^{0.5} \quad \text{...(7.15)}$$
$$P_{\max} = \left[\frac{2(c-1)}{c+1}\right]\left(\frac{T_o - T_0}{3}\right)\left[\frac{T_o}{3m}\right]^{0.5}$$
$$= \left(\frac{4}{3}\right)\left[\frac{c-1}{c+1}\right]\frac{T_o^{15}}{(3m)^{0.5}} \quad \text{...(7.16)}$$

While starting, $v = 0$ and $T_c = 0$. Hence
$$T_1 = \left[\frac{2c}{c+1}\right]T_o$$

While running,
$$T_1 = \left[\frac{2c}{c+1}\right](T_o - T_c)$$

Hence maximum belt tension,
$$T_1 = \left[\frac{2c}{c+1}\right]T_o \quad \text{...(7.17)}$$

7.2.13 Belt Creep

The tension on the tight side is more than the tension on the slack side. As a result of this, the belt is stretched more on the tight side as compared to the slack side. Therefore, the driver pulley receives more length of the belt and delivers less. Hence, the belt creeps (or jumps) forward. The reverse occurs on the follower pulley. The follower pulley receives less length of the belt and delivers more. As a result of it, the belt creeps backward. This phenomenon is called *creeping of the belt.*

$$\text{Creep} = \frac{T_1 - T_2}{bt\,E} \quad \text{...(7.18)}$$

where E is the modulus of elasticity of belt material.

The velocity ratio becomes,

$$\frac{n_2}{n_1} = \left(\frac{d_1}{d_2}\right)\left[\frac{E + \sqrt{\sigma_2}}{E + \sqrt{\sigma_1}}\right] \quad \text{...(7.19)}$$

where $\sigma_1 = \dfrac{T_1}{bt}$ and $\sigma_2 = \dfrac{T_2}{bt}$

7.2.14 Crowning of Pulleys

The pulley face is given a convex cu~vature and is never kept flat. This is called crowing of the pulleys. This helps in running the belt in the centre of the pulley width. Crowing prevents any tendency of the belt to fall off the pulley face.

7.2.15 Cone Pulleys

Consider the cone pulley block shown in Fig. 7.6. Let N be the speed of the driver block, and R_1, R_2, R_3, etc. the radii of its pulleys. Let the radii of the driven pulleys be r_1, r_2, r_3, etc. and speeds n_1, n_2, n_3, etc. The centre distance between the pulleys is C.

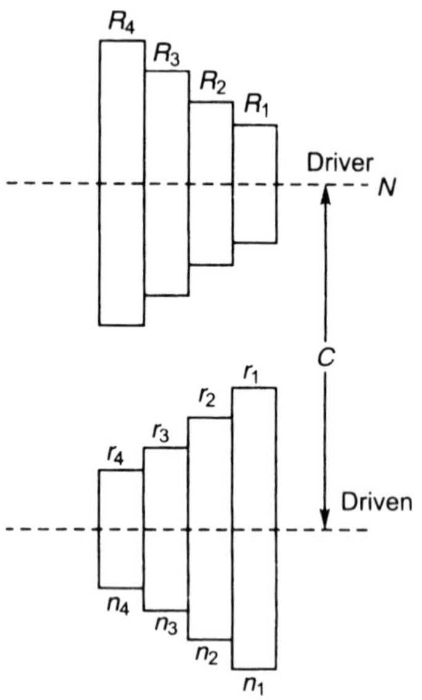

Fig. 7.6 Cone pulley system.

Now $\quad R_1 N = r_1 n_1$

or $\quad \dfrac{N}{n_1} = \dfrac{r_1}{R_1}$

Similarly $\quad \dfrac{N}{n_2} = \dfrac{r_2}{R_2}$

$\dfrac{N}{n_3} = \dfrac{r_3}{R_3}$ and so on.

Let $\quad n_2 = k n_1$

Therefore $\quad \dfrac{n_2}{N} = \dfrac{k n_1}{N} = k\left(\dfrac{R_1}{r_1}\right)$

Similarly $\quad \dfrac{n_3}{N} = k^2\left(\dfrac{R_1}{r_1}\right)$

or in general,

$$\frac{n_i}{N} = k^{(i-1)}\left(\frac{R_1}{r_1}\right) = \frac{R_i}{r_i} \quad \text{...(7.20)}$$

and

$$\frac{n_i}{n_1} = k^{(i-1)} \quad \text{...(7.21)}$$

The length of the belt is same for all the cones, therefore

$$R_i + r_i = \text{constant}, \; i = 1, 2, 3 \text{ etc.}$$

7.2.16 Compound Belt Drive

In compound belt drive, the driven pulley of the first set is mounted on the same shaft on which the driver of the second set is mounted. Let pulley 1 be the driver for the first set and pulley 2 its follower. The driver of the second set, pulley 3, is mounted on the same shaft on which pulley 2 is mounted. The follower of second set is pulley 4.

$$\frac{n_2}{n_1} = \frac{d_1}{d_2}$$

$$\frac{n_4}{n_3} = \frac{d_3}{d_4}$$

Hence

$$\frac{n_4}{n_1} = \left(\frac{d_1 d_2}{d_2 d_4}\right)$$

or in general,

$$\frac{\text{Speed of last driver}}{\text{Speed of first driver}} = \frac{\text{Product of diameter of drivers}}{\text{Produt of diameters of drivers}} \quad \text{...(7.23)}$$

■ **Example 7.1** Two pulleys of diameters 450 mm and 150 mm are mounted on two parallel shafts 2 m apart and are connected by a flat belt drive. Find the power which can be transmitted by the belt when the larger pulley rotates at 180 rpm. The maximum permissible tension in the belt is 1 kN, and the coefficient of friction between the belt and the pulley is 0.25. Also find the length of the cross belt required and the angle of arc of contact between the belt and the pulleys.

Solution :

$$v_2 = \frac{\pi \times 450 \times 180}{60 \times 1000} = 4.24 \text{ m/s}$$

$$n_1 = \frac{180 \times 450}{150} = 540 \text{ rpm}$$

Now

$$\sin \alpha = \frac{d_1 + d_2}{2C} = \frac{150 + 450}{4000} = 0.15$$

$$\theta = 8.627°$$

Angle of contact, $\theta = \pi + 2\alpha = 197.254°$ or 3.443 rad

$$\frac{T_1}{T_2} = \exp(\mu\theta)$$

$$= \exp(0.25 \times 3.443) = 2.365$$

$$T_2 = \frac{T_1}{2.365} = \frac{1000}{2.365} = 422.83 \text{ N}$$

Power transmitted, $P = (T_1 - T_2)\dfrac{v_2}{1000}$

$$= (1000 - 422.83) \times \frac{4.24}{1000} = 2.447 \text{ kW}$$

Length of cross belt, $L_c = 2C + \dfrac{\pi(d_1 + d_2)}{2} + \dfrac{(d_1 + d_2)^2}{4C}$

$$= 4 + \frac{\pi(0.450 + 0.150)}{2} + \frac{(0.40 + 0.150)^2}{8}$$

$$= 4987 \text{ m}$$

■ **Example 7.2** A pulley is driven by a flat 100 mm wide and 6 mm thick. The density of belt material is 1000 kg/m^3. The angle of lap is 120° and the coefficient of friction 0.3. The maximum stress in the belt does not exceed 2 MPa. Find the maximum power that can be transmitted and the corresponding speed of the belt.

Solution : Maximum tension in the belt,

$$T = 2 \times 10^6 \times 0.1 \times 0.0006 = 1200 \text{ N}$$

Mass of the belt per metre length, $m = 0.1 \times 0.006 \times 1 \times 1000 \times 0.6$ kg/m

Speed of the belt for maximum power, $v = \left[\dfrac{T}{3m}\right]^{0.5}$

$$= \left[\frac{200}{3 \times 0.6}\right]^{0.5}$$

$$= 25.82 \text{ m/s}$$

For maximum power to be transmitted, the centrifugal tension,

$$T_c = \frac{T}{3} = \frac{1200}{3} = 400 \text{ N}$$

$$\frac{T_1}{T_2} = \exp(0.3 \times 120 \times \pi/180) = 1.874$$

$$T_1 = T - T_c = 1200 - 400 = 800 \text{ N}$$

$$T_2 = \frac{T_1}{1.874} = \frac{800}{1.874} = 426.8 \text{ N}$$

Maximum power transmitted $= (T_1 - T_2)\dfrac{v}{1000}$

$$= (800 - 426.8) \times \frac{25.82}{1000}$$

$$= 9.636 \text{ kW}$$

■ **Example 7.3** An open belt drive is used to connect two parallel shafts 4 m apart. The diameter of bigger pulley is 1.5 m and that of the smaller pulley 0.5 m. The mass of the belt is 1 kg/m length. The maximum tension is not to exceed 1500 N. The coefficient of friction is 0.25. The bigger pulley, which is the driver, runs at 250 rpm. Due to slip, the speed of the driven pulley is 725 rpm. Calculate the power transmitted, power lost in friction, and the efficiency of the drive.

Solution :

$$v_1 = \pi \times 1.5 \times \frac{250}{60} = 19.635 \text{ m/s}$$

$$v_2 = \pi \times 0.5 \times \frac{725}{60} \; 19.98 \text{ m/s}$$

$$T_c = mv_1^2 = 1 \times (19.635)^2 = 385.53 \text{ N}$$

$$T_1 = T - T_c = 1500 - 385.53 = 1114.47 \text{ N}$$

$$\sin\alpha = \frac{d_2 - d_1}{2C} = \frac{1.5 - 0.5}{8} = 0.125$$

$$\alpha = 7.18°$$

$$\theta = 180° - 2\alpha = 165.64° = 2.89 \text{ rad}$$

$$\frac{T_1}{T_2} = \exp(\mu\theta) = \exp(0.25 \times 2.59) = 2.06$$

$$T_2 = \frac{T_1}{2.06} = \frac{1114.47}{2.06} = 541.0 \text{ N}$$

$$T_1 = T_2 = 1114.47 - 541.00 = 573.47 \text{ N}$$

Torque on bigger pulley, $M_1 = 573.47 \times 0.75 = 430.1$ Nm

Torque on smaller pulley, $M_2 = 573.47 \times 0.25 = 143.37$ Nm

Power transmitted,

$$P = (T_1 - T_2)\frac{v_1}{1000} = 573.47 \times \frac{19.635}{1000} = 11.26 \text{ kW}$$

Input power, $\quad P_1 = M_1\omega_1 = 430.1 \times \dfrac{\left(2\pi \times \dfrac{250}{60}\right)}{1000}\ 11.26$ kW

Output power, $\quad P_2 = M_2\omega_2 = 143.37 \times \dfrac{\left(2\pi \times \dfrac{725}{60}\right)}{1000} = 10.885$ kW

Power lost in friction, $P_f = 11.26 - 10.885 = 0.375$ kW

Efficiency of the drive $= \dfrac{10.885}{11.26} = 96.67\%$

7.3 V-BELT DRIVE

The V-belt derive is more positive than the flat belt drive. It is a short centre drive and is preferred for power transmission from the prime mover. The belt touches the sides of the grooved pulley only. The V-belts are classified as A, B, C, D, E.

7.3.1 Ratio of Belt Tensions

A V-belt in a grooved pulley is shown in Fig. 7.7.

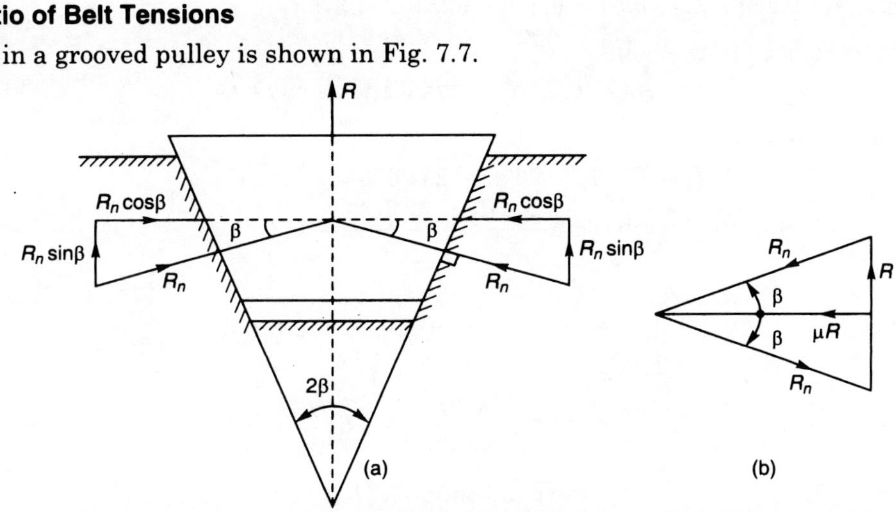

Fig. 7.7 Forces on V-belt.

Let 2β = pulley groove angle;

R = total reaction on the pulley;

R_n = normal reaction between the belt and the sides of the groove and

μ = coefficient of friction between the belt and the groove sides.

Then $$R = 2R_n \sin\beta$$

or $$R_n = \frac{R}{2\sin\beta}$$

Force of friction, $$F = 2\mu R_n = \left(\frac{\mu}{\sin\beta}\right)R = \mu_e R$$

where μ_e is called the virtual, apparent or equivalent coefficient of friction.

Ratio of tensions, $$\frac{T_1}{T_2} = \exp(\mu_e\theta) \qquad ...(7.24)$$

■ **Example 7.4** A compressor requires 100 kW to run at 240 rpm from an electric motor of speed 750 rpm, by means of a V-belt drive. The diameter of the compressor shaft pulley should not be more than 1 m while the centre distance between the shafts is 2 m. The belt speed should not exceed 25 m/s.

Determine the number of V-belts required to transmit the power if each belt has a cross-sectional area of 375 mm^2, density 100 kg/m^3, and an allowable tensile stress of 2.5 MPa. The pulley groove angle is 40 and coefficient of friction between the belt and the pulley sides is 0.25.

Solution : Diameter of motor pulley, $d_1 = \frac{240 \times 1}{750} = 0.32$ m

Mass of belt per metre length,

$$m = 375 \times 10^{-6} \times 1 \times 1000 = 0.375 \text{ kg/m}$$

Velocity of belt, $v = 25$ m/s

Centrifugal tension, $T_c = mv^2 = 0.375 \times 625 = 234.375$ N

Maximum tension in the belt,

$$T = \sigma A = 2. \times 10^6 \times 375 \times 10^{-6} = 937.5 \text{ N}$$

Tight side tension,

$$T_1 = T - T_c = 937.5 - 234.375 = 703.125 \text{ N}$$

$$\sin\alpha = \frac{d_2 - d_1}{2C} = \frac{1 - 0.32}{4} = 0.17$$

$$\alpha = 9.78°$$

$$\theta = 180° - 2\alpha = 160.44° = 2.8 \text{ rad}$$

$$T_1/T_2 = \exp(\mu_e\theta)$$

$$= \exp\left[\left(\frac{0.25}{\sin 20°}\right) \times 2.8\right]$$

$$= \exp(2.0466) = 7.74$$

$$T_2 = \frac{703.125}{7.74} = 90.74 \text{ N}$$

Power transmitted per belt

$$= (T_1 - T_2)\frac{v}{1000}$$

$$= (703.125 - 90.84) \times \frac{25}{1000} = 15.31 \text{ kW}$$

Number of V-belts required

$$= \frac{100}{15.31} = 6.53 \approx 7$$

7.4 CHAIN DRIVE

Chains are mostly used to transmit power without slipping and with better efficiency than belts. They are commonly used in motor cycles, bicycles, road rollers, and agricultural machinery. A chain on the sprocket is shown in Fig. 7.8. The pitch of the chain is the distance between the hinge centers of the adjacent links. The pitch circle diameter is the diameter of the circle on which the hinge centres of the chain like lie, when chain is wrapped round the sprocket.

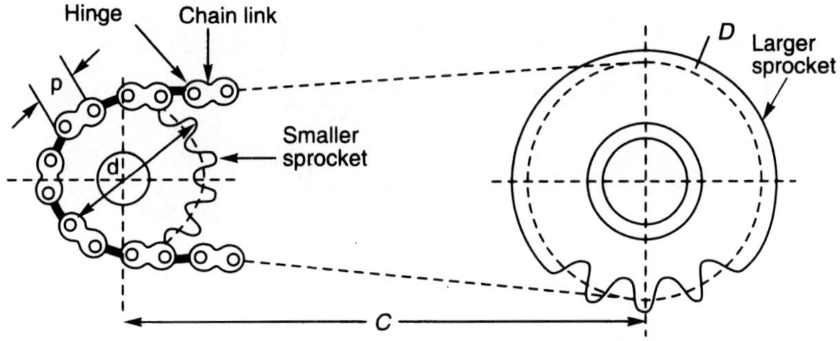

Fig. 7.8 Chain drive.

7.4.1 Chain Pitch

As shown in Fig. 7.9, the chain pitch is,

$$p = \frac{2 \times D}{2} \times \sin\left(\frac{\theta}{2}\right) = D \sin\left(\frac{\theta}{2}\right) \quad \text{...(7.25a)}$$

where $\theta = \frac{360°}{z}$ and z = Number of teeth

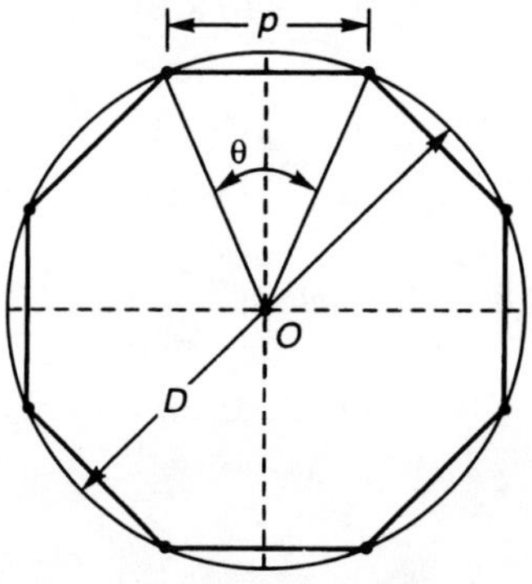

Fig. 7.9 Chain pitch.

Hence,
$$p = D \sin\left(\frac{180°}{z}\right) \qquad \text{...(7.25b)}$$

7.4.2 Chain Length

As shown in Fig. 7.10, the chain length

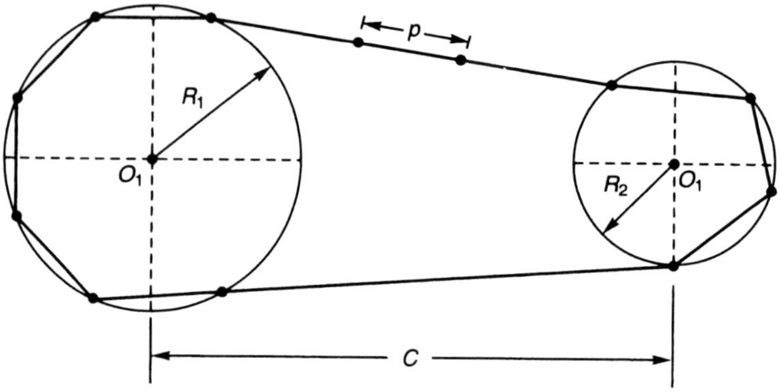

Fig. 7.10 Chain length.

$$L = 2C + \pi(R_1 + R_2) + \frac{(R_1 - R_2)^2}{C} \qquad \text{...(7.26a)}$$

$$\pi(R_1 + R_2) = \frac{p(z_1 + z_2)}{2}$$

$$R_1 = \left(\frac{p}{2}\right) \text{cosec}\left(\frac{180°}{z_1}\right)$$

$$R_2 = \left(\frac{p}{2}\right) \text{cosec}\left(\frac{180°}{z_2}\right)$$

$$L = 2C + \frac{p(z_1 + z_2)}{2} + \frac{p^2\left[\text{cosec}\left(\frac{180°}{z_1}\right) - \text{cosec}\left(\frac{180°}{z_2}\right)\right]^2}{4C} \qquad \text{...(7.26b)}$$

■ **Example 7.5** A chain drive is used for reduction of speed from 240 rpm to 120 rpm. The number of teeth on the driving sprocket is 24. Find the number of teeth on the driven sprocket. If the pitch circle diameter of the driven sprocket is 600 mm and centre distance is 1 m, determine the pitch and length of the chain.

Solution :

$$z_2 = \frac{n_1 z_1}{n_2} = \frac{240 \times 24}{120} = 48$$

$$R_2 = \left(\frac{p}{2}\right) \text{cosec}\left(\frac{180°}{z_2}\right)$$

$$0.3 = \left(\frac{p}{2}\right) \text{cosec}\left(\frac{180}{48}\right)$$

$$p = 0.0392 \text{ m or } 39.2 \text{ mm}$$

$$L = 2C + \frac{p(z_1 + z_2)}{2} + \left(\frac{p^2}{4C}\right)\left[\text{cosec}\left(\frac{180°}{z_1}\right) - \text{cosec}\left(\frac{180°}{z_2}\right)\right]^2$$

$$= 2 \times 1 + \left(\frac{0.0392}{2}\right)(24 + 48) + \frac{(0.0392)^2}{4 \times 1}\left[\text{cosec}\left(\frac{180°}{24}\right) - \text{cosec}\left(\frac{180°}{48}\right)\right]^2$$

$$= 2 + 1.4112 + 0.0223 = 3.4335 \text{ m}$$

7.5 ROPE DRIVE

Ropes are used for power transmission over long distances. They are commonly used in hoisting equipment, drilling rigs, and textile industry. Ropes are either made of fibre of steel. Ropes are generally of circular cross-section and require grooved sheaves or pulleys.

7.5.1 Ratio of Tensions

The ratio of tensions, $$\frac{T_1}{T_2} = \exp(\mu_e \theta) \quad \text{...(7.27)}$$

where $\mu_e = \dfrac{\mu}{\sin \beta}$

μ = coefficient of friction between rope and sheave

β = semi-groove angle of the sheave

θ = angle of contact.

■ **Example 7.6** A pulley of groove angle 45°, diameter 4 m and having 15 grooves is used to transmit power. The angle of contact is 170° and the coefficient of friction between the ropes and the groove sides is 0.30. The maximum possible tension in the ropes is 1 kN and the mass of the rope is 1.5 kg/m length. Determine the speed of the pulley for maximum power conditions.

Solution : For maximum power to be transmitted,

$$v = \left[\frac{T}{3m}\right]^{0.5} = \left[\frac{1000}{3 \times 1.5}\right]^{0.5} = 14.91 \text{ m/s}$$

Speed of the pulley, $$n = \frac{60 \times 14.91}{4\pi} = 71.2 \text{ rpm}$$

For maximum power to be transmitted,

$$T_c = \frac{T}{3} = \frac{1000}{3} = 333.3 \text{ N}$$

Tension on the side, $$T_1 = T - T_c = 1000 - 333.33 = 666.67 \text{ N}$$

$$\frac{T_1}{T_2} = \exp(\mu_e \theta)$$

$$= \exp\left[0.3 \times \text{cosec}(22.5°) \times 170 \times \frac{\pi}{180}\right]$$

$$= \exp(2.326) = 10.23$$

$$T_2 = \frac{T_1}{10.23} = \frac{666.67}{10.23} = 65.12 \text{ N}$$

Power transmitted per rope

$$= (T_1 - T_2)\frac{v}{1000}$$

$$= (666.67 - 65.12) \times \frac{14.91}{1000} = 8.97 \text{ kW}$$

Total power transmitted, $P = 8.97 \times 15 = 134.5$ kW.

■ **Example 7.7** An open belt drive transmits power from a 300 mm diameter pulley running at 240 rpm to a pulley 450 mm diameter. Angle of lap on smaller pulley is 160°. The belt is on the point of slipping when 3 kW is being transmitted. The coefficient of friction between belt and pulley is 0.3.

It is desired to increase the power transmitted. State which of the following methods suggested would be more effective?

(a) Initial tension in the belt is increased by 10%.

(b) Suitable dressing is given to the belt surface to increase the coefficient of friction by 10%. Assume that initial tension is kept the same.

Solution : Linear velocity of belt,

$$v = \pi d_1 n_1 /(60 \times 10^{3)} = \pi \times 300 \times 240/(60 \times 10^{3)} = 2.77 \text{ m/s}$$

Power transmitted, $$P = (T_1 - T_2)v/10^3$$

$$3 = (T_1 - T_2) \times 3.77/10^3$$

$$T_1 - T_2 = 3 \times 10^3 / 3.77 = 795.76 \text{ N}$$

$$T_1/T_2 = \exp(\mu\theta) = \exp(0.3 \times 160 \times \pi/180) = 2.3112$$

$$T_2(2.3112 - 1) = 795.76$$

$$T_2 = 606.89 \text{ N}$$

$$T_1 = 1402.65 \text{ N}$$

(a) Initial tension,

$$T_i = (T_1 + T_2)/2 = (1402.65 + 606.89)/2 = 1004.77 \text{ N}$$

When the initial tension is increased by10%,

$$T_1 + T_2 = 2 \times 1.1 \times 1004.77 = 2210.494 \text{ N}$$

$$T_2(2.3112 + 1) = 2210.494$$

$$T_2 = 667.58 \text{ N}$$

$$T_1 = 1542.91 \text{ N}$$

$$P = (1542.91 - 667.58) \times 3.77 / 10^3 = 3.30 \text{ kW}$$

(b) When coefficient of friction is increased by 10%, then $\mu = 1.1 \times 0.3 = 0.33$

$$T_1 / T_2 = \exp(\mu\theta) = \exp(0.33 \times 160 \times \pi / 180) = 2.51314$$

When the initial tension is kept constant.

$$T_1 + T_2 = 2 \times 1004.77 = 2009.54 \text{ N}$$

$$T_2(2.5134 + 1) = 2009.54.464$$

$$T_2 = 572 \text{ N}$$

$$T_1 = 1437.5 \text{ N}$$

$$P = (1437.5 - 572) \times 3.77/10^3 = 3.26 \text{ kW}$$

■ **Example 7.8** A rope drive transmits 120 kW at 220 rpm by ropes 25 mm in diameter and 0.6 kg/m length mass. The maximum rope tension is 1360 N and it is designed for maximum power transmission conditions. The angle of contact is 160° and the coefficient of friction is 0.25. The sheave groove angle is 45°. Determine the diameter of thc sheave and the number of ropes.

Solution : For maximum power transmission,

$$T_{max} = 3T_c$$

$$1360 = 3mv^2$$

$$v^2 = 1360/(3 \times 0.6) = 755.56$$

$$v = 27.487 \text{ m/s}$$

Diameter of sheave,

$$D = 60v/(\pi N) = 60 \times 27.487/(\pi \times 220) = 2.386 \text{ m}$$

$$T_c = mv^2 = 0.6 \times (27.487)^2 = 453.32 \text{ N}$$

$$T_1 = 1360 - 453.32 = 906.68 \text{ N}$$

$$T_1/T_2 = \exp\,[(0.25/\sin 22.5°)(\pi \times 160/180)] = 6.1985$$

$$T_2 = 906.68/6.1985 = 146.27 \text{ N}$$

Power transmitted, $P = n(906.68 - 146.27) \times 27.487/10^3) = 120$

Number of ropes, $n = 5.74 \approx 6$

SOLVED QUESTIONS FROM UNIVERSITY EXAMINATIONS

■ **Example 7.9** Derive an expression for the ratio of tensions on the two sides of vee belt. [KU, Dec. 2006]

Solution : Refer to art. 7.2.7 and art. 7.3.1.

■ **Example 7.10** A rope drive transmits 140 H.P. at 225 rpm by ropes, each 2.5 cm diameter and density 0.0068 kg/cm^3. The maximum rope tension is 150 kg and it is designed for maximum power conditions. The angle of contact is 160° and $\mu = 0.2$. The groove angle is 45°. Determine the diameter of the pulley and the number of ropes. [KU, Dec. 2007]

Solution : Given : $P_t = 140\,\text{HP}$, $N = 225\,\text{rpm}$, $d_r = 2.5\,\text{cm}$, $\rho = 0.0068\,\text{kg/cm}^3$,

$$T_{max} = 150\,\text{kg or } 1471.5\text{ N},\ \theta = 160°,\ \mu = 0.2,\ 2\beta = 45°,\ T_{max} = 3T_c$$

Mass of rope per metre length,

$$m = \frac{\pi}{4}d_r^2 \times l \times \rho = \frac{\pi}{4}(2.5)^2 \times 100 \times 0.0068 = 3.338\,\text{kg/m}$$

$$T_{max} = 3 \times mv^2$$

$$1471.5 = 3 \times 3.338 \times v^2$$

$$v = 12.12 \text{ m/s}$$

Now

$$\frac{\pi DN}{60} = v$$

Diameter of pulley,

$$D = \frac{60 \times 12.12}{\pi \times 225} = 1.029 \text{ m}$$

$$T_c = \frac{1471.5}{3} = 490.5 \text{ N}$$

$$T_1 = T_{max} - T_c = 1471.5 - 490.5 = 981 \text{ N}$$

$$\mu_e = \frac{\mu}{\sin\beta} = \frac{0.2}{\sin 22.5°} = 0.5226$$

$$\frac{T_1}{T_2} = e^{\mu_e\theta} = \exp\left[0.5226 \times \frac{\pi}{180} \times 160\right] = 1.459$$

$$T_2 = \frac{981}{1.459} = 672.2 \text{ N}$$

Power transmitting capacity of one rope,

$$P = \frac{(T_1 - T_2)v}{10^3} = \frac{(981 - 672.2) \times 12.12}{10^3} = 3.743 \text{ kW}$$

Number of ropes required,

$$n = \frac{140 \times 0.7357}{3.743} = 27.5 \simeq 28 \qquad [1 \text{ HP} = 0.7357 \text{ kW}]$$

■ **Example 7.11** Two pulleys of 30 cm and 90 cm diameter are on parallel shafts 180 cm apart. Determine the necessary length of an open belt to connect the two shafts. Show that if coefficient of limiting friction between the belt and pulley is 0.2 and initial tension is same, the maximum power which can be transmitted by the open and crossed belts are in the ratio of about 1 to 1.35. [KU, Dec, 2006]

Solution : Given $d_1 = 30$ cm, $d_2 = 90$ cm, $C = 180$ cm, $\mu = 0.2$

Length of open belt,

$$L_0 = 2C + \frac{(d_2 - d_1)^2}{4C} + \frac{\pi}{2}(d_1 + d_2)$$

$$= 2 \times 180 + \frac{(90 - 30)^2}{4 \times 180} + \frac{\pi}{2}(30 + 90) = 553.49 \text{ cm}$$

Open belt :

$$\sin\alpha_0 = \frac{d_2 - d_1}{2C} = \frac{90 - 30}{2 \times 180} = \frac{60}{2 \times 180} = 0.16667$$

$$\alpha = 9.594°$$

$$\theta_0 = 180° - 2\alpha_0 = 180° - 2 \times 9.594° = 160.81°$$

$$\frac{T_1}{T_2} = e^{\mu\theta_0} = e^{(0.2 \times \pi/180 \times 160.81)} = e^{0.56133} = 1.753$$

Initial tension,

$$T_0 = \frac{1}{2}(T_1 + T_2) = \frac{1}{2}(1.753 + 1)T_2 = 1.3765\, T_2$$

$$T_2 = 0.7265T_0$$

$$T_1 - T_2 = (1.7530 - 1)T_2 = 0.7530T_2 = 0.5470T_0$$

Power transmitted,

$$P_0 = \frac{(T_1 - T_2)v}{10^3} = \frac{0.547T_0 v}{10^3} \text{ kW}$$

Cross belt:

$$\sin \alpha_c = \frac{(d_1 + d_2)}{2C} = \frac{30 + 90}{2 \times 180} = \frac{120}{2 \times 180} = \frac{1}{3} = 0.33333$$

$$\alpha_c = 19.47°$$

$$\theta_c = 180° + 2\alpha_c = 180° + 2 \times 19.47° = 218.94°$$

$$\frac{T_1}{T_2} = \exp(\mu\theta_c) = \exp\left(0.2 \times \frac{\pi}{180} \times 218.94\right) = \exp(0.7642) = 2.147$$

$$T_0 = \frac{1}{2}(T_1 + T_2) = \frac{1}{2}(2.147 + 1)T_2 = 1.5737T_2$$

or $$T_2 = 0.6354T_0$$

$$T_1 - T_2 = (2.147 - 1)T_2 = 1.147T_2 = 0.7288T_0$$

Power transmitted,

$$P_c = \frac{(T_1 - T_2)v}{10^3} = \frac{0.7288T_0 v}{10^3} \text{ kW}$$

Ratio of power transmitted for same T_0,

$$\frac{P_c}{P_0} = \frac{0.7288}{0.5470} = 1.33, \text{ which lies between 1 to 1.35.}$$ Hence proved.

■ **Example 7.12** What is initial tension in belts? How is it determined? [KU, Dec. 2007]

Solution : The initial tension T_0 in the belt is the initial stretching given to the belt when the driving and driven pulleys are stationary so that the belt is not having unnecessary sag. As the driving pulley starts rotating the pulley pulls the belt from one side and delivers the belt to the other side. The tension on tight side increase to T_1 and on the slack side decreases to T_2.

For equilibrium of the belt,

$$T_1 - T_0 = T_0 - T_2$$

$$\therefore \quad T_0 = \frac{1}{2}(T_1 + T_2)$$

■ **Example 7.13** A leather belt transmits 8 kW of power from a pulley 1.1 m in diameter running at 200 rpm. The angle of lap is 160° and the coefficient of friction between belt and pulley is 0.25. The maximum safe working stress in the belt is 2.2 N/mm^2. Thickness of belt is 8 mm and the density of leather is 0.001 g/mm^3. Find the width of the belt taking centrifugal tension into account. [KU, Dec. 2008]

Solution : Given : $P = 8\,\text{kW}$, $d_1 = 1.1$ m, $n_1 = 200$ rpm, $\theta = 160°$, $\mu = 0.25$, $\sigma_w = 2.2\,\text{N/mm}^2$, $t = 8$ mm, $\rho = 0.001\,\text{g/mm}^3$ or $1000\,\text{kg/m}^3$

Let b = width of belt, mm

Mass of belt per metre length,

$$m = bt \times l \times \rho = b \times 8 \times 10^{-6} \times 1 \times 10^3 = 8b \times 10^{-3} \text{ kg/m}$$

Centrifugal tension,

$$T_c = mv^2$$

$$v = \frac{\pi d_1 n_1}{60} = \frac{\pi \times 1.1 \times 200}{60} = 11.519 \text{ m/s}$$

$$T_c = 8b \times 10^{-3}(11.519)^2 = 1.062 \times b \text{ N}$$

$$T_{max} = \sigma_w \times bt = 2.2 \times b \times 8 = 17.6\, b \text{ N}$$

$$T_1 = T_{max} - T_c = 17.6b - 1.062\text{b} = 16.538\, b \text{ N}$$

Now

$$\frac{T_1}{T_2} = \exp(\mu\theta) = \exp\left(0.25 \times \frac{\pi}{180} \times 160\right) = \exp(0.69813) = 2.01$$

Power transmitted, $P = \frac{(T_1 - T_2)v}{10^3}$

$$8 = \frac{T_1\left(1 - \frac{1}{2.01}\right) \times 11.519}{10^3} = 5.788 \times T_1$$

$$T_1 = \frac{8 \times 10^3\ 2.01}{11.519 \times 1.01} = 1382 \text{ N} = 16.538b$$

$$b = 83.6 \text{ mm}$$

■ **Example 7.14** Explain the following with the help of sketches :

(i) Idler puleys.

(ii) Intermediate pulleys. [KU, Dec. 2008]

Solution :

(i) Idler pulley : An idler pulley is used to increase the angle of contact of belt on the driven pulley when the power is being transmitted to two or more pulleys, as shown in Fig. 7.11.

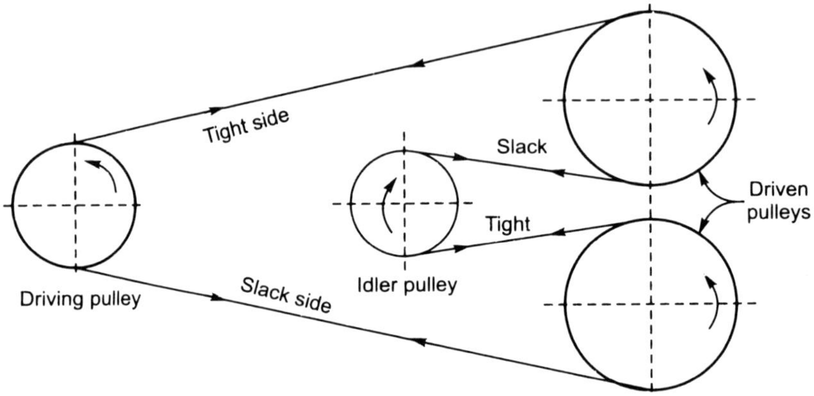

Fig. 7.11

(ii) Intermediate pulley : A pulley mounted on a shaft between the driving and driven shaft pulleys is called the intermediate pulley. The shaft on which this pulley is mounted is called the intermediate shaft. The intermediate pulley is used to increase the speed ratio as shown in Fig. 7.12.

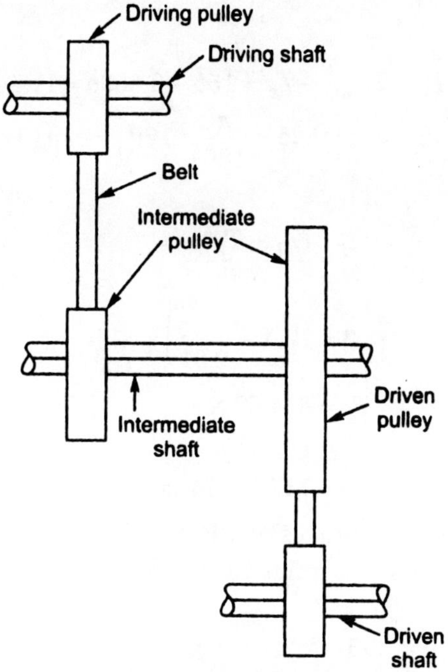

Fig. 7.12

■ **Example 7.15** Differentiate between Cross and Open belt drives. [KU, Dec. 2009]

Solution :

Parameter	*Open belt drive*	*Cross belt drive*
1. Direction of rotation of driven pulley	Same	Opposite
2. Angle of contact on smaller pulley	Less than 180°	More than 180°
3. Power transmitting capacity	Less	More
4. Slip	More	Less

■ **Example 7.16** A leather belt is required to transmit 8 kW from a pulley 1.5 m diameter running at 240 rpm. The angle of contact is160 degrees and the coefficient of friction between belt and pulley is 0.25. The safe working stress for leather is 1.5 MPa and density of leather is 1000 kg/m^3. Determine the width of the belt if its thicknes is 10 mm. Take into account the effect of centrifugal tension. [KU, Dec. 2009]

Solution : Given : $P = 8\,\text{kW}$, $d = 1.5$ m, $n = 240$ rpm, $\theta = 160°$, $\mu = 0.25$,

$$\sigma_w = 1.6\,\text{MPa}, \rho = 1000\,\text{kg/m}^3, t = 10\text{ mm}$$

Let b = width of belt, mm

Mass of belt per metre length, $m = btl\rho$

$$= b \times 10 \times 10^{-6} \times 1 \times 1000 = b \times 10^{-2}\text{ kg/m}$$

Belt speed, $$v = \frac{\pi dn}{60} = \frac{\pi \times 1.5 \times 240}{60} = 6\pi\text{ m/s}$$

Centrifugal tension,

$$T_c = mv^2 = b \times 10^{-2} \times (6\pi)^2 = 3.553\text{b N.}$$

$$T_{max} = \sigma_w \times bt = 1.5 \times b \times 10 = 15b\text{ N}$$

Tight side tension,

$$T_1 = T_{max} - T_c = 15b - 3.553b = 11.447b \text{ N}$$

$$\frac{T_1}{T_2} = \exp(\mu\theta) = \exp\left(0.25 \times \frac{\pi}{180} \times 160\right) = \exp(0.6981) = 2.01$$

Power transmitted,

$$P = \frac{(T_1 - T_2)v}{10^3}$$

$$8 = T_1\left(1 - \frac{1}{2.01}\right) \times \frac{6\pi}{10^3}$$

$$T_1 = 844.62 \text{ N}$$

$$= 11.447b$$

$$b = 73.78 \simeq 74 \text{ mm}$$

■ **Example 7.17** (a) Give the advantages of a V-belt drive over flat belt drive.

(b) The following data relate to a flat belt drive :

Power transmitted = 20 kW
Pulley diameter = 200 cm
Angle of contact = 175°
Speed of pulley = 300 rpm
Coefficient of friction between belt and pulley surfaces = 0.30
Permissible stress for belt = 250 N/cm^2
Thickness of belt = 8 mm
Density of belt material = 0.95×10^{-3} kg/cm^3.

Determine the width of belt required taking centrifugal tension into account.

[MDU, Dec. 2008]

Solution :

(a) The advantages of V-belt drive over flat belt drive are :
   (i) More positive drive because of negligible slip.
   (ii) There is no joint problem as the belt is endless.
   (iii) Longer life.
   (iv) More power transmission capacity
   (v) Easy to instal and remove.

(b) Given : $P = 20$ kW, $D = 200$ cm, $\theta = 175°$, $N = 300$ rpm, $\mu = 0.30$, $\sigma_w = 250$ N/cm^2, $t = 8$ mm, $\rho = 0.95 \times 10^{-3}$ kg/cm^3

Belt speed,
$$v = \frac{\pi DN}{10^2 \times 60} = \frac{\pi \times 200 \times 300}{10^2 \times 60} = 31.4 \text{ m/s}$$

$$P = \frac{(T_1 - T_2)v}{10^3}$$

$$T_1 - T_2 = \frac{20 \times 10^3}{31.4} = 636.62 \text{ N}$$

Let b = width of belt, cm

Mass of belt per metre length,

$$m = bt \times 10^2 \times \rho = b \times 0.8 \times 10^2 \times 0.95 \times 10^{-3}$$

$$= 0.076 \times b \text{ kg/m}$$

Centrifugal tension,

$$T_c = mv^2 = 0.076 \times b \times (31.4)^2 = 74.93 \times b \text{ N}$$

Maximum tension in belt,

$$T_{max} = b \times 0.8 \times 250 = 200 \times b \text{ N}$$

Tight side belt tension,

$$T_1 = 200b - 74.93b = 125.07b \text{ N}$$

$$\frac{T_1}{T_2} = \exp(\mu\theta) = \exp\left(0.30 \times \frac{\pi \times 175}{180}\right) = 2.5$$

$$T_2 = \frac{125.07\ b}{2.5} = 50.03\ b \text{ N}$$

$$(125.07 - 50.03)b = 636.62$$

$$b = \frac{636.62}{75.04} = 8.48 \simeq 8.5 \text{ cm}$$

SUMMARY

(a) Flat Belt Drive

1. Belts are used for short drive to transmit motion on parallel shafts.
2. Angular velocity ratio :

$$n_2/n_1 = d_1/d_2 = (d_1 + t)/(d_2 + t) = (1 - s/100) \cdot (d_1 + t)/(d_2 + t)$$

where $s = (s_1 + s_2 - 0.01 s_1 s_2)$ is the total fractional slip.

t = thickness of belt.

3. Law of Belting : The law of belting states that the centre line of the belt as it approaches the pulley must lie in a plane perpendicular to the axis of that pulley, or must lie in the plane of the pulley, otherwise the belt will run off the pulley.
4. Length of the open belt,

$$L_o = 2C + (d_2 - d_1)^2/(4C) + \pi(d_1 + d_2)/2$$

5. Length of cross belt,

$$L_c = 2C + \pi(d_1 - d_2)^2/2 + (d_1 + d_2)^2/(4C)$$

6. Angle of arc of contact

$$\theta = \pi - 2\sin^{-1}[(d_2 - d_1)/(2C)] \quad \text{for open type belt}$$

$$= \pi + 2\sin^{-1}[(d_1 + d_2)/(2C)] \quad \text{for cross type belt}$$

7. Ratio of belt tensions

$$T_1/T_2 = \exp(\mu\theta)$$

8. Power transmitted

$$P = (T_1 - T_2)v/10^3 = T_1 v[1 - \exp(-\mu\theta)]/10^3 \text{ kW}$$

9. Centrifugal tension

$$T_c = mv^2$$

10. Effective tension on tight side $= T_1 + T_c$

    Effective tension on slack side $= T_2 + T_c$

11. Condition for maximum power transmission

$$v = [T/(3m)]^{0.5} \text{ or } T_c = \frac{T}{3}$$

12. Initial belt tension

$$T_o = (T_1 + T_2)/2$$
$$= (T_1 + T_2 + 2T_c))/2 \text{ considering centrifugal tension.}$$

13. The jumping of the belt forward and backward on the driving and driven pulleys respectively due to unequal tension on the two sides of the belt, is called creep.
14. Creep $= (T_1 - T_2)/(btE)$.
15. Velocity ratio after accounting for creep.

$$(n_2/n_1) = (d_1 / d_2)\ [(E + \sqrt{\sigma_2})/(E + \sqrt{\sigma_1})],\ \sigma_1 = \frac{T_1}{bt},\ \sigma_2 = \frac{T_2}{bt}$$

16. The convex curvature given to the pulley rim is called crowing.
17. Crowning helps in running the belt in the centre of the pulley width.
18. Cone pulleys are used to obtain a range of speeds between the connected shafts.
19. Fast pulley is keyed to the shaft and loose pulley is free on the shaft.

(b) *V*-Belt Drive

20. Virtual (or apparent) coefficient of friction,

$$\mu_e = \mu/\sin\beta$$

    where β = semi-groove angle of pulley.

21. Ratio of belt tensions,

$$T_1/T_2 = \exp(\mu_e\theta).$$

(c) Chain Drive

22. Chains are used for short centre drive.
23. Chain pitch, $p = D\sin(180°/z)$.
24. Chain length, $L = 2C + \pi(R_1 + R_2) + (R_1 - R_2)^2/C$

$$R = (p/2)\ \text{cosec}\ (180°/z).$$

(d) Rope Drive

25. Ropes are used for long distance drive.
26. Ratio of rope tensions, $T_1/T_2 = \exp(\mu_e\theta)$, $\mu_e = \mu/\sin\beta$, $\beta =$ semi-groove angle of sheave.

SHORT ANSWER TYPE QUESTIONS

1. State the law of belting.
2. What is the effect of belt thickness and slip on speed ratio?
3. What is the effect of centrifugal tension on power transmission?
4. What is the role of initial tension in flat belt drive?
5. Write the expression for ratio of belt tensions.
6. What is centrifugal tension in a belt?
7. What is the condition for maximum power transmission by a belt?
8. What is belt creep?
9. Why crowing of pulley is done?
10. Differentiate between a fast and loose pulley.
11. What is virtual coefficient of friction in a *V*-belt drive?
12. Define chain pitch.
13. When rope drive is preferred?
14. When chain drive is preferred?
15. When *V*-belt drive is preferred?

EXERCISES

1. A flat belt drive is required to transmit 20 kW at 300 rpm of 2 m diameter pulley. The angle of contact is 170° and coefficient of friction between belt and pulley is 0.30. The permissible stress for belt material is 3 MPa. Thickness of belt is 8 mm and density of its material is 0.95 kg/m^3. Find the width of belt required taking centrifugal tension into account. **[Ans.** 45 mm]
2. A shaft running at 500 rpm carries a pulley 1 m diameter and drives another pulley by means of ropes with a speed ratio of 2 : 1. The drive transmits 200 kW. Angle of groove is 40° and distance between pulley centres is 2 m. The coefficient of friction between rope and pulley is 0.20. The mass of rope is 0.12 kg/m and has a safe stress of 1.75 MPa. The initial tension in the rope should not exceed 800 N. Calculate the number of ropes required and rope diameter. **[Ans.** 8, 30 mm]
3. If the difference between tight and slack side tensions for a leather belt does not exceed 100 N/cm of width for a belt 5 mm thick, find the maximum stress in the belt. Assume the following data: Angle of lap = 170°, coefficient of friction = 0.25, density of leather = 10^{-3} kg/cm^3, belt speed = 1000 m /min. **[Ans.** 4.1 MPa]
4. The initial tension in a flat drive is 1800 N and angle of lap on the smaller pulley is 170°. The coefficient of friction between belt and pulley surface is 0.25. The pulley diameter is 0.9 m and runs at 540 rpm. Neglecting centrifugal tension, determine the power that can be transmitted. **[Ans.** 36.59 kW]
5. It is required to reduce speed from 360 to 120 rpm by the use of chain drive. The driving sprocket has 10 teeth. Calculate : (a) the number of teeth on the follower, (b) the pitch of chain if pitch circle diameter of follower is 0.5 m, (c) the pitch circle diameter of driver, and (d) the length of chain, if centre distance is 0.4 m. **[Ans.** 30, 52.5 mm, 85, 1.95 m]
6. A rope pulley having a mean diameter of 1.5 m rotates at 90 rpm; angle of lap of ropes = 190°; angle of groove = 45°, safe tension per rope = 750 N, and coefficient of friction between the ropes and sides of groove = 0.25. Calculate the number of ropes for transmitting 50 kW. **[Ans.** 18]

7. A belt of density 1000 kg/m^3 has maximum permissible stress of 2.5 MPa. Calculate the maximum power that can be transmitted by a belt of 200 mm × 12 min size if the ratio of tensions is 2. [**Ans.** 57.18 kW]

8. Determine the maximum power that can be transmitted by a belt of 100 mm × 10 mm size with an angle of lap of 160°. The belt density is 1000 kg/m^3 and coefficient of friction is 0.25. The tension in the belt should not exceed 1.5 MPa. [**Ans.** 11.18 kW]

9. The included angle of a *V*-grooves pulley is 30°. The belt is 20 mm deep and maximum width is 20 mm. The mass of belt is 0.35 kg per metre length and maximum allowable stress is 1.4 MPa. Determine the maximum power that can be transmitted if angle of lap is 140° and coefficient of friction is 0.15. [**Ans.** 4.09 kW]

10. A pulley used to transmit power by means of ropes has a diameter of 3.6 m and has 15 grooves of 45° angle. The angle of contact is 170° and coefficient of friction between ropes and groove side is 0.28. The maximum possible tension in the ropes is 960 N and rope mass is 1.5 kg per metre length. Calculate the pulley speed in rpm and maximum power transmitted. [**Ans.** 77.45 rpm, 124.16 kW]

❑❑❑

Chapter 8

Cams

Inside This Chapter

8.1 INTRODUCTION

A cam is a rotating or a reciprocating element of a mechanism which imparts rotating, reciprocating or oscillating motion to another element, called the follower. There is line contact between the cam and the follower, and thereby forms a higher pair. Cams are used in clocks, printing machines, automatic screw cutting machines, internal combustion engines for operating the valves, and shoe making machines etc. In this chapter, we shall study the various types of planar cams from the point of view of drawing their profile and motion analysis.

The three essential components of a cam mechanism are :

(i) the cam,

(ii) the follower, and

(iii) the frame.

The cam revolves at a constant speed and drives the follower. The motion of the follower depends upon the profile of the cam. The frame supports and guides the follower and the cam.

8.2 CLASSIFICATION OF CAMS

The planar cams can be classified according to their shape as follows :

1. **Wedge and flat cams :** Such a cam has a wedge A to which transnational motion is given to actuate the follower *B* in order to either reciprocate or oscillate it, as shown in Fig. 8.1(a) and (b). The follower is guided in the guides C. In Fig. 8.1(c), the cam is stationary and the follower guide *C* cause the relative motion of the cam *A* and followr *B*. Fig. 8.1(d) shows a flat plate with a groove in which the follower is held to obtain the desired motion.

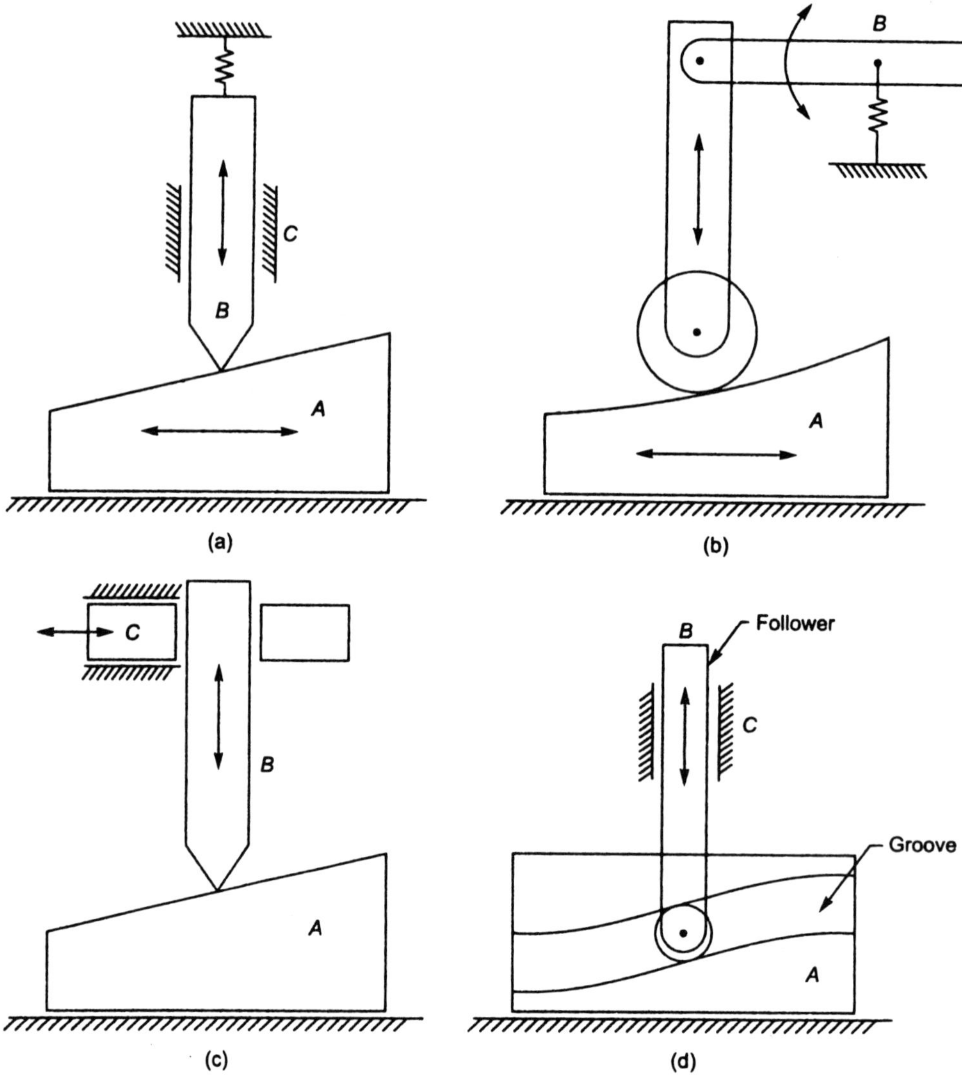

Fig. 8.1 Wedge and flat cams.

2. **Radial and Offset Cams :** A cam in which the follower moves radially from the centre of rotation of the cam is called a radial (or disc or plate) cam, as shown in Fig. 8.2(a) and (b). In such a cam, the axis of the follower passes through the axis of the cam.

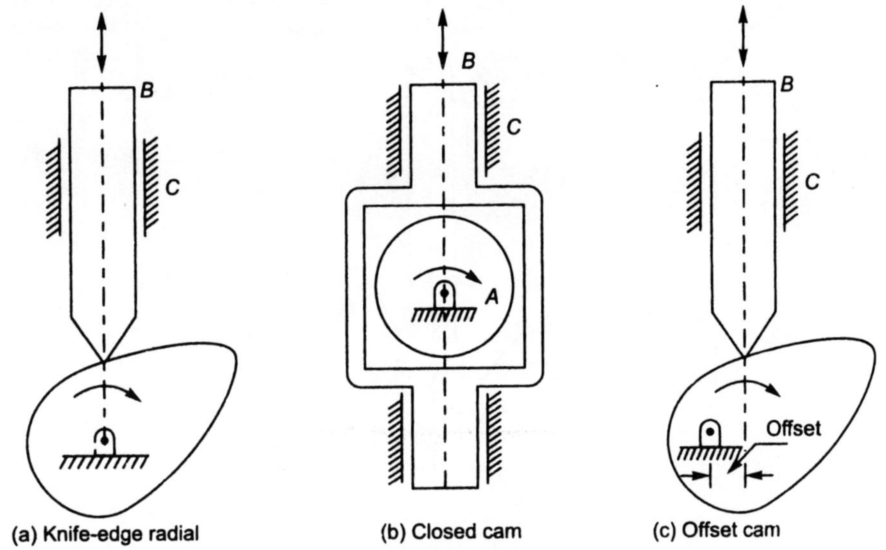

Fig. 8.2 Radial and offset cams.

If the axis of the follower does not pass through the axis of the cam, it is called an offset cam, as shown in Fig. 8.2(c).

3. **Cylindrical Cams :** In a cylindrical cam, a cylinder which has a circumferential groove cut in the surface, rotates about its axis. The follower motion can be either oscillaotory or reciprocating type, as shown in Fig. 8.3 (a) and (b). They are also called barrel or drum cams.

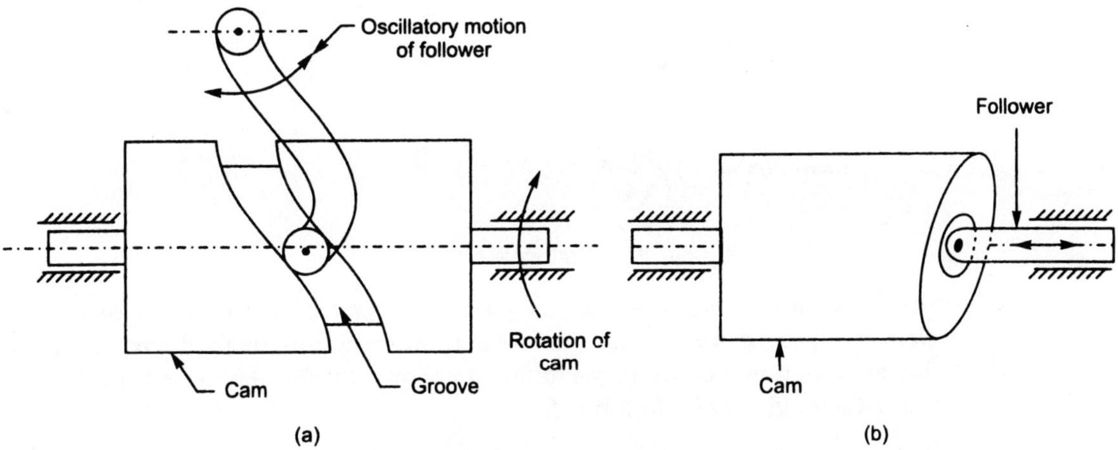

Fig. 8.3 Cylindrical cams.

4. **Spiral Cams :** A spiral cam is a face cam in which a groove is cut in the form of a spiral, as shown in Fig. 8.4. The spiral groove consists of teeth which mesh with a pin gear follower.
5. **Conjugave Cams :** It is a double disc cam in which the two discs keyed together are in constant touch with the two rollers of a follower, as shown in Fig. 8.5. Such a cam gives low wear, low noise, better control of follower, high speed, and high dynamic loads, etc.

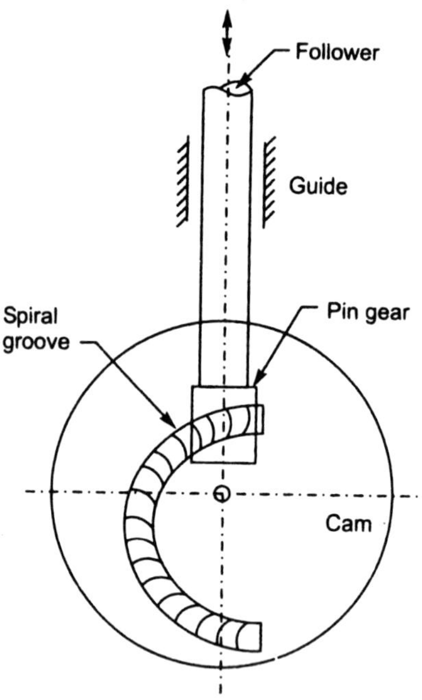

Fig. 8.4 Spiral cam.

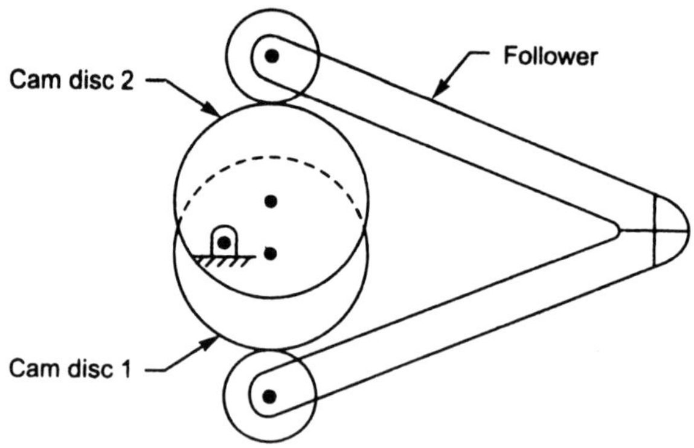

Fig. 8.5 Conjugate cam.

6. **Globoidal Cams :** A globoidal cam has either a convex or a concave surface. A circumferential groove is cut on the surface of rotation of the cam to impart motion to the follower, which has oscillatory motion, as shown in Fig. 8.6(a) and (b). This is used where the angle of oscillation is large.

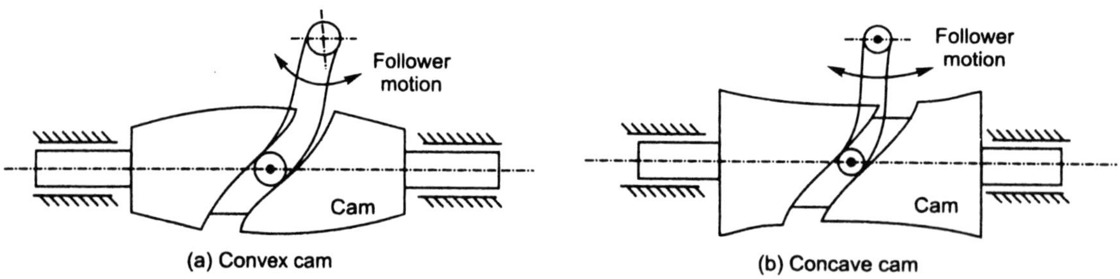

Fig. 8.6 Globoidal cam.

7. **Spherical Cams :** In a spherical, cam, the follower ocillates about an axis perpendicular to the axis of rotation of the cam. The spherical cam is in the form of a spherical surface, as shown in Fig. 8.7.

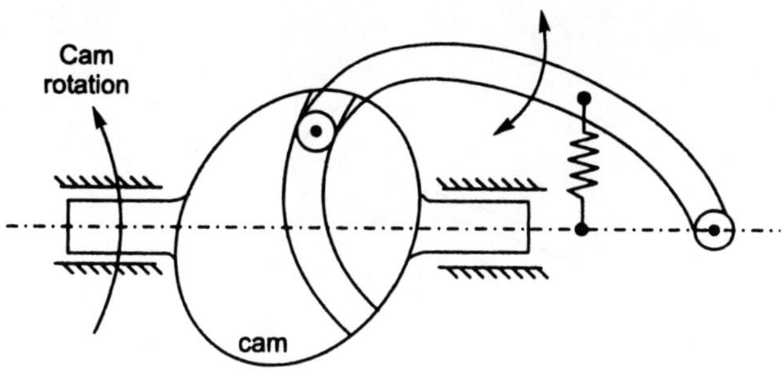

Fig. 8.7 Spherical cam.

8.3 TYPES OF FOLLOWERS

The followers can be of the following types :

1. **Based on the surface in contact :** The followers based on the types of surface in contact can be classified as : knife edge, roller, flat faced (or mushroom), and spherical faced follower, as shown in Fig. 8.8(a) to (g).
2. **Based on the motion of the follower :** Depending upon the motion, the follower could be of the reciprocating or translating, oscillating or rotating types.
3. **Based on the path of motion of follower :** When the axis of the follower passes through the centre of rotation of the cam, it is called a radial follower, and when the axis of the follower does not pass through the axis of the cam, it is called an offset follower.

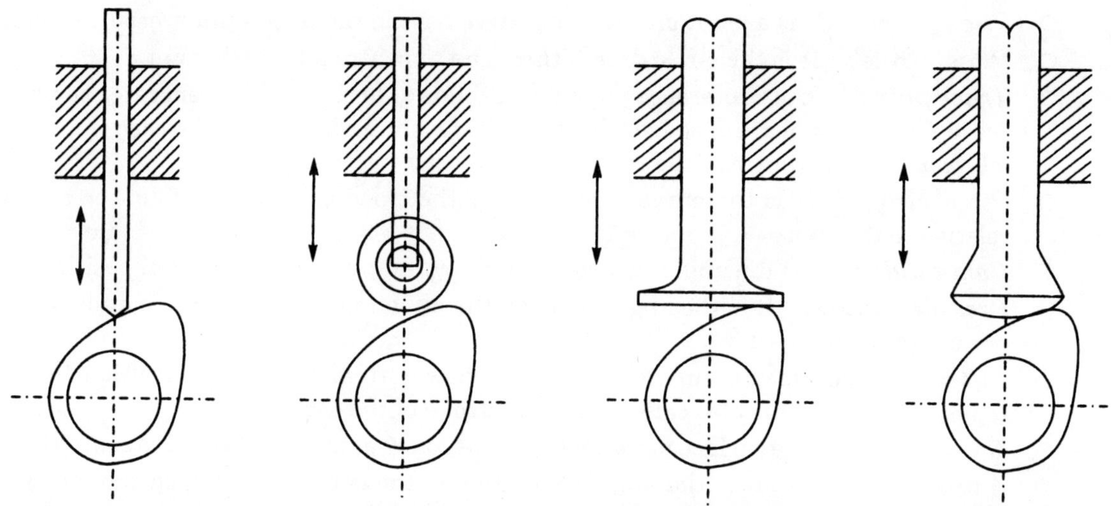

(a) Knife-edge follower (b) Roller follower (c) Flat faced (plain/ mushroom) follower (d) Spherical-faced follower (mushroom)

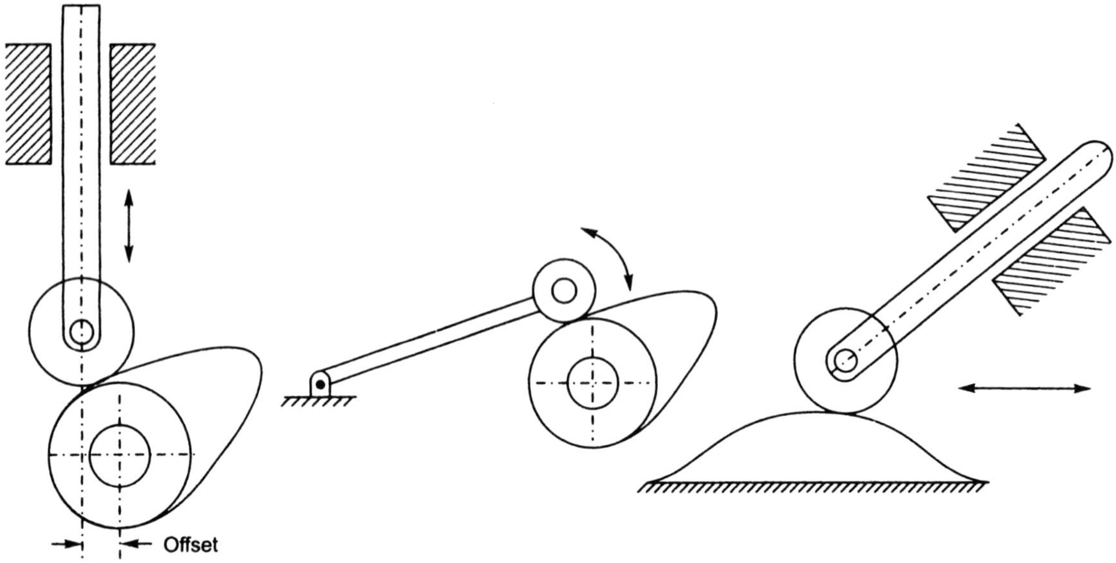

(e) Offset follower (f) Oscillating follower (g) Translating follower

Fig. 8.8 Types of fo!lowers.

8.4 CAM NOMENCLATURE

A radial cam with reciprocating roller follower is shown in Fig. 8.9. The following nomenclature is used in reference to planar cam mechanisms :

- ***Base circle :*** It is the smallest circle that can be drawn to the cam profile from the centre of rotation.
- ***Prime circle :*** It is the smallest circle drawn to the pitch curve from the centre of rotation of the cam.
- ***Pitch point :*** It is a point on the pitch curve having the maximum pressure angle.
- ***Pitch circle :*** It is the circle drawn through the centre and pitch point.
- ***Trace point :*** It is a reference point onthe follower and is used to generate the pitch curve. In the case of a knife edge follower, it is the knife edge, and in the case of a roller follower, it is the centre of the roller.
- ***Pitch curve :*** It is the curve generated by the trace point as the follower moves relative to the cum.
- ***Cam angle :*** It is the angle turned through by the cam from the initial position.
- ***Pressure angle :*** It is the angle between the direction of the follower motion and a normal to the pitch curve.
- ***Lift :*** It is the maximum travel of the follower from the lowest position to the topmost position. It is also called throw or stroke of the cam.
- ***Cam profile :*** The curface in contact with the follower is known as the cam profile.
- ***Angle of ascent :*** It is the angle of rotation of the cam during which the follower rises up.
- ***Angle of descent :*** It is the angle of rotation of the cam during which the follower lowers down.

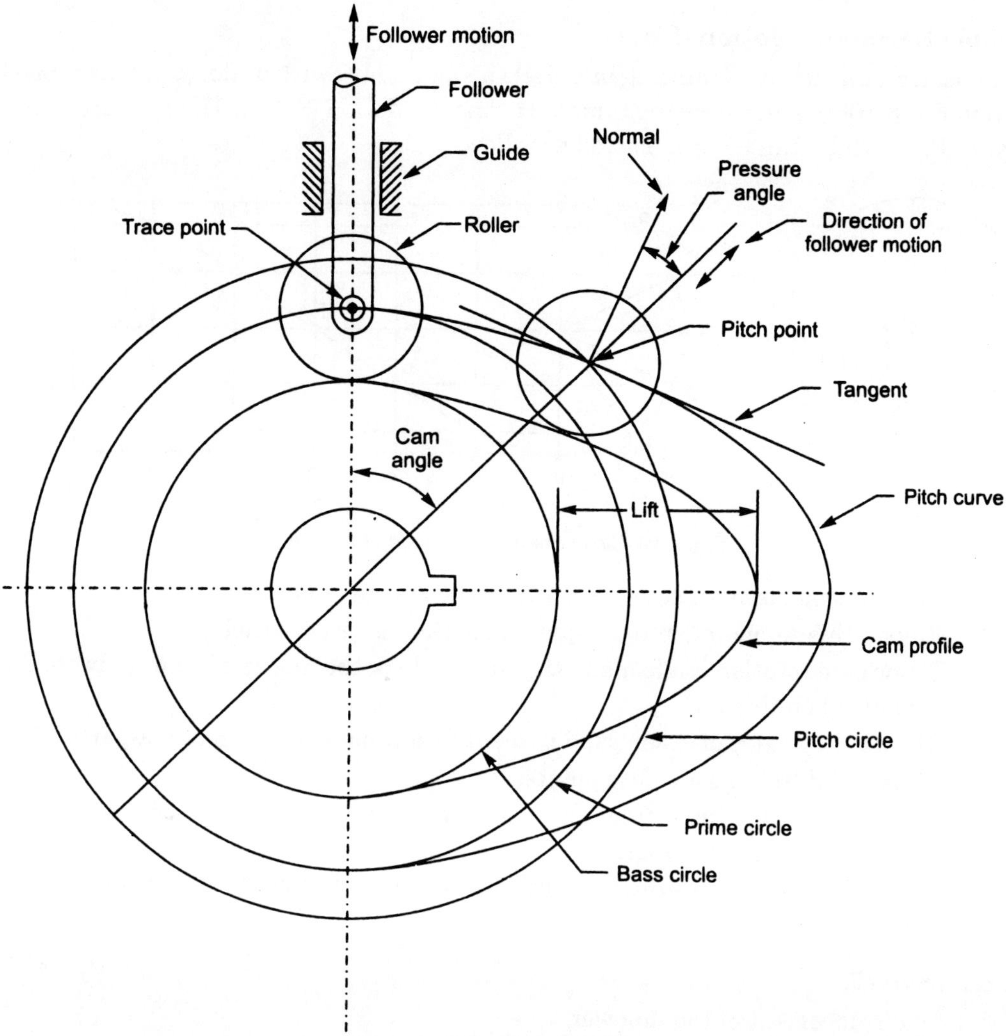

Fig. 8.9 Cam nomenclature.

- ***Angle of dwell:*** It is the angle of rotation of cam during which the follower remains stationary.

8.5 FOLLOWER MOTIONS

The follower can have following type of motions :

1. Simple harmonic motion (SHM),
2. Uniform acceleration and deceleration,
3. Uniform velocity,
4. Parabolic motion, and
5. Cycloidal motion

We shall discuss all these motions.

1. Simple Harmonic Motion (SHM)

Consider a particle at A rotating in a circle about point O with uniform angular speed ω, as shown in Fig. 8.10(a), and executing Simple Harmonic Motion (SHM). The displacement curve shown in Fig. 8.10(b), can be constructed as follows :

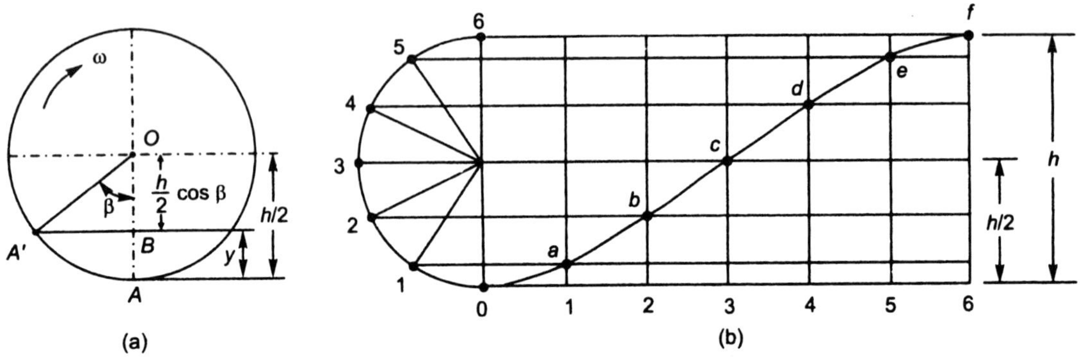

Fig. 8.10 Simple harmonic motion of follower.

1. Draw a semicircle with follower lift as the vertical diameter.
2. Divide this semicircle into n equal parts (say 6, *i.e.*, 30° each).
3. Draw cam rotation angle along the x-axis. Mark the angles of ascent, dwell, descent, and dwell on this line.
4. Divide the angles of ascent and descent into same equal number of parts.
5. Draw vertical lines at these points.
6. Draw horizontal lines from the points on the circumference of the semicircle to intersect the vertical lines.
7. Mark the points of intersection and join by a smooth curve to obtain the displacement diagram.

Motion Analysis

Let y = displacement of the follower

h = lift of the follower (maximum displacement or rise)

θ = angle turned through by the crank OA from given datum

ϕ = cam rotation angle for the maximum follower displacement

β = angle on harmonic circle.

Then

$$y = OA - OB$$

$$= \left(\frac{h}{2}\right) \times (1 - \cos\beta) \qquad \text{...(1)}$$

For the ascent or desecnt h of the follower dispalcement, the cam is rotated through an angle ϕ, whereas a point on the harmonic semicircle traverses an angle π radians. Thus, the cam rotation is proportional to the angle turned through by the point on the harmonic semicircle, *i.e.*,

$$\beta = \frac{\pi\theta}{\phi}$$

Thus, Eq. (1) becomes,

$$y = \left(\frac{h}{2}\right)\left[1 - \cos\left(\frac{\pi\theta}{\phi}\right)\right] \qquad \text{...(8.1a)}$$

Now $\theta = \omega t$

$$y = \left(\frac{h}{2}\right)\left[1 - \cos\left(\frac{\pi\omega t}{\phi}\right)\right] \qquad \text{...(8.1.b)}$$

Velocity, $$v = \frac{dy}{dt}$$

Differentiating Eq. (8.1b), we get

$$v = \left(\frac{h}{2}\right)\left(\frac{\pi\omega}{\phi}\right)\sin\left(\frac{\pi\omega t}{\phi}\right)$$

$$= \left(\frac{h}{2}\right)\left(\frac{\pi\omega}{\phi}\right)\sin\left(\frac{\pi\theta}{\phi}\right) \qquad \text{...(8.2)}$$

Let θ_1 = angle of ascent
θ_2 = angle of dwell
θ_3 = angle of descent

Then velocity during ascent,

$$v_a = \left(\frac{h}{2}\right)\left(\frac{\pi\omega}{\theta_1}\right)\sin\left(\frac{\pi\theta}{\theta_1}\right) \qquad \text{...(8.3a)}$$

Velocity during descent,

$$v_d = \left(\frac{h}{2}\right)\left(\frac{\pi\omega}{\theta_3}\right)\sin\left(\frac{\pi\theta}{\theta_3}\right) \qquad \text{...(8.3b)}$$

The velocity is maximum when $\theta = \frac{\phi}{2}$, and from Eq. (8.2), we get

$$v_{max} = \left(\frac{h}{2}\right)\left(\frac{\pi\omega}{\phi}\right) \qquad \text{...(8.4)}$$

Maximum velocity of follower during ascent $= \left(\frac{h}{2}\right)\left(\frac{\pi\omega}{\theta_1}\right)$ ...(8.5a)

Maximum velocity of follower during descent $= \left(\frac{h}{2}\right)\left(\frac{\pi\omega}{\theta_3}\right)$ ...(8.5b)

Acceleration, $$f = \frac{dv}{dt}$$

Differentiating Eq. (8.2), we get

$$f = \left(\frac{h}{2}\right)\left(\frac{\pi\omega}{\phi}\right)^2 \cos\left(\frac{\pi\theta}{\phi}\right) \qquad \text{...(8.6)}$$

Acceleration during ascent,

$$f_a = \left(\frac{h}{2}\right)\left(\frac{\pi\omega}{\theta_1}\right)^2 \cos\left(\frac{\pi\theta}{\theta_1}\right) \qquad \text{...(8.7a)}$$

Acceleration during descent,

$$f_d = \left(\frac{h}{2}\right)\left(\frac{\pi\omega}{\theta_3}\right)^2 \cos\left(\frac{\pi\theta}{\theta_3}\right) \quad ...(8.7b)$$

The acceleration is maximum when $\theta = 0°$

$$f_{max} = \left(\frac{h}{2}\right)\left(\frac{\pi\omega}{\phi}\right)^2 \quad ...(8.8)$$

Maximum acceleration of follower during ascent $= \left(\frac{h}{2}\right)\left(\frac{\pi\omega}{\theta_1}\right)^2$ ...(8.9a)

Maximum acceleration of follower during descent $= \left(\frac{h}{2}\right)\left(\frac{\pi\omega}{\theta_3}\right)^2$ ...(8.9b)

The motion of the follower is shown in Fig. 8.11.

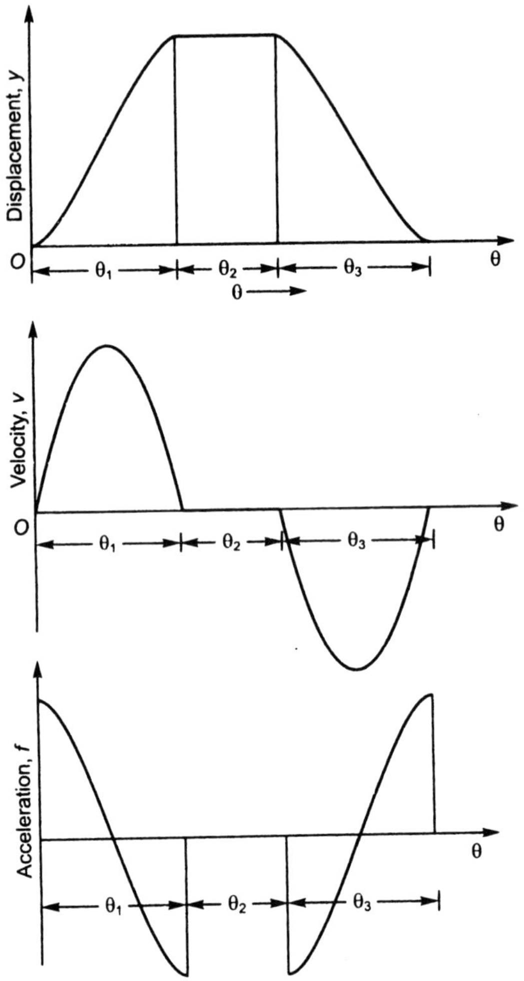

Fig. 8.11 Displacement, velocity and acceleration distribution in SHM of follower.

It can be observed from Fig. 8.11 that there is an abrupt change of acceleration from zero to maximum at the beginning of the follower motion and also from maximum (negative) to zero at the end of the follower motion. The same pattern is repeated during descent. This leads to jerk, vibration and noise, etc. Therefore, SHM should be adopted only for low and moderate cam speeds.

2. Motion with Uniform Acceleration and Deceleration

In such a motion, there is acceleration during the first half of the follower motion and deceleration during the later half. The magnitude of both acceleration and decleration is the same in the two halves.

The displacement diagram, as shown in Fig. 8.12, can be constructed as follows :

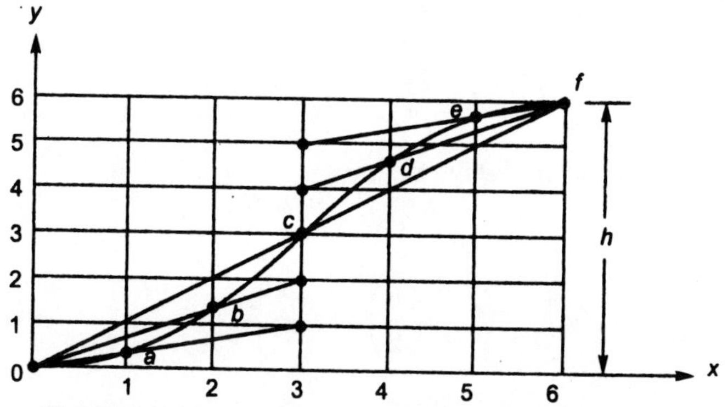

Fig. 8.12 Uniform acceleration and deceleration of follower.

1. Draw the cam rotation angle along the x-axis and the follower lift along the y-axis.
2. Mark the angles of ascent, dwell, descent, and dwell on the horizontal line and the lift on the vertical line at the origin.
3. Divide the angles of ascent and descent into equal number of parts (say 6). Also divide the lift line into same equal number of parts.
4. Draw horizontal and vertical lines at these points.
5. Join the origin with the points of intersection on the middle line of ascent upto half the lift. Also join the points on the other half of the middle line of lift with the topmost point of the last line.
6. Mark the points of intersection of these lines with the vertical lines.
7. Join these points with a smooth curve to get the displacement diagram.

Motion Analysis

Let f = uniform acceleration or deceleration of the follower

y = displacement of follower

Then
$$\frac{dv}{dt} = f$$

Integrating, we have
$$v = ft + C_1$$

where C_1 is a constant of integration

If at $t = 0$, $v = 0$, then $C_1 = 0$. Hence

$$v = ft$$

Now $$v = \frac{dy}{dt} = ft$$

Integrating again, we have

$$y = \frac{ft^2}{2} + C_2$$

where C_2 is another constant of integration.

As $y = 0$ at $t = 0$, therefore $C_2 = 0$. Hence

$$y = \frac{ft^2}{2}$$

$$f = \frac{2y}{t^2} = \text{constant} \qquad \text{....(8.10a)}$$

Considering the follower at midway, we have

$$y = \frac{h}{2} \text{ and } t = \frac{(\phi/2)}{\omega}$$

$$f = 2\left(\frac{h}{2}\right) \times \left[\frac{4\omega^2}{\phi^2}\right]$$

$$= \frac{4h\omega^2}{\phi^2} \qquad \text{...(8.10b)}$$

Maximum acceleration during ascent,

$$f_{max} = \frac{4h\omega^2}{\theta_1^2} \qquad \text{...(8.11a)}$$

Maximum acceleration during descent,

$$f_{max} = \frac{4h\omega^2}{\theta_3^2} \qquad \text{...(8.11b)}$$

Velocity, $$v = ft = \left[\frac{4h\omega^2}{\phi^2}\right]\left(\frac{\theta}{\omega}\right)$$

$$= \left(\frac{4h\omega}{\phi^2}\right)\theta \qquad \text{...(8.12)}$$

The velocity is maximum when the follower is at midway position, *i.e.*, $\theta = \frac{\phi}{2}$

Maximum velocity, $$v_{max} = \left(\frac{4h\omega}{\phi^2}\right)\left(\frac{\phi}{2}\right)$$

$$= \frac{2h\omega}{\phi} \qquad \text{...(8.13)}$$

Maximum velocity during ascent $= \frac{2h\omega}{\theta_1}$ ...(8.14a)

$$\text{Maximum velocity during descent} = \frac{2h\omega}{\theta_3} \quad ...(8.14b)$$

The motion of the follower is shown in Fig. 8.13.

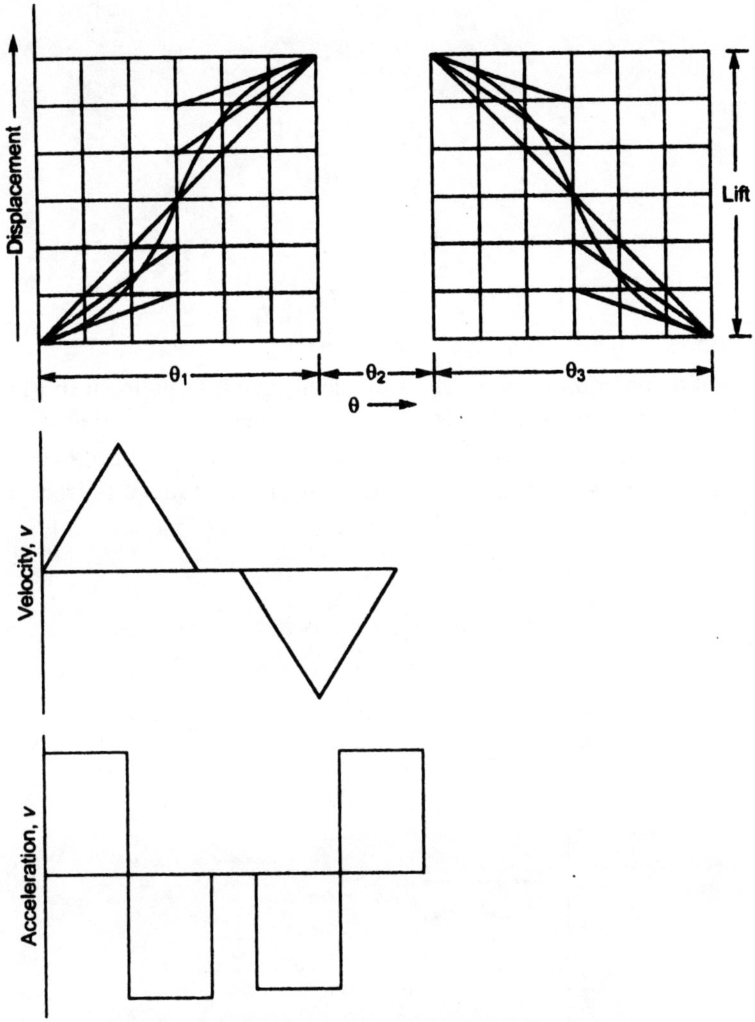

Fig. 8.13 Motion with uniform acceleration and deceleration.

It may be observed from Fig. 8.13 that there are abrupt changes in the acceleration at the beginning, midway and at the end of the follower motion. At the midway, an infinite jerk is produced. Therefore, this motion is adopted for moderate speeds only.

3. Motion with Uniform Velocity

In this case, the displacement of the follower is proportional to the angle of cam rotation. Therfore, slope of the displacement curve is constant.

Let $$y = c\theta$$

where c = constant of proportionality, and θ = angle of cam rotation.

If h = follower rise

ϕ = angle through which the cam is to rotate to lift the follower by h

Then $$h = c\phi$$

so that $$c = \frac{h}{\phi}$$

$\therefore$ $$y = \frac{h\theta}{\phi} \quad \text{...(8.15)}$$

Velocity, $$v = \left(\frac{h}{\phi}\right)\cdot\left(\frac{d\theta}{dt}\right)$$

$$= \frac{h\omega}{\phi} \quad \text{...(8.16)}$$

Acceleration, $$f = \left(\frac{h}{\phi}\right)\cdot\left(\frac{d\omega}{dt}\right) = 0 \quad \text{...(8.17)}$$

The variation of dispalcement, velocity and acceleration are shown in Fig. 8.14. It may be observed that although the acceleration is zero during ascent or descent of the follower, it is infinite at the beginning and end of the motion. There are abrupt changes in velocity at these points. This results in infinite inertia forces and is therfore unsuitable from practical point of view.

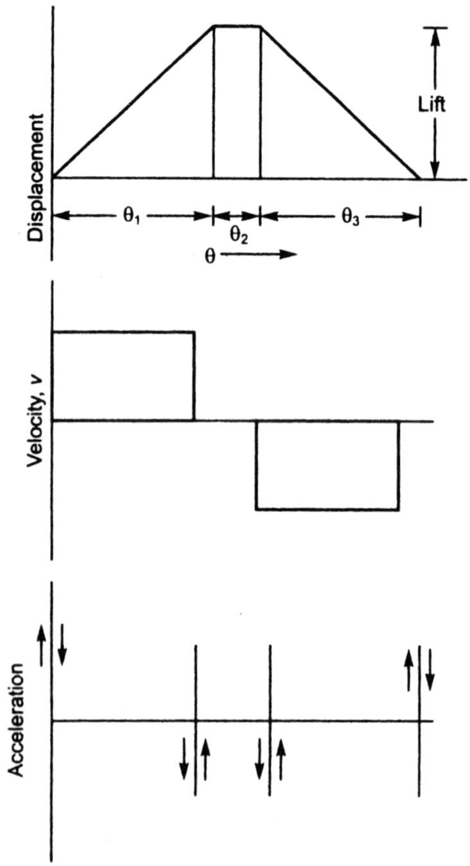

Fig. 8.14 Motion with uniform velocity.

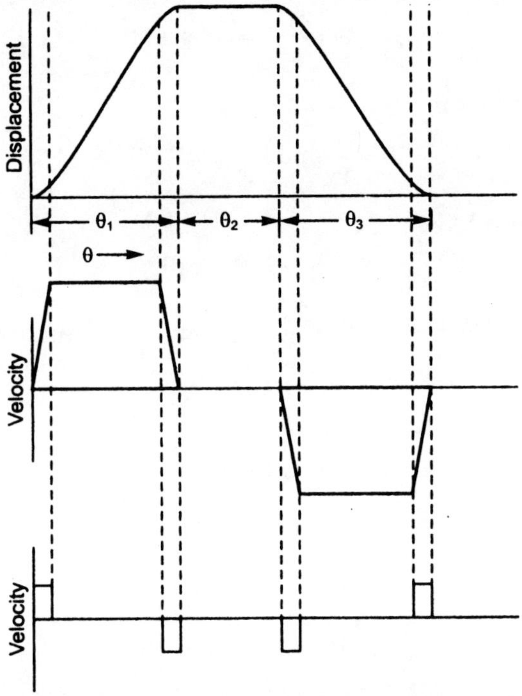

Fig. 8.15 Modified motion with uniform velocity.

This can be avoided by rounding the sharp corners of the displacement curve so that the velocity changes are gradual at the beginning and end of the follower motion. During these periods the acceleration may be assumed to be constant and of finite values. A modified uniform velocity motion is shown in Fig. 8.15.

4. Parabolic Motion

In the parabolic motion, the displacement of the follower is proportional to the square of the angle of cam rotation.

Let the parabolic motion, for the first half, be represented by,

$$y = c\theta^2$$

where c = a constant of proportionality

For $y = \frac{h}{2}$, $\theta = \frac{\phi}{2}$

$$\therefore \qquad \frac{h}{2} = \frac{c\phi^2}{4}$$

or

$$c = \frac{2h}{\phi^2}$$

Hence,

$$y = 2h\left(\frac{\theta}{\phi}\right)^2 \qquad \text{...(8.18)}$$

Velocity,

$$v = \frac{4h\omega\theta}{\phi^2} \qquad \text{...(8.19)}$$

Accleration, $$f = 4h\left(\frac{\omega}{\phi}\right)^2 = \text{constant.} \quad \text{...(8.20)}$$

For velocity to be maximum,

$$\theta = \frac{\phi}{2}, \text{ and } v_{\max} = \frac{2h\omega}{\phi} \quad \text{...(8.21)}$$

For the second half, we have

$$y = c_1 + c_2\theta + c_3\theta^2$$

For $\theta = \beta$, $y = h$ and $v = 0$. Also for $\theta = \frac{\phi}{2}$, $v = v_{max}$.

Therefore, $$h = c_1 + c_2\phi + c_3\phi^2$$

$$\theta = c_2\omega + 2c_3\omega\phi$$

or $$0 = c_2 + 2c_3\phi$$

Hence, $$c_1 = -h,\ c_2 = \frac{4h}{\phi},\ c_3 = -\frac{2h}{\phi^2}$$

$$y = h\left[1 - 2\left(1 - \frac{\theta}{\phi}\right)^2\right] \quad \text{...(8.22)}$$

$$v = \left(\frac{4h\omega}{\phi}\right)\left(1 - \frac{\theta}{\phi}\right) \quad \text{...(8.23)}$$

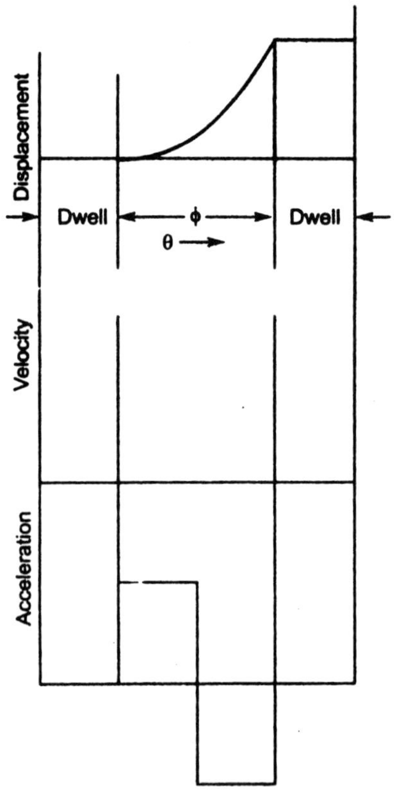

Fig. 8.16 Parabolic motion.

$$f = -4h\left(\frac{\omega}{\phi}\right)^2 \qquad \text{...(8.24)}$$

The variation of displacement, velocity and acceleration are shown in Fig. 8.16.

In this case, the displacement of the follower is proportional to the angle of cam rotaiton.

5. Cycloidal Motion

A cycloid is the locus of a point on a circle rolling on a straight line. The cycloidal curve, as shown in Fig. 8.17, can be constructed by adopting the following steps :

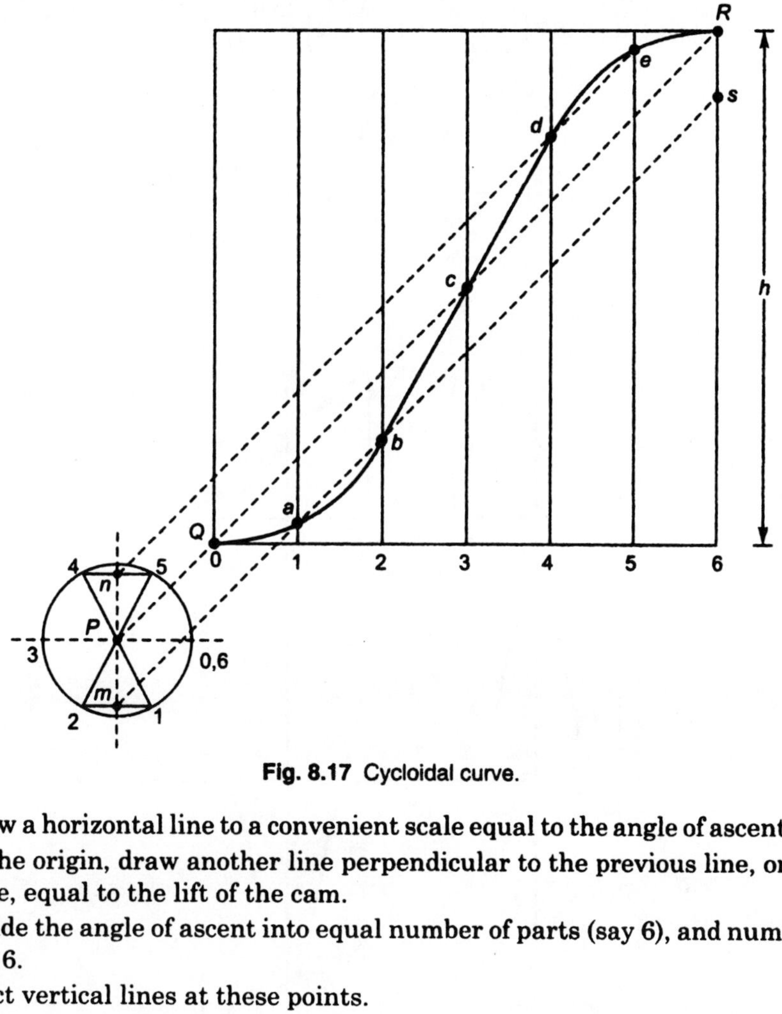

Fig. 8.17 Cycloidal curve.

1. Draw a horizontal line to a convenient scale equal to the angle of ascent of the cam.
2. At the origin, draw another line perpendicular to the previous line, on a convenient scale, equal to the lift of the cam.
3. Divide the angle of ascent into equal number of parts (say 6), and number them from 0 to 6.
4. Erect vertical lines at these points.
5. Draw the diagonal *QR* passing through the origin and produce it backward.
6. Calculate the radius of the circle, $r = \dfrac{h}{2\pi}$, where h = lift.
7. Select a convenient point *P* on the diagonal produced backwards and draw a circle with radius equal to *r*.
8. Divide the circle into six equal parts and number the ends of the diameters from 0 to 6.

9. Join points 1–2 and 4–5, intersecting the vertical diameter at points m and n respectively.
10. From points m and n draw lines parallel to QR intersecting vertical lines at 1 and 2 at 'a' and b and lines at 4 and 5 at d and e. The diagonal PQR shall intersect lines at 3 at c.
11. Join the points a, b, c, d, e and R by a smooth curve to get the cyclodidal curve.

Motion Analysis

A cycloid is expressed by,

$$y = \left(\frac{h}{\pi}\right)\left[\frac{\pi\theta}{\phi} - \left(\frac{1}{2}\right)\sin\left(\frac{2\pi\theta}{\phi}\right)\right] \quad \ldots(8.25)$$

$$v = \frac{dy}{dt} = \left(\frac{dy}{d\theta}\right)\left(\frac{d\theta}{dt}\right)$$

$$= \left[\frac{h}{\phi} - \left\{\frac{h}{2\pi}\right\}\left(\frac{2\pi}{\phi}\right)\cos\left(\frac{2\pi\theta}{\phi}\right)\right]\omega$$

$$= \frac{h\omega}{\phi} - \left(\frac{h\omega}{\phi}\right)\cos\left(\frac{2\pi\theta}{\phi}\right)$$

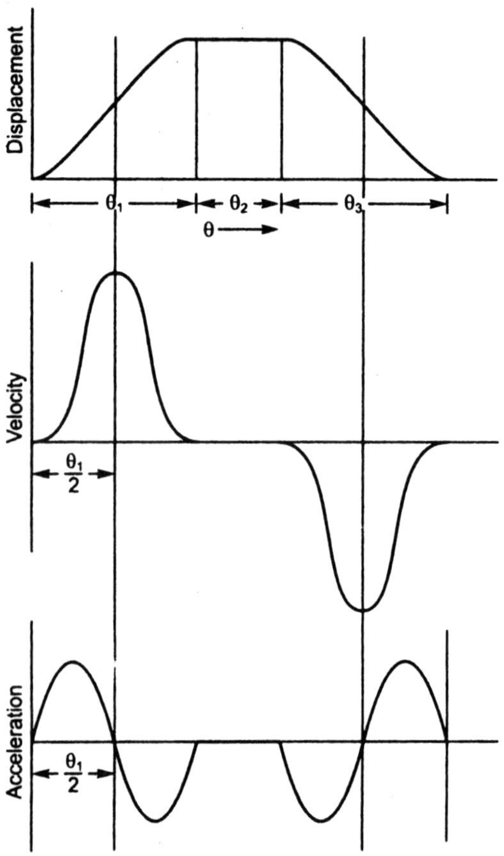

Fig. 8.18

$$= \left(\frac{h\omega}{\phi}\right)\left[1 - \cos\left(\frac{2\pi\theta}{\phi}\right)\right] \qquad ...(8.26)$$

$$v_{max} = \frac{2h\omega}{\phi} \text{ at } \theta = \frac{\phi}{2} \qquad ...(8.27)$$

$$f = \frac{dv}{dt} = \left(\frac{dv}{d\theta}\right)\frac{d\theta}{dt}$$

$$= \left(\frac{h\omega}{\phi}\right)\left(\frac{2\omega}{\phi}\right) \times \left[\sin\left(\frac{2\pi\theta}{\phi}\right)\right]\omega$$

$$= (2h\pi) \times \left(\frac{\omega}{\phi}\right)^2 \sin\left(\frac{2\pi\theta}{\phi}\right) \qquad ...(8.28)$$

$$f_{max} = (2h\pi) \times \left(\frac{\omega}{\phi}\right)^2 \text{ at } \theta = \frac{\phi}{4} \qquad ...(8.29)$$

The variation of cycloidal motion is shown in Fig. 8.18. It may be observed that there are no abrupt changes in the velocity and acceleration at any stage of the motion. Therefore, cycloidal motion is the most ideal for high speed follower motion.

8.6 CAM PROFILE WITH KNIFE-EDGE FOLLOWER

The following procedure may be adopted to draw the cam profile with knife edge follower :

1. Draw the displacement diagram for follower motion.
2. Consider that cam remains stationary and that the follower moves round it in a direction opposite to the direction of cam rotation.
3. Draw the cam base circle and divide its circumference into number of divisions depending upon the divisions used in the displacement diagram.
4. Draw various positions of follower with dotted lines corresponding to different angular dispalcement from the radius from which ascent is to commence.
5. Draw a smooth curve tangential to the contact surface in different positions.

■ **Example 8.1** A disc cam is to give SHM to a knife edge follower during out stroke of 50 mm. The angle of ascent is 120°, dwell 60°, and angle of descent 90°. The minimum radius of cam is 50 mm. Draw the profile of the cam when the axis of the follower passes through the axis of the camshaft.

Also calculate the maximum velocity and acceleration during ascent and descent when the camshaft revolves at 240 rpm.

Solution : Given $h = 50$ mm, $N = 240$ rpm. $\theta_1 = 120°$, $\theta_3 = 90°$

Angular velocity of camshaft,

$$\omega = \frac{2\pi N}{60} = \frac{2\pi \times 240}{60} = 25.133 \text{ rad/s}$$

With SHM, maximum velocity during ascent.

$$v_{max} = \left(\frac{\pi h}{2}\right)\left(\frac{\omega}{\theta_1}\right)$$

$$= \left(\frac{\pi \times 50 \times 10^{-3}}{2}\right) \times \left(\frac{25.133}{\frac{\pi}{180} \times 120}\right) = 0.942 \text{ m/s}$$

and during descent, $v_{max} = \left(\frac{\pi h}{2}\right)\left(\frac{\omega}{\theta_3}\right)$

$$= \left(\frac{\pi \times 50 \times 10^{-3}}{2}\right) \times \left(\frac{25.133}{\frac{\pi}{180} \times 90}\right) = 1.257 \text{ m/s}$$

Maximum acceleration during ascent,

$$f_{max} = \left(\frac{\pi\omega}{\theta_1}\right)^2 \times \left(\frac{h}{2}\right) = \left[\frac{\pi \times 25.133}{\frac{\pi}{180} \times 120}\right]^2 \times \left(\frac{50 \times 10^{-3}}{2}\right) = 35.53 \text{ m/s}^2$$

and during descent,

$$f_{max} = \left(\frac{\pi\omega}{\theta_3}\right)^2 \times \left(\frac{h}{2}\right) = \left[\frac{\pi \times 25.133}{\frac{\pi}{180} \times 90}\right]^2 \times \left(\frac{50 \times 10^{-3}}{2}\right) = 63.17 \text{ m/s}^2$$

Displacement Diagram

The dispalcement diagram shown in Fig. 8.19(a) may be drawn as follows :

1. Draw a vertical line 0-6 equal to the lift of 50 mm.
2. Draw a semicircle on this line and divide the semicircle into six equal parts of 30° each.
3. Draw a horizontal line 00′ perpendicular to 0-6 line representing cam rotation angle θ on a scale of 1 cm = 20°.
4. Divide the ascent angle of 120° into six equal parts and also the descent angle of 90° into six equal parts.
5. Erect perpendiculars at points 0 to 6 and 6′ to 0′.
6. Draw horizontal lines from points 1 to 5 on the semicircle to intersect vertical lines drawn previously, as shown in the figure.
7. Join the points of intersection with a smooth curve to get the displacement diagram.

Cam Profile

The cam profile shown in Fig. 8.19(b) may be drawn as follows :

1. Draw a circle of radius equal to the base circle radius of 50 mm with centre *O*.
2. Draw angles of ascent, dwell and descent of 120°, 60° and 90° respectively. Divide the angles of ascent and descent into six equal parts. Draw radial lines for these angles.
3. Mark points 0 to 6 and 6′ to 0′ on the base circle in the angles of ascent and descent respectively.

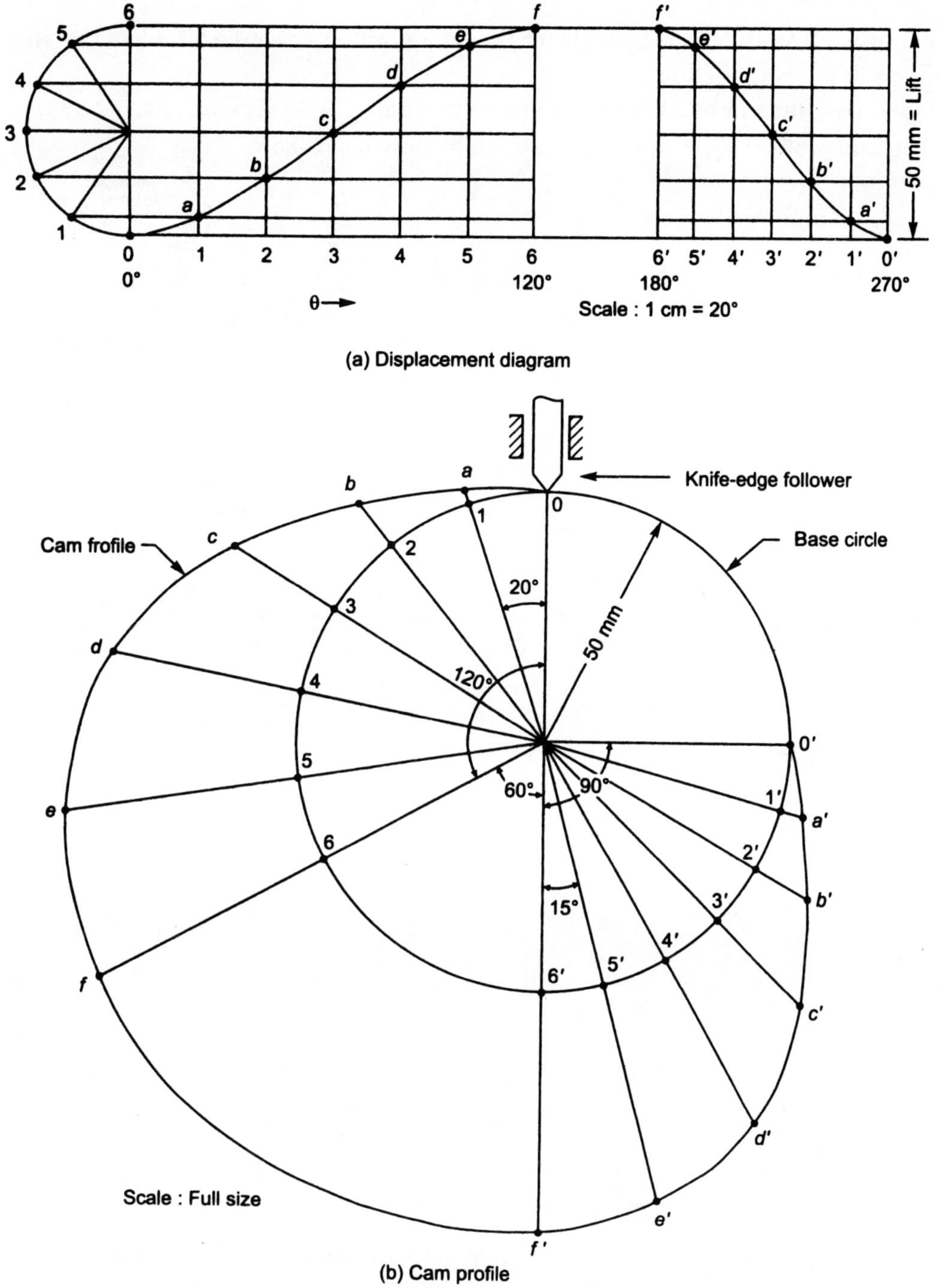

(b) Cam profile

Fig. 8.19 Cam profile with radial knife-edge follower having SHM.

4. Measure distance 1*a*, 2*b*, 3*c*, 4*d*, 5*e* and 6*f* from the displacement diagram for ascent and cut-off corresponding distances on the radial lines drawn in the cam profile. Repeat the same process during descent.
5. Join the points so obtained by a smooth curve to get the cam profile.

■ **Example 8.2** Draw the cam profile for the data given in Example 8.1 when the knife-edge follower is offset by 20 mm.

Solution : Displacement diagram remains the same as in Example 8.1.

The cam profile as shown in Fig. 8.20 can be drawn as follows :

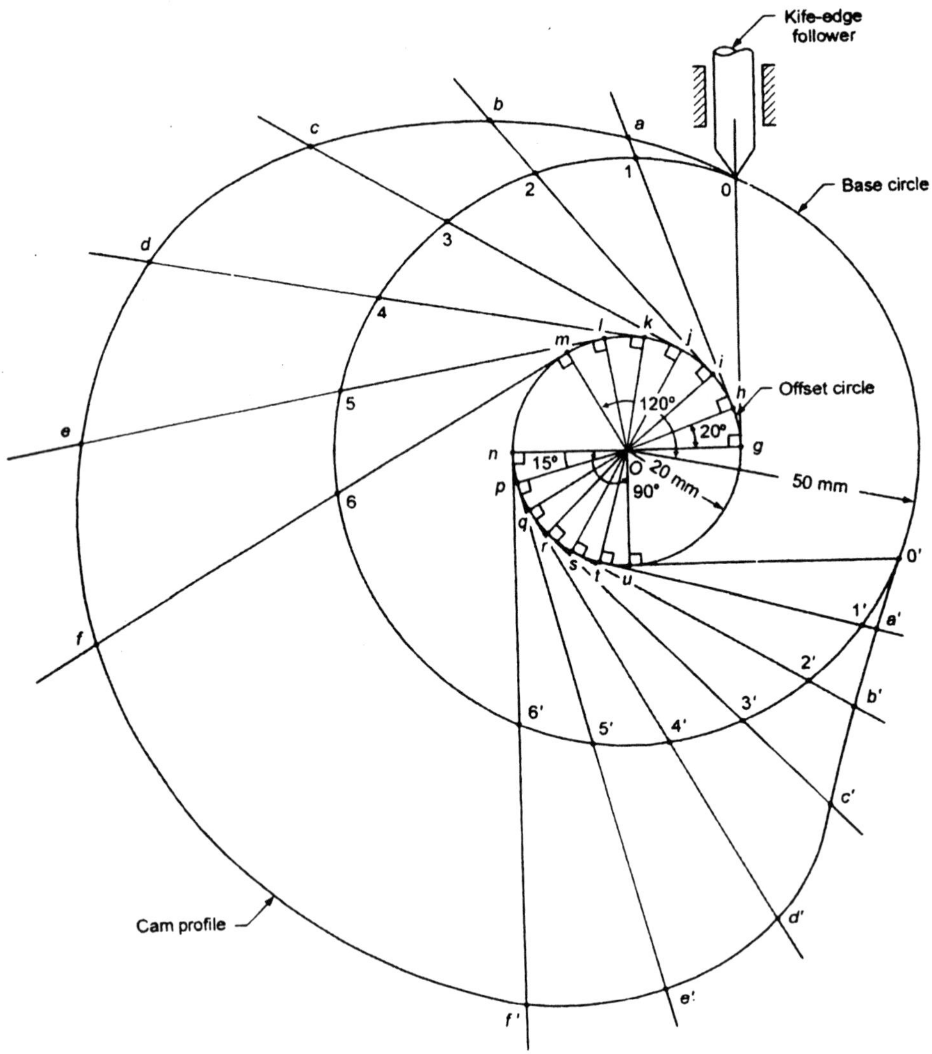

Fig. 8.20 Cam profile with offset knife-edge follower having SHM.

1. Draw the offset circle with radius equal to the offset of 20 mm and the base circle with radius 50 mm at centre O.
2. Divide the angles of ascent and descent in the offset circle into six equal parts. Draw radial lines intersecting the offset circle from g to m and n to u in the angle of ascent and descent respectively.
3. Draw tangents at points g to u intersecting the base circle at 0 to 6 and 6′ to 0′ in the angle of ascent and descent respectively.

4. Measure distances 1*a*, 2*b*, 3*c*, 4*d*, 5*e* and 6*f* on the displacement diagram for ascent and cut-off corresponding distances on the tangential lines drawn in the cam profile. Repeat the same process for the angle of descent.
5. Join the points so obtained by a smooth curve to get the cam profile.

8.7 CAM PROFILE WITH ROLLER FOLLOWER

8.7.1 Radial Roller Follower

The profile of a cam with radial roller follower has been shown in Fig. 8.21. The following steps may be used to draw the cam profile :

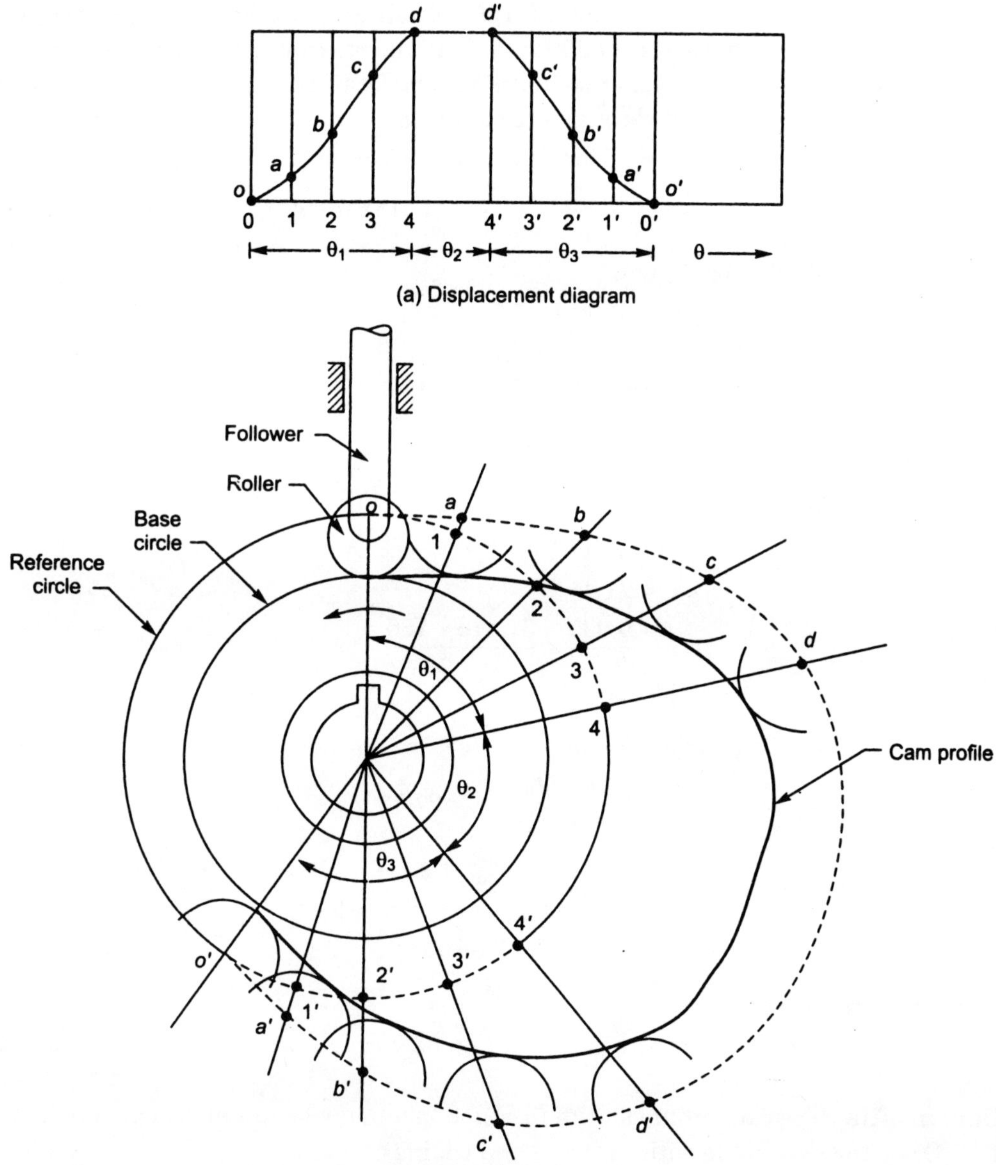

(a) Displacement diagram

(b) Cam profile

Fig. 8.21 Radial cam with roller follower.

1. Draw the base circle.
2. Draw the follower in its 0° position, tangent to the base circle.
3. Draw the refernce circle through the centre of the follower in its 0° position.
4. Draw radial lines from the centre of the cam, corresponding to the vertical lines in the displacement diagram.
5. Transfer displacements 1*a*, 2*b*, 3*c*, ..., etc. from the dispalcement diagram to the appropriate radial lines, measuring from the reference circle.
6. Draw in the follower outline on the various radial lines.
7. Draw a smooth curve tangent to these follower outlines.

■ **Example 8.3** A disc cam with base circle radius of 50 mm is operating a roller follower with SHM. The lift is 25 mm, angle of ascent 120°, dwell 90°, return 90°, and dwell during remaining period. The cam rotates at 300 rpm. Find the maximum velocity and acceleration during ascent and descent. The roller radius is 10 mm. Draw the cam profile when the line of reciprocation of follower passes through the cam axis.

Solution : Given : $N = 300$ rpm, $h = 25$ mm, $\theta_1 = 120°$, $\theta_3 = 90°$

$$\omega = \frac{2\pi N}{60} = \frac{2\pi \times 300}{60} = 31.416 \text{ rad/s.}$$

Maximum velocity during ascent,

$$v_a = \left(\frac{\pi h}{2}\right)\left(\frac{\omega}{\theta_1}\right)$$

$$= \frac{\pi \times 0.025}{2} \times \frac{31.416 \times 180}{\pi \times 120}$$

$$= 0.589 \text{ m/s}$$

Maximum velocity during descent,

$$v_d = \left(\frac{\pi h}{2}\right)\left(\frac{\omega}{\theta_3}\right)$$

$$= \frac{\pi \times 0.025}{2} \times \frac{31.416 \times 180}{\pi \times 90}$$

$$= 0.785 \text{ m/s}$$

Maximum acceleration during ascent

$$= \frac{2v_a^2}{h} = \frac{2 \times (0.589)^2}{0.025} = 27.75 \text{ m/s}^2$$

$$\text{Maximum acceleration during descent} = \frac{2v_d^2}{h} = \frac{2 \times (0.785)^2}{0.025}$$

$$= 49.35 \text{ m/s}^2$$

Displacement Diagram

Draw the displacement diagram shown in Fig. 8.22(a) following the procedure explained in Example 8.1.

Cam profile : The cam profile shown in Fig. 8.22(b) may be drawn as explained below :

1. Draw the base circle with centre *O* and radius 50 mm.

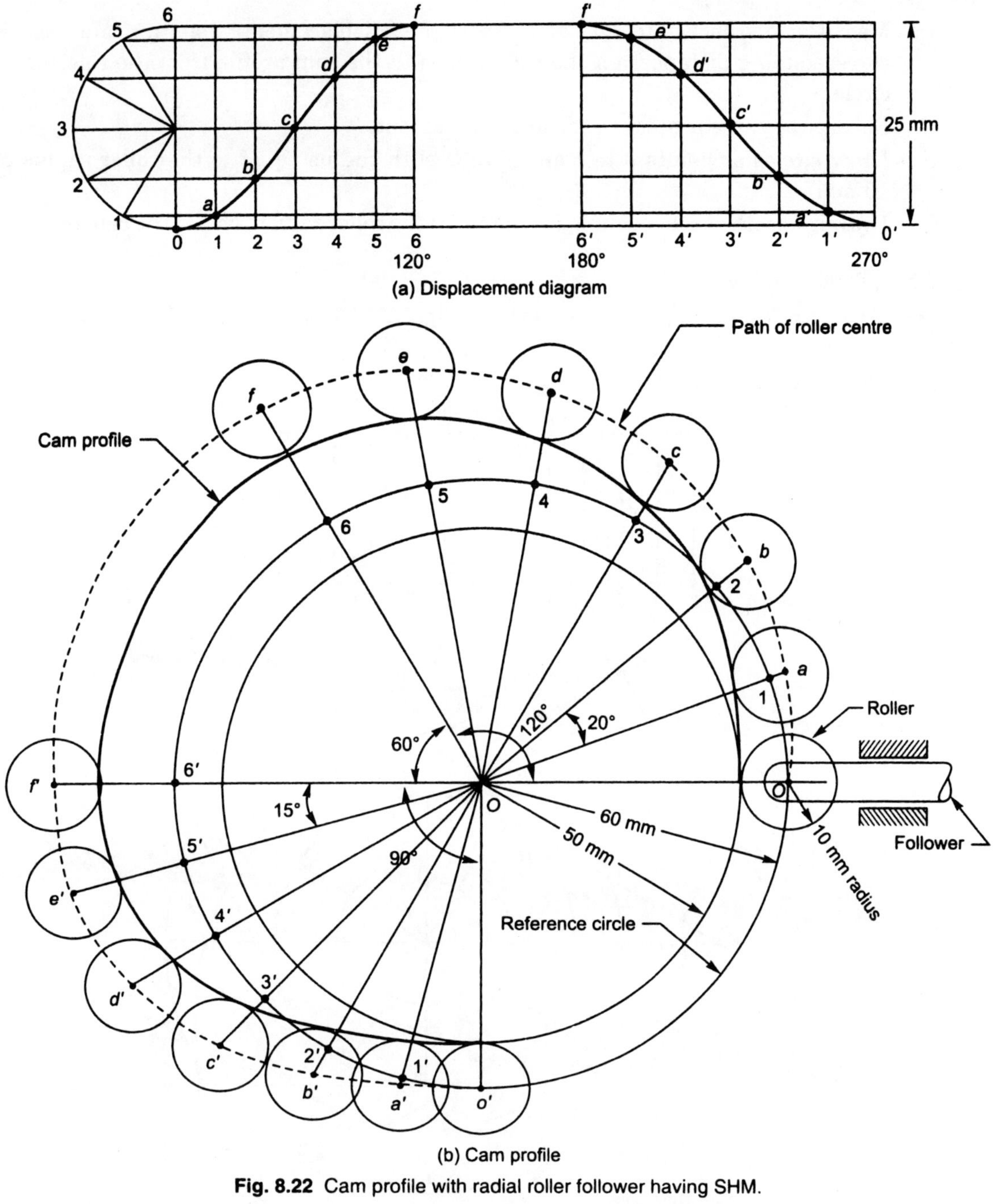

Fig. 8.22 Cam profile with radial roller follower having SHM.

2. Draw the reference circle with radius equal to the sum of the radius of base circle and roller radius, *i.e.*, 60 mm. Draw angles of ascent, dwell, and descent of 120°, 60° and 90° respectively.
3. Divide the angle of ascent and descent into six equal parts.
4. Draw radial lines intersecting the reference circle at point 0 to 6 in the angle of ascent and 6′ to 0′ in the angle of descent.

5. Measure distances $1a$, $2b$, etc. from the displacement diagram and mark corresponding distances on the radial lines in the cam profile from the refernence circle.
6. Repeat the same process in the angle of descent.
7. Draw circles at points 0 to f and f' to $0'$ with radius equal to the roller radius of 10 mm.
8. Draw a smooth curve touching (asymptotic) the roller radii to obtain the cam profile.

8.7.2 Offset Roller Follower

A cam profile with offset roller follower is shown in Fig. 8.23. The following steps may be used to draw the cam profile :

1. Draw the base circle.

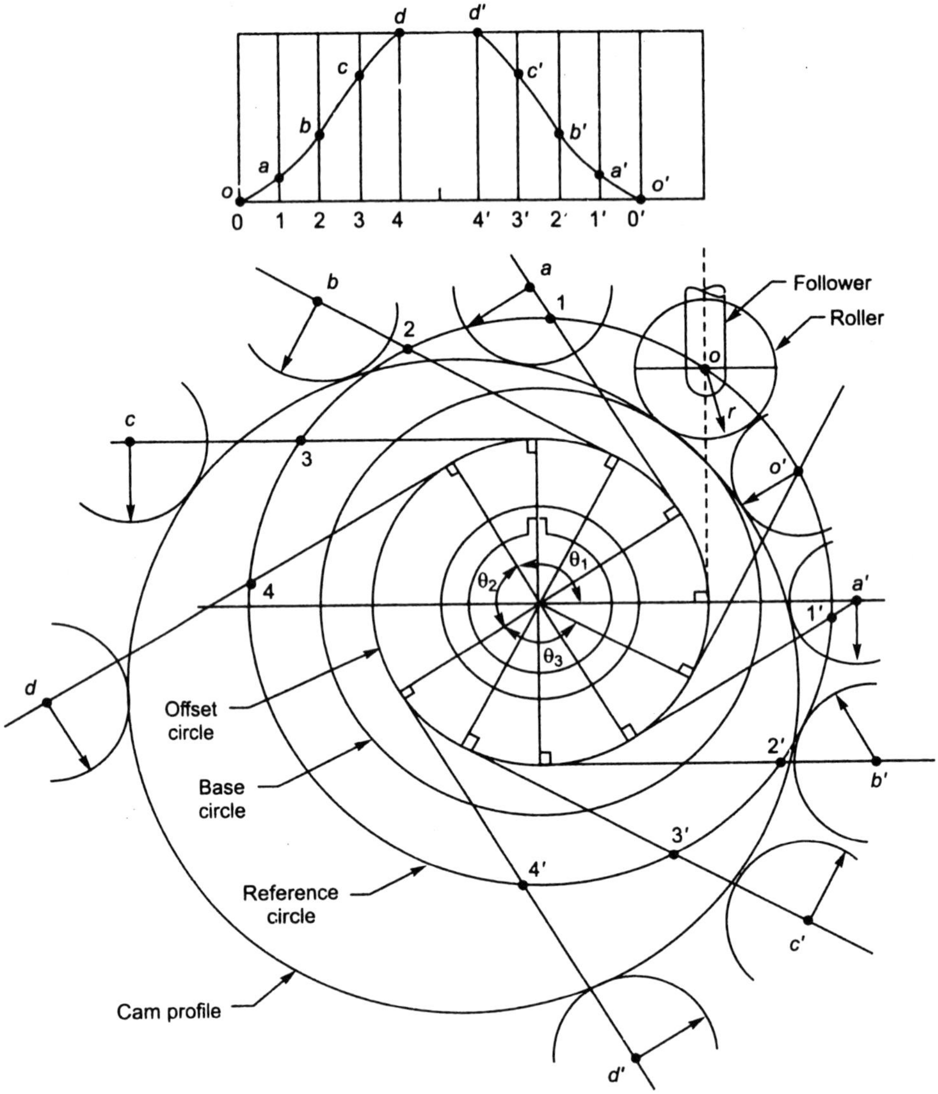

Fig. 8.23 Cam with offset roller follower.

2. Draw the follower in its 0° position, tangent to the base circle.
3. Draw the reference circle through the centre of the follower in its 0° position.
4. Draw the offset circle tangent to the follower centre line.
5. Divide the offset circle into a number of divisions corresponding to the divisions in the displacement diagram.
6. Draw tangents to the offset circle at each number.
7. Lay off various displacement 1*a*, 2*b*, 3*c*, ..., etc. along the appropriate tangent lines, measuring from the reference circle.
8. Draw in the follower outlines on the various tangent lines.
9. Draw a smooth curve to these follower outlines.

■ **Example 8.4** Draw the cam profile for the data given in Examle 8.3 when the roller follower is offset by 20 mm.

Solution : The displacement diagram remains the same as in Example 8.3.

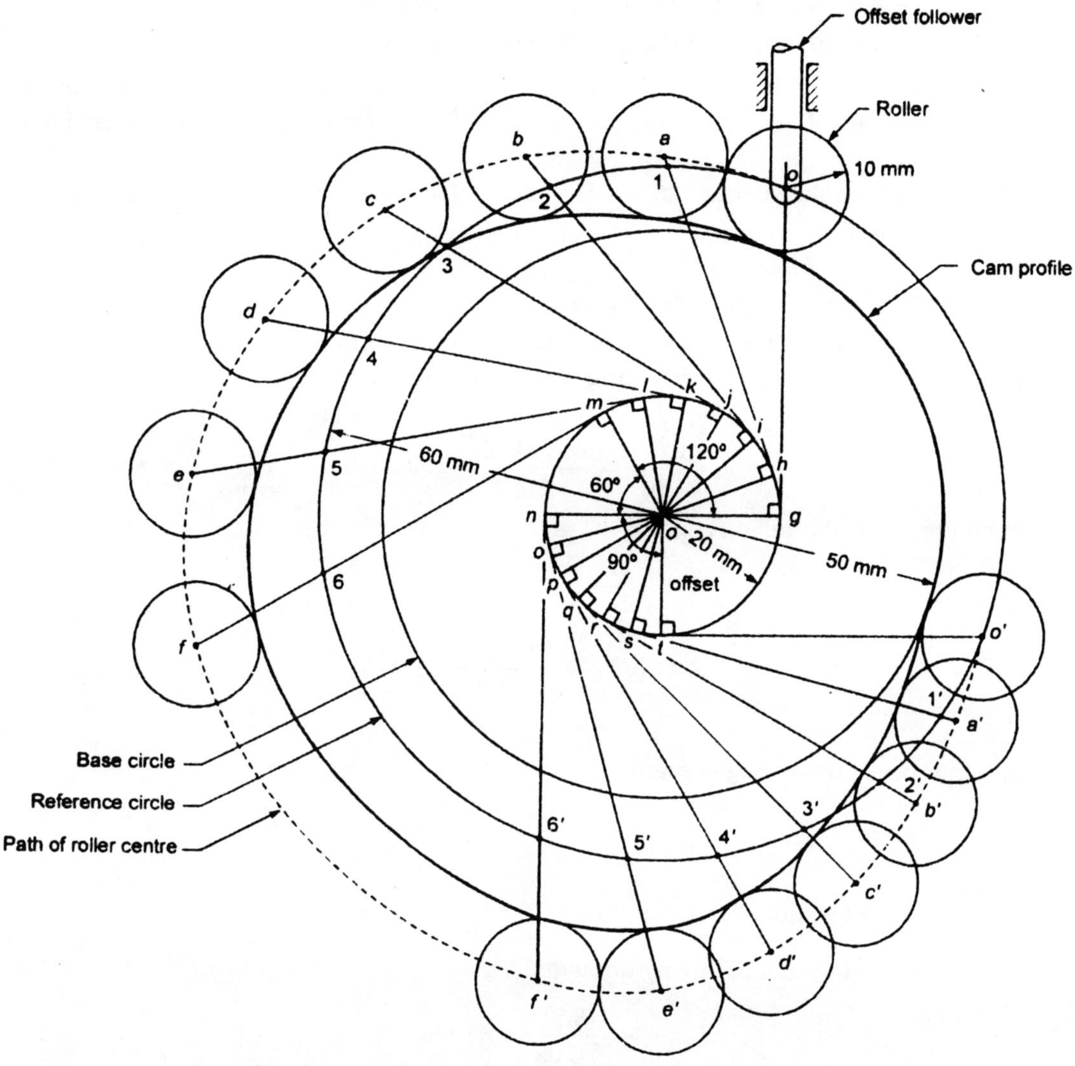

Fig. 8.24 Cam profile with off-set roll3r follower having SHM.

The cam profile shown in Fig. 8.24 may be drawn as explained below :

1. Draw the base circle at centre O with radius 50 mm and reference circle with radius 60 mm.
2. Draw the offset circle with radius 20 mm.
3. Draw angles of ascent, dwell, and descent of 120°, 60° and 90° respectively in the offset circle.
4. Divide the angle of ascent and descent into six equal parts intersecting the offset circle at points g to m and n to u respectively.
5. Draw tangents at points g to u intersecting the refernece circle at points 0 to 6 and 6′ to 0′ in the angle of ascent and descent respectively.
6. Measure off distances 1a, 2b, etc. from the displacement diagram in the angle of ascent and measure the corresponding distances in the cam profile. Repeat the same process for the angle of descent.
7. Draw circles with radius of roller at points 0 to f and f' to 0′.
8. Draw a smooth curve touching these circles to get the cam profile.

■ **Example 8.5** A cam with 30 mm as minimum diameter and 20 mm lift is rotating clockwise at a uniform speed of 1200 rpm and has to given the following motion to a roller follower 10 mm in diameter :

1. Outward stroke during 120° with equal uniform acceleration and deceleration.
2. Dwell for 60°.
3. Return during 90°.
4. Dwell during the remaining period.

Draw the cam profile if the cam axis coincides with the follower axis. Also calculate the maximum velocity and acceleration during ascent and return strokes.

Solution : Given $N = 12000$, rpm, $h = 20$ mm, $\theta_1 = 120°$, $\theta_3 = 90°$

$$\omega = \frac{2\pi N}{60} = \frac{2\pi \times 1200}{60} = 40\pi \text{ rad/s}$$

Maximum velocity during ascent,

$$v_a = 2h \times \left(\frac{\omega}{\theta_1}\right)$$

$$= 2 \times 0.02 \times \left[\frac{40\pi \times 180}{\pi \times 90}\right]$$

$$= 2.4 \text{ m/s}$$

Maximum velocity during descent,

$$v_d = 2h \times \left(\frac{\omega}{\theta_3}\right) = 2 \times 0.02 \times \left[\frac{40\,\pi \times 180}{\pi \times 90}\right] = 3.2 \text{ m/s}$$

Maximum acceleration during ascent $= \dfrac{v_a^2}{h} = \dfrac{(2.4)^2}{0.02} = 288 \text{ m/s}^2$

Maximum acceleration of during descent

$$= \frac{v_d^2}{h} = \frac{(3.2)^2}{0.02} = 512 \text{ m/s}^2$$

Displacement Diagram

The displacement diagram is shwon in Fig. 8.25(a) for uniform acceleration and deceleration may be drawn as follows :

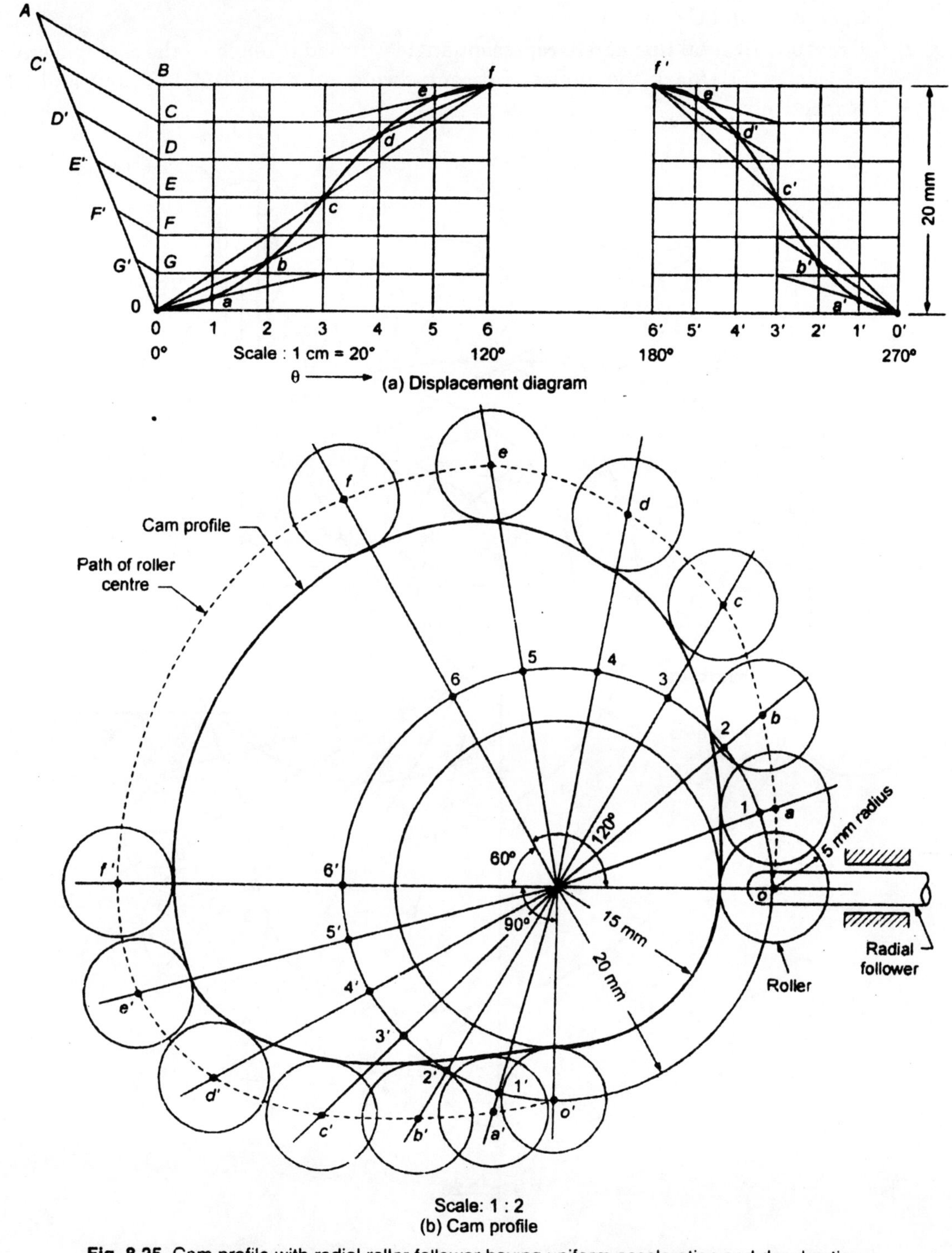

Fig. 8.25 Cam profile with radial roller follower having uniform acceleration and deceleration.

1. Draw a vertical line OB representing the lift of the follower equal to 20 mm to a scale of 1 : 2. To divide this line into six equal parts, draw a line OA inclined at any angle with OB such that OA = 60 mm. Divide the line OA into six equal parts. Join AB and draw parallel lines from C', D', E', F', and G', to line AB to meet OB at C, D, E, F, and G respectively.
2. Draw a horizontal line at 0 to represent angle θ turned through by the cam to a scale of 1 cm = 20°. Mark the angles of ascent, dwell and descent of 120°, 60° and 90° respectively.

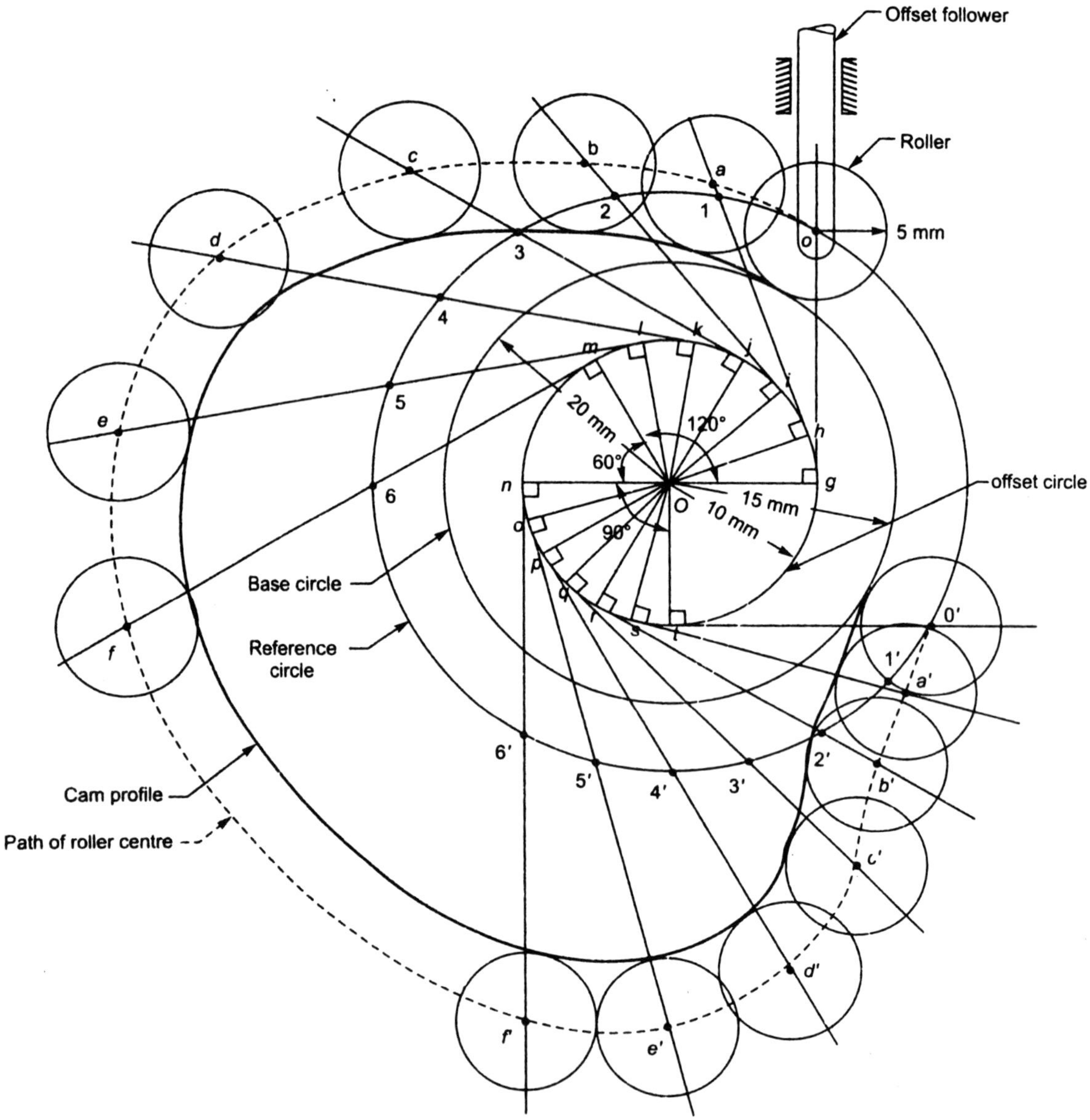

Fig. 8.26

3. Divide the angles of ascent and descent into six equal parts and erect perpendiculars at these points to intersect the horizontal lines from points B to G.
4. Join the points o and f with the points of intersection on the middle vertical line in the angle of ascent. Join the points of intersection of these lines with the vertical lines by a smooth curve $o\ a\ b\ c\ d\ e\ f$. Repeat the similar process for the angle of descent to get the curve $f'\ e'\ d'\ c'\ b'\ a'\ o'$.

Cam Profile

The cam profile has been drawn in Fig. 8.25(b) to a scale of 1 : 2. The procedure described in Example 8.3 may be followed to draw the profile.

■ **Example 8.6** Draw the cam profile for the data given in Examples 8.5 when the roller follower is offset by 10 mm from the axis of the cam.

Solution : The displacement diagram remains the same as drawn in Example 8.5. The cam profile has been drawn in Fig. 8.26 following the same procedure as explained in Example 8.4.

9. Draw a smooth curve to these follower outlines.

8.8 CAM PROFILE WITH TRANSLATIONAL FLAT-FACED FOLLOWER

A cam with flat-faced follower is shown in Fig. 8.28. The following steps may be adopted to draw the cam profile :

1. Draw the base circle, which also serves the reference circle in this case.
2. Draw the follower in the home position, tangent to the base circle.
3. Draw radial lines corresponding to the divisions in the displacement diagram, and number them accordingly
4. Draw in the follower outline on the various radial lines by laying off the appropriate displacements and drawing lines perpendicular to the radial lines.
5. Draw a smooth curve tangent to the follower lines.

■ **Example 8.7** A cam is to operate a flat faced follower having uniform acceleration and deceleration during ascent and descent. The least radius of the cam is 50 mm. During descent, the deceleration period is half of the acceleration period. The ascent life is 37.5 mm. The ascent is for 1/4 the period, dwell for 1/4th, descent for 1/3rd, and dwell for the remaining 1/6th period. The cam rotates at 600 rpm. Find the maximum velocity and acceleration during ascent and descent. Draw the cam profile.

Solution : Acceleration period during descent $= \frac{1}{3} \times 300 \times \frac{2}{3} = 80°$

Deceleration period during descent = 40°

Distances moved during acceleration period $= \frac{80}{120} \times 37.5 = 25$ mm

Distance moved during deceleration period = 12.5 mm

Angular velocity of cam,

$$\omega = \frac{2\pi N}{60} = \frac{2\pi \times 600}{60} = 20\pi \text{ rad / s}$$

Maximum velocity during ascent,

$$v_{max} = \frac{2\omega h}{\theta_1}$$

$$= \frac{2 \times 20\pi \times 37.5 \times 10^{-3}}{\left(\frac{\pi}{2}\right)} = 3 \text{ m/s}$$

Maximum velocity during descent,

$$v_{max} = \frac{2\omega h}{\theta_3} = \frac{2 \times 20\pi \times 37.5 \times 10^{-3} \times 180}{80 \times \pi} = 3.375 \text{ m/s}$$

$$\text{Maximum acceleration during ascent} = \frac{4\omega^2 h}{\theta_1^2}$$

$$= \frac{4 \times (20\pi)^2 \times 37.5 \times 10^{-3}}{\left(\frac{\pi}{2}\right)^2}$$

$$= 240 \text{ m/s}^2$$

$$\text{Maximum acceleration during decent} = \frac{v_{max}}{t}$$

$$= \frac{3.375 \times 360 \times 600}{80 \times 60}$$

$$= 151.875 \text{ m/s}^2$$

$$\text{Maximum deceleration during decent} = \frac{v_{max}}{t}$$

$$= \frac{3.375 \times 360 \times 600}{40 \times 60}$$

$$= 303.75 \text{ m/s}^2$$

The velocity and acceleration diagrams are shown in Fig. 8.27.

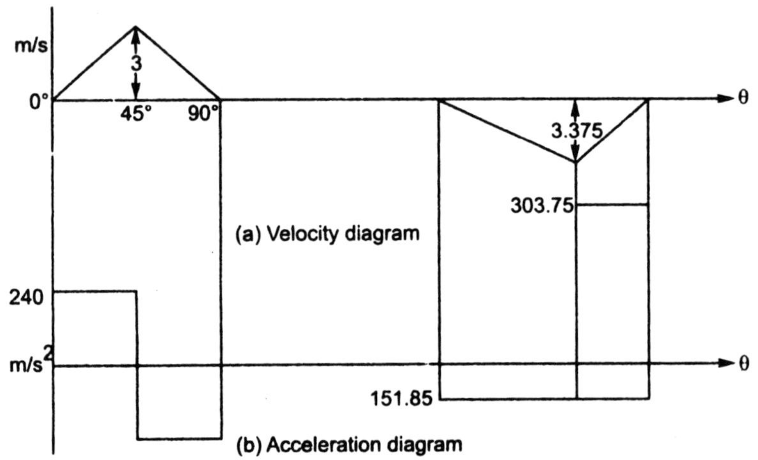

Fig. 8.27 Velocity and acceleration diagram for a flat-faced follower having uniform acceleration and deceleration.

Displacement Diagram

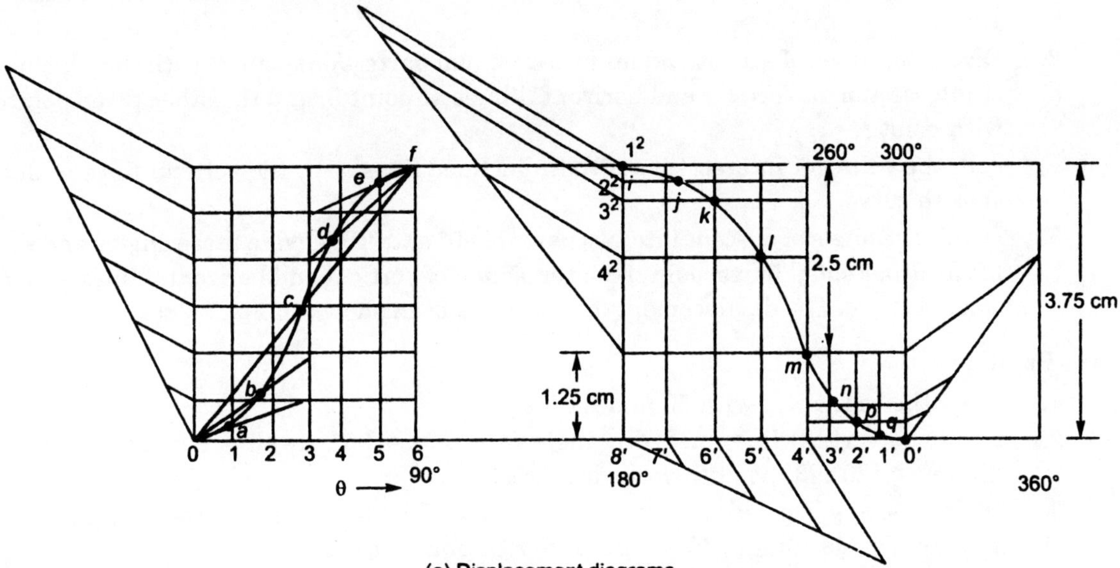

(a) Displacement diagrame

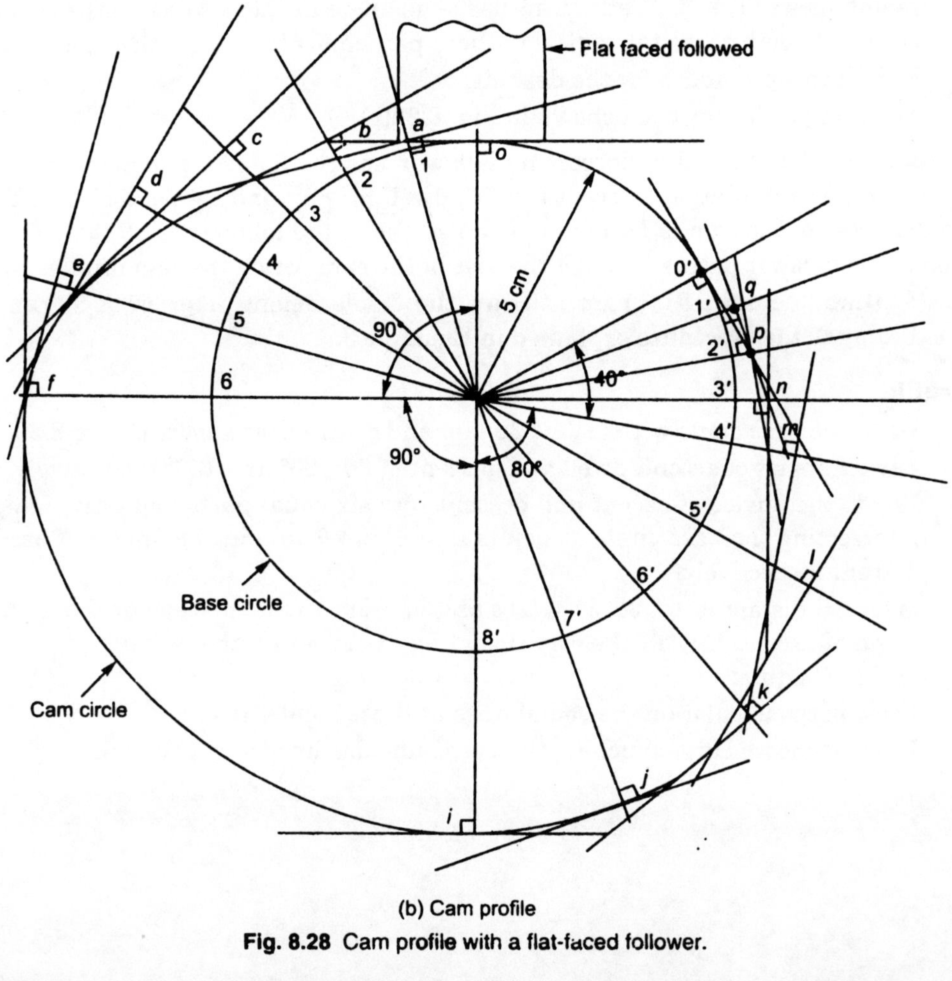

(b) Cam profile

Fig. 8.28 Cam profile with a flat-faced follower.

1. Draw a vertical line equal to the lift of 37.5 mm and a horizontal line perpendicular to it representing the cam angle. Mark angle of ascent = 90°, dwell = 90°, and descent = 120°.
2. Divide the angle of ascent and lift into six equal parts. Join point *O* with three points of intersection of vertical and horizontal lines at point 3, and the other three points with point *f*.

   Join the points of intersection of these inclined lines with the vertical lines with a smooth curve.
3. Divide the angle of descent into two parts of 80° and 40°. Divide these angles and the lift into four equal parts. Join the intersection of vertical and horizontal lines with a smooth curve. The displacement diagram has been shown in Fig. 8.28(a).

Cam Profile

1. Draw the base circle with 50 mm radius.
2. From the vertical line, measure angle of ascent = 90°, dwell = 90°, and angle of descent = 120°, in the anti-clockwise direction.
3. Divide the angle of ascent into six equal parts. Divide the angle of descent into 80° and 40°. Divide 80° and 40° angles into four equal parts.
4. From the circumference of the base circle, mark distances 1*a*, 2*b*, 3*c*, ...,etc. along the radial lines at 1, 2, 3, ... etc. Draw perpendiculars at these points to the radial lines. Draw a smooth curve tangential to these perpendiculars to get the cam profile.
5. Repeat this procedure for the descent.

   The cam profile has been shown in Fig. 8.28(b).

■ **Example 8.8** A flat-faced reciprocating follower has the following motion with uniform acceleration and retardation : Ascent for 80°, dwell for 80°, and return for 120°. The base circle diameter of the cam is 60 mm and the stroke of the follower is 20 mm. The line of motion of the follower passes through the axis of the cam. Draw the cam profile.

Solution : Displacement diagram : Draw the displacement diagram as shown in Fig. 8.29(a) according to the procedure explained in Example 8.5.

Cam Profile

1. Draw the base circle with radius 30 mm and centre *O* as shown in Fig. 8.29(b).
2. Lay off angles of ascent, dwell and descent of 80°, 80° and 120° respectively.
3. Divide the angles of ascent and descent into six equal parts and draw radial lines intersecting the base circle at points 0 to 6 and 6′ to 0 in the angles of ascent and descent respectively.
4. Measure distances 1*a*, 2*b*, 3*c*, 4*d*, 5*e* and 6*f* from the displacement diagram in the angle of ascent. Cut-off the corresponding distances on the redial lines in the cam profile diagram.
5. Draw perpendicular on the radial lines at these points '*a*' to *f*.
6. Draw a smooth curve touching the perpendicular lines to get the cam profile.

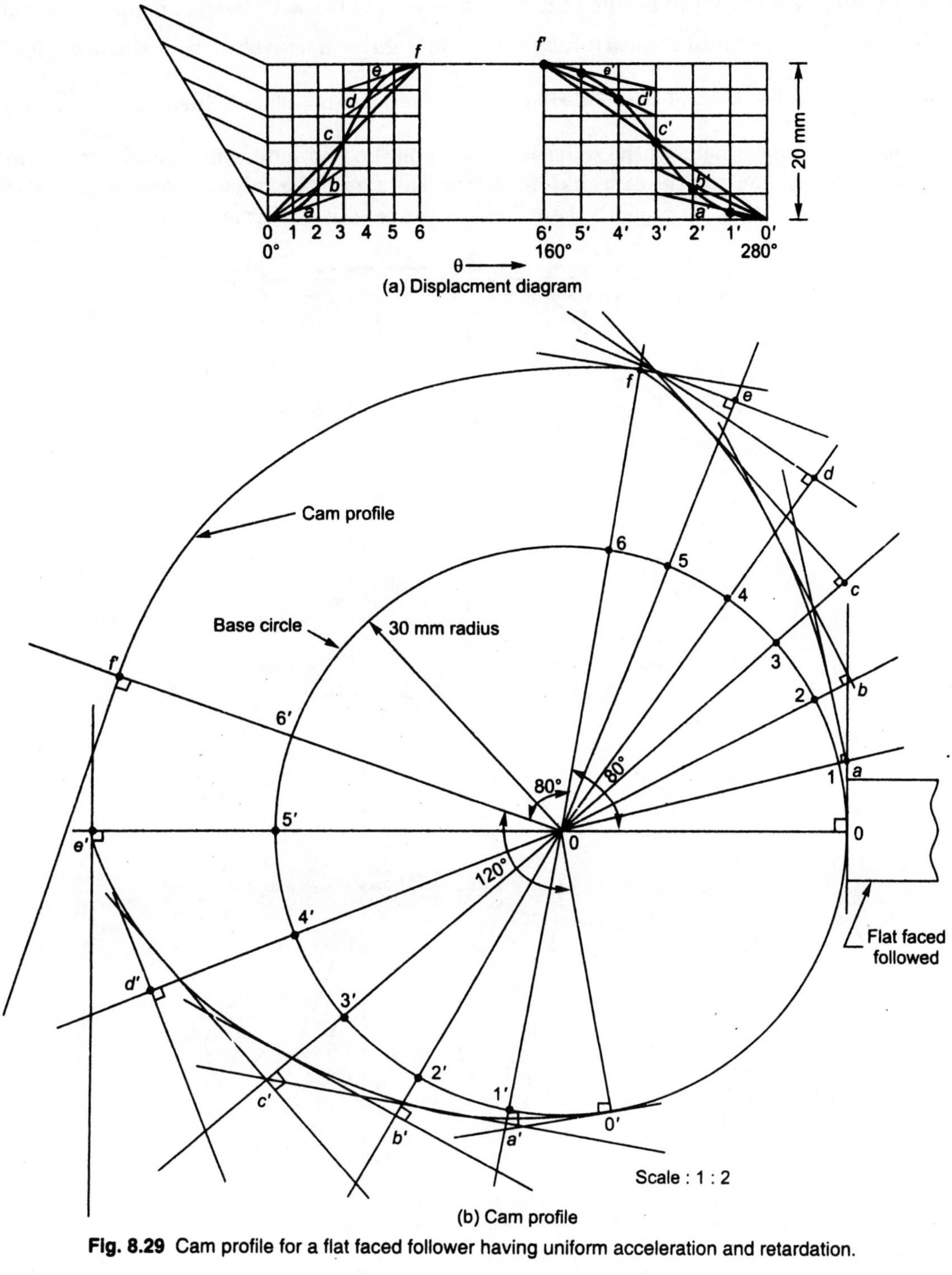

(a) Displacment diagram

(b) Cam profile

Fig. 8.29 Cam profile for a flat faced follower having uniform acceleration and retardation.

■ **Example 8.9** Draw the profile of a cam to raise a valve with harmonic motion through 50 mm in $\frac{1}{3}$ of a revolution, keep it fully raised through $\frac{1}{12}$ of a revolution, and to lower it with harmonic motion in $\frac{1}{6}$ of a revolution. The valve remains closed during the rest of the revolution. The diameter of the roller is 20 mm and the minimum radius of the cam is to be 25 mm. The diameter of the cam shaft is 25 mm. The axis of the valve rod passes through the axis of the cam shaft. Assume the camshaft to rotate with a uniorm velocity.

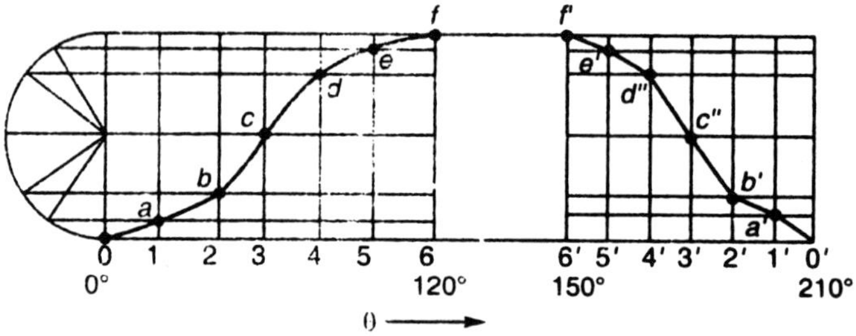

(a) Displacement diagram

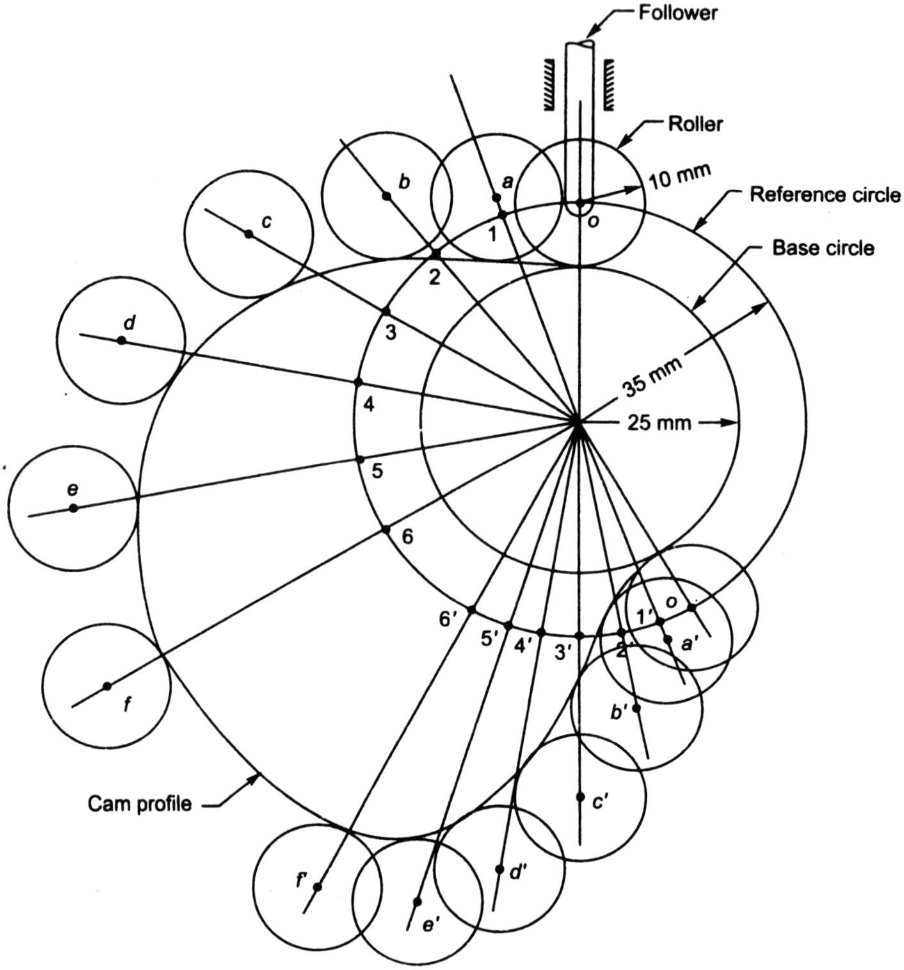

Fig. 8.30

Solution : Given life = 50 mm, angle of ascent = $\frac{1}{3} \times 360° = 120°$, dwell = $\frac{1}{12} \times 360° = 30°$, descent = $\frac{1}{6} \times 360° = 60°$

Roller diameter = 20 mm, minimum radius of cam = 25 cm

Diameter of cam shaft = 25 mm.

Displacement diagram : Draw the displacement diagram shown in Fig. 8.30(a) according to the procedure explained in Example 8.1.

Cam profile : Draw the cam profile shown in Fig. 8.30(b) according to the procedure described in Example 8.3

8.9 CAM SIZE

The cam size is define by the following parameters :

1. Pressure angle
2. Radius of curvature of cam profile
3. Hub size.

The following methods may be used to reduce the pressure angle :

1. Increase the diameter of the base circle.
2. Increase the angle of rotation of the cam, thereby lengthening the pitch curve for the specified follower displacement. The cam profile becomes flatter and the pressure angle becomes smaller.
3. Select the motion curve for a smaller pressure angle.
4. By changing the offset of the follower.

SOLVED QUESTIONS FROM UNIVERSITY EXAMINATIONS

■ **Example 8.10** Explain the different types of cams and followers commonly used.

[KU, Dec., 2007]

Solution : Refer to art. 8.2 and art. 8.3.

■ **Example 8.11** Draw the profile of a cm which is to give oscillatory motion to the follower with uniform angular velocity about its pivot. The base circle diameter is 50 mm, angle of oscillation of the follower 60 mm. The oscillating lever is 60 mm long with a roller of 8 mm diameter at the end. One oscillation of the follower is completed in one revolution of the cam.

[KU, Dec., 2008]

Solution : Roller diameter = 8 mm

Distance between can centre and pivot of follower = 60 mm

Length of oscillating lever = 60 mm

Base circle diameter = 50 mm

Angle of oscillation of follower = 30°

Length of circular arc during movement of follower, as shown in Fig. 8.31(a)

$$= 60 \times \frac{\pi}{180} \times 30 = 31.416 \text{ mm}$$

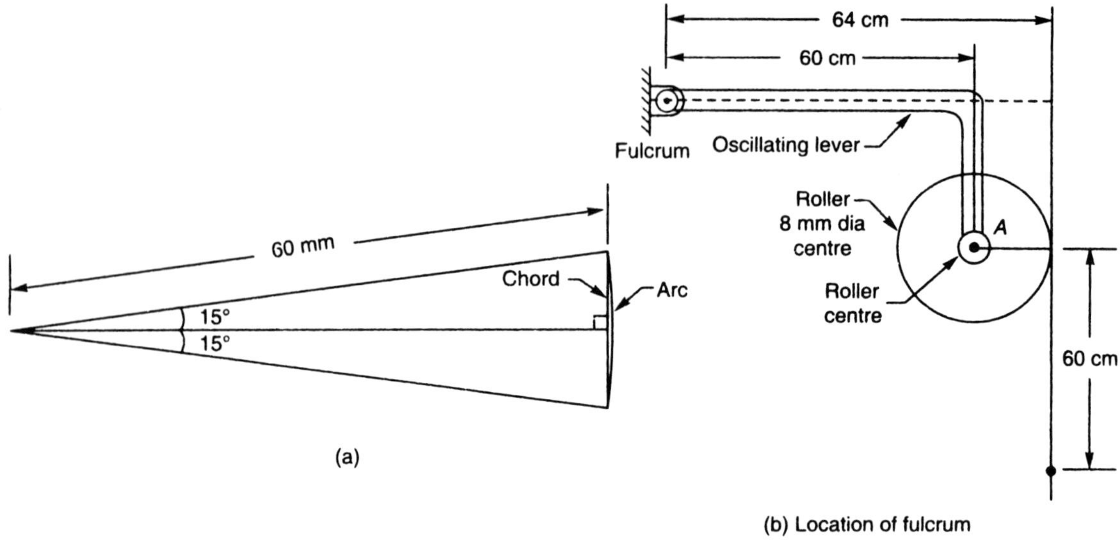

Fig. 8.31

Length of chord joining the roller centres in exterme positions

$$= 2 \times 60 \sin 15° = 31.058 \text{ mm}$$

For small angle, length of arc $\simeq$ length of chord.

We take the follower displacement = 31.058 mm $\simeq$ 31 mm

Let angle of ascent = 90°

Angle of dwell = 60°

Angle of descent = 90°

Angle of dwell = 120°

Motion : SHM

Procedure :

1. Draw the displacement diagram with follower lift = 31 mm, as shown in Fig. 8.32(a).
2. Draw base circle of cam, with O as centre and 25 mm as radius (Fig. 8.32b).
3. Draw roller centre path circle with radius 29 mm.
4. Draw another circle with 64 mm radius, which is equal to the distance between cam centre and lever fulercum.
5. Draw a vertical line OA = 29 mm. With A as centre and radius equal to 60 mm draw an arc cutting the 64 mm circle at point B. Join AB and OB.
6. With OB as the reference line, draw angle of ascent $\angle BOf_1 = 90°$ and divide it into six equal parts. Draw angle of dwell $f_1of_1' = 60°$ and angle of descent $f_1'\,oo_1' = 90°$. Divide the angle descent into six equal parts.
7. With $a_1, b_1, \ldots$ as centres draw arcs of 60 mm radius between the 29 mm and 64 mm radii circles to meet the 29 mm circle at 0, 1, 2, ... points.
8. Mach displacements $1a$, $2b$, ...etc along the chords of displacement arcs.
9. Repeat the above procedure for the angle of descent.
10. From the points 0, 1, 2, ... etc draw the roller radius circles.
11. The envelope of the roller circle radius gives the cam profile.

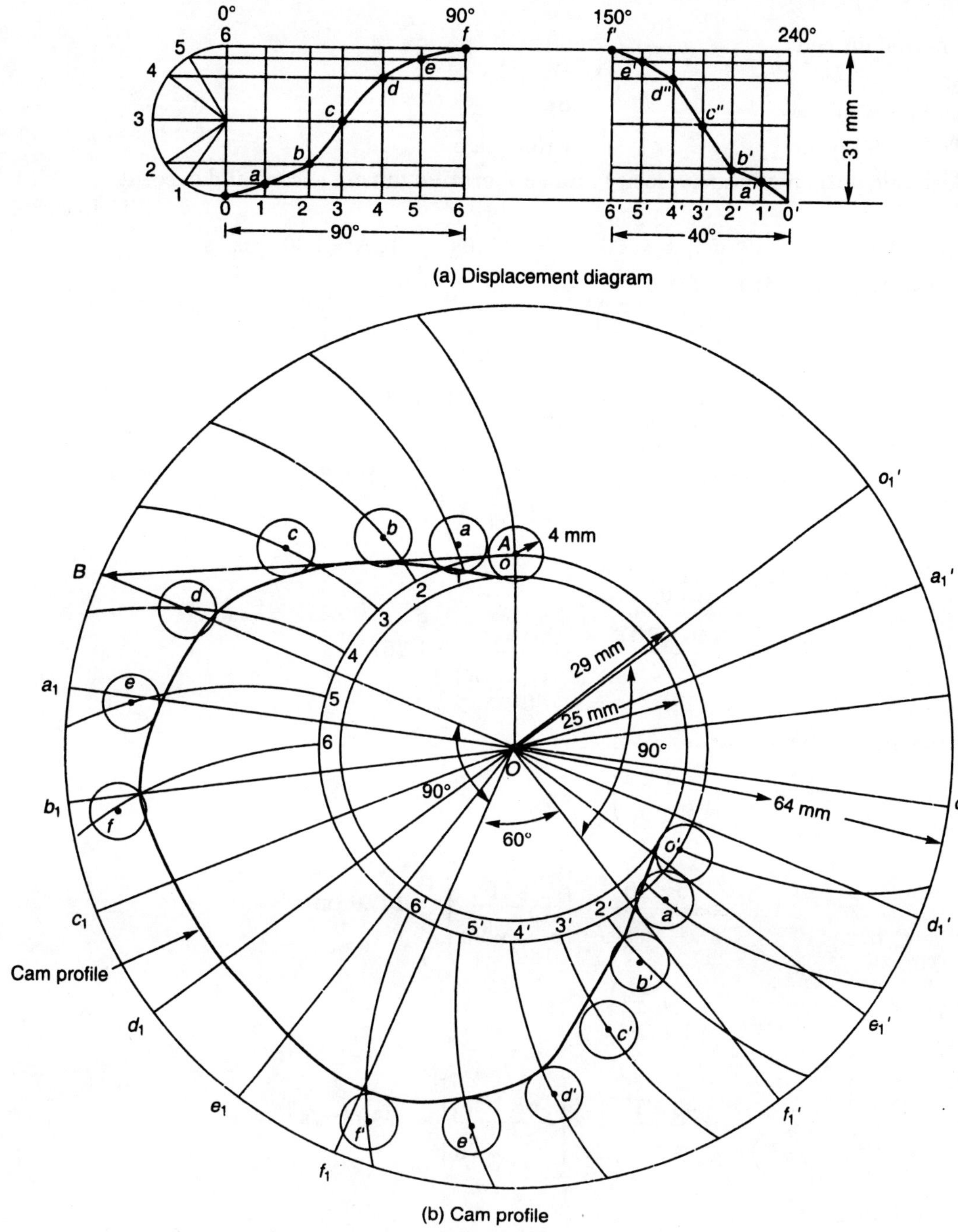

Fig. 8.32

■ **Example 8.12** Draw the cam profiel for the data given below :

Base circle radius of cam	= 50 mm
Lift	= 40 mm
Angle of ascent	= 60°,
Angle of dwell	= 40°

Angle of descent = 90°
Speed of cam = 300 rpm
Motion of follower = SHM
Type of follower = Knife edge.
Also calculate maximum velocity and acceleration during ascent and descent.

[KU, Dec., 2007]

Solutin : Given : $h = 40$ mm, $\theta_1 = 60°$, $\theta_2 = 40°$, $\theta_3 = 90°$, $N = 300$ rpm

Maximum velocity of follower :

During ascent, $$v_{max} = \frac{h}{2}\left(\frac{\pi\omega}{\theta_1}\right)$$

$$\omega = \frac{2\pi N}{60} = \frac{2\pi \times 300}{60} = 10\ \pi\text{rad/s}$$

$$v_{max} = \frac{40 \times 10^{-3}}{2} \times \frac{\pi \times 10\pi}{\left(\frac{\pi}{180} \times 60°\right)} = 1.885 \text{ m/s}$$

During desent, $$v_{max} = \frac{h}{2}\left(\frac{\pi\omega}{\theta_3}\right)$$

$$= \frac{40 \times 10^{-3}}{2} \times \frac{\pi \times 10\pi}{\left(\frac{\pi}{180} \times 90\right)} = 1.257 \text{ m/s}$$

Maximum acceleration :

During ascent, $$f_{max} = \left(\frac{h}{2}\right)\left(\frac{\pi\omega}{\theta_1}\right)$$

$$= \frac{40 \times 10^{-3}}{2} \times \left(\frac{\pi \times 10\pi}{\frac{\pi}{180} \times 90}\right)^2 = 177.65 \text{m/s}^2$$

During ascent, $$f_{max} = \left(\frac{h}{2}\right)\left(\frac{\pi\omega}{\theta_3}\right)$$

$$= \frac{40 \times 10^{-3}}{2} \times \left(\frac{\pi \times 10\pi}{\frac{\pi}{180} \times 90}\right)^2 = 78.96 \text{m/s}^2$$

Produce :

1. Draw the displacement diagram with SHM as shown in Fig. 8.33(a).
2. Draw the cam profile with knife-edge follower as shown in Fig. 8.33 (b). Follow all the steps as explained in example 8.1. The cam profile is shown in Fig. 8.33(b).

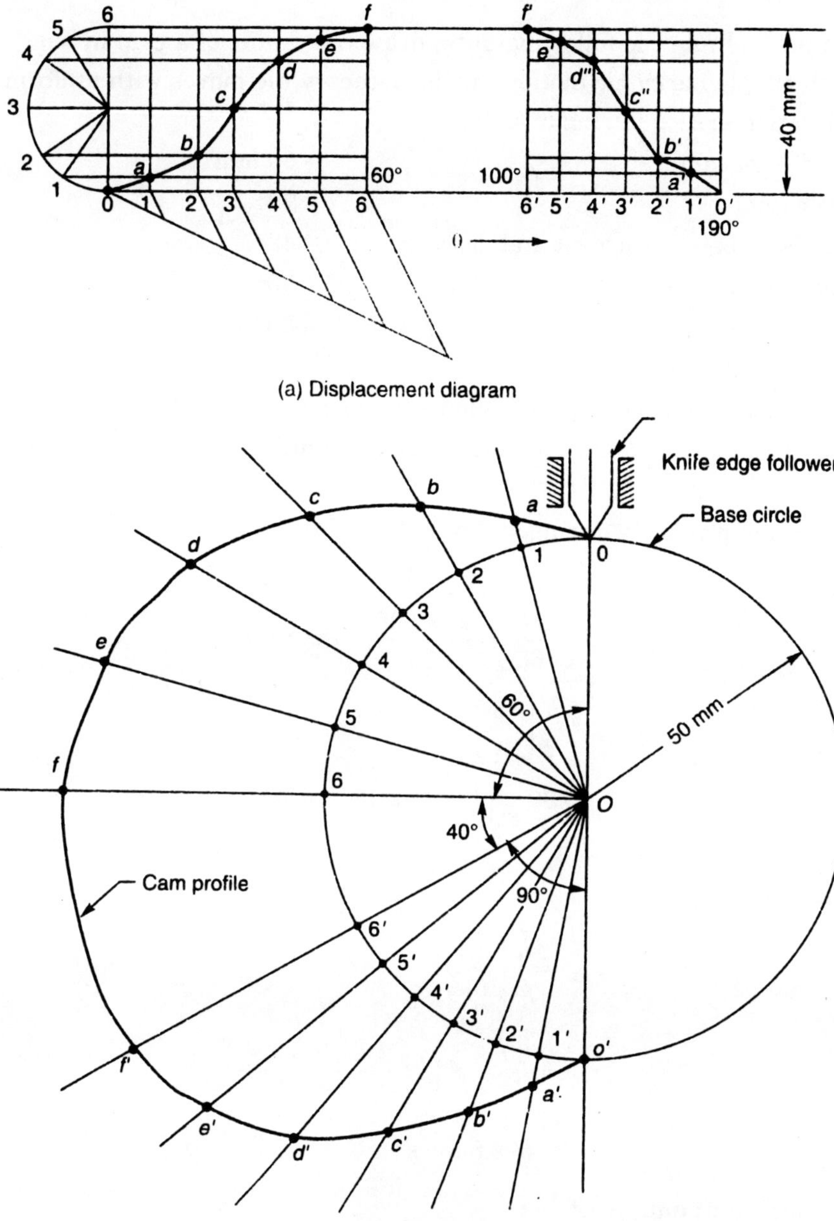

(a) Displacement diagram

(b) Cam profile

Fig. 8.33

■ **Example 8.13** Briefly discuss the displacement, velocity and acceleration diagram of the follower for the following types of follower motions in a cam-follower system :

(a) SHM,

(b) constant acceleration [MDU, Dec., 2007]

Solution : Refer to art 8.5.

■ **Example 8.14** From the following data, draw the profile of a cam in which the follower moves with simple harmonic motion during ascent while moves with uniform acceleration and deceleration motion during descent.

Least radius of cam	= 60 mm
Angle of ascent	= 72°
Angle of dwell between ascent and descent	= 42°
Angle of descent	= 90°
Lift of follower	= 60 mm
Diameter of roller	= 40 mm

Distance between line of action of follower and axis of cam = 30 mm. If the cam rotates at 540 rpm anti-clockwise, find the maximum velocity and acceleration of the follower during ascent. [MDU, May, 2008]

Solution : Maximum velocity during ascent with SHM

$$h = 60 \text{ mm}, \theta_1 = 72°, N = 540 \text{ rpm}$$

$$\omega = \frac{2\pi N}{60} = \frac{2\pi \times 540}{60} = 18\pi \text{ rad/s}$$

$$v_{max} = \frac{h}{2}\left(\frac{\pi\omega}{\theta_1}\right)$$

$$= \frac{60}{2} \times 10^{-3} \frac{(\pi \times 18\pi \times 180)}{\pi \times 72} = 4.24 \text{ m/s}$$

Maximum acceleration,

$$f_{max} = \frac{h}{2}\left(\frac{\pi\omega}{\theta_1}\right)^2$$

$$= \frac{60}{2} \times 10^{-3}\left(\frac{\pi \times 18\pi \times 180}{\pi \times 72}\right)^2$$

$$= 599.6 \text{ m/s}^2$$

Construction of cam profile :

1. Draw the displacement diagram as shown in Fig. 8.34(a). The ascent is with SHM and descent with uniform acceleration and deceleration.
2. Draw the cam profile as shown in Fig. 8.34(b). Follow the procedure as explained in example 8.6.

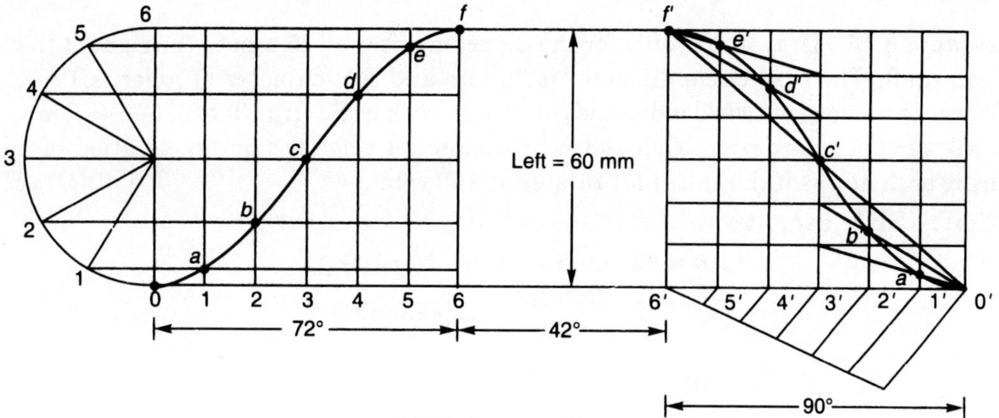

(a) Displacement diagram

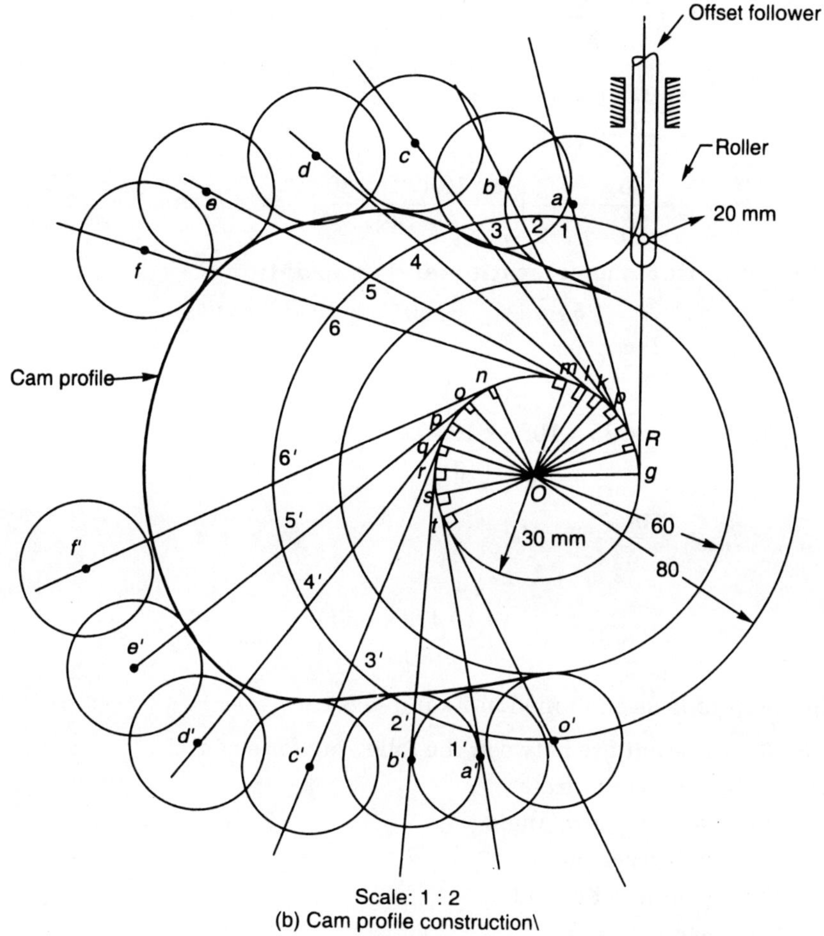

(b) Cam profile construction\

Fig. 8.34

■ **Example 8.15** Draw the profile for the disc cam offset by 15 mm to the right of the centre of camshaft. The base circle diameter is 70 mm and the diameter of roller is 12 mm. The follower is to move outward a distance of 35 mm with SHM in 140° of cam rotation to dwell for 40° of cam retardation. Calculate the maximum velocity and acceleration of follower during each storks if the camshaft rotates at 100 rpm. [MDU, Dec., 2008]

Solution : **SHM ascent**

$$h = 35 \text{ mm}, \theta_1 = 140°, N = 100 \text{ rpm}$$

$$\omega = 2\pi N / 60 = \frac{2\pi \times 100}{60} = 10.472 \text{ rad/s}$$

$$v_{max} = \frac{h}{2}\left(\frac{\pi\omega}{\theta_1}\right)$$

$$= \frac{35 \times 10^{-3}}{2}\left(\frac{\pi \times 10.472 \times 180}{\pi \times 140}\right) = 0.236 \text{ m/s}$$

$$f_{max} = \frac{h}{2}\left(\frac{\pi\omega}{\theta_1}\right)^2$$

$$= \frac{35 \times 10^{-3}}{2}\left(\frac{\pi \times 10.472 \times 180}{\pi \times 140}\right)^2 = 3.172 \text{ m/s}^2$$

Descent with uniform acceleration and retardation :

$$h = 35 \text{ mm}, \theta_3 = 150°, \omega = 10.472 \text{ rad / s}$$

$$v_{max} = \frac{2h\omega}{\theta_3}$$

$$= \frac{2 \times 35 \times 10^{-3} \times 10.472 \times 180}{\pi \times 150} = 0.28 \text{ m / s}$$

$$f_{max} = \frac{4h\omega^2}{\theta_3^2}$$

$$= 4 \times 35 \times 10^{-3}\left(\frac{10.472 \times 180}{180 \times \pi}\right)^2 = 2.24 \text{ m/s}^2$$

Draw the comprofile as explained in Example 8.15.

■ **Example 8.16** Differentiate between the following for a cam :

(i) Base circle and prime circle
(ii) Cam angle and pressure angle
(iii) Pitch point and trace point
(iv) Period of ascent and descent. [MDU, May, 2009]

Solution : Refer to art. 8.4.

■ **Example 8.17** Draw the profile of a cam operating a knife edge follower (when the axis of follower passes through the axis of cam shaft) from the following data :

(a) Follower to move outward through 30 mm with SHM during 120° of cam rotation.
(b) Follower to dwell for next 60°

(c) Follower to return to its original position with uniform velocity during 90° of cam rotation.

(d) Follower to dwell for the rest of the cam rotation.

The least radius of the cam is 20 mm and the cam rotates at 240 rpm.

Determine : Maximum velocity and maximum acceleration during return strok.

[MDU, May, 2009]

Solution : Given : $h = 30$ mm, $\theta_1 = 120°$, SHM, $\theta_2 = 60°, \theta_3 = 90°$, uniform velocity, $r_1 = 20$ mm, $N = 240$ rpm

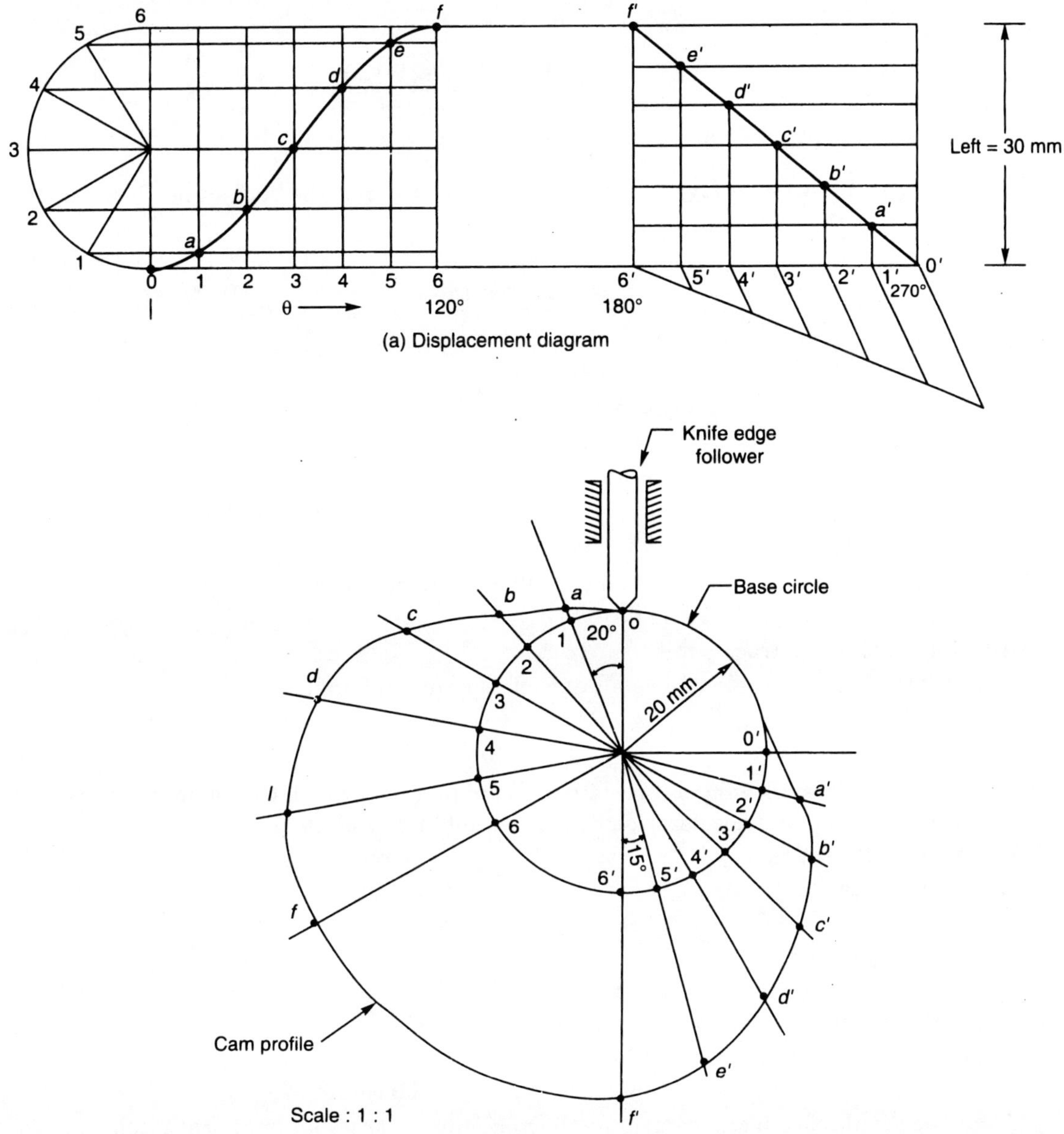

(a) Displacement diagram

(b) Cam profile

Fig. 8.35

Ascent (SHM) :

$$\omega = \frac{2\pi N}{60} = \frac{2\pi \times 240}{60} = 8\pi \text{ rad / s}$$

$$v_{max} = \frac{h}{2}\left(\frac{\pi\omega}{\theta_1}\right)$$

$$= \frac{30 \times 10^{-3}}{2}\left(\frac{\pi \times 8\pi \times 180}{\pi \times 120}\right) = 0.565 \text{ m/s}$$

$$f_{max} = \frac{h}{2}\left(\frac{\pi\,\omega}{\theta_1}\right)^2$$

$$= \frac{30 \times 10^{-3}}{2}\left(\frac{\pi \times 8\pi \times 180}{\pi \times 120}\right)^2$$

$$= 21.318 \text{ m/s}^2$$

Descent (uniform velocity) : Maximum velocity occurs at the beginning of descent.

$$v_{max} = \frac{h\omega}{\theta_3}$$

$$= \frac{30 \times 10^{-3} \times 8\pi \times 180}{\pi \times 90}$$

$$= 0.48 \text{ m/s}$$

$$F_{max} = 0$$

Construction :

1. Draw the displacement diagram as shown in Fig. 8.35(a).
2. Draw the cam profile as shown in Fig. 8.35(b). Follow the procedure as described in example 8.1.

■ **Example 8.18** Differentiate between the following in reference to a cam :

(i) Cam angle and pressure angle

(ii) Pitch point and trace point

(iii) Angle of ascent, angle of dwell, angle of descent, and angle of action.

[MDU, Dec., 2009]

Solution : Refer to art. 8.4.

■ **Example 8.19** The following data relate to a cam profile in which the follower moves with uniform acceleration and deceleration during ascent and descent.

Minimum radius of cam	= 25 mm
Roller diameter	= 7.5 mm
Lift	= 28 mm
Offset of follower axis	= 12 mm towards right
Angle of ascent	= 60°
Angle of descent	= 90°
Angle of dwell between ascent and descent	= 45°
Speed of cam	= 200 rpm.

Draw the profile of cam and determine the maximum velocity and uniform acceleration of the follower during the outstroke and the instroke. [MDU, Dec., 2009]

Solution : Given : $h = 28\,\text{mm}$, $\theta_1 = 60°$, $\theta_2 = 45°$, $\theta_3 = 90°$, $N = 200$ rpm

$$\omega = \frac{2\pi N}{60}$$

$$= \frac{2\pi \times 200}{60} = 20.944 \text{ rad/s}$$

In-stroke :

$$v_{max} = \frac{2h\omega}{\theta_1} = \frac{2 \times 28 \times 10^{-3} \times 20.944 \times 180}{\pi \times 60}$$

$$= 1.12 \text{ m/s}$$

$$f = 4h\left(\frac{\omega}{\theta_1}\right)^2$$

$$= 4 \times 28 \times 10^{-3}\left(\frac{20.944 \times 180}{\pi \times 60}\right)^2 = 44.8 \text{ m/s}^2$$

Outside stroke :

$$v_{max} = \frac{2h\omega}{\theta_3} = 2 \times 28 \times 10^{-3} \times \left(\frac{20.944 \times 180}{\pi \times 90}\right)^2$$

$$= 0.747 \text{ m/s}$$

$$f = 4h\left(\frac{\omega}{\theta_3}\right)^2$$

$$= 4 \times 28 \times 10^{-3}\left(\frac{20.944 \times 180}{\pi \times 90}\right)^2 = 19.91 \text{ m/s}^2$$

Draw the cam profile as described in Example 8.6.

■ **Example 8.20** (a) Explain the path of contact and arc of contact.

(b) Differentiate the following :

(i) Cam angle and pressure angle

(ii) Base circle and prime circle of cam

(iii) Circular pitch and diametral pitch

(iv) Period of ascent and period of descent. [MDU, Dec., 2010]

Solution : (a) Refer to art. 8.3

(b) Refer to art 8.4

■ **Example 8.21** Draw the profile of a cam operating a roller reciprocating follower and with the following data.

Minimum radius of cam	= 25 mm
Lift	= 30 mm
Roller diameter	= 15 mm

The cam lifts the follower for 120° with SHM followed by a dwell-period of 30°. Then, the follower lowers down during 150° of the cam rotation with uniform acceleration and deceleration followed by a dwell-period. If the cam rotates at a uniform speed of 150 rpm, calculate the max. velocity and acceleration of the follower during the descent period.

[MDU, Dec., 2010]

Solution : Given $r_1 = 25$ mm, $h = 30$ mm, $r_2 = 7.5$ mm, $\theta_1 = 120°$, SHM, $\theta_2 = 30°$, $\theta_3 = 150°$, f = constant, $N = 150$ rpm.

During descent period there is uniform acceleration and deceleration.

$$\omega = \frac{2\pi N}{60} = \frac{2\pi \times 150}{60} = 5\pi \text{ rad/s}$$

$$v_{max} = \frac{2h\omega}{\theta_3} = \frac{2 \times 30 \times 10^{-3} \times 5\pi \times 180}{\pi \times 150}$$

$$= 0.36 \text{ m/s}$$

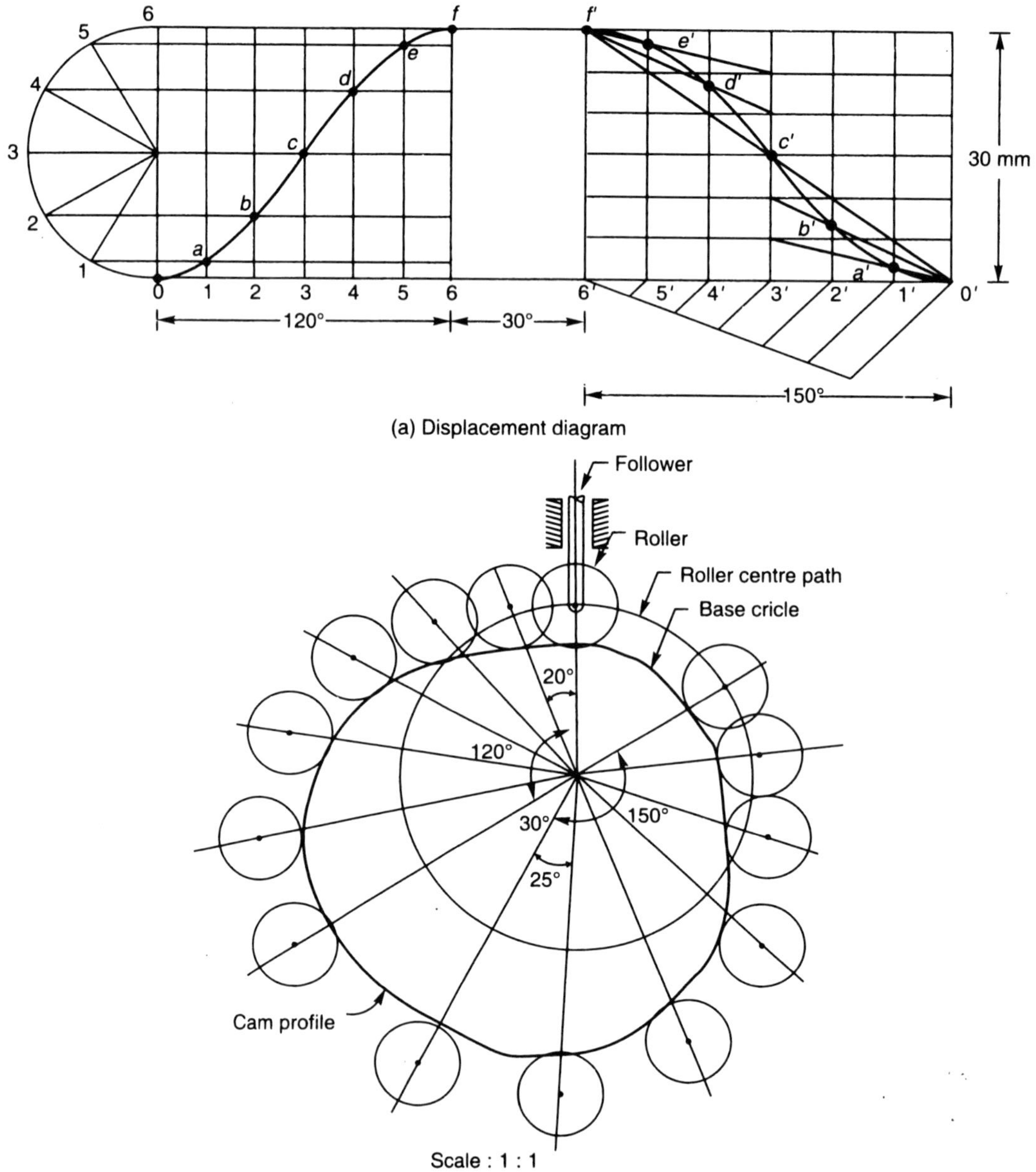

(a) Displacement diagram

(b) Cam profile

Fig. 8.36

$$f_{max} = 4h\left(\frac{\omega}{\theta_3}\right)^2$$

$$= 4 \times 30 \times 10^{-3}\left(\frac{5\pi \times 180}{\pi \times 150}\right)$$

$$= 4.32 \text{ m/s}^2$$

Construction :

1. Draw the displacement diagram as shown in Fig. 8.36(a).
2. Draw the cam profile as shown in Fig. 8.36(b).
   To draw the cam profile, follow the procedure of art. 8.7.

SUMMARY

1. Follower Motions

(a) Simple Harmonic Motion :

Displacement, $y = \left(\frac{h}{2}\right) \times \left[1 - \cos\left(\frac{\pi\theta}{\phi}\right)\right]$

Velocity, $v = \left(\frac{h}{2}\right) \times \left(\frac{\pi\omega}{\phi}\right)\left(\frac{\pi\theta}{\phi}\right)$

Maximum velocity, $v_{max} = \left(\frac{h}{2}\right) \times \left(\frac{\pi\omega}{\phi}\right)$ where $\theta = \frac{\phi}{2}$

Maximum velocity of follower during ascent $= \left(\frac{h}{2}\right)\left(\frac{\pi\omega}{\theta_1}\right)$

Maximum velocity of follower during descent $= \left(\frac{h}{2}\right)\left(\frac{\pi\omega}{\theta_3}\right)$

Acceleration, $f = \left(\frac{h}{2}\right)\left(\frac{\pi\omega}{\phi}\right)^2 \cos\left(\frac{\pi\theta}{\phi}\right)$

Maximum acceleration, $f_{max} = \left(\frac{h}{2}\right)\left(\frac{\pi\omega}{\phi}\right)^2$, when $\theta = 0°$

Maximum acceleration of follower during ascent $= \left(\frac{h}{2}\right)\left(\frac{\pi\omega}{\theta_1}\right)^2$

Maximum acceleration of follower during descent $= \left(\frac{h}{2}\right)\left(\frac{\pi\omega}{\theta_3}\right)^2$

where θ_1 = angle of ascent, and θ_3 = angle of descent.

(b) Motion with uniform acceleration of deceleration :

Displacement, $y = \frac{ft^2}{2}$

Maximum velocity during ascent, $v_{max} = \frac{2h\omega}{\theta_1}$

Maximum velocity during descent, $v_{max} = \frac{2h\omega}{\theta_3}$

Maximum acceleration during ascent, $f_{max} = \frac{4\omega^2 h}{\theta_1^2}$

Maximum acceleration during descent, $f_{max} = \frac{4\omega^2 h}{\theta_3^2}$

(c) Motion with uniform velocity :

Displacement, $y = \frac{h\theta}{\phi}$

Velocity, $v = \frac{h\omega}{\phi}$

Acceleration, $f = 0$

(d) Parabolic motion :

For the first half,

$$y = \left(\frac{2h\theta}{\phi^2}\right)$$

$$4h = \left(\frac{4\omega\theta}{\phi^2}\right)$$

$$f = 4h\left(\frac{\omega}{\phi}\right)^2 = \text{constant.}$$

For velocity to be maximum,

$$\theta = \frac{\phi}{2} \text{ and } v_{max} = \frac{2h\omega}{\phi}$$

For the second half, we have

$$y = h\left[1 - 2\left(1 - \frac{\theta}{\phi}\right)^2\right]$$

$$v = \left(\frac{4h\omega}{\beta}\right)\left(1 - \frac{\theta}{\phi}\right)$$

$$f = -4h\,(\omega/\phi)^2$$

(e) Cycloidal motion :

$$y = \left(\frac{h}{\pi}\right)\left[\frac{\pi\theta}{\phi} - \left(\frac{1}{2}\right)\sin\left(\frac{2\pi\theta}{\phi}\right)\right]$$

$$v = \frac{h\omega}{\phi}\left[1 - \cos\left(\frac{2\pi\theta}{\phi}\right)\right]$$

$$v_{max} = \frac{2h\omega}{\phi} \text{ at } \theta = \frac{\phi}{2}$$

$$f = (2h\pi)\left(\frac{\omega}{\phi}\right)^2 \sin\left(\frac{2\pi\theta}{\phi}\right)$$

$$f_{max} = (2h\pi)\left(\frac{\omega}{\phi}\right)^2 \text{ at } \theta = \frac{\phi}{4}$$

Cycloidal motion is the most ideal for high speed follower motion.

2. Radius of Curvature

The cam profile must be convex everywhere for a flat-face follower. In the case of a roller follower, the concave portion of the cam profile must have a radius of curvature greater than that of the roller to ensure proper contact along the cam profile.

3. Pressure Angle

Pressure angle is the angle between the direction of the follower motion and a normal to the pitch curve. The following methods may be used to reduce the pressure angle :

1. Increase the diameter of the base circle.
2. Increase the angle of rotation of the cam, thereby lengthening the pitch curve for the specified follower dispalcement. The cam profile becomes flatter and the pressure angle becomes smaller.
3. Select the motion curve for a smaller pressure angle.
4. By changing the offset of the follower.

4. Cam Size

The cam size is defined by the following parameters :

1. Pressure angle
2. Radius of curvature of cam profile
3. Hub size.

SHORT ANSWER TYPE QUESTIONS

1. What is a cm? What is its use?
2. How cams can be classified?
3. What are the various types of followers?
4. Name the different motions that a follower can have.
5. Differentiate between (a) base circle and prime circle, (b) can angle and pressure angle.
6. Compare the knife-edge follower with roller follwer.
7. What is a tangent cam?
8. Compare various types of follower motions.
9. Define pressure angle of a cm.
10. Differntiate between trace point and pitch point.

11. How the cam size is defined?
12. What are the methods for reducing pressure angle of a cam?
13. What is an offset follower?

EXERCISES

1. Draw the profile of a cam operating a knife edge follower when the axis of the following passes through the axis of the cam shaft. The following data is given : Lift = 40 mm, angle of ascent = 60°, dwell = 45°, angle of descent = 90°, and dwell for the remaining period of cam rotation.
   The motion of the cam is simple harmonic during both ascent. The least radius of cam is 50 mm. If the cam rotates at 300 rpm, determine the maximum velocity and acceleration of the followr during ascent and descent. [**Ans.** 1.88 m/s, 1.26 m/s; 177.47 m/s^2, 78.87 m/s^2]
2. If the follower in Exercise 8.1 is offset by 25 mm, then draw the cam profile.
3. A cam with 30 mm as minimum diameter is rotating clockwise at a uniform speed of 1200 rpm and operates a roller follower of 10 mm diameter as given below :
   (i) Outward stroke of 30 mm during 120° of cam rotation with equal uniform acceleration and retardation.
   (ii) Follower is to dwell for 50° of cam rotation.
   (iii) Inward stroke during 90° of cam rotation with equal uniform acceleration and retardation.
   (iv) Follower is to dwell for the remaining period of cam rotation.
   Draw the cam profile if the axis of follower passes through the axis of the cam. Determine the maximum velocity and acceleration during outward and inward strokes.
   [**Ans.** 3 m/s, 4 m/s; 359.6 m/s^2, 639.35 m/s^2]
4. Draw the cam profile from the following data if the radial follower moves with simple harmonic motion during ascent and uniform acceleration and deceleration during descent; Lift = 40 mm, Least radius of cam = 60 mm, Angle of ascent = 54°, Dwell = 40°, Angle of descent = 72°, Roller diameter = 20 mm.
5. If the roller follower is offset by 20 mm in Exercise 8.5, then draw the cam profile.
6. Draw the profile of a cam to give reciprocating motion to a flat-faced follower for the following data;
   Lift = 25 mm, Ascent = 120°, Dwell = 30°, Descent = 120°, Dwell = 90°
   Minimum radius of cam = 25 mm
   The ascent and descent is to take place with SHM. The line of movement of follower passes through the cam centre.

❑❑❑

Chapter 9

Gears

Inside This Chapter

9.1 INTRODUCTION

A gear may be defined as any toothed member designed to transmit or receive motion from another member by successively engaging each tooth. The smaller gear is called the **pinion** and the bigger one the **gear wheel.** They are used in metal cutting machine tools, automobiles, tractors, hoisting and transporting machinery, rolling mills, etc.

The gears provide many advantages over other modes of power transmission likes belts, ropes and chains, etc. Some of their advantages are :

1. They occupy lesser space.

2. There is no slip between the gears in mesh and provide exact speed ratio.
3. They can transmit higher power.
4. Their efficiency is higher. However, the error in tooth meshing may cause undesirable vibration and noise during operation.

9.2 CLASSIFICATION OF GEARS

The gear may be classifies as follows :

1. **Spur gears :** A spur gear is a cylindrical gear whose tooth traces are straight line generators of the reference cylinder. A spur gear pair is mesh in shown in Fig. 9.1
2. **Helical gears :** This is similar to the spur gear in which the tooth traces are helices.
3. **Double helical (or herringbone) gears :** It is a cylindrical gear in which a part of the face width is right hand and the other left hand, with or without a gap between them.
4. **Spiral gears :** In spiral gears, the tooth trances are curved lines other than helices.
5. **Bevel gears :** The reference surface is a cone in bevel gears. The bevel gears may be straight, spiral, zerol, the face gears. In zerol belvel gears, the teeth are curved in the lengthwise direction and are arranged in such a manner that the effective spiral angle is zero. In face gears, the bevel gear teeth are cut on the flat-face of the blank. A crown gear is a bevel gear with a reference cone angle of 90°.

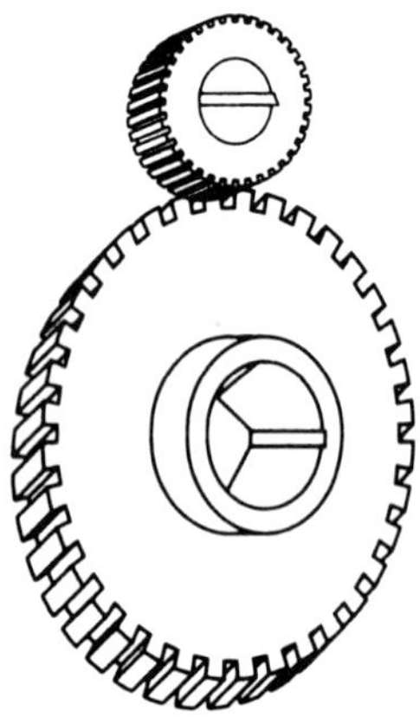

Fig. 9.1 Spur gears.

6. **Hypiod gears :** They are similar to the spiral bevel gear with the difference that the axes of the shafts do not intersect.
7. **Worm gears :** In these gears, the are screw threads on the worm and teeth on the worm wheel.
8. **Planetary gears :** A gear pair or a gear train one of whose axes, instead of being fixed in position in the mechanism of which the gear pair is apart, moves around the other is called planetary gear train.

Gears may also be classified based on the orientation of the shafts as :

1. Parallel shafts—spur, helical, and double helical gears.
2. Intersecting shafts—straight bevel, spiral bevel, zerol bevel, and face gears.
3. Non-parallel and no-intersecting—spiral, hypoid, and worm gears.

Gears may be of the external, internal and rack and pinion type. In external gears, the teeth of gears mesh externally, whereas in internal gears the teeth of the two gears mesh internally. A rack is a gear of infinite radius.

9.3 GEAR TERMINOLOGY

A spur gear pair in mesh is shown in Fig. 9.2. The various terms related to gears are defined as follows :

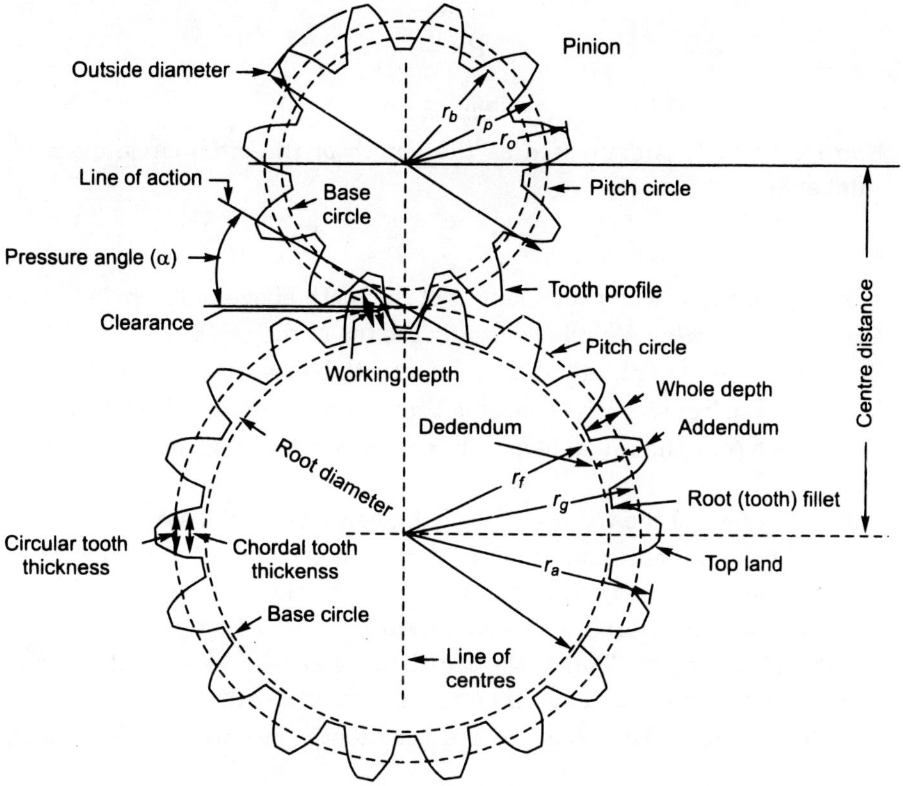

Fig. 9.2 Terminology of spur gears.

- ***Pitch circle :*** It is an imaginary circle which by pure rolling action, would give the same motion as the actual gear.
- ***Pitch circle diameter (d) :*** It is the diameter of a circle which by pure rolling action would produce the same motion as the toothed gear wheel.
- ***Base circle :*** It is the circle from which involute form is generated.
- ***Pitch surface :*** It is the surface of the disc which the toothed gear has replaced at the pitch circle.
- ***Pitch point :*** It is the pitch of the tangency or the point of contact of the two pitch circles of the mating gears.
- ***Circular pitch (p) :*** It is the distance measured along the circumference of the pitch circle from a point on one tooth to a corresponding point on the adjacent tooth.

$$p = \frac{\pi d}{z} \quad \text{...(9.1)}$$

where z = number of teeth.

- ***Base pitch (p_b) :*** It is the distance measured along the circumference of the base cirlce from a point on one tooth to a corresponding point on the adjacent tooth.

Base pitch, $p_b = p \times \cos \alpha$ ...(9.2)

where α = pressure angle of gear tooth profile.

- **Diametral pitch (*P*) :** It is expressed as the number of teeth per unit pitch circle diameter.

$$P = \frac{z}{d} \quad \text{...(9.3)}$$

$$Pp = \pi \quad \text{...(9.4)}$$

- ***Module (m) :*** It is expressed as the length of the pitch circle diameter per unit number of teeth.

$$m = \frac{d}{z} = \frac{1}{P} \quad \text{...(9.5)}$$

- ***Addendum (h_a) :*** The radial height of the tooth above pitch circle.
- ***Addendum circle :*** A circle bounding the top of the teeth.
- ***Dedendum (h_f) :*** The radial depth of a tooth below the pitch circle.
- ***Dedendum circle :*** A circle passing through the roots of all the teeth.
- ***Clearance (c) :*** The radial height difference between addendum and dedendum of a teeth.
- ***Face :*** The part of the tooth surface lying below the pitch surface.
- ***Backlash :*** The minimum distance between the non-driving side of a tooth and adjacent side of the mating tooth at the pitch circle.
- ***Profile :*** The curve forming face and flank.
- ***Tooth thickness (t) :*** This is the arc distance measured along the pitch circle from its intercept with one flank to its intercept with the other flank of the same tooth.
- ***Face width (b) :*** The width of the gear tooth measured axially along the pitch surface.
- ***Top land :*** It is the surface of the top of the tooth.
- ***Tooth fillet :*** The radius that connects the root circle to the profile of the tooth.
- ***Tooth space :*** It is the width of the space between two teeth measured on the pitch circle.
- ***Pressure angle (α) :*** The angle between the common normal at the point of contact and the common tangent at the pitch point. The pressure angle is either 14.5° or 20°.
- ***Path of contact :*** It is the locus of the point of contact of two mating teeth from the beginning of engagement to the end of engagement. It is a straight line.
- ***Path of approach :*** It is the portion of the path of contact from the beginning of engagement to the pitch point.
- ***Angle of approach :*** The angle turned by gears during path of approach.
- ***Path of recess :*** It is the portion of the path of contact from the pitch point to the end of engagement of the two mating teeth.
- ***Angle of recess :*** The angle turned through during path of recess.
- ***Arc of contact :*** It is the locus of a point on the pitch circle, from the beginning of engagement to the end of engagement of pair of teeth in mesh.
- ***Involute :*** The involute of a circle is the curve traced by the end of a thread as it is unwound from a stationary cylinder.
- ***Base Circle Diameter (d_b) :*** It is the diameter of the base circle.

$$d_b = d \cos a \quad \text{...(9.6)}$$

- ***Cycloid :*** It is the locus of a point on the circumference of a circle which rolls without slipping on a fixed straight line.

- ***Centre distance* (C) :** It is the distance between the centres of rotation of the two gears in mesh.

$$C = \frac{d_1 + d_2}{2} = \frac{m\,(z_1 + z_2)}{2} \qquad \ldots(9.7)$$

9.4 FUNDAMENTAL LAW OF GEARING

Let us consider two curved bodies 1 and 2 rotating about their centres O_1 and O_2 and contacting at point A, as shown in Fig. 9.3. A_1 A_2 are two coincident points, A_1 lying on body 1 and A_2 lying on body 2. TT' and NN' represent common tangents and normals respectively at point A. Let ω_1 and ω_2 be the angular velocities of body 1 and 2 respectively. Let v_{a1} and v_{a2} be the linear velocities of point A. $v_{a1}$is perpendicular to O_1A_1 and v_{a2} is perpendicular to O_2A_2. Let the common normal intersect the line of centres at point P. Let O_1G and O_2H be perpendiculars to AP.

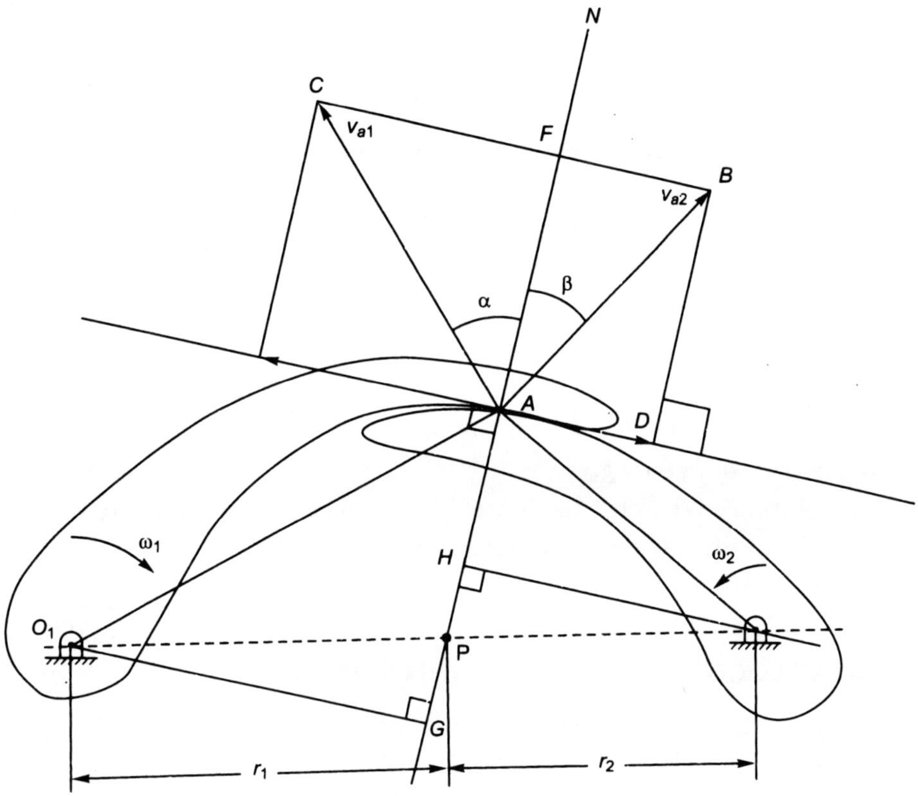

Fig. 9.3 Low of gearing.

If the two bodies are to remain in contact, the component of velocities of A_1 and A_2 along the common normal must be equal.

Therefore
$$v_{a1} \cos \alpha = v_{a2} \cos \beta$$

or
$$\omega_1 . O_1A_1 \cos \alpha = \omega_2 . O_2A_2 \cos \beta$$

$$\omega_1 \,.\, O_1A_1 \cdot \frac{A_1F}{A_2C} = \omega_2 \,.\, O_2A_2 \cdot \frac{A_2F}{A_2B}$$

The condition for pure rolling is that the point of contact shall lie on the line of centres.

$$\frac{\omega_1}{\omega_2} = \frac{A_1C}{O_1A_1} \cdot \frac{O_2A_2}{A_2B}$$

Triangles A_1CF and $O_1A_1G_1$ are similar. Also triangles A_2FB and O_2A_2H are similar.

Therefore
$$\frac{A_1C}{O_1A_1} = \frac{A_1F}{O_1G}$$

and
$$\frac{A_2B}{O_2A_2} = \frac{A_2F}{O_2H}$$

Hence
$$\frac{\omega_1}{\omega_2} = \frac{A_1F}{O_1G} \cdot \frac{O_2H}{A_2F}$$

But
$$A_1F = A_2F$$

$$\frac{\omega_1}{\omega_2} = \frac{O_2H}{O_1G}$$

Now triangles O_1PG and O_2PH are similar. Hence

$$\frac{O_2H}{O_1G} = \frac{O_2P}{O_1P}$$

$$\frac{\omega_1}{\omega_2} = \frac{O_2P}{O_1P} \quad ...(9.8)$$

Therefore, *for constant angular velocity ratio of the two gears in contact the common normal at the point of contact must always intersect the line of centres at a fixed point (pitch point) and divide this line in the inverse ratio of the angular velocities of the two gears.* ***This is the fundamental law of gearing.***

Conjugate Action : When the tooth profiles are so shaped so as the produce a constant angular velocity ratio during meshing, then the surfaces are said to be conjugate. Two conjugate surfaces in contact always satisfy the law of gearing.

9.5 SLIDING VELOCITY BETWEEN GEAR TEETH

The relative velocity along the common tangent is called the sliding velocity, v_s. Considering Fig. 9.3 again, the sliding velocity the common tangent,

$$v_s = v_{a1} \sin\alpha + v_{a2} \sin\beta$$
$$= \omega \,.\, O_1A_1 \cdot \frac{FC}{A_1C} + \omega_2 \,.\, O_2A_2 \cdot \frac{FB}{A_2B}$$

Now
$$\frac{A_1C}{O_1A_1} = \frac{A_1F}{O_1G} = \frac{FC}{AG}$$

and $$\frac{A_2B}{O_2A_2} = \frac{A_2F}{O_2H} = \frac{FB}{AH}$$

Hence, $$v_s = \omega_1 \,.\, AG + \omega_2 \,.\, AH$$
$$= \omega_1\,(AP + PG) + \omega_2\,(AP - PH)$$
$$= (\omega_1 + \omega_2)\,AP + \omega_1 \,.\, PG - \omega_2 \,.\, PH$$

Now $$\frac{\omega_1}{\omega_2} = \frac{O_2P}{O_1P} = \frac{PH}{PG}$$

Hence $$v_s = (\omega_1 + \omega_2)\,AP \qquad ...(9.9)$$

v_s = (sum of the angular velocities) × distance of the point of contact from the pitch point.

Thus, we find that the ***velocity of sliding is proportional to the distance of the pitch point from the point of contact.***

9.6 GEAR TOOTH FORMS

There are two types of the gear tooth forms : involute and cycloidal. The involute of a circle is the curve traced by the end of a thread as it is unwound from a stationary cylinder. Cycloid is the locus of a point on the circumference of a circle which rolls without slipping on a fixed straight line. If the circle, instead of rolling without slipping on a straight line, rolls on the outside of another circle, the locus of the point on the circumference is called as epicycloid. Conversely, if the circle rolls on the outside of another circle, the corresponding locus of the point on the circumference of the rolling circle is called hypo-cyloid.

9.6.1 Comparison between Involute and Cycloidal Tooth Profiles

The comparison of involute and cycloidal tooth profiles is given in Table 9.1. The cycloidal profile is not commonly used for gear tooth, due to the reasons given in Table 9.1.

Table 9.1 : Comparison between Involute and Cycloidal Tooth Profiles

Characteristics	*Involute gears*	*Cycloidal gears*
1. Pressure angle	Constant throughout the engagement	Varies from commencement to end
2. Ease of manufacture	Easy to manufacture	Difficult to manufacture
3. Centre distance	Do not require exact center distance	Requires exact centre distance
4. Interference	May occur	No interference
5. Strength	Less	More
6. Wear	More wear and tear	Less wear and tear
7. Operation	Smooth	Less smooth

9.7 CONSTRUCTION OF AN INVOLUTE

The construction of an involute tooth profile is shown in Fig. 9.4. The following steps may be followed to draw the involute :

1. Draw the base circle.

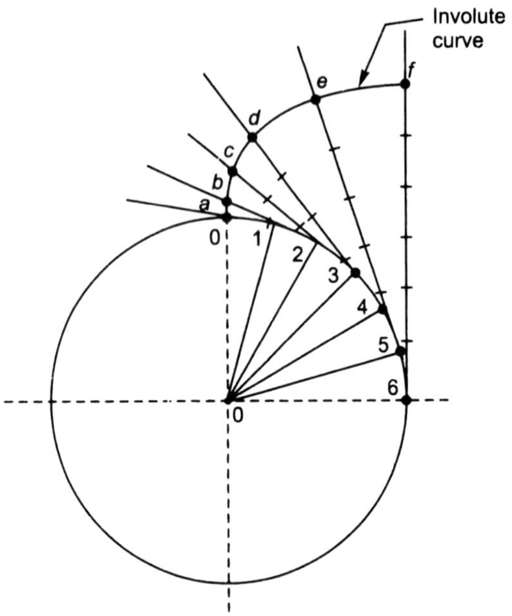

Fig. 9.4 Involute profile.

2. Divide the base circle quadrant into equal number of parts (say 6). Mark the points 0 to 6 on the circumference of the circle.
3. Draw tangents at points 1 to 6.
4. Cut-off 1 a = 01 on tangent at 1; 2b = 02 on tangent at 2, and so on.
5. Join o, a, b, c, etc. by a smooth curve to obtain the involute profile.

9.8 INVOLUTE FUNCTION

Consider the involute of a circle shown in Fig. 9.5 (a)

Let l = length of the thread unwrapped

$= AB$ arc

$= r_b\ (\beta + \delta)$

$= r_b\ \tan \alpha$

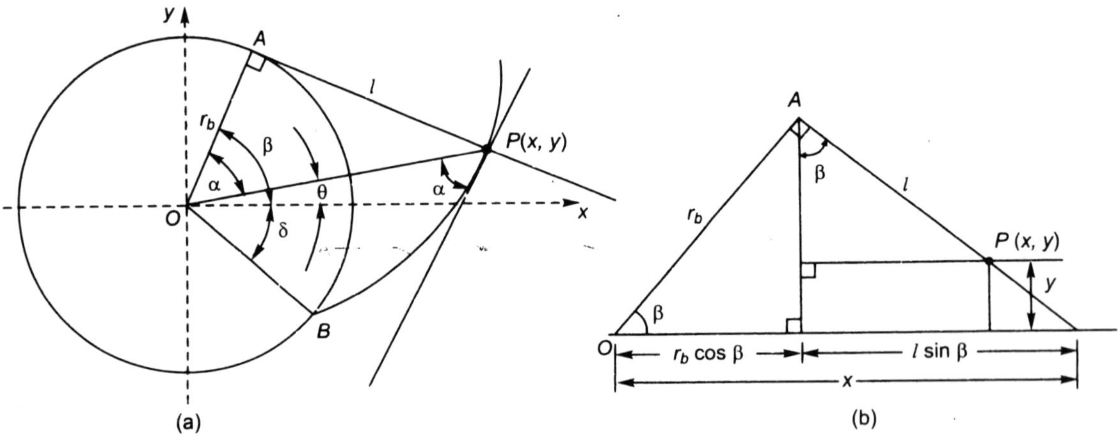

Fig. 9.5 Involute function.

Thus $$\beta + \delta = \tan \alpha$$

where r_b = base radius.

Also from Fig. 9.5 (b), we have

$$x = r_b \cos \beta + l \sin \beta$$
$$= r_b [\cos \beta + (\beta + \delta) \sin \beta]$$

and $$y = r_b [\sin \beta - (\beta + \delta) \cos \beta]$$

Also $$\theta + \alpha = \beta = \tan \alpha - \delta$$

or $$\theta = \tan \alpha - \alpha - \delta$$
$$= \text{inv}(\alpha) - \delta$$

where $$\text{inv}(\alpha) = \tan \alpha - \alpha \quad \text{...(9.10)}$$

Eq. (9.10) represents the involute function. It values are given in standard tables.

9.9 CHARACTERISTICS OF INVOLUTE ACTION

The characteristics of the involute action are :

- arc of contact,
- length of path of contact, and
- the contact ratio.

As shown in Fig. 9.6, the contact of two gear teeth beings at *A* and ends at *B*.

Addendum radius of pinion, $r_{a1} = r_1 + h_{a1}$

Base circle radius of pinion, $r_{b1} = r_1 \cos \alpha$

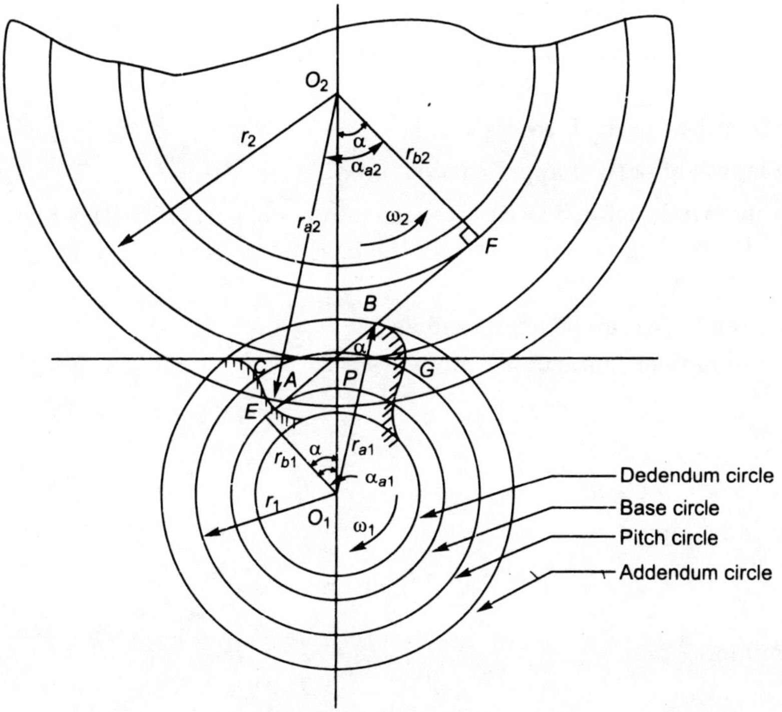

Fig. 9.6 Angles of action.

Addendum radius of gear, $r_{a2} = r_2 + h_{a2}$

Base circle radius of gear, $r_{b2} = r_2 \cos \alpha$

where r_1 = pitch circle radius of pinion

r_2 = pitch circle radius of gear

h_{a1} = addendum of pinion

h_{a2} = addendum of gear

r_{b1} = base circle radius of pinion

r_{b2} = base circle radius of gear

Length of path of recess, $L_r = PB = EB - EP$

$$= (r_{a1}^2 - r_{b1}^2)^{0.5} - O_1P \sin \alpha$$

$$= (r_{a1}^2 - r_{b1}^2)^{0.5} - r_1 \sin \alpha \qquad \text{...(9.11)}$$

Length of path of approach,

$$L_a = AP = AF - PF$$

$$= (r_{a2}^2 - r_{b2}^2)^{0.5} - O_2P \sin \alpha$$

$$= (r_{a2}^2 - r_{b2}^2)^{0.5} - r_2 \sin \alpha \qquad \text{...(9.12)}$$

Length of path of contact,

$$AB = L_p = L_r + L_a$$

$$= (r_{a1}^2 - r_{b1}^2)^{0.5} + (r_{a2}^2 - r_{b2}^2)^{0.5} - (r_1 + r_2) \sin \alpha \qquad \text{...(9.13)}$$

Length of arc of contact, L_c = arc CG

$$= \frac{AB}{\cos \alpha} \qquad \text{...(9.14)}$$

Maximum length of path of recess $= r_2 \sin \alpha$ ...(9.15)

Maximum length of path of approach $= r_1 \sin \alpha$ ...(9.16)

The **contact ratio** is defined as the average number of pairs of teeth which are in contact. This can be found by noting how many times the base pitch fits into the length of the path of contact.

The contact ratio (CR) can be expressed as :

CR = length of path of contact/base pitch

$$= \frac{L_p}{p_b} \qquad \text{...(9.17)}$$

where $p_b = p \cos \alpha = \pi m \cos a$

For a rack and a pinion,

$$L_P = (r_{a1}^2 - r_{b1}^2)^{0.5} - r_1 \sin \alpha + \frac{a}{\sin a} \qquad \text{...(9.18)}$$

where a = addendum.

9.10 INTERFERENCE AND UNDERCUTTING IN INVOLUTE GEAR TEETH

An involute starts at the base circle and is generated outwards. It is therefore impossible to have an involute inside the base circle. The line of action is tangent to the two base circles of a pair of gears in mesh, and these two points represent the extreme limits of the length of action.These two points are called **interference points.** If the teeth are of such proportion that the beginning of contact occurs before the interference point is met, then the involute portion of the driven gear will mate with a non-involute portion of the driving gear, and **involute interference** is said to occur.This condition is shown in Fig. 9.7; E_1 and E_2 show the interference points that should limit the length of action. A shows the beginning of contact, and *B* shows the end of contact. It can be seen that the beginning of contact occurs before the interference point E_1 is met; therefore, interference is present. The tip of the driven tooth will gauge out or undercut the flank of the driving tooth as shown by the dotted line.

There are several ways of eliminating interference. Interference can be avoided by undercutting, making stub teeth, increasing the pressure angle, and cutting the gears with long and short addendum gear teeth. The method of undercutting is to limit addendum of the driven gear so that it passes through the interference point E_1, thus giving a new beginning of contact. Interference and the resulting undercutting not only weaken the pinion tooth but may also remove a small portion of the involute adjacent to the base circle, which may cause a serious reduction in the length of action.

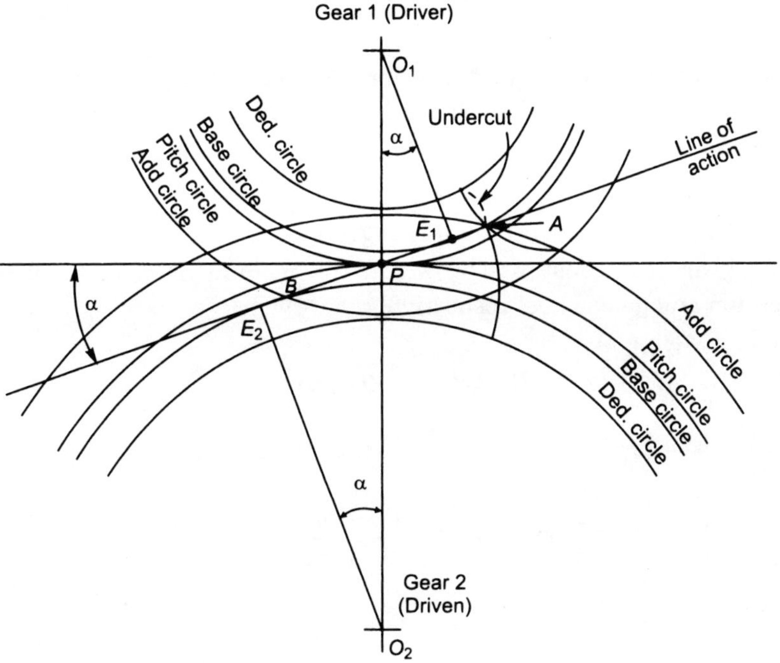

Fig. 9.7 Interference in rack and pinion.

Fig. 9.8 shows a rack and a pinion in mesh. The point of tangency of the line of action and the base circle of the pinion is labelled as the interference point *E*, which fixes the maximum addendum for the rack. The contact begins at *A*, and undercutting will occur as shown by the dotted line. If the addendum of the rack extends only to the line that passes through the

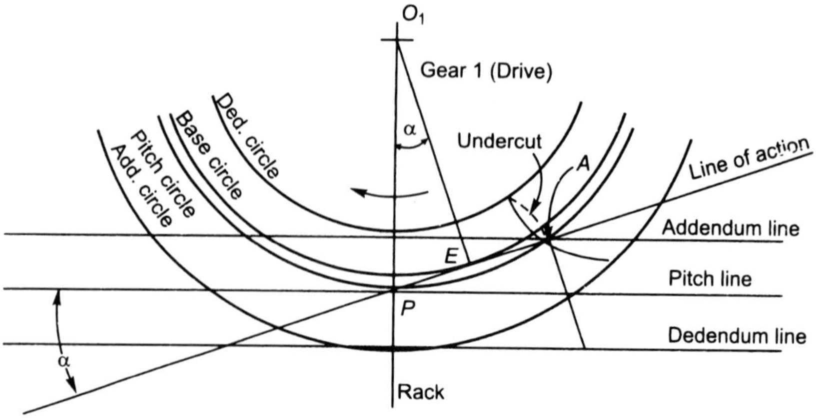

Fig. 9.8 Interference in rack and pinion.

interference point E, then the interference point becomes the beginning of contact, and interference is eliminated. If the number of teeth on the pinion is such that it will mesh without interference, it will mesh without interference with any other gear having the same or a larger number of teeth.

9.11 MINIMUM NUMBER OF TEETH

9.11.1 Gear Wheel

For minimum number of teeth to avoid interference, the common tangent to the base circles cuts the addenda circles at A and B, as shown in Fig. 9.9.

Let speed ratio, $i = z_2/z_1$

Addendum of pinion, $h_{a1} = a_p m$

Addendum of gear wheel, $h_{a2} = a_w m$

where a_p and a_w are the constants by which the module must be multiplied to get the addendum of pinion and gear wheel respectively.

From $\Delta\, AO_2P$, we have

$$O_2A^2 = O_2P^2 + AP^2 - 2 . O_2P . AP . \cos \angle O_2PA$$

$$= \left(\frac{mz_2}{2}\right)^2 + (O_1P \sin \alpha)^2 - 2 \cdot \frac{mz_2}{2} \cdot O_1P \sin \alpha . \cos (90° + \alpha)$$

$$= \left(\frac{mz_2}{2}\right)^2 + \left(\frac{mz_1}{2}\right)^2 \sin \alpha + 2 \cdot \left(\frac{mz_2}{2}\right) \cdot \frac{mz_1}{2} \sin^2 \alpha$$

$$= \left(\frac{mz_2}{2}\right)^2 \left[1 + \left(\frac{1}{i^2}\right) \sin^2 \alpha + \left(\frac{2}{i}\right) \sin^2 \alpha\right]$$

$$O_2A = \frac{mz_2}{2} \left[1 + \left(\frac{1}{i^2}\right) \sin^2 \alpha + \left(\frac{2}{i}\right) \sin^2 \alpha\right]^{0.5} \quad ...(1)$$

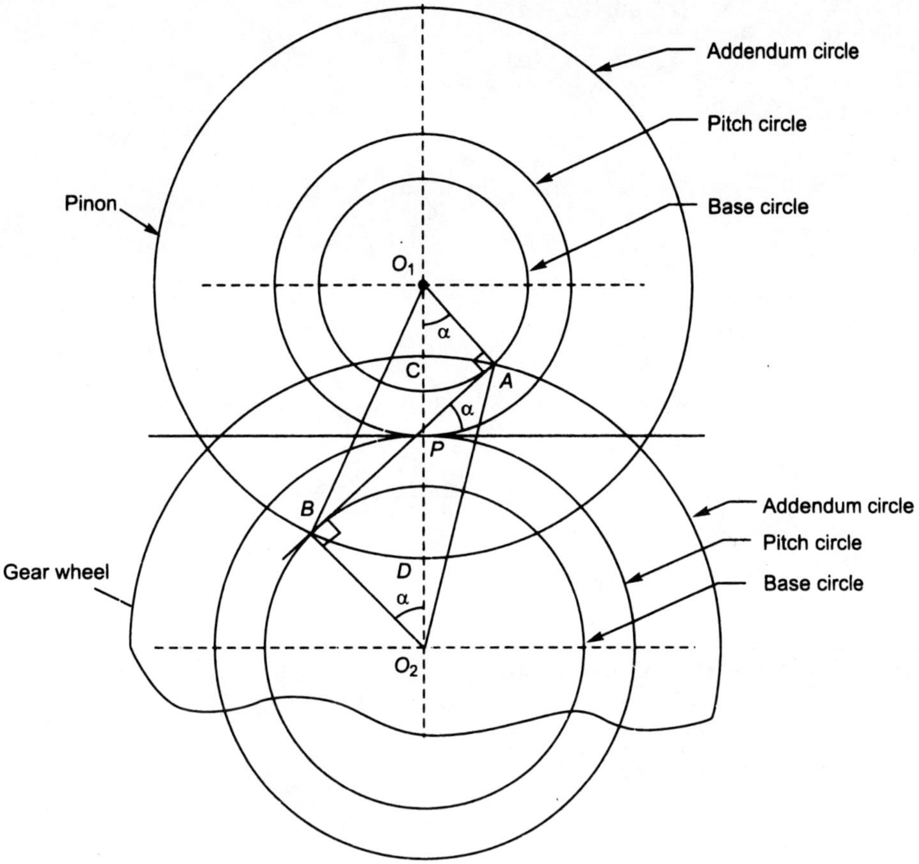

Fig. 9.9 Calculating minimum number of teeth on pinion and gear wheel.

Also
$$O_2A = O_2P + PC$$
$$= \frac{mz_2}{2} + h_{a2}$$
$$= \frac{mz_2}{2} + a_w m \qquad \ldots(2)$$

From Eqs. (1) and (2), we get

$$\frac{mz_2}{2} + a_w m = \left(\frac{mz_2}{2}\right)\left[1 + \left(\frac{1}{i^2}\right)\sin^2\alpha + \left(\frac{2}{i}\right)\sin^2\alpha\right]^{0.5}$$

or
$$a_w = \left(\frac{z_2}{2}\right)\left[\left\{1 + \left(\frac{1}{i^2}\right)\sin^2\alpha + \left(\frac{2}{i}\right)\sin^2\alpha\right\}^{0.5} - 1\right]$$

or
$$z_2 = \frac{2a_w}{\left[\left\{1 + \left(\frac{1}{i^2}\right)\sin^2\alpha + \left(\frac{2}{i}\right)\sin^2\alpha\right\}^{0.5} - 1\right]} \qquad \ldots(9.19)$$

or
$$z_2 = \frac{[z_1^2 \sin^2\alpha - 4a_w^2]}{[4a_w - 2z_1 \sin^2\alpha]} \qquad \ldots(9.20)$$

For $a_w = 1$, $$z_2 = \frac{[z_1^2 \sin^2 \alpha - 4]}{[4 - 2z_1 \sin^2 \alpha]} \quad ...(9.21)$$

For $i = 1$, $a_w = a_p$ and

$$z_2 = \frac{2a_w}{[\{1 + 3 \sin^2 a\}^{0.5} - 1]} \quad ...(9.22)$$

9.11.2 Pinion

From ΔBO_1P, we have

$$O_1B^2 = O_1P^2 + BP^2 - 2 . O_1P . BP \cos \angle O_1PB$$

$$= \left(\frac{mz_1}{2}\right)^2 + (O_2P \sin \alpha)^2 - 2 \cdot \frac{mz_1}{2} \cdot O_2P \sin \alpha . \cos (90° + \alpha)$$

$$= \left(\frac{mz_1}{2}\right)^2 + \left(\frac{mz_2}{2}\right)^2 \sin^2 \alpha + 2 \cdot \left(\frac{mz_1}{2}\right)^2 \cdot \left(\frac{mz_2}{2}\right)^2 \sin^2 \alpha$$

$$= \left(\frac{mz_1}{2}\right)^2 [1 + i^2 \sin^2 \alpha + 2i \sin^2 \alpha]$$

$$O_1B = \left(\frac{mz_1}{2}\right)^2 [1 + i^2 \sin^2 \alpha + 2i \sin^2 \alpha]^{0.5} \quad ...(1)$$

Also $$O_1B = O_1P + PD$$

$$= \frac{mz_1}{2} + h_{a1}$$

$$= \frac{mz_1}{2} + a_m \quad ...(2)$$

From Eqs. (1) and (2), we get

$$\frac{mz_1}{2} + a_p m = \left(\frac{mz_1}{2}\right) [1 + i^2 \sin^2 \alpha + 2i \sin^2 \alpha]^{0.5}$$

or $$a_p = \left(\frac{z_1}{2}\right) [\{1 + i^2 \sin^2 \alpha + 2i \sin^2 \alpha\}^{0.5} - 1]$$

or $$\frac{2a_p}{[\{1 + i^2 \sin^2 \alpha + 2i \sin^2 \alpha\}^{0.5} - 1]} \quad ...(9.23)$$

or $$z_1 = \frac{[z_2^2 \sin^2 \alpha - 4a_p^2]}{[4a_p - 2 z_2 \sin^2 \alpha]} \quad ...(9.24)$$

For $$a_p = 1, z_1 = \frac{[z_2^2 \sin^2 \alpha - 4]}{[4 - 2 z_2 \sin^2 \alpha]} \quad ...(9.25)$$

For $\alpha = 14.5°$

$$(z_1)_{min} = \left(\frac{z_2^2 - 63.8}{63.8 - 2z_2}\right) \quad ...(9.26)$$

For α = 20°

$$(z_1)_{min} = \left(\frac{z_2^2 - 34.2}{34.2 - 2z_2} \right) \qquad ...(9.27)$$

9.11.3 Rack and Pinion

A rack is a gear of infinite pitch radius. Thus its pitch circle is straight line, called the **pitch line.** The line of action is tangent to the base circle at infinity;thence the involute profile of the rack is a straight line and is perpendicular to the line of action. Fro the rack and pinion shown in Fig. 9.10, let

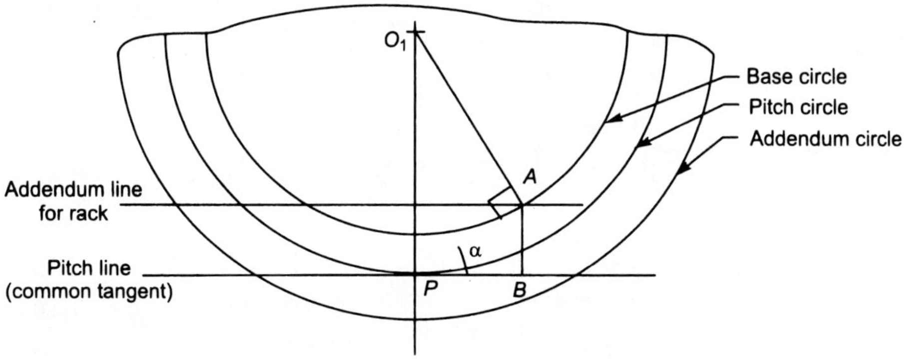

Fig. 9.10 Minimum number of teeth on rack and pinion.

Addendum of rack, $h_r = a_r m$

$$h_r = AB = AP \sin \alpha$$
$$= O_1P \sin \alpha \sin \alpha$$
$$= O_1P \sin a \sin \alpha$$
$$= r_1 \sin^2 \alpha$$

or
$$a_r m = \left(\frac{mz_1}{2} \right) \sin^2 \alpha$$

or
$$a_r = \left(\frac{z_1}{2} \right) \sin^2 \alpha$$

or
$$z_1 = \frac{2a_r}{\sin^2 \alpha} \qquad ...(9.28)$$

For $a_r = 1$, the minimum number of teeth on the pinion are given in Table 9.2.

Table 9.2 : Minimum number of Teeth on the Pinion for a Rack

α	14.5°	20°	20° stub	25°
$(z_1)_{min}$	32	18	14	12

9.12 GEAR STANDARDIZATION

A set of gears is interchangeable when any two gears selected from the set will mesh with each other and satisfy the fundamental law of gearing. For interchangeability all gear of the set must have the same circular pitch, module, diametral pitch, pressure angle, addendum and dedendum; and tooth thickness must be one-half of the circular pitch. Standard tooth forms ensure readily availability of gears.

The standard pressure angles are 14.5° and 20°. The 20° full-depth system has several advantages when compared with the 25° or 30° full-depth system. The lower pressure angle gives a higher contact ratio which results in quieter operation, reduced wear, reduced tooth load, and reduced bearing loads.

The larger pressure angle tooth forms results in boarder teeth at the base and hence are stronger in bending. Also fewer teeth may be used on the pinion without undercutting the teeth. The proportions of standard tooth forms are given in Table 9.3.

Table 9.3 : Standard Tooth Forms

	14.5° full-depth	*20° full-depth*	*20° stub*
Addendum, h_a	m	m	0.800 m
Dedendum, h_f	1.157 m	1.250	m
Clearance, c	0.157 m	0.250	0.200 m
Fillet radius, r	0.209 m	0.300 m	0.304 m
Tooth thickness, t	1.5708 m	1.5708 m	1.5708 m

9.13 EFFECT OF CENTRE DISTANCE VARIATION ON VELOCITY RATIO

Consider a pair of teeth in contact at L, as shown in Fig. 9.11. The angular velocity ratio is,

$$\frac{\omega_1}{\omega_2} = \frac{O_2P}{O_1P}$$

Let the centre distance of rotation of gear 2 be shifted from O_2 to O'_2. As a result of this change, the contact point will shift to L'. Common normal at the point of contact L' is tangent to the base circle, because it is a contact between two involute curves, and they are generated from base circle. Let the tangent to the base circle $M'N'$ intersect the line joining the centres of rotation O' and O_1 and P'.

Triangles O_1PN and O_2MP are similar. Also triangles $O_1N'P'$ and $O'_2M'P'$ are similar. Therefore,

$$\frac{MO_2'}{N'O_1} = \frac{O_2'P'}{O_1P'}$$

and

$$\frac{MO_2}{NO_1} = \frac{O_2P}{O_1P}$$

But

$$NO_1 = N'O \text{ and } O_2M = O_2'M'$$

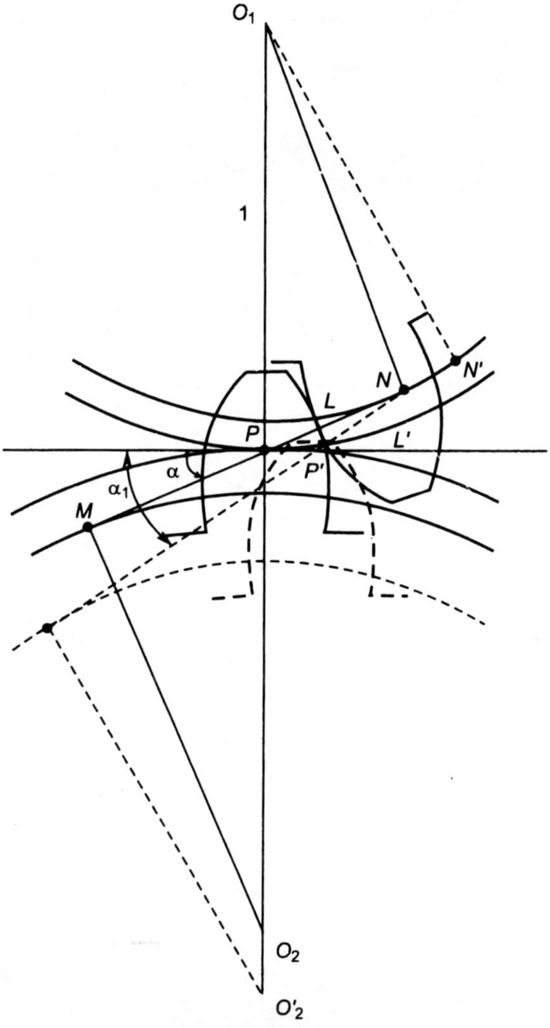

Fig. 9.11 Effect of centre distance variation.

Therefore,
$$\frac{O_2P}{O_1P} = \frac{O_2'P'}{O_1P'} \quad \text{...(9.29)}$$

Hence, *the variation in the centre distance, within limits, does not affect the angular velocity ration. But the length of arc of contact is decreased, and the pressure angle is increased.*

9.14 INTERNAL GEARS

A pinion in mesh with an internal gear is shown in Fig. 9.12. Internal gears have some advantages over the external gears. The most important advantages is the compactness of the drive. Other advantages are : greater length of contact, greater tooth strength, and lower relative sliding velocity between meshing teeth.

The tooth profile is concave in internal gears instead of convex as in external gears. Because of this a type of interference called **fouling** may occur in internal gears. Fouling occurs

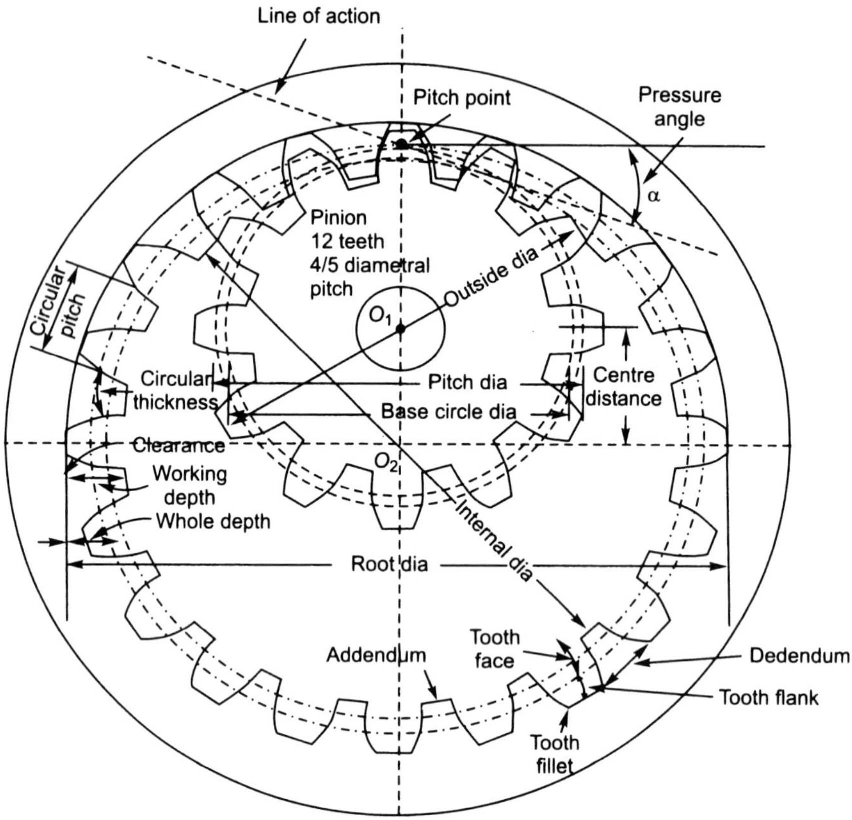

Fig. 9.12 Internal gears.

between inactive profiles as the teeth go in and out of the mesh and the there is not sufficient difference between the numbers of teeth on the internal gear and the pinion.

■ **Example 9.1** A 0.2 mm module, 20° pinion of 42 teeth drives a gear of 90 teeth. Calculate the contact ratio.

Solution : Given $= 0.2\text{ mm}, \alpha = 20°, z_1 = 42, z_2 = 90$

$$d_1 = mz_1 = 0.2 \times 42 = 8.4 \text{ mm}, d_2 = mz_2 = 0.2 \times 90 = 18\text{ mm}$$

$$r_{b1} = r_1 \cos \alpha = 4.2 \cos 20° = 3.95 \text{ mm}, r_{b2} = r_2 \cos \alpha$$

$$= 9 \cos 20° = 8.46 \text{ mm}$$

$$r_{a1} = r_1 + m = 4.2 + 0.2 = 4.4 \text{ mm}$$

$$r_{a2} = r_2 + m = 9 + 0.2 = 9.2 \text{ mm}$$

Length of patch of contact,

$$L_p = (r_{a1}^2 - r_{b1}^2)^{0.5} + (r_{a2}^2 - r_{b2}^2)^{0.5} - (r_1 + r_2) \sin \alpha$$

$$= [(4.4)^2 - (3.95)^2]^{0.5} + [(9.2)^2 - (8.46)^2]^{0.5} - (4.2 + 9) \sin 20°$$

$$= 1.9384 + 3.6150 - 4.5147 = 1.0387 \text{ mm}$$

Base pitch, $p_b = \dfrac{2\pi r b_1}{z_1} = 2\pi \times \dfrac{3.95}{42} = 0.5909 \text{ mm}$

Contact ratio $= \dfrac{L_p}{p_b} = \dfrac{1.0387}{0.5909} = 1.758 \simeq 2$

■ **Example 9.2** For a 20° pressure angle, calculate the minimum number of teeth in a pinion to mesh with a rack without involute interference. Also calculate the number of teeth in a pinion to mesh with a gear of equal size without involute interference, the addendum equal the module.

Solution : Given : $z_1 = \dfrac{2a_r}{\sin^2 \alpha}$

Here $a_r = 1$, $z_1 = \dfrac{2}{\sin^2 20°} = 17$

$$z_1 = \frac{2a_p}{[\{1 + i^2 \sin^2 \alpha + 2i \sin^2 \alpha\}^{0.5} - 1]}$$

Here, $i = 1$, $a_p = 1$

$$z_1 = \frac{2}{[\{1 + \sin^2 20° + 2 \sin^2 20°\}^{0.5}] - 1}$$

$$= 12.32 \cong 13$$

■ **Example 9.3** A pair of a spur gears has 16 and 18 teeth, a module of 13 mm, addendum of 13 mm, and pressure angle of 14.5°. Show that the gears have interference. Determine the amount by which the addendum must be reduced to eliminate the interference.

Solution : Given : $z_1 = 16$, $z_2 = 18$, $m = 13$ mm, $h_1 = 13$ mm, $\alpha = 14.5°$

$$r_1 = \frac{mz_1}{2} = 13 \times \frac{16}{2} = 014 \text{ mm}, \; r_2 = 13 \times \frac{18}{2} = 117 \text{ mm}$$

$$r_{b1} = r_1 \cos \alpha = 104 \cos 14.5° = 100.7 \text{ mm},$$

$$r_{b2} = 117 \cos 14.5° = 113.3° \text{ mm}$$

$$r_{a1} = r_1 + h_a = 104 + 13 = 117 \text{ mm},$$

$$r_{a2} = 117 + 13 = 130 \text{ mm}$$

Let pinion be the driver,

Length of approach, $L_a = (r_{a2}^2 - r_{b2}^2)^{0.5} - r_2 \sin \alpha$

$$= [(130)^2 - (113.3)^2]^{0.5} - 117 \sin 14.5°$$

$$= 63.74 - 211.3 = 34.44 \text{ mm}$$

$$(L_a)_{max} = r_1 \sin \alpha = 104 \sin 14.5° = 26.04 \text{ mm.}$$

Since length of approach is more than the maximum length of approach, therefore interference will occur. To eliminate interference, we make $L_1 = (L_a)_{max}$.

$$26.04 = (r_{a2} - r_{b2}^2)^{0.5} - r_2 \sin a$$

$$= [r_{a2} - (113.3)^2]^{0.5} - 117 \sin 14.5°$$

$$= [r_{a2}^2 - (113.3)^2]^{0.5} - 211.3$$

$$[r_{a2}^2 - (113.3)^2] = (26.04 + 211.3)^2$$

$$= 126.09 \text{ mm}$$

$$h'_a = r_{a2} - r_2 = 126.09 - 117 = 11.09 \text{ mm}$$

Decrease in addendum $= h'_a - h_a = 13 - 11.09 = 3.91$ mm

■ **Example 9.4** Two spur gears of 24 teeth and 36 teeth of 8 mm module and 20° pressure angle are in mesh. Addendum of each gear is 8 mm. The teeth are of involute form and the pinion rotates at 450 rpm. Determine the velocity of sliding when the pinion is at a radius of 102 mm.

Solution : $z = 24$, $z_2 = 36$, $m = 8\,\text{mm}$, $h_a = 8\,\text{mm}$, $\alpha = 20°$, $n_1 = 450\,\text{rpm}$, $r = 102\,\text{mm}$

$$r_1 = \frac{mz_1}{2} = 8 \times \frac{24}{2} = 96\,\text{mm},\ r_2 = 8 \times \frac{36}{2} = 144\,\text{mm}$$

$$r_{b1} = r_1 \cos\alpha = 96 \cos 20° = 90.21\,\text{mm},\ r_{b2} = 144 \cos 20° = 135.31\,\text{mm}$$

$$r_{a1} = r_1 + h_a = 96 + 8 = 104\,\text{mm},\ r_{a2} = 144 + 8 = 152\,\text{mm}$$

Let pinion be the driver,

Length of recess at 102 mm radius,

$$L_r = (r_{a1}^2 - r_{b1}^2)^{0.5} - r_1 \sin\alpha$$

$$= [(102)^2 - (90.21)^2]^{0.5} - 96 \sin 20°$$

$$= 47.60 - 32.83 = 14.77\,\text{mm}$$

$$n_2 = \frac{n_1 z_1}{z_2} = 450 \times \frac{24}{36} = 300\,\text{rpm}$$

$$\omega_1 = 2\pi \times \frac{450}{60} = 47.124\,\text{rad/s}$$

$$\omega_2 = 2\pi \times \frac{300}{60} = 31.416\,\text{rad/s}$$

Velocity of sliding at 102 mm radius $= (\omega_1 + \omega_2)\, L_r$

$$= (47.124 + 31.416) \times \frac{14.77}{1000} = 1.16\,\text{m/s}$$

■ **Example 9.5** A pair of 20° pressure angle gears in mesh has the following data :

Speed of pinion	= 400 rpm
Number of teeth on pinion	= 24
Number of teeth on gear	= 28
Module	= 10 mm

Determine the addendum of the gears if the pat of aproach and recess is half the maximum value. Determine also the arc of contact and the maximum velocity of sliding between the mating surfaces.

Solution : Given $z_1 = 24$, $z_2 = 28$, $m = 10\,\text{mm}$, $h_a = ?$ $\alpha = 20°$, $n_1 = 400\,\text{rpm}$

$$r_1 = \frac{mz_1}{2} = 10 \times \frac{24}{2} = 120\,\text{mm},\ r_2 = 10 \times \frac{28}{2} = 140\,\text{mm}$$

$$r_{b1} = r_1 \cos\alpha = 120 \cos 20° = 112.76\,\text{mm},$$

$$r_{b2} = 140 \cos 20° = 131.56\,\text{mm}$$

$$r_{a1} = r_1 + h_{a1} = 120 + h_{a1},\ r_{a2} = 140 + h_{a2}$$

Let pinion be the driver,

Length of approach, $L_a = 0.5\,(L_a)_{max}$

$$(r_{a2}^2 - r_{b2}^2)^{0.5} - r_2 \sin\alpha = 0.5\, r_1 \sin\alpha$$

$$[r_{a2}^2 - (131.56)^2]^{0.5} - 140 \sin 20° = 0.5 \times 120 \times \sin 20°$$

$$r_{a2}^2 - (131.56)^2 = (200)^2 \sin^2 20°$$

$$r_{a2}^2 = 21987,\ r_{a2} = 148.3 \text{ mm}$$

$$h_{a2} = 148.3 - 140 = 8.3 \text{ mm}$$

Length of recess, $L_r = 0.5 \times (L_r)_{max}$

$$(r_{a1}^2 - r_{b1}^2)^{0.5} - r_1 \sin\alpha = 0.5\, r_2 \sin\alpha$$

$$[r_{a1}^2 - (112.76)^2]^{0.5} - 120 \sin 20° = 0.5 \times 140 \times \sin 20°$$

$$r_{a1}^2 - (112.76)^2 = (190)^2 \sin^2 20°$$

$$r_{a1}^2 = 16938,\ r_{a1} = 130.14 \text{ mm}$$

$$h_{a1} = 130.14 - 120 = 10.14 \text{ mm}$$

$$\text{Arc of contact} = \frac{L_p}{\cos\alpha} = 0.5\,(r_1 + r_2) \tan\alpha$$

$$= 0.5\,(120 + 140) \tan 20° = 47.3 \text{ mm}$$

Path of recess, $L_r = (r_{a1}^2 - r_b 1^2)^{0.5} - r_1 \sin\alpha$

$$= [(130.14)^2 - (112.76)^2]^{0.5} - 120 \sin 20°$$

$$= 23.93 \text{ mm}$$

$$\omega_1 = 2\pi \times \frac{400}{60} = 41.88 \text{ rad/s},\ \omega_2 = 41.88 \times \frac{24}{28}$$

$$= 35.9 \text{ rad/s}$$

Maximum velocity of sliding $= (\omega_1 + \omega_2) \times L_r$

$$= (41.88 + 35.9) \times \frac{23.91}{1000} = 1.86 \text{ m/s}$$

9.15 NONSTANDARD GEARS

The term nonstandard gear apply to such gears which are modified by changing some standard parameters like pressure angle, addendum, tooth depth or centre distance. These changes are made to improve the performance of the gear operation or from economical point of view.

The recent trend these days is to make the designs of machines are complete as possible to reduce their size and weight which also results in reduction in the costs.

For a given module, if the number of teeth on pinion are decreased, the pinion size decreases. For a given speed ratio, the size of the gear also decreases. Also the size of other components associated with the gear set such as shafts, casings and bearings are also reduced. The only way to have a smaller size of gears is to reduce the number of teeth. However, for a typical type of teeth, it is observed that if the number of teeth is reduced from a certain number, the problems of interefrence, undercutting and contact ratio hamper the smooth running of the gears. Therefore, the main reason to employ nonstandard gears is to prevent interference and undercutting and to maintain a reasonable contact ratio.

It should be remembered that as an involute is generated, its radius of curvature go on becoming larger and larger, being zero at the base circle. As far as possible, the curve near the base should be avoided because high stresses are developed in the region of sharp curvature.

1. Centre-distance Modifications

The number of teeth on a pinion can be reduced from the minimum allowable number by increasing the centre distance marginally and by changing the tooth proportions and the pressure angle of the gears. A reduction in the interference and improvement in the contact ratio is brought this way. The teeth can be generated with rack cutters of standard pressure angles by displaying the pitch line of the rack from the pitch circle of the gear. This action produces teeth which are thicker than before. As the teeth are cut with a displaced or offset cutter, they will engage at a new pressure angle and at new centre distance.

2. Clearance Modifications

If the clearance between mating teeth is increased to 0.3 m or 0.4 m instead of the usual value of 0.25 m to have a larger fillet at the root of the tooth, the fatigue strength of the tooth is increased. This way some extra depth is available to smoothen the tooth profile. Interchangeability is not lost this way.

3. Addendum Modification

In cases where it is not possible to change the centre distances, modifications can be made to the addendum. In such cases, there has to be no change in the pitch circles and the pressure angles. However, the contact region is shifted away from the pinion centre towards the gear centre, decreasing the approach action and increasing the recess action.

9.16 HELICAL GEARS

A helical gear has teeth in the form of a helix around the gear. The helix may be right handed on one gear and left handed on the other gear. The pitch surfaces are cylindrical like spur gears but the teeth wind around the cylinder helically like screw threads. Helical gears are used to transmit power between parallel shafts.

If a plane is rolled on a base cylinder, a line in the plane parallel to the axis of the cylinder will generate the surface of an involute spur gear tooth. If the generating line is inclined to the

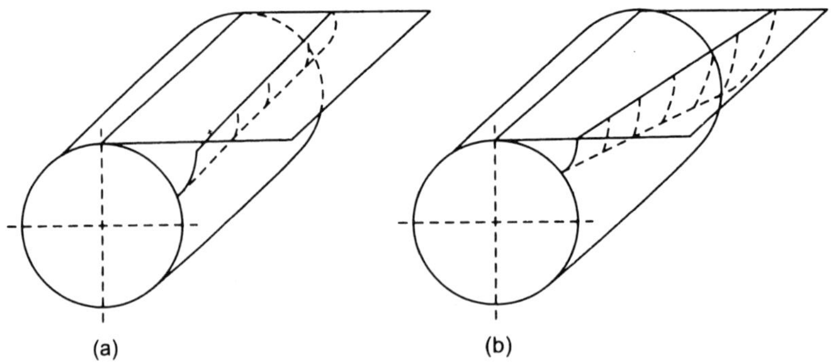

Fig. 9.13 Generation of helical gears.

axis, the surface of a helical gear tooth will be generated. These two conditions are shown in Fig. 9.13.

Helical gears are used to connect parallel shafts and non-intersecting shafts. The former are known as parallel helical gears and the latter as crossed helical gears, as shown in Fig. 9.14.

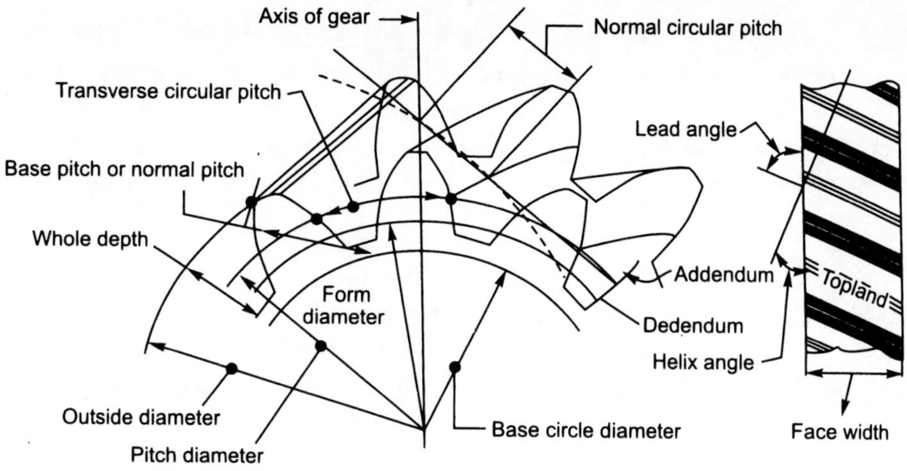

(a) Helical gear

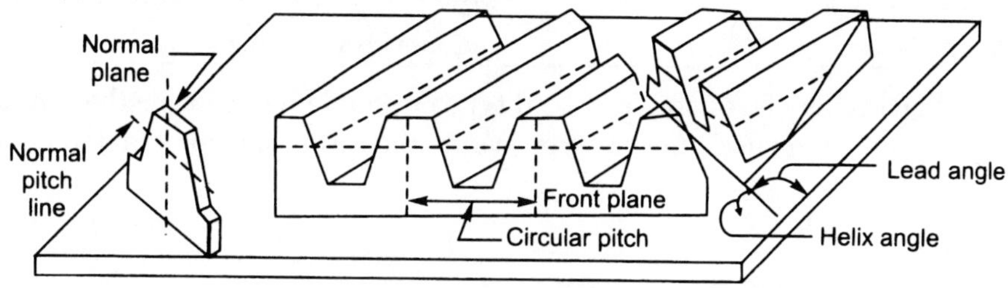

(b) Helical rack

Fig. 9.14 Helical gear and rack terminology.

In determining the tooth proportions of helical gear for either crossed or parallel shafts,it is necessary to consider the manner in which the teeth are to be cut. If the gear is to be hobbed, all dimensions are figured in a plane that is normal to the tooth pitch element, and the diametral pitch and the pressure angle are standard values in that plane. As the cutting action of the hob occurs in the normal plane, it is possible to use the same hob to cut both helical and spur gears of a given pitch; in a spur gear the normal plane and the plane of rotation (or transverse plane) are identical.

Helical gears connecting parallel shafts have line contact which runs diagonally across the face of the teeth. Parallel helical gears have smoother action and hence less noise and vibration than spur gears. Also the tooth contact is gradual, beginning at one end of the tooth and processing across the tooth surface, whereas in spur gears, contact takes place simultaneously over the entire face width. However, helical gears give rise to end thrust.

9.17 COMPARISON BETWEEN SPUR AND HELICAL GEARS

The comparison between spur and helical gears is given in Table 9.4

Table 9.4 : Comparison of Spur and Helical Gears

Spur Gears	Helical Gears
1. Teeth are cut parallel to the axis of the shaft.	Teeth are cut in the form of a helix on the pitch cylinder between meshing gears.
2. Contact between meshing teeth occurs along the entire face width of the tooth.	Contact between meshing gears beings with a point on the leading edge of the tooth and gradually extends along the diagonal line across the tooth.
3. Load application is sudden resulting into impact conditions and generating noise in high speed applications.	Pick up of load by the tooth is gradual, resulting in smooth engagement and quite operation even at high speeds.
4. Used for parallel shafts only.	Crossed helical gears are used on shafts with crossed axes.
5. Speed is limited to about 20 m/s.	Used in automobiles, turbines and high speed applications upto 50 m/s.
6. Imposes radial load only.	Imposes radial and axial thrust loads.
7. Contact ratio is low.	Contact ratio is high.

The helical gears may be of single helical type or double helical type (herringbone gears) The axial thrust present in single helical type is automatically cancelled in double helical type gears.

9.18 HELICAL GEAR TERMINOLOGY

1. **Helix angle (β) :** It is the angle between a line drawn through one of the teeth and the centre line of the shaft (Fig. 9.15) on which the gear is mounted. It varies from 15° to 30°.

2. **Normal circular pitch (p_n) :** The normal circular pitch is the distance between corresponding points of adjacent teeth as measured in a plane perpendicular (or normal) to the helix. It is the perpendicular distance between two adjacent teeth.

$$p_n = p_t \cos\beta = \pi m \cos\beta \quad ...(9.30)$$

3. **Normal diametral pitch (P_n) :** The normal diametral pitch is the diametral pitch measured in the plane normal to the helix. It is equal to the diametral pitch of the hob.

$$P_n = \frac{P_t}{\cos\beta} \quad ...(9.31)$$

4. **Transverse circular pitch (p_t) :** The transverse circular pitch is the distance measured in a plane perpendicular to the shaft axis (or plane of rotation) between corresponding points of adjacent teeth.

$$p_t = \frac{\pi d}{z} = \pi m \quad ...(9.32)$$

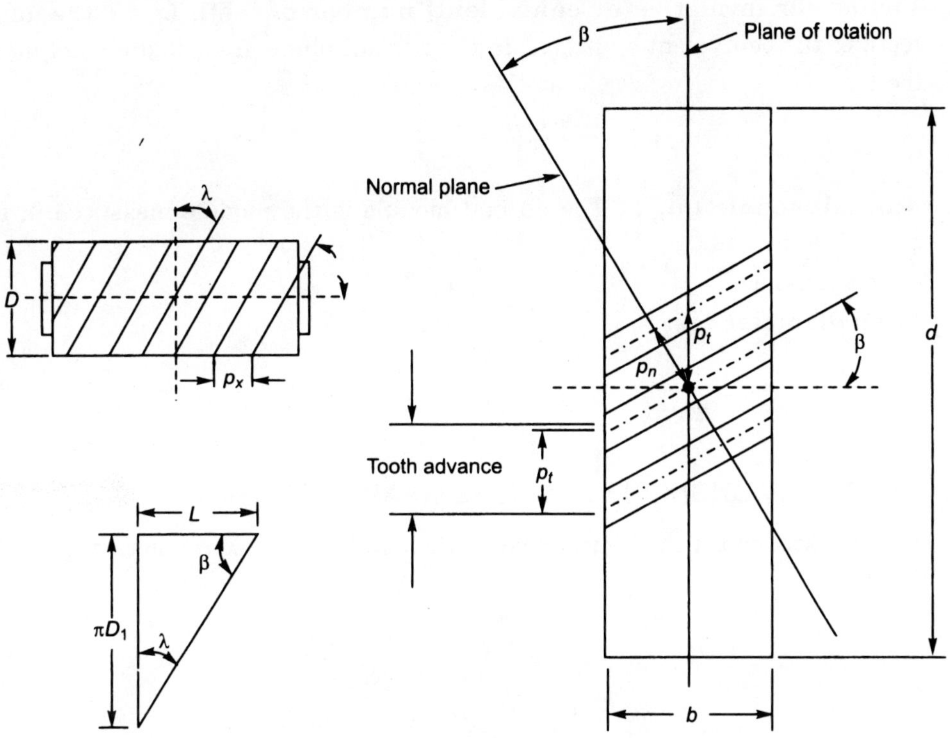

(b) (c)

Fig. 9.15 Definition of various pitches.

5. **Transverse diametral pitch (P_t):** The transverse diametral pitch is the diametral pitch measured in the plane of rotation,that is, transverse to the axis of rotation.

$$P_t = \frac{z}{d} = \frac{1}{m} \quad ...(9.33)$$

$$P_t\, p_t = P_n\, p_n = \pi \quad ...(9.34)$$

6. **Transverse pressure angle (α_t) :** It is the pressure angle measured in the transverse plane or plane of rotation.
7. **Normal pressure angle (α_n) :** It is the pressure angle measured in the normal plane or plane perpendicular to the teeth.
8. **Axial pitch (p_x):** The axial pitch is the distance measured in a plane parallel to the shaft axis between corresponding points of adjacent teeth.

$$p_x = p_t \cot \beta \quad ...(9.35)$$

9. **Lead :** The lead is the distance measured parallel to the axis to represent the distance advanced by each tooth per revolution.
10. **Lead angle :** The lead angle is the acute angle between the tangent to the helix and a plane perpendicular to axis of cylinder.

11. **Virtual (or formative or equivalent) number of teeth (z_v) :** The number of teeth of the equivalent spur gear in the normal plane are called virtual number of teeth.

$$z_v = \frac{z}{\cos^3 \beta} \quad \text{...(9.36)}$$

12. **Normal module (m_n) :** The normal module is the module measured in a plane normal to the helix.

Normal module, $m_n = m_t \cos \beta$ ...(9.37)

13. **Pitch diameter :**

$$d = \frac{z m_n}{\cos \beta} \quad \text{...(9.38)}$$

9.19 ANGLE RELATIONSHIPS IN HELICAL GEARS

Consider the cross-sections of the helical gear in the axial plane x-x and normal plane y-y, as shown in Fig. 9.16.

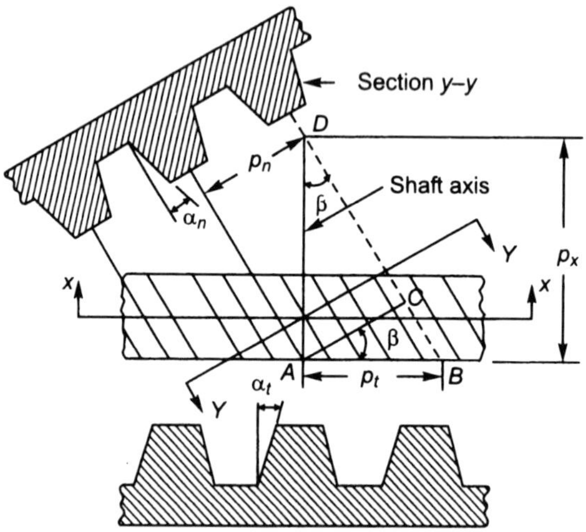

Fig. 9.16 Angle relationships.

In ΔABC, we have

$$\frac{AC}{AB} = \frac{p_n}{p_t} = \cos \beta$$

∴ Normal circular pitch, $p_n = p_t \cos \beta$ ...(9.39)

Transverse circular pitch, $p_t = \frac{\pi d}{z}$ ...(9.40)

Transverse diametral pitch, $P_t = \frac{z}{d}$

Transverse module, $m_t = \frac{1}{P_t} = \frac{d}{z}$

$$p_t P_t = \pi = p_n P_n$$

Normal diametral pitch, $$P_n = \frac{P_t}{\cos\beta} = \frac{z}{d\cos\beta} \qquad ...(9.41)$$

Normal module, $$m_n = m_t \cos\beta \qquad ...(9.42)$$

From ΔABD, Axial pitch, $$p_x = \frac{p_t}{\tan\beta} = \left(\frac{\pi d}{z}\right)\cot\beta \qquad ...(9.43)$$

Also $$\cos\beta = \frac{\tan\alpha_n}{\tan\alpha_t} \qquad ...(9.44)$$

$$d = \frac{zp_t}{\pi} = zm_t = \frac{zm_n}{\cos\beta} \qquad ...(9.45)$$

Centre distance, $$C = \left(\frac{1}{2}\right)(d_1 + d_2) = \frac{m_n}{2\cos\beta}(z_1 + z_2) \qquad ...(9.46)$$

Speed ratio, $$i = \frac{z_2}{z_1} = \frac{n_1}{n_2}$$

If the gear is to be cut by a gear shaper method, the dimensions are considered in the plane of rotation and the diametral pitch and the pressure angle are standard values in that plane. When a helical gear is cut by a gear shaper, the circular pith p_t of Fig. 9.15 becomes equal to the circular pitch of the cutter so that the following relations apply :

$$p_t = \frac{\pi d}{z} = \frac{\pi}{P_t} = \pi m \qquad ...(9.47)$$

$$P_t = \frac{z}{d} \qquad ...(9.48)$$

$$m_t = \frac{d}{z} \qquad ...(9.49)$$

9.20 VIRTUAL NUMBER OF TEETH

In helical gears, the plane *x-x* normal to the gear teeth intersects the pitch cylinder to form an ellipse, as shown in Fig 9.17 . The gear tooth profile generated in this plane, using the radius of curvature of the ellipse, would be a spur gear having the same properties as the actual helical

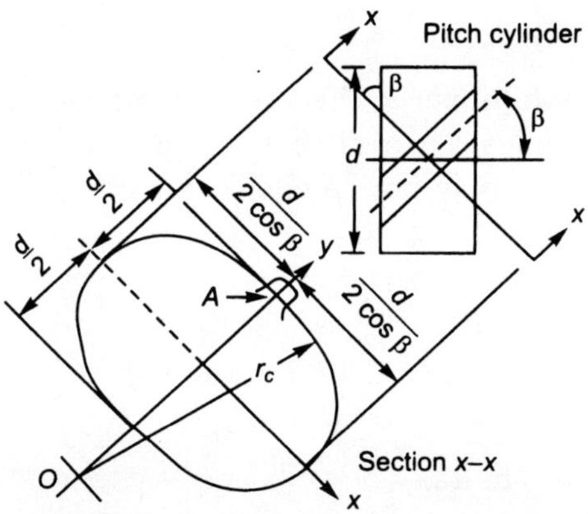

Fig. 9.17 Concept of virtual number of teeth.

gear. The semi-major and semi-minor axes of this ellipse are : $a = \dfrac{d}{2\cos\beta}$ and $b = \dfrac{d}{2}$ respectively.

The radius of curvature r_c at point A is,

$$r_c = \frac{a^2}{b} = \frac{d}{2\cos^2\beta}$$

where d = pitch circle diameter.

In the design of helical gears, an imaginary spur gear is considered in the plane x-x with a pitch circle radius r_c and module m_n. It is called a "*formative*" or "*virtual*" spur gear.

Pitch circle diameter of virtual gear,

$$d_c = \frac{d}{\cos^2\beta}$$

The number of teeth of the equivalent (or virtual) spur gear in the normal plane are called the virtual number of teeth, z_v

$$z_v = \frac{\pi d_c}{p_n} = \frac{\pi d}{\cos^2\beta} \times \frac{1}{\pi m_n} = \frac{d}{m_n \cos^2\beta} = \frac{z}{\cos^3\beta}$$

9.21 PARALLEL HELICAL GEARS

For parallel helical gears to mesh properly, the following conditions must be satisfied :

1. Equal helix angles
2. Equal pitches or modules
3. Opposite hand of helices

Velocity ratio,

$$\frac{\omega_1}{\omega_2} = \frac{z_2}{z_1} = \frac{d_2 \cos\beta_2}{d_1 \cos\beta_1} \quad \text{...(9.50a)}$$

$$= \frac{d_2}{d_1} \quad (\because \beta_1 = \beta_2) \text{ ...(9.50b)}$$

Centre distance,

$$C = (z_1 + z_2)\frac{m_t}{2} \quad \text{...(9.51)}$$

9.22 CROSSED HELICAL GEARS

For crossed helical gears to mesh properly, there is only one requirement that they must have common normal pitches or modules. Their pitches in the plane of rotation are not necessarily equal. Their helix angles may or may not be equal and the gears may be of the same or of opposite hand.

Velocity ratio,

$$i = \frac{\omega_1}{\omega_2} = \frac{z_2}{z_1}$$

$$= \frac{d_2 \cos\beta_2}{d_1 \cos\beta_1} \quad \text{...(9.52)}$$

The angle between the two shafts,

$$\Sigma = \beta_1 \pm \beta_2 \quad \text{...(9.53)}$$

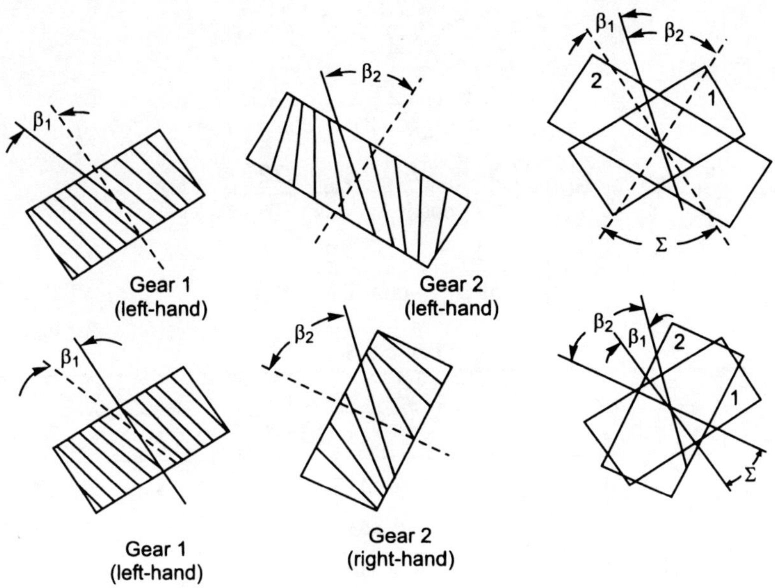

Fig. 9.18 Helical gear in and out of mesh.

The plus and minus signs apply respectively, when the gears have the same or the opposite hand. Fig. 9.18 illustrated pairs of crossed helical gears in and out of mesh.

Centre distance, $$C = \left(\frac{m_n}{2}\right)\left[\frac{z_1}{\cos\beta_1} + \frac{z_2}{\cos\beta_2}\right] \qquad ...(9.54)$$

■ **Example 9.6** A pair of crossed helical gears connect two shafts at an angle of 60° with a velocity ratio of 1.5 : 1. The pinion has a normal diametal pitch of 0.25, a pitch diameter of 200 mm and a helix angle of 35°. Determine the helix angle and the pitch diameter of the gear and the number of teeth on both the pinion and the gear.

Solution : $$\Sigma = \beta_1 + \beta_2$$

Here $\Sigma = 60°$ and $\beta_1 = 35°$. Therefore, $\beta_2 = 25°$.

Now $$\frac{\omega_1}{\omega_2} = \frac{z_2}{z_1} = \frac{d_2 \cos\beta_2}{d_1 \cos\beta_1}$$

$$d_2 = 200 \times \frac{1.5 \cos 35°}{\cos 2.5°} = 271.15 \text{ mm}$$

$$z_1 = p_n d_1 \cos\beta_1 = 0.25 \times 200 \times \cos 35° = 40.95 \approx 41$$

$$z_2 = 41 \times 1.5 = 61.6 \approx 62$$

■ **Example 9.7** Two left handed helical gears connect two shafts inclined at 60°. The normal module is 8 mm. The larger gear has 72 teeth an the velocity ratio is 1 : 2. If centre distance is 500 mm, calculate the helix angles of the two gears.

Solution : Here $\Sigma = 60°$, $m_n = 8$ mm, $z_2 = 72$, $C = 500$ mm

$$\Sigma = \beta_1 + \beta_2$$

$$\beta_2 = 60° - \beta_1$$

$$i = \frac{n_2}{n_1} = \frac{z_1}{z_2}$$

$$z_1 = \frac{72}{2} = 36$$

$$C = \left(\frac{m_n}{2}\right)\left[\frac{z_1}{\cos\beta_1} + \frac{z_2}{\cos\beta_2}\right]$$

$$500 = \left(\frac{8}{2}\right)\left[\frac{36}{\cos\beta_1} + \frac{72}{\cos(60° - \beta_1)}\right]$$

or
$$3.472 = \frac{1}{\cos\beta_1} + \frac{2}{\cos(60° - \beta_1)}$$

Table 9.5

β_1, *deg*	$\frac{1}{\cos\beta_1}$	$\frac{2}{\cos(60° - \beta_1)}$	$\frac{1}{\cos\beta_1} + \frac{2}{\cos(60° - \beta_1)}$
25	1.1034	2.4415	3.5449
27	1.223	2.3847	3.5070
28	1.1325	2.3583	3.4908
29	1.1433	2.3332	3.4765
30	1.1547	2.3094	3.4641

We take $\beta_1 = 29°$, then $\beta_2 = 31°$.

■ **Example 9.8** A pair of helical gears having 30 and 48 teeth and a 23° helix angle transmits power between parallel shafts. The module in the normal plane is 3 mm, and the pressure angle in this plane is 20°. Determine (a) the module in the plane of rotation, (b) pitch diametres, (c) center distance, (d) circular pitch in the normal plane, and (e) circular pitch in the plane of rotation.

Solution : Given : $z_1 = 30, z_2 = 48, m_n = 3$ mm, $\alpha_n = 20°, \beta = 23°$

(a) $m_t = m_n/\cos\beta = 3/\cos 23° = 3.259$ mm

(b) $d_1 = z_1 m_n/\cos^3\beta = 30 \times 3/\cos^3 23° = 115.4$ mm

$$d_2 = z_2 m_n/\cos^3\beta = 48 \times 3/\cos^3 23° = 184.64 \text{ mm}$$

(c) $C = 0.5\, m_t\,(z_1 + z_2) = 0.5 \times 3.259\,(30 + 48) = 127.1$ mm

(d) $p_n = \pi m_n = \pi \times 3 = 11.42$ mm

(e) $p_t = \pi m_t = \pi \times 3.259 = 10.238$ mm

■ **Example 9.9** A pair of crossed helical gears connected shafts making angle of 45°. The right-hand pinion has 36 teeth and a helix angle of 20°. The right-hand gear had 48 teeth and its module in the normal plane is 2.5 mm. Determine (a) The helix angle of the gear, (b) circular pitch in the normal plane, (c) module of the pinion in its plane of rotation, (d) module of the gear in its plane of rotation, and (e) centre distance.

Solution : Given : $z_1 = 36, z_2 = 48, m_n = 2.5$ mm, $\Sigma = 45°, \beta_1 = 20°$

(a) $\Sigma = \beta_1 + \beta_2 = 45°, \beta_2 = 45 - 20 = 25°$

(b) $p_n = \pi m_n = \pi \times 2.5 = 7.854$ mm

(c) $m_{t1} = m_n/\cos\beta_1 = 2.5/\cos 20° = 2.66$ mm

$$m_{t2} = m_n/\cos\beta_2 = 2.5/\cos 25° = 2.758 \text{ mm}$$

(d) $C = 0.5\ m_n\ [z_1/\cos\beta_1 + z_1/\cos\beta_2]$

$= 0.5 \times 2.5\ [36/\cos 20° + 36/\cos 25°]$

$= 114.1$ mm

9.23 HERRINGBONE GEARS

It is a gear, half of whose width is cut with a tooth helix in one direction and the other half in the opposite direction. It is in effect a double helical gear cut on a blank. The advantage of herringbone gears is that the end thrust is automatically eliminated.

9.24 BEVEL GEARS

Bevel gears are used to connect shafts whose axes intersect. The shaft angle is defined as the angle between the centre lines which contains the engaging teeth. Fig. 9.19 shows the details of a pair of bevel gears.

- **Pitch cone :** The pitch cone is the pitch surface of a bevel gear in a gear pair.
- **Cone centre :** The cone centre is the apex of the pitch cone.
- **Pitch cone radius (r) :** The pitch cone radius is the length of the pitch cone element.
- **Pitch angle (δ) :** The pitch angle is the angle that the pitch line makes with the axis of the gear.
- **Reference cone angle :** The reference cone angle is the angle between the gear axis and the reference cone generator containing the root cone generator.
- **Tip (or face) angle (δ_a) :** The tip angle is the angle between the tip cone generator and the axis of the gear.
- **Root (or cutting) angle (δ_f) :** The root angle is angle between the root cone generator and the axis of the gear.
- **Back cone :** The back cone is an imaginary cone the elements of the pitch cone at the larger end of the tooth.
- **Gear diameter :** The gear diameter is the diameter of which are perpendicular to the elements of the largest pitch circle.
- **Virtual number of teeth :** The virtual number of teeth is number of teeth on an imaginary spur gear laid out on a pitch radius equal to the back cone radius.

$$z_v = \frac{z}{\cos\delta} \quad \text{...(9.55)}$$

- **Crown gears :** The crown gear is a gear pair for which pitch cone angle is 90°.
- **Miter gears :** Miter gears are two bevel gears of the same size having pitch cone angle of 90°.
- **Angular bevel gears :** The shaft angle is greater or less than 90°.

Let θ_a = addendum angle

θ_f = dedendum angle

Tip angle, $\delta_a = \delta + \theta_a$

$\delta_f = \delta - \theta_f$

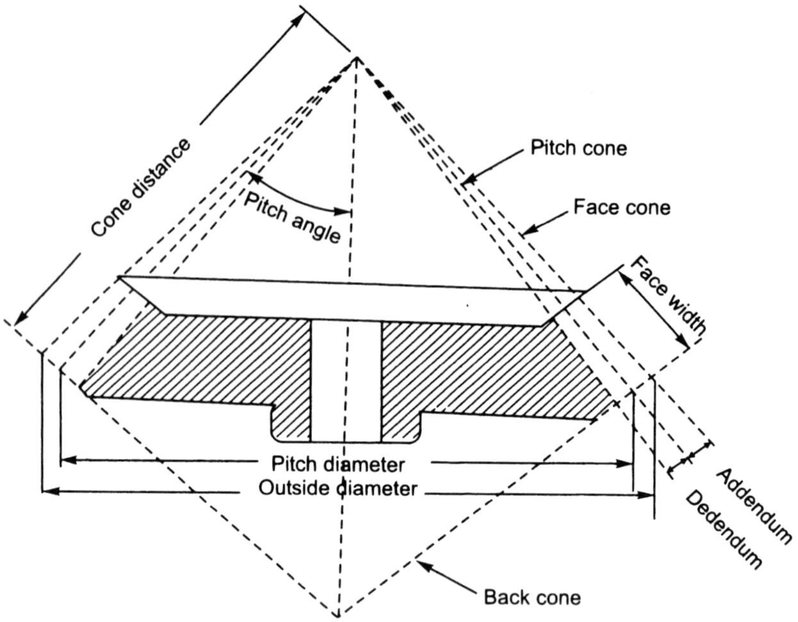

(a) Bevel gear terminology

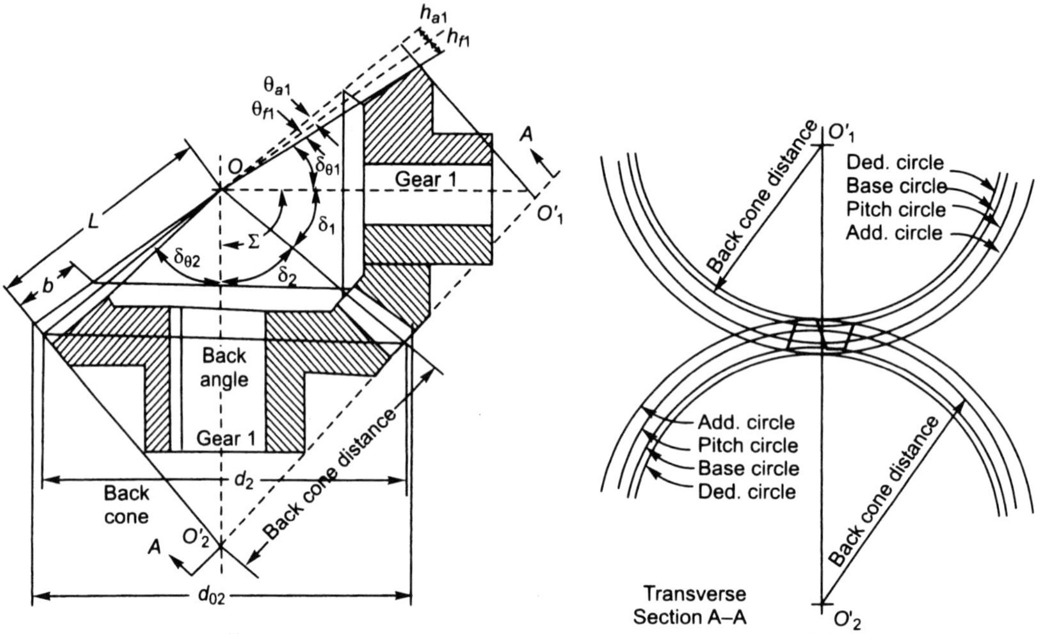

(i) (ii)

(b)

Fig. 9.19 Bevel gear nomenclature.

Velocity ratio,
$$\frac{\omega_1}{\omega_2} = \frac{d_2}{d_1} = \frac{z_2}{z_1}$$

Module
$$m = \frac{d}{z}$$

Angle relationships :

Let L = length of the pitch cone

$$= \left[\frac{d_1^2 + d_2^2}{4}\right]^{0.5} \qquad \text{...(9.56)}$$

$$\sin \delta_1 = \frac{d_1}{2L} = \sin(\Sigma - \delta_2) = \sin\Sigma \cos\delta_2 - \cos\Sigma \sin\delta_2$$

$$\frac{\sin\delta_1}{\sin\Sigma \sin\delta_2} = \frac{\cos\delta_2}{\sin\delta_2} - \frac{\cos\Sigma}{\sin\Sigma}$$

$$\left(\frac{1}{\sin\Sigma}\right)\left[\frac{\sin\delta_1}{\sin\delta_2} + \cos\Sigma\right] = \frac{1}{\tan\delta_2} \qquad \text{...(9.57)}$$

Also
$$\frac{\sin\delta_1}{\sin\delta_2} = \frac{d_1}{d_2}$$

Therefore
$$\tan\delta_2 = \frac{\sin\Sigma}{\cos\Sigma + \dfrac{d_1}{d_2}} = \frac{\sin\Sigma}{\cos\Sigma + \dfrac{z_1}{z_2}}$$

Similarly
$$\tan\delta_1 = \frac{\sin\Sigma}{\cos\Sigma + \dfrac{d_2}{d_1}}$$

$$= \frac{\sin\Sigma}{\cos\Sigma + \dfrac{z_2}{z_1}} \qquad \text{...(9.58)}$$

Addendum angle,
$$\tan\theta_a = 2h_a \sin\frac{\delta_1}{d_1}$$

$$= 2h_a \sin\frac{\delta_2}{d_2} \qquad \text{...(9.59)}$$

Dedendum angle,
$$\tan\theta_f = 2hf_f \sin\frac{\delta_1}{d_1}$$

$$= 2h_f \sin\frac{\delta_2}{d_2} \qquad \text{...(9.60)}$$

Outside (or tip) diameter of pinion,

$$d_{a1} = d_1 + 2h_a \cos\delta_1 \qquad \text{...(9.61)}$$

Outside diameter of gear,

$$d_{a2} = d_2 + 2h_a \cos\delta_2 \qquad \text{...(9.62)}$$

For right angle gears,

$$\Sigma = 90°\text{, and } \tan\delta_1 = \frac{z_1}{z_2}$$

$$\tan\delta_2 = \frac{z_2}{z_1}$$

For obtuse angle gears, Σ is grater than 90°, and

$$\tan \delta_2 = \frac{\sin (180° - \Sigma)}{\frac{d_1}{d_2} - \cos (180° - \Sigma)} \quad ...(9.63a)$$

$$= \frac{\sin (180° - \Sigma)}{\frac{z_1}{z_2} - \cos (180° - \Sigma)} \quad ...(9.63b)$$

Similarly,

$$\tan \delta_1 = \frac{\sin (180° - \Sigma)}{\frac{d_2}{d_1} - \cos (180° - \Sigma)}$$

$$= \frac{\sin (180° - \Sigma)}{\frac{z_2}{z_1} - \cos (180° - \Sigma)} \quad ...(9.64)$$

Forces on straight tooth bevel gears are :

Normal force, $$F_n = \frac{F_t}{\cos \alpha_n} \quad ...(9.65)$$

Radial force, $$F_r = F_t \tan \alpha_n \sin \delta \quad ...(9.66)$$

Axial force, $$F_a = F_t \sin \alpha_n \cos \delta$$

where $F = \frac{10^3 P}{v_m}$, $v_m = \frac{\pi d_m n}{10^3 \times 60}$ m/s, ...(9.67)

and $d_m = d - b \sin \delta$, b = face width of gear tooth.

■ **Example 9.10** A crown bevel gear of 48 teeth and a module of 2 mm in driven by 24-tooth pinion. Calculate the pitch angle of the pinion and the shaft angle.

Solution : $m = 2$ mm, $z_1 = 24, z_2 = 48$

$$i = z_2/z_1 = 48/24 = 2$$

$$d_1 = mz_1 = 2 \times 24 = 48 \text{ mm}, \; d_2 = 2 \times 48 = 96 \text{ mm}$$

$$L = 0.5 [d_1^2 + d_2^2]^{0.5} = 0.5 (48^2 + 96^2)^{0.5} = 53.66 \text{ mm}$$

$$\sin \delta_1 = d_1/(2L) = 48/(2 \times 53.66) = 0.4472$$

$$\delta_1 = 26.56°$$

$$\tan \delta_1 = \sin \Sigma/(\cos \Sigma + i)$$

$$\tan 26.56° = \sin \Sigma/(\cos \Sigma + 2)$$

$$0.5 = \sin \Sigma/(\cos \Sigma + 2)$$

$$0.5 = \cos \Sigma + 1 = \sin \Sigma$$

Both sides are equal, if Σ = 90°

Hence shaft angle is 90°.

■ **Example 9.11** A 6.35 mm module, straight bevel pinion of 14 teeth drives a gear of 20 teeth. The shaft angle is 90°. Calculate the addendum and dedendum, circular tooth thickness for each gear, and the pitch and base radii of the equivalent spur gear.

Solution : $m = 6.25$ mm, $z_1 = 14$, $z_2 = 20$, $\Sigma = 90°$, $h_a = h_f = ?$

$$i = z_2/z_1 = 20/14 = 1.4286$$

$$d_1 = mz_1 = 6.25 \times 14 = 88.9 \text{ mm}, \; d_2 = 6.25 \times 20 = 127 \text{ mm}$$

$$L = 0.5\,[d_1^2 + d_2^2]^{0.5}\; [(88.9)^2 + (127)^2]^{0.5} = 77.51 \text{ mm}$$

$$\sin \delta_1 = d_1/(2L) = 48/(2 \times 53.66) = 0.4472\, \delta_1 = 26.56°$$

$$\tan \delta_1 = \sin \Sigma/(\cos \Sigma + i)$$

$$= \sin 90°/(\cos 90° + 1.4286) = 0.7$$

$$\delta_1 = 35°, \; \delta_2 = 90 - 35 = 55°$$

$$h_a = m = 6.35 \text{ mm}, \; h_f = 1.25 \text{ m} = 1.25 \times 6.35 = 7.94 \text{ mm}$$

For a bevel gear, $z_v = z/\cos \delta$

$$z_{v1} = z_1/\cos \delta_1 = 14/\cos 35° = 17, \; z_{v2} = 20/\cos 55° = 35$$

For the equivalent spur gear, $d = mz_v$. The pitch diameters are :

$$d_1 = 6.35 \times 17 = 108 \text{ mm}, \; d_2 = 6.35 \times 35 = 222.25 \text{ mm}$$

Circular tooth thickness = $\pi \times 108/17 = 111.96$ mm

9.25 SPIRAL GEARS

Spiral gears are used to connect non-parallel and non-intersecting shafts, as shown in Fig. 9.20. The shaft angle may be less than or greater than 90°, as show in figure below.

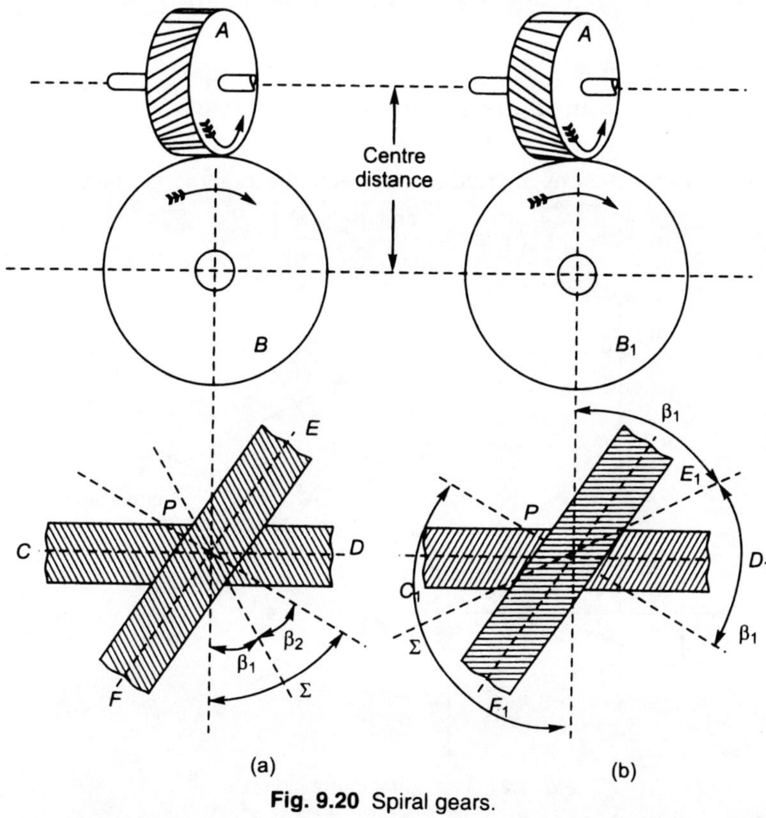

Fig. 9.20 Spiral gears.

Let β_1 = spiral angle of gear 1

β_2 = spiral angle of gear 2

$$\Sigma = \beta_1 + \beta_2$$

Gear ratio, $$i = \frac{z_2}{z_1}$$

$$p_1 = \frac{p_n}{\cos\beta_1} \text{ and } p_2 = \frac{p_n}{\cos\beta_2}$$

Also $$d_1 = \frac{z_1 p_1}{\pi} = \frac{z_1 p_n}{\pi\cos\beta_1}$$

and $$d_2 = \frac{z_2 p_2}{\pi} = \frac{z_2 p_n}{\pi\cos\beta_2}$$

Centre distance, $$C = \frac{d_1 + d_2}{2}$$

$$= \left(\frac{z_1 p_n}{2\pi}\right)\left[\frac{1}{\cos\beta_1} + \frac{i}{\cos\beta_2}\right] \quad \text{...(9.68)}$$

Let $d_v = \frac{z_1 p_n}{\pi}$ be the virtual pitch circle diameter of a spur gear with the same number of teeth and normal pitch as the spiral gear 1.

Thus $$C = \left(\frac{d_v}{2}\right)\left[\frac{1}{\cos\beta_1} + \frac{i}{\cos\beta_2}\right] \quad \text{...(9.69)}$$

9.25.1 Efficiency of Spiral Gears

Two spiral gears in mesh at point P are shown in Fig. 9.21. Gear 1 is the driver and 2 is the driven gear.

Let $\mu = \tan\phi$ be the coefficient of friction between the mating surfaces.

Driving force, $$F_1 = F_2\left[\frac{\cos(\beta_1 - \phi)}{\cos(\beta_2 + \phi)}\right]$$

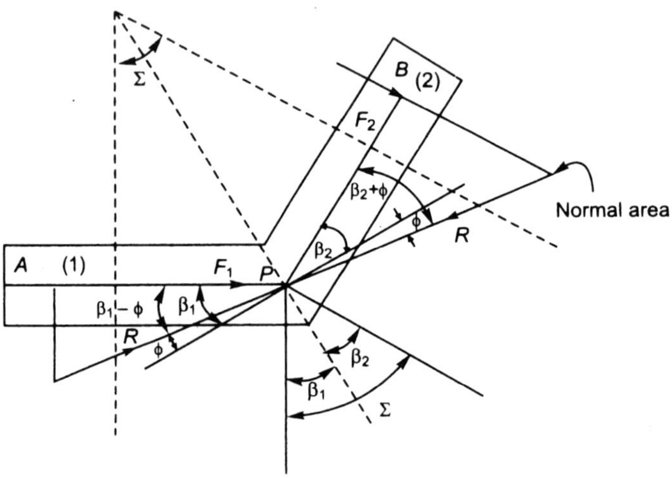

Fig. 9.21 Two spiral gears in mesh.

Without friction, $$F_{10} = \frac{F_2 \cos\beta_1}{\cos\beta_2}$$

Efficiency of the drive,

$$\eta = \frac{F_{10}}{F_1} = \frac{\cos\beta_1 \cos(\beta_2 + \phi)}{\cos(\beta_1 - \phi)\cos\beta_2}$$

$$= \frac{\cos\beta_1 \cos(\Sigma - \beta_1 + \phi)}{\cos(\Sigma - \beta_1)\cos(\beta_1 - \phi)}$$

$$= \frac{\cos(\Sigma + \phi) + \cos(2\beta_1 - \Sigma - \phi)}{\cos(\Sigma - \phi) + \cos(\Sigma + \phi - 2\beta_1)}$$

$$= \frac{\cos(\Sigma + \phi) + \cos(2\beta_1 - \Sigma - \phi)}{\cos(\Sigma - \phi) + \cos(2\beta_1 - \Sigma - \phi)}$$

$$= \frac{\cos(\Sigma + \phi) + \cos(\beta_1 - \beta_2 - \phi)}{\cos(\Sigma - \phi) + \cos(\beta_1 - \beta_2 - \phi)} \quad \text{...(9.70)}$$

where $\Sigma = \beta_1 + \beta_2$.

For efficiency to be maximum,

$$\cos(2\beta_1 - \Sigma - \phi) = 1$$

or $$2\beta_1 - \Sigma - \phi = 0$$

or $$\beta_1 = \frac{\Sigma + \phi}{2}$$

$$\eta_{max} = \frac{1 + \cos(\Sigma + \phi)}{1 + cso(\Sigma - \phi)} \quad \text{...(9.71)}$$

■ **Example 9.12** Two shafts connected by spiral gears are 500 mm apart. The speed ratio is 3 and the angle between the shafts is 60°. The normal pitch is 20 mm.The spiral angles for the driving and driven gears are equal. Find (a) number of teeth on each gear, (b) exact centre distance and (c) efficiency of the drive. Take friction angle equal to 6°.

Solution : $$\beta_1 = \beta_2 = \frac{\Sigma}{2} = \frac{60}{2} = 30°$$

Centre distance, $$C = \left(\frac{p_n z_1}{2\pi}\right)\left[\frac{1}{\cos\beta_1} + \frac{i}{\cos\beta_2}\right]$$

$$= \left(\frac{20z_1}{2\pi}\right)\left[\frac{1}{\cos 30°} + \frac{1}{\cos 30}\right]$$

$$500 = 14.7\, z_1$$

or $$z_1 = 34$$

$$z_2 = 34 \times 3 = 102$$

Exact centre distance,

$$C = \left(\frac{20 \times 34}{2\pi}\right)\left[\frac{1}{\cos 30°} + \frac{3}{\cos 30°}\right] = 499.87 \text{ mm}$$

Efficiency of the drive,

$$\eta = \frac{\cos(\beta_2 + \phi)\cos\beta_1}{\cos(\beta_1 - \phi)\cos\beta_2}$$

$$= \frac{\cos(30° + 6°)\cos 30°}{\cos(30° - 6°)\cos 30°}$$

$$= 88.56\%$$

■ **Example 9.13** Two spiral gears in mesh have the following data :

Angle of friction	= 6°
Normal pitch	= 20 mm
Shaft angle	= 55°
Speed ratio	= 3
Approximate centre distance	= 400 mm
Spiral angle of pinion	= 25°

Determine (a) the exact distance, (b) the number of teeth in each wheel, and (c) the efficiency of the drive.

Solution :

$$C = \left(\frac{z_1 p_n}{2\pi}\right)\left[\frac{1}{\cos\beta_1} + \frac{i}{\cos\beta_2}\right]$$

$$400 = \left(\frac{z_1 \times 20}{2\pi}\right)\left[\frac{1}{\cos 25°} + \frac{3}{\cos 30°}\right]$$

$$z_1 = 27.51 \approx 28$$

$$z_2 = 84$$

Exact centre distance,

$$C = \left(\frac{z_1 p_n}{2\pi}\right)\left[\frac{1}{\cos\beta_1} + \frac{i}{\cos\beta_2}\right]$$

$$= \left(\frac{28 \times 20}{2\pi}\right)\left[\frac{1}{\cos 25°} + \frac{3}{\cos 30°}\right]$$

$$= 407.08 \text{ mm}$$

Efficiency of drive,

$$= \frac{\cos(\beta_2 + \phi)\cos\beta_1}{\cos(\beta_1 - \phi)\cos\beta_2}$$

$$= \frac{\cos(30° + 6°)\cos 25°}{\cos(25° - 6°)\cos 30°} = 81.54\%$$

■ **Example 9.14** The angle between two shafts is 90°. They are joined by two spiral gears having a normal circular pitch of 8 mm and gear ratio of 3. If the approximate centre distance between the shafts is 250 mm and friction angle 6°, determine for the maximum efficiency of the drive (a) the number of teeth,(b) the exact centre distance, (c) pitch diameters, and (d) the efficiency.

Solution : (a) $\Sigma = 90°$, $C = 250$ mm, $p_n = 8$ mm, i 3, $\phi = 6°$

For maximum efficiency, $\beta_1 = 0.5(\Sigma + \phi) = 0.5(90 + 6) = 48°$,

$$\beta_2 = 42°$$

$$C = [z_1 p_n/(2\pi)]\,[1/\cos\beta_1 + i/\cos\beta_2]$$

$$250 = [z_1 \times 8/(2\pi)]\,[1/\cos 48° + 3/\cos 42°]$$
$$z_1 = 35.49 \cong 36$$
$$z_2 = 36 \times 3 = 108$$

(b) Exact centre distance

$$= [36 \times 8/(2\pi)]\,[1/\cos 48° + 3/\cos 42°]$$
$$= 253.54 \text{ mm}$$

(c) $d_1 = z_1 p_n/(\pi \cos \beta_1) = (36 \times 8)/(\pi \times \cos 48°) = 137 \text{ mm}$

$$d_2 = z_2 p_n/(\pi \cos \beta_2) = (36 \times 8)/(\pi \times \cos 42°) = 370 \text{ mm}$$

(d) $\eta = [1 + \cos(\Sigma + \phi)]/[1 + \cos(\Sigma - \phi)]$

$$= [1 + \cos 96°]/[1 + \cos 84°] = 81\%$$

9.26 WORM GEARS

A worm and worm gear is used to provide a high angular velocity reduction between non-intersecting shafts which are usually at right angles. The pinion or worm has a small

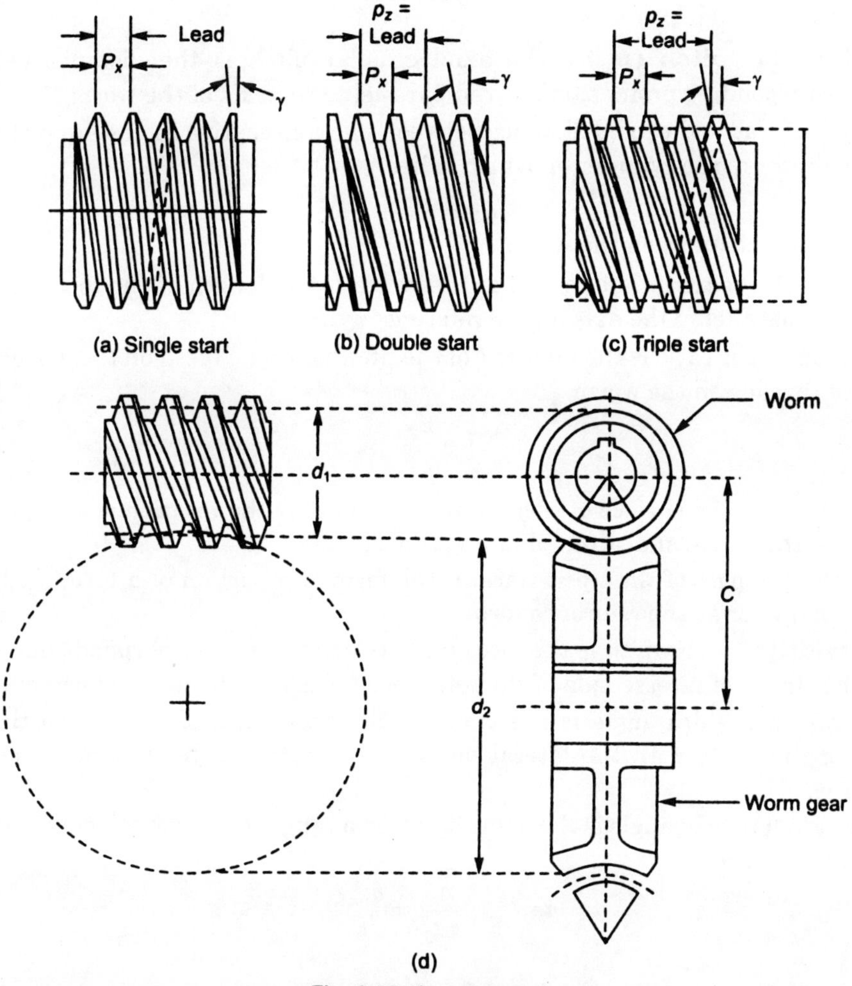

Fig. 9.22 Worm gears.

number of teeth (threads), usually one to four. Its mating gear is called the worm wheel. There is a line contact between the worm threads and the worm wheel teeth. Because of this, worm gears can transmit high tooth loads. However, the high sliding velocities give rise to high heating of the worm. The geometry of a worm and worm gear is shown in Fig. 9.22. For a worm and worm gear with shafts at right angles to mesh properly, the following conditions must be satisfied :

1. Lead angle of worm = helix angle of worm gear
2. Axial pitch of worm = circular pitch of worm gear.

Axial diametral pitch (P_x): The axial diametral pitch is the quotient of the number π by the axial pitch.

Diametral quotient (q): The diametral quotient is the ratio of the reference diameter to the axial module,

$$q = \frac{d}{m} \qquad \text{...(9.72)}$$

Axial module (m_x): The axial module is the quotient of the axial pitch by the number π.

$$m_x = \frac{p_x}{\pi} \qquad \text{...(9.73)}$$

Axial circular pitch (p_x) : The axial circular pitch is the distance between two consecutive corresponding profiles, measured parallel to the axis of the worm.

Lead (p_z): Lead is the axial distance between two consecutive intersection of a helix and a straight generator of the cylinder on which it lies.

$$p_z = p_x z_1 \qquad \text{...(9.74)}$$

where z_1 = number of starts on the worm.

Length of the worm : The length of the worm is the length of the toothed part of the worm measured parallel to the axis on the reference cylinder.

Gear ratio : Gear ratio is the quotient of the number of the teeth on the wheel divided by the number of threads on the worm.

$$i = \frac{z_2}{z_1} \qquad \text{...(9.75)}$$

Torus : Torus is the surface of revolution generated by the rotation of a circle around an axis external to this circle and situated in its plane.

Gorg : Gorg is part of the tip surface in the form of a portion of a torus with the same middle circle diameter as the reference torus.

Tooth width : Tooth width is the distance between two planes perpendicular to the axis containing the circles of intersection of the reference torus and the lateral faces of the teeth.

Width angle : Width angle is the angle at the centre included between the points of intersection of the circle with the lateral faces of the teeth, in the generating circle of the reference torus.

Lead angle (γ): Lead angle is the angle between a tangent to the pitch helix and the plane of rotation of the worm.

$$\tan \gamma = \frac{p_z}{\pi d_1} = \frac{m_x z_1}{d_1} \qquad \text{...(9.76)}$$

$$\frac{\omega_1}{\omega_2} = \frac{z_2}{z_1} = \frac{\pi d_2}{p_z} \quad ...(9.77)$$

$$p_n = p_x \cos\gamma \quad ...(9.78)$$

$$m_x = \frac{m_n}{\tan\gamma} \quad ...(9.79)$$

Centre distance, $C = \frac{d_1 + d_2}{2} = \left(\frac{m_n}{2}\right)\left[\frac{z_1}{\cos\beta_1} + \frac{z_2}{\cos\beta_2}\right]$

For 90° shafts, $\beta_2 = \gamma$ and $\beta_1 = 90° - \gamma$.

Hence $C = \left(\frac{m}{2}\right)[z_1 \cot\gamma + z_2]$

$$\left(\frac{p_z}{2\pi}\right)[\cot\gamma + i] \quad ...(9.80)$$

where, velocity ratio, $i = \frac{z_2}{z_1} = \frac{\pi d_2}{p_x} = \frac{n_1}{n_2}$

9.26.1 Efficiency of Worm Gears

The efficiency of the worm gears when worm is the driver,

$$\eta = \frac{\cos\beta_1 \cos(\beta_2 + \phi)}{\cos(\beta_1 - \phi)\cos\beta_2}$$

For $\beta_1 + \beta_2 = 90°$; $\beta_2 = \gamma$ and $\beta_1 = 90° - \gamma$, we get

$$\eta = \frac{\sin\gamma \cos(\gamma + \phi)}{\cos\gamma \sin(\gamma + \phi)}$$

$$= \frac{\tan\gamma}{\tan(\gamma + \phi)} \quad ...(9.81)$$

When worm wheel is the driver, then

$$\eta = \frac{\tan(\gamma - \phi)}{\tan\gamma} \quad ...(9.82)$$

A worm and worm gear may be considered self-locking when the lead angle of the worm is less than 5°.

■ **Example 9.15** A triple-threaded worm drives a worm gear of 60 teeth, the shaft angle is 90°. The circular pitch of the worm gear is 30 mm and the pitch diameter of the worm is 95 mm. Determine the lead angle of the worm, the helix angle of the worm gear and the distance between shaft centres.

Solution : Lead, $p_z = p_x z_1 = 30 \times 3 = 90$ mm

$$\tan\gamma = \frac{p_z}{\pi d_1} = \frac{90}{\pi \times 95} = 0.30155$$

$$\gamma = 16.78°$$

Helix angle of worm gear = lead angle of worm

Hence $\beta = 16.78°$

$$d_2 = \frac{p_x z_2}{2} = \frac{30 \times 60}{\pi} = 571.95 \text{ mm}$$

$$\text{Centre distance, } C = \frac{d_1 + d_2}{2}$$

$$= \frac{95 + 572.96}{2} = 333.98 \text{ mm}$$

■ **Example 9.16** A two start worm rotating at 900 rpm drives a 27 tooth worm gear. The worm has a pitch diameter of 60 mm and a pitch of 20 mm. The coefficient of friction is 0.05. Find (a) the helix angle of worm, (b) the speed of gear, (c) center distance, (d) efficiency and (e) maximum efficiency.

Solution : $N_1 = 900 \text{ rpm}, z_2 = 27, d_1 = 60 \text{ mm}, p_x = 20 \text{ m}, z_1 = 2$

$$\phi = \tan^{-1} 0.05 = 2.86°, \; p_z = p_x z_1 = 20 \times 2 = 40 \text{ mm}$$

(a) $\tan \gamma = \dfrac{p_z}{\pi d_1} = \dfrac{40}{\pi \times 60} = 0.2122$

$$\beta_2 = \gamma = 11.98°$$

$$\beta_1 = 90° - \beta_2 = 78.02°$$

(b) $d_2 = \dfrac{p_x z_2}{\pi} = \dfrac{20 \times 27}{\pi} = 171.88 \text{ mm}$

$$i = \frac{\omega_1}{\omega_2} = \frac{n_1}{n_2} = \frac{\pi d_2}{p_z} = \frac{\pi \times 171.98}{40} = 13.5$$

$$n_2 = \frac{900}{13.5} = 66.67 \text{ rpm}$$

(c) $C = \left(\dfrac{p_z}{2\pi}\right)(\cot \gamma + i) = \left(\dfrac{40}{2\pi}\right)(\cot 11.98° + 13.5) = 115.95 \text{ mm}$

(d) $\eta = \dfrac{\tan \gamma}{\tan(\gamma + \phi)} = \dfrac{\tan 11.98°}{\tan(11.98° + 2.86°)} = 0.8 \text{ or } 80\%$

(e) $\eta_{max} = \dfrac{1 - \sin \phi}{1 + \sin \phi} = \dfrac{1 - \sin 2.86°}{1 + \sin 2.86°} = 90.49\%$

SOLVED QUESTIONS FROM UNIVERSITY EXAMINATIONS

■ **Example 9.17** Draw and label the following gears and discuss their application :

(a) Helical gears (b) Spiral bevel gears (c) Work gears.

[MDU, Dec., 2007]

Solution : (a) Refer to art. 9.16

(b) Refer to art. 9.25

(c) Refer to art. 9.26

■ **Example 9.18** State and prove the law of gear tooth action for contact velocity ratio and show how the involute tooth profile satisfies the condition. [MDU, May 2008]

Solution : Refer to art. 9.4.

■ **Example 9.19** Calculate : (i) Length of path of approach, (ii) Arc of contact, and (iii) The contact ratio when a pinion having 23 teeth drives a gear having 57 teeth. The profile of the gears is involute with 20° pressure angle, module 8 mm and addendum equal to one module.

[MDU, May 2008]

Solution : Given : $z_1 = 23, z_2 = 57, \alpha = 20°, m = 8\,mm, h_a = 1\,m = 8\,mm$

$$d_1 = mz_1 = 8 \times 23 = 184\,mm, r_1 = 92\,mm$$

$$d_2 = mz_2 = 8 \times 57 = 456\,mm, r_2 = 228\,mm$$

Addendum radius of pinion, $r_{a1} = r_1 + h_a = 92 + 8 = 100$ rpm

Base circle radius of pinion, $r_{b1} = r_1 \cos\alpha = 92 \cos 20° = 86.45$ mm

Addendum radius of gear, $r_{a2} = r_2 + h_a = 228 + 8 = 236$ mm

Base circle radius of gear, $r_{b2} = r_2 \cos\alpha = 228 \cos 20° \;\; 214.25$ mm

(i) Length of path of approach,

$$L_a = \sqrt{r_{aw}^2 - r_{b2}^2} - r_2 \sin\alpha$$

$$= \sqrt{(236)^2 - (214.25)^2} - 228 \sin 20°$$

$$= 20.98\,mm$$

Length of path of recess,

$$L_r = \sqrt{r_{a1}^2 - r_{b1}^2} - r_1 \sin\alpha$$

$$= \sqrt{(100)^2 - (86.45)^2} - 92 \sin 20°$$

$$= 18.79\,mm$$

Length of path of contact,

$$L_p = L_r + L_a = 18.79 + 20.98 = 39.77\,mm$$

(ii) Arc of contact, $L_c = \dfrac{L_p}{\cos\alpha} = \dfrac{39.77}{\cos 20°} = 42.32$ mm

(iii) Contact ratio, $CR = \dfrac{L_p}{\pi m \cos\alpha} = \dfrac{39.77}{\pi \times 8 \times \cos 20} = 1.68 \simeq 2$

■ **Example 9.20** Determine the minimum number of teeth required on a pinion to avoid interference, which is to gear with a wheel to given gear ratio 3 : 1. The pressure angle is 20° and a standard addendum of 1 module may be assumed. [MDU, Dec., 2008]

Solution : Given : $i = 3, \alpha = 20°, a_w = 1$

$$z_2 = \frac{2a_w}{\left[1 + \frac{1}{i^2}\sin^2\alpha + \frac{2}{i}\sin^2\alpha\right]^{0.5} - 1}$$

$$= \frac{2 \times 1}{\left[1 + \frac{1}{9}\sin^2 20° + \frac{2}{3}\sin^2 20°\right]^{0.5} - 1} = 44.94 \simeq 45$$

$$z_1 = \frac{z_2}{i} = \frac{45}{3} = 15$$

Minimum number of teeth on pinion, $z_1 = 15$

■ **Example 9.21** What do you mean by pitch point, circular pitch, module, addendum and dedendum of a gear? Explain with the help of a neat sketch. What is the the relation between circular pitch, diametral pitch and number of teeth? [MDU, May, 2009]

Solution : Refer to art. 9.3.

■ **Example 9.22** Prove that minimum number of teeth (z_1) required on the pinion in order to avoid interference is given by

$$z_1 = \frac{2a_p}{\sqrt{1 + G\,(G + 2)\sin^2\phi} - 1}$$

where a_p = fraction by which the standard addendum of one module for the pinion should be multiplied in order to avoid interference.

G = gear ratio

ϕ = pressure angle [MDU May 2009]

Solution : Refer to art 9.11.2.

■ **Example 9.23** What do you mean by pitch point, circular pitch, module, diametral pitch, backlash of a gear? Explain with the help of neat sketches. What is relation between module and circular pitch? [MDU, Dec., 2009]

Solution : Refer to art 9.3.

■ **Example 9.24** Two 20° involute spur gear mesh extanally and give a velocity ratio of 3. The module is 3 mm and addendum = 1.1 module. If the pinion rotates at 120 rpm, determine :

(i) The minimum number of teeth on each wheel to avoid interference.

(ii) Contact ratio. [MDU, Dec., 2009]

Solution :

Given : $\alpha = 20°$, $i = 3$, $m = 3\text{ m}$, $h_a = 1.1\text{ mm} = 1.1 \times 3 = 3.3\text{ mm}$, $a_w = 1.1$, $n_1 = 120\text{ rpm}$

(i)

$$z_2 = \frac{2a_w}{\left[1 + \frac{1}{i}\left(\frac{1}{i} + 2\right)\sin^2\alpha\right]^{0.5} - 1}$$

$$= \frac{2 \times 1.1}{\left[1 + \frac{1}{3}\left(\frac{1}{3} + 2\right)\sin^2 20°\right]^{0.50} - 1} = 49.4 \approx 51$$

$$z_1 = \frac{z_2}{i} = \frac{51}{3} = 17$$

(ii)

$$r_1 = \frac{mz_1}{2} = \frac{3 \times 17}{2} = 25.5\text{ mm}$$

$$r_2 = \frac{mz_2}{2} = \frac{3 \times 51}{2} = 76.5\text{ mm}$$

$$r_{a1} = r_1 + h_a = 25.5 + 3.3 = 28.8\text{ mm}$$

$$r_{b1} = r_1\cos\alpha = 25.5\cos 20° = 23.96\text{ mm}$$

$$r_{a2} = r_2 + h_a = 76.5 + 3.3 = 79.8\text{ mm}$$

$$r_{b2} = r_2\cos\alpha = 46.5\cos 20° = 71.88\text{ mm}$$

$$L_a = \sqrt{r_{a2}^2 - r_{b2}^2} - r_2\sin\alpha = \sqrt{(79.8)^2 - (71.88)^2} - 76.5\sin 20° = 8.49\text{ mm}$$

$$L_r = \sqrt{r_{a1}^2 - r_{b1}^2} - r_1\sin\alpha = \sqrt{(28.8)^2 - (23.96)^2} - 25.5\sin 20° = 7.26\text{ mm}$$

$$L_p = L_a + L_r = 8.49 + 7.26 = 15.75 \text{ mm}$$

$$CR = \frac{L_p}{\pi\, m \cos \alpha} = \frac{15.75}{\pi \times 3 \cos 20^\circ} = 1.778 \approx 2$$

■ **Example 9.25** prove that minimum number of teeth (T) required on the gear in order to avoid interference is given by :

$$T = \frac{2a_w}{\left(\sqrt{1 + \frac{1}{G}\left(\frac{1}{G} + 2\right) \sin^2 \phi} - 1\right)}$$

where $a_w \to$ Addendum co-efficient

$G \to$ Gear-Ratio

$\phi \to$ pressure-angle [MDU, Dec., 2010]

Solution : Refer to art. 9.11.1

SUMMARY

1. A gear may be defined as a toothed member designed to transmit or receive motion from one shaft to another by successively engaging tooth.
2. Gears occupy less space, no slip, transmit higher power, and gives higher efficiency.
3. Gears may be classified as : spur, helical double helical, spiral, bevel, worm, hypoid and planetary.
4. Pressure angle is the angle between the common normal at the point of contact and the common tangent at the pitch point.
5. Involute of a circle is the curve traced by the end of a thread as it is unwound from a stationary cylinder.
6. Cyloid is the locus of a point on the circumference of a circle which rolls without slipping on a fixed straight line.
7. Basic relations for a spur gear.

   Circular pitch, $p = \pi d / z$

   where z = number of teeth, d = pitch circle diameter.

   Base pitch, $p_b = p \times \cos \alpha$

   Diametral pitch, $P = z / d$

   Relationship between circular and diametral pitch, $P_p = \pi$

   Module, $m = d / z = 1/P$

   Base Circle Diameter, $d_b = d \cos \alpha$

   Center distance, $C = (d_1 + d_2)/2 = m\,(z_1 + z_2)/2$
8. Fundamental law of gearing : For constant angular velocity ratio of the two gears in contact the common normal at the point of contact must always intersect the line of centres at a fixed point (pitch point) and divide this line in the inverse ratio of the angular velocities of the two gears.

9. Relative velocity between gear teeth :

$v_r = (\omega_1 + \omega_2)\, AP$

$= $ (sum of the angular velocities) $\times$

distance of the point of the contact from the pitch point.

10. Involute function inv (α) = tan $\alpha - \alpha$

11. Characteristics of involute action :

Addendum radius of pinion, $r_{a1} = r_1 + h_{a1}$

Base circle of radius of pinion, $r_{b1} = r_1 \cos\alpha$

Addendum radius of gear, $r_{a2} = r_2 + h_{a2}$

Base circle radius of gear, $r_{a2} = r_2 \cos\alpha$

where r_1 = pitch circle radius of gear

r_2 = pitch circle radius of gear

h_{a1} = addendum of pinion

h_{a2} = addendum of gear

r_{b1} = base circle radius of pinion

r_{b2} = base circle radius of gear

Length of path of recess, $L_r = (r_{a1}^2 - r_{b1}^2)^{0.5} - r_1 \sin\alpha$

Length of path of approach, $L_a = (r_{a2}^2 - r_{b2}^2)^{0.5} - r_2 \sin\alpha$

Length of path of contact,

$$AB = L_p = L_p = L_r + L_2$$
$$= (r_{a1}^2 - r_{b1}^2)^{0.5} + (r_{a2}^2 - r_{b2}^2)^{0.5} - (r_1 + r_2)\sin\alpha$$

Length of arc of contact, $L_c = AB/\cos\alpha$

Maximum length of path of recess $= r_2 \sin\alpha$

Maximum length of path of approach $= r_1 \sin\alpha$

Contact ratio,

$$m_c = \text{length of path of contact/base pitch} = L_p / p_b$$

where $p_b = p\cos\alpha = \pi\, m \cos\alpha$

For a rack and a pinion,

$$L_p = (r_{a1}^2 - r_{b1}^2) - r_1 \sin\alpha + a/\sin\alpha$$

where 'a' = addendum

12. Interference in gears is a phenomenon in which the tip of the driven gear tooth will dig into the flank of the driving gear tooth.

13. Interference can be avoided by undercutting, making stub tooth, increasing the pressure angle, and cutting the gears with long and short addendum gear teeth.

14. Minimum number of teeth on gear wheel.

Speed ratio, $i = z_2/z_1$, addendum of pinion, $h_{a1} = a_p m$, addendum of gear wheel, $h_{a2} = a_w m$

where a_p and a_w are the constants by which the module must be multiplied to get the addendum of pinion and gear wheel respectively.

$$z_2 = 2a_w / [\{1 + (1/i^2)\sin^2\alpha + (2 + i)\sin^2\alpha\}^{0.5} - 1]$$

For $a_w = 1$, $\quad z_2 = [z_1^2 \sin^2\alpha - 4] / [4 - 2z_1 \sin^2\alpha]$

For $i = 1$, $\quad a_w = a_p$ and $z_2 = 2a_w / [\{1 + 3\sin^2\alpha\}^{0.5} - 1]$

15. Minimum number of teeth on pinion.

$$z_1 = 2a_p / [\{1 + i^2 \sin^2\alpha + 2i \sin^2\alpha\}^{0.5} - 1]$$

For $a_p = 1$, $\quad z_1 = [z_2^2 \sin^2\alpha - 4] / [4 - 2z_2 \sin^2\alpha]$

16. Minimum number of teeth on rack pinion.

$$z_1 = 2a_r / \sin^2\alpha$$

$$z_{min} = 32 \text{ for } \alpha = 14.5° \text{ and } 18 \text{ for } \alpha = 20°$$

17. Effect of centre distance variation on velocity ratio.

The variation in the centre distance, within limits, does not affect the angular velocity ratio. But the length of arc of contact is decreased, and the pressure angle is increased.

18. Helical gears have teeth in the form of a helix around the gear.

19. Helical gears have line contact which takes place gradually and smoothly, gives less noise as compared to spur gears. However they give rise to end thrust.

20. Basic relations for a helical gear :

Transverse circular pitch, $\quad p_t = \pi d / z = \pi m$

Axial pitch, $\quad p_x = p_t \cot\beta$,

where β = helix angle

Virtual number of teeth, $\quad z_v = z / \cos\beta$

Normal module, $\quad m_n = m_t \cos\beta$

Normal circular pitch, $\quad p_n = p_t \cos\beta$

Transverse diametral pitch, $\quad P_t = z / d = 1/m$

Normal diametral pitch, $\quad P_n = P_t / \cos\beta$

Speed ratio, $\quad i = z_2 / z_1 = n_1 / n_2$

Centre distance, $\quad C = (d_1 + d_2)/2 = (m_n / \cos\beta)\,[(z_1 + z_2)/2]$

21. Crossed helical gears :

Velocity ratio, $i = n_1 / n_2 = z_2 / z_1 = (d_2 \cos\beta_2)/(d_1 \cos\beta_1)$

Angle between the two shafts, $\Sigma = \beta_1 \pm \beta_2$ (use + ve sign for same hand and – ve sign for opposite hand gears)

Centre distance, $\quad C = (m_n / 2)\,[z_1 / \cos\beta_1 + z_2 / \cos\beta_2]$

22. In herringbone gears, the end thrust is automatically eliminated.

23. Bevel gears are used for shafts whose axes intersect.

24. Crown gears is a gear pair for which pitch cone angle is 90°.

25. Miter gears are two bevel gears of the same in mesh whose pitch cone angle is 90°.

26. Basic relations for a bevel gears :

Length of the pitch cone, $L = [(d_1^2 + d_2^2)/4]^{1/2}$

Pitch cone angle of pinion, $\delta_1 = \sin^{-1} [d_1/(2L)]$

Angle between the shafts, $\Sigma = \delta_1 + \delta_2$

$$\tan \delta_1 = \sin \Sigma / [\cos \Sigma + (z_1 + z_2)]$$

27. Spiral gears are used to connect non-parallel and non-intersecting shafts.

28. Basic relations for spiral gears :

Angle between the shafts, $\Sigma = \beta_1 + \beta_2$

Circular pitch, $p = p_n / \cos \beta$

Pitch circle diameter, $d = z p_n / (\pi \cos \alpha_n)$

Centre distance, $C = [(z_1 p_n)/(2\pi)] [1/\cos \beta_1 + 1/\cos \beta_2]$

Efficiency of spiral gears = $[\cos (\Sigma + \phi) + \cos (\beta_1 - \beta_2 - \phi)] / [\cos (\Sigma - \phi) + \cos (\beta_1 + \beta_2 - \phi)]$

Where $\phi = \tan^{-1} \mu$, μ = coefficient of friction

For maximum efficiency, $\beta_1 = (\Sigma + \phi)/2$

Maximum efficiency = $[1 + \cos (\Sigma + \phi)] / [1 + \cos (\Sigma - \phi)]$

29. Worm gears are used to provide high angular velocity reduction between non-intersecting shafts.

30. Basic relations for worm gears :

Lead angle of worm = Helix angle of worm gear

Axial pitch of worm = Circular pitch of worm gear

Axial diametral pitch, $P_x = \pi / p_x$

Diametral quotient, $q = d/m$

Axial module, $m_x = p_x / \pi$

Lead, $p_z = z_1 p_x$,

where z_1 = number of starts on the worm

Gear ratio, $i = z_2 / z_1$

Lead angle, $\gamma = \tan^{-1} [p_z / (\pi d_1)]$

Axial module, $m_x = m_n / \tan \gamma$

Centre distance, $C = (m_n/2) [z_1/\cos \beta_1 + z_2/\cos \beta_2]$

Efficiency of worm gears

$$= [\cos \beta_1 \cos (\beta_2 + \phi)] / [\cos (\beta_1 - \phi) \cos \beta_2]$$

SHORT ANSWER TYPE QUESTIONS

1. Define pressure angle of a gear.
2. What is the relationship between circular pitch and diametral pitch?

3. State the law of gearing.
4. Write expression for velocity of sliding in a gear pair.
5. What is the law for velocity of sliding between gear pair?
6. Differentiate between path of approach and path of recess.
7. Compare involute and cyloidal tooth profiles.
8. When interference will occurring gear teeth?
9. Define contact ratio.
10. How undercutting can be reduced in gears?
11. What are stub gears?
12. How many are the minimum number of teeth on a pinion for a rack?
13. What is the effect of centre distance variation on speed ratio in gears?
14. What is a rack?
15. What are standard gears? What are their advantages?
16. What are nonstandard gears?
17. Differentiate between helical and herrigbone gears.
18. What are the advantages of helical gears over spur gears?
19. Define virtual member of teeth on helical gear.
20. Write the expression for virtual number of teeth on helical gear.
21. What is the advantage of herringbone gear over helical gear.
22. Differentiate between crown and miter bevel gears.
23. Write the expression for virtual number of teeth on bevel gear.
24. Where do we use spiral gears?
25. When do we use worm gear?

EXERCISES

(a) Spur Gears

1. A pair of 20° full depth involute spur gears having 40 and 60 teeth of module 4 mm are in mesh. The smaller gear rotates at 1440 rpm. Find (a) sliding velocity at engagement and disengagement of the pair of teeth, and (b) contact ratio. [**Ans.** 6 m/s, 10.315 m/s; 6]
2. Calculate the minimum number of teeth on a pinion to avoid interference to have a speed ratio of 2.5 : 1. The pressure angle is 20° and addendum of one module of gear may be used. [**Ans.** 15]
3. A pinion of involute profile has 25 teeth and 150 mm pitch circle diameter. It drives a rack. The addendum of both pinion and rack is 6.25 mm. Calculate the pressure angle to avoid interference. [**Ans.** 16.78°]
4. The following data refer to two meshing involute gears of 20° pressure angle : Number of teeth on pinion = 20, speed ratio 2, speed of pinion = 250 rpm, module = 12 m.
   The addendum of each wheel is such that the path of approach and path of recess on each side are half of the maximum possible length. Calculate (a) addendum for both the wheels, (b) the length of arc of contact, and (c) the maximum possible sliding velocity during approach and recess.
   [**Ans.** 19.47 mm, 7.77 mm; 65.51 mm; 0.806 m/s, 1.612 m/s]

5. Two spur gears wheels of 80 mm and 120 mm pitch diameters have involute teeth of standard addenda of 3 mm and 20° pressure angle. The module is 1 mm. Determine (a) the length of path of contact, (b) contact ratio, and (c) angle turned through by pinion, while any pair of teeth is in contact. [**Ans.** 14.79 mm, 5, 22.49°]

6. A pair of involute profile spur gears is to give a speed ratio of 3. The arc of approach is not to be less than the circular pitch. The pressure angle is 20° and pinion is the driver. The module is 4 mm. Calculate (a) minimum number of teeth on gear, and (b) addendum of gear wheel. [**Ans.** 57, 5.07 mm]

7. A pinion having 30 teeth drives a gear of 80 teeth. The profile is involute with 20° pressure angle, 12 mm module, and 10 mm addendum. Find (a) the length of path of contact, (b) arc of contact, and (c) contact ratio. [**Ans.** 52.26 mm, 55.61 mm, 1.47]

8. Find the minimum number of teeth on gear wheel to avoid undercutting when the addendum for stub teeth is 0.84 module, if (a) gear ratio is 3 : 1, and (b) the wheel is used to engage a rack. [**Ans.** 16.18]

9. A pair of involute spur gears having 20 and 40 teeth are in mesh, the speed of smaller wheel being 2000 rpm. Calculate the sliding velocity between gear teeth faces (a) at the point of engagement, (b) at the pitch point, and (c) at the point of disengagement, if smaller wheel is the driver. Pressure angle is 20°, addendum = 5 mm, and module = 5 mm.
   Also find the angle turned through by the pinion while any one pair of teeth is in contact. [**Ans.** 3.97 m/s, 0, 3.61 m/s; 29.43°]

(b) Helical Gears

10. The centre distance between the two shafts connected by two left handed helical gears is 0.4 m. The shaft angle is 60° and normal modules is 6 mm. The gear ratio is 2 and 12 larger gear is having to teeth. Find the helix angles of the two gears. [**Ans.** 165°, 44°]

11. Two left-handed helical gears connect two shafts 60° apart. The normal module is 6 mm. The larger gear has 60 teeth and the velocity ration is 0.5. The centre distance is 320 mm find the helix angles of the two gears. [**Ans.** 25°, 35.5°]

12. Two right-handed helical gears connect two shafts 70° apart. The larger gear has 50 teeth and the smaller 20. The centre distance is 167 mm. Determine the helix angle of the gears. The normal module is 4 mm. [**Ans.** 28°, 42°]

(c) Bevel Gears

13. A pair of bevel gears is connected on two intersecting shafts at an angle of 72°. The velocity of the gears is 2. Calculate the pitch angles. [**Ans.** 22.39°, 49.61°]

14. A 6 mm module, straight bevel pinion of 17 teeth drives a gear of 25 teeth. The shaft angle is 90°. Calculate the virtual number of teeth on the pinion and gear. [**Ans.** 21, 44]

(d) Spiral Gears

15. Two spiral gears A and B have 45 and 15 teeth at spiral angels of 20° and 50° respectively. Both gears are of same hand. A is 150 mm in diameter. Find the centre distance between the shafts and the angle between the shafts. If the teeth are of 20° involute from and coefficient of friction is 0.08, find the efficiency of A is the driver. [**Ans.** 111.5 mm, 70°, 82.59%]

16. Two spiral gears of diameter ratio 1.5 are used on a machine tool. The angle between the shafts is 76° and approximate centre distance is 115 mm. Speed ratio is 1.5 and normal pitch is 10 mm. Calculate the number of teeth on each gear and spiral angles. [**Ans.** 23, 35, 38°, 38°]

17. A pair of spiral gears is required to connect two shafts, the angle between the non-intersecting axes is 90°. The speed ratio is 3 and pitch circle diameters of the gears are equal. If approximate centre distance is 250 mm, estimate the helix angle of gears and number of teeth on each gear. Take module = 3 mm. **[Ans.** 71.56°, 18.44°; 26, 78]

18. A drive on a machine tool is to be made by two spiral gears of the same hand and normal pitch of 12.5 mm. The gears are of equal diameters and centre distance between the shafts is approximately 134 mm. The angle between the two shafts is 80° and speed ratio 1.25. Calculate
(a) spiral angle of each wheel,
(b) number of teeth on each wheel
(c) efficiency of drive, and
(d) maximum efficiency
The friction angle is 6°. **[Ans.** 32.46°, 47.54°, 30, 24, 82.98%, 83.85%]

(e) Worm Gears

19. A three-start worm has a pitch diameter of 80 mm and a pitch of 20 mm. It rotates at 600 rpm and drives a 40 tooth worm gear. If coefficient of friction is 0.05, find the
(a) helix angle of worm,
(b) speed of the gear,
(c) centre distance, and
(d) efficiency and maximum efficiency. **[Ans.** 76-56, 45 rpm, 167.3 mm, 81.7%, 90.5%]

❑❑❑

Chapter 10

Gear Trains

Inside This Chapter

10.1 INTRODUCTION

A gear train is composed of two or more gears in mesh for the purpose of transmitting motion from one shaft to another. A gear train enables to have larger centre distance between the driving and driven shafts, provides control on the direction of rotation of the driven gear, and facilitates increased transmission ratio.

10.2 TYPES OF GEAR TRAINS

There are four types of gear trains :

1. Simple gear train
2. Compound gear train
3. Reverted gear train
4. Planetary (or epicyclic) gear train.

1. Simple Gear Train

A simple gear train is one in which there is only one gear on each shaft. A simple gear train is often used to change the direction of rotation of a gear without changing its angular velocity. This can be done by placing an idler gear between the driving and driven gears.

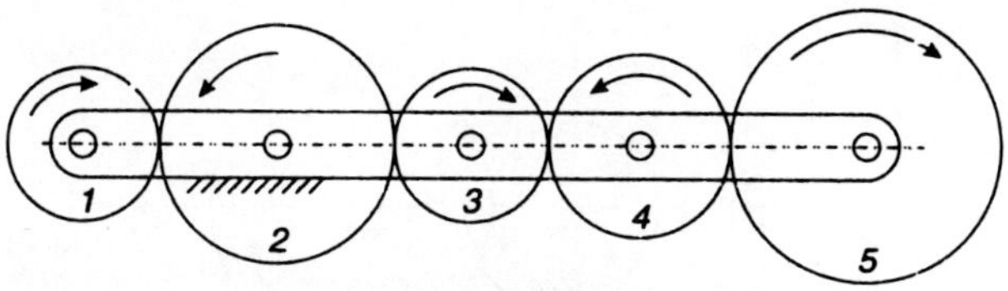

Fig. 10.1 Simple gear train.

Consider the simple gear train shown in Fig. 10.1

$$\frac{n_1}{n_2} = \frac{z_2}{z_1}$$

$$\frac{n_2}{n_3} = \frac{z_3}{z_2}$$

$$\frac{n_{m-1}}{n_m} = \frac{z_m}{z_{m-1}}$$

Multiplying, we get

$$\frac{n_1}{n_m} = \frac{z_m}{z_1} \quad ...(10.1)$$

Therefore, the velocity ratio of a simple gear train is the ratio of the angular velocity of the first gear in the train to the angular velocity of the last gear. We find that the intermediate gears do not in any way affect the velocity ratio. These gears are called **idler gears.** If the number of gears in the train are even then the direction of rotation of the last gear is reversed and if the number of gears in the train are odd then the direction of rotation of the last gear remains the same. Idler gears are used for two purposes : to connect gears where a large centre distance is required, and to control the directional relationship between gears.

2. Compound Gear Train

A pair of gears is compound if they have a common axis and are integral. A compound gear train is a gear train containing compound gears. Consider the compound gear train shown in Fig. 10.2. Gears 2 and 3 are on the same shaft.

$$\frac{n_2}{n_1} = \frac{-z_1}{z_2}$$

$$\frac{n_4}{n_3} = \frac{-z_3}{z_4}$$

Multiplying, we get

$$\frac{n_2}{n_1} \times \frac{n_4}{n_3} = \frac{z_1}{z_2} \times \frac{z_3}{z_4}$$

But

$$n_2 = n_3$$

Thus

$$\frac{n_4}{n_1} = \frac{z_1}{z_2} \times \frac{z_3}{z_4}$$

$$= \frac{z_1 z_3}{z_2 z_4}$$

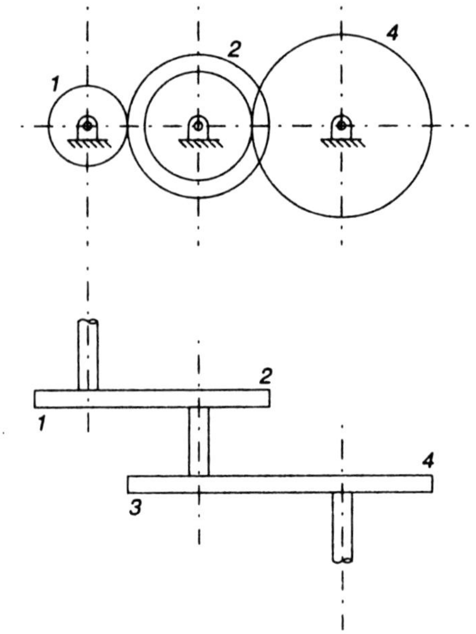

Fig. 10.2 Compound gear train.

Speed of driven gear/speed of driving gear

= Product of teeth of driving gears/Product of teeth of driven gears ...(10.2)

The ratio given by Eq. (10.2) is called the **train value.**

3. Reverted Gear Train

In a reverted gear train the first and the last gear are coaxial (same axis). Reverted gear trains are used in automotive transmission, lathe back gears, industrial speed reducers, and in clocks where the minute and hour hand shafts are coaxial. In the reverted gear train shown in Fig. 10.3, we have

Centre distance, $C = r_1 + r_2 = r_3 + r_4$

For same module ($m = d/z$) of all gears, the pitch radius of gears is proportional to the number of teeth. Hence

$$z_1 + z_2 = z_3 + z_4 \quad \text{...(10.3)}$$

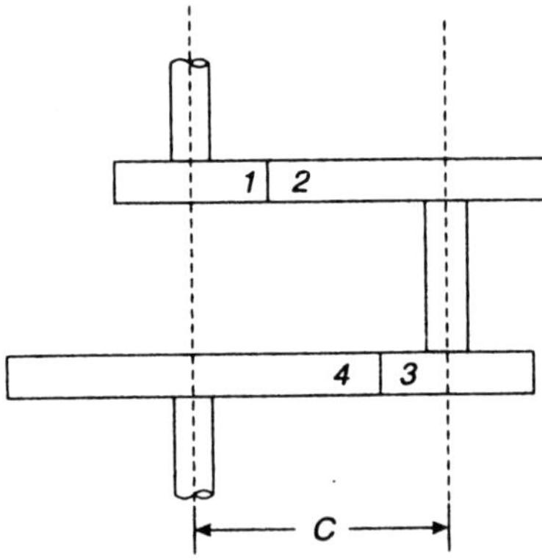

Fig. 10.3 Reverted gear train.

4. Planetary Gear Trains

These are gear trains in which the axis of one or more gears move relative to the frame. The gear at the centre is called the sun, and the gears whose axes move are called the **planets.**

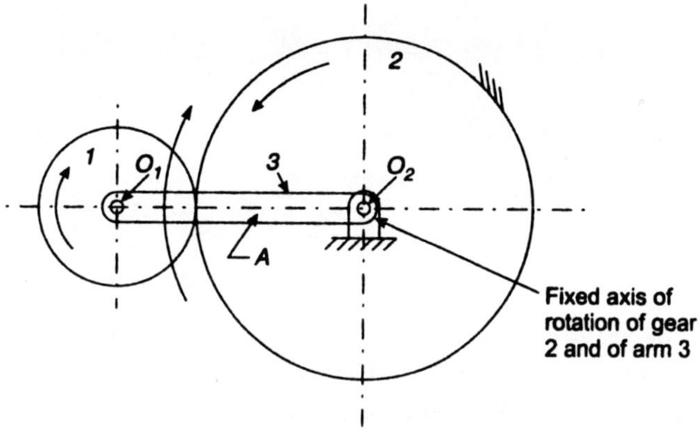

Fig. 10.4 Planetary gear train.

In Fig. 10.4, arm 3 drives gear 1 about gear 2, which is a fixed external gear. Gear 1 rotates about its centre *d* while this centre rotates about centre O_2 of the fixed gear. As gear 1 rolls on the outside of gear 2, a point on its surface will generate an epicycloid. If gear 2 happens to be an internal gear and gear 1 rolls on the inside of gear 2, then a point on the surface of gear 1 will generate a hypocycloic. Because of the curves generated, a planetary gear train is often called as an **epcyclic, or cyclic,** gear train.

10.3 DETERMINATION OF SPEED RATIO OF PLANETARY GEAR TRAINS

The speed ratio of a planetary gear train can be determined by the following methods :

1. Relative velocity method
2. Algebraic or tabular method.

1. Relative Velocity Method

Consider the planetary gear train shown in Fig. 10.4.

$$\omega_{13} = \omega_{14} - \omega_{34}$$

$$\omega_{23} = \omega_{24} - \omega_{34}$$

and

$$\frac{\omega_{13}}{\omega_{23}} = \frac{\omega_{14} - \omega_{34}}{\omega_{24} - \omega_{34}}$$

Link 3 is the arm, link 1 is the driven gear, link 2 is the driving gear, and link 4 is fixed. If subscript '*a*' is used for the arm, then

$$\frac{\omega_{1a}}{\omega_{2a}} = \frac{\omega_1 - \omega_a}{\omega_2 - \omega_a}$$

or

$$\frac{n_{1a}}{n_{2a}} = \frac{n_1 - n_a}{n_2 - n_a}$$

Now

$$\frac{n_{1a}}{n_{2a}} = \frac{-z_1}{z_2}$$

Hence, $$\frac{n_1 - n_a}{n_2 - n_a} = \frac{-z_1}{z_2}$$

If gear 2 is fixed, then $n_2 = 0$. Thus,

$$\frac{n_1}{n_a} = \left(1 + \frac{z_1}{z_2}\right) \qquad \ldots(10.4)$$

If arm is fixed, then $n_a = 0$. Thus

$$\frac{n_1}{n_2} = \frac{-z_1}{z_2} \ldots \qquad (10.5)$$

2. Tabular Method

This method is based on the principle of superposition, which states that the resultant revolutions or turns of any gear may be found by taking the number of turns it makes with the arm plus the number of turns it makes relative to the arm. The following steps may be followed for this method :

1. Assume the arm to be fixed and determine the revolutions of different gears of the train for one revolution of a particular convenient gear.
2. Multiply all columns in the first row by x and write in the second row.
3. To account for rotation of arm, add y to the various quantities of second row.
4. Out of the three quantities involved in the last row, two of them are given. From these the values of x and y can be determined. On substituting in the third, its magnitude can be determined.

The following table may be conveniently used for this purpose :

Operation	*Revolutions of*		
	Arm A	*Gear 1, z_1*	*Gear 2, z_2*
1. Arm A fixed, + 1 revolution to gear 1, cw	0	+ 1	$\frac{-z_1}{z_2}$
2. Multiply by x	0	$+x$	$\frac{-xz_1}{z_2}$
3. Add y	$+y$	$x+y$	$\frac{-xz_1}{z_2}+y$

10.4 SUN AND PLANET GEARS

Fig. 10.5 shows the sun and planet gears in which gear P is the planet, gear S the sun, arm A, and the annular (or internal) gear I. The annular gear is fixed and the planet gear rolls over the sun and the annular gear. O_1 is the moving axis of rotation of planet gear P and O_2 is the fixed axis of rotation of sun gear 2 and arm A. The axes O_1 and O_2 are coupled by the arm A. Let z_p, z_s, and z, be the number of teeth on the planet, sun, and annular gears respectively. To find the speed of any gear, the following table may be used :

Operation	*Revolution of*			
	Arm A	*Sun gear S, z_s*	*Planet gear P, z_p*	*Annular gear I, z_i*
1. Arm *A* fixed, + 1 revolution to *S*, cw	0	+ 1	$\frac{-z_s}{z_p}$	$\frac{-z_s}{z_i}$
2. Multiply by x	0	$+x$	$\frac{-z_s x}{z_p}$	$\frac{-z_s x}{z_i}$
3. Add y	y	$x+y$	$\frac{-z_s x}{z_p}+y$	$\frac{-z_s x}{z_i}+y$

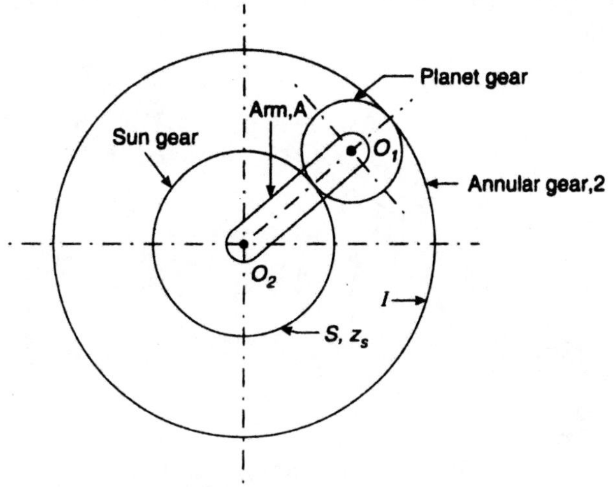

Fig. 10.5 Sun and planet gears.

Let n_s and n_i be the speed of sun and annular gears respectively. Then

$$x + y = n_s - \left(\frac{z_s}{z_i}\right) \cdot x + y = n_i$$

Solving for x and y, we get

$$x = \frac{z_i n_s - z_i n_i}{z_i + z_s}$$

$$y = \frac{z_s n_s + z_i n_i}{z_i + z_s}$$

(a) When sun gear is fixed, then $n_s = 0$. Thus

$$y = \frac{-z_i n_i}{z_i + z_s} \qquad \text{...(10.6)}$$

$$= \frac{-n_i}{\left(1 + \frac{z_s}{z_i}\right)}$$

(b) When annular gear is fixed, then $n_i = 0$. Thus

$$y = \frac{z_i n_s}{z_i + z_s}$$

$$= \frac{n_s}{\left(1 + \frac{z_i}{z_s}\right)} \quad \text{...(10.7)}$$

10.5 TORQUE IN EPICYCLIC GEAR TRAINS

Consider the rotating parts of an epicyclic gear train shown in Fig. 10.6, if the rotating parts have no angular acceleration, then the gear train is kept in equilibrium by the following three externally applied torques :

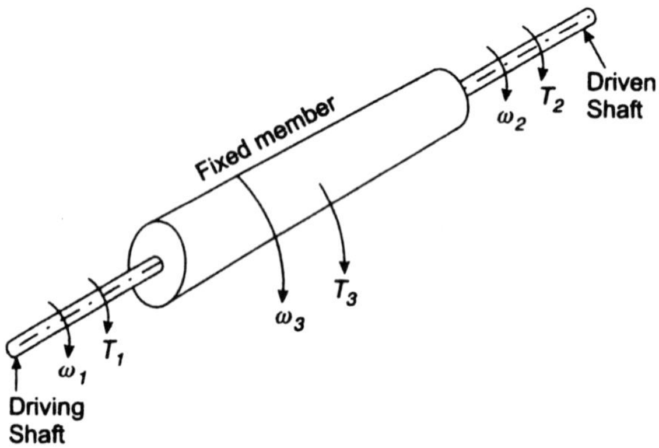

Fig. 10.6 Torque in epicyclic gear trains.

1. Input torque (T_1) on the driving member.
2. Output torque (T_2) on the driven member.
3. Braking torque (T_3) on the fixed member.

The net torque applied on the gear train must be zero, *i.e.*,

$$T_1 + T_2 + T_3 = 0$$

If F_1, F_2, and F_3 are the externally applied forces at radii r_1, r_2 and r_3, then

$$F_1 r_1 + F_2 r_2 + F_3 r_3 = 0$$

If ω_1, ω_2, and ω_3 are the angular velocities of the driving, driven, and fixed members respectively, and friction is neglected, then the net kinetic energy dissipated by the gear train must be zero.

$$T_1\omega_1 + T_2\omega_2 + T_3\omega_3 = 0$$

For a fixed member, $\omega_3 = 0$. Therefore

$$T_1\omega_1 + T_2\omega_2 = 0$$

$$T_2 = -\left(\frac{\omega_1}{\omega_2}\right) T_1$$

$$T_3 - (T_1 + T_2) = T_1\left(\frac{\omega_1}{\omega_2} - 1\right) = T_1\left(\frac{n_1}{n_2} - 1\right) \quad \text{...(10.8)}$$

■ **Example 10.1** In a reverted gear train, as shown in Fig. 10.7, two shafts *A* and *B* are in the same straight line and are geared together through an intermediate parallel shaft *C*. The gears connecting the shafts *A* and *C* have a module of 3 mm and those connecting the shafts *C* and *B* have a module of 4.5 mm. The speed of shaft *A* is to be about but greater than 12 times the speed of shaft *B*. The ratio of each reduction is same. Find suitable number of teeth on all gears. The minimum number of teeth is 18. Also find the exact velocity ratio and the distance of shaft *C* from *A* and *B*.

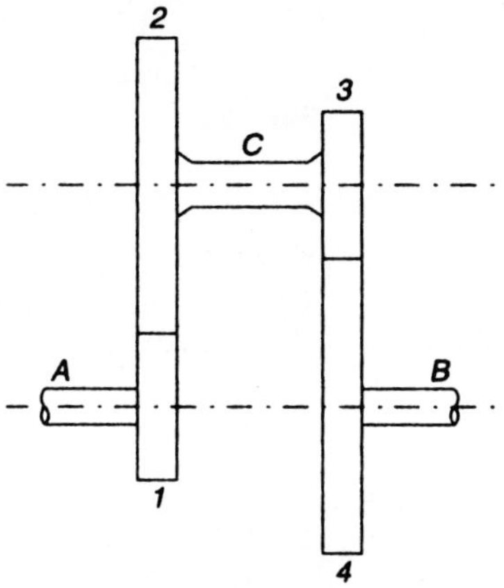

Fig. 10.7

Solution : Given $m_1 = m_2 = 3$ mm, $m_3 = m_4 = 4.5$ mm, $z_{min} = 18, n_A > 12n_B$

Now $$\frac{n_1}{n_2} = \frac{n_3}{n_4}$$

But $$n_2 = n_3$$

Therefore, $$n_2^2 = n_1\, n_4 \text{ or } n_2 = \sqrt{n_1 n_4} = \sqrt{12}\, n_4$$

$$\frac{n_2}{n_4} = \sqrt{12} = 3.464 \cong 3.5$$

For same centre distance,

$$3(z_1 + z_2) = 4.5\,(z_3 + z_4)$$

$$(z_1 + z_2) = 1.5\,(z_3 + z_4)$$

$$z_1(1 + 3.5) = 1.5 z_3(1 + 3.5)$$

$$z_1 = 1.5 z_3$$

Let $z_3 = 20$, then $z_4 = 70, z_1 = 30, z_2 = 105$

$$\frac{n_A}{n_B} = (3.5)^2 = 12.25$$

Centre distance, $$C = 3\,(30 + 105) = 405 \text{ mm}$$

■ **Example 10.2** The speed ratio of the reverted gear train shown in Fig. 10.8 is to be 12. The module of gears *A* and *B* is 3.125 mm and of gears *C* and *D* is 2.5 mm. Calculate the suitable number of teeth for the gears. No gear is to have less than 24 teeth.

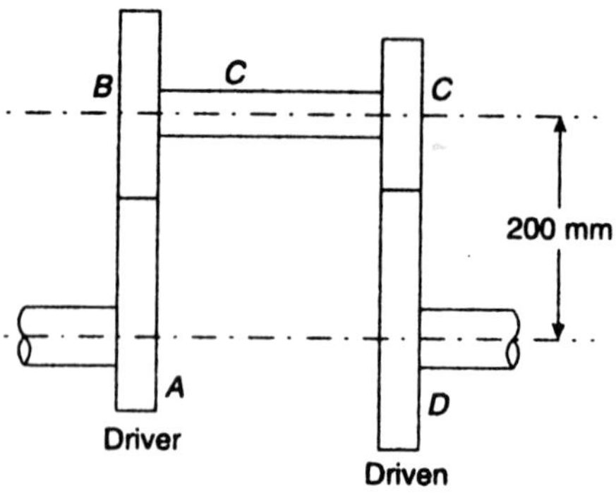

Fig. 10.8

Solution : Given : $m_a = m_b = 3.125$ mm, $m_c = m_d = 2.5$ mm, $z_{min} = 24$

$$3.125\,(z_a + z_b) = 2.5(z_c + z_d) \quad ...(1)$$

$$(z_c + z_d) = 1.25(z_a + z_b)$$

$$3.125\,(z_a + z_b)/2 = 200$$

$$z_a + z_b = 128 \quad ...(2)$$

Let $z_a = 24$, so that $z_b = 104$

Eq. (1) becomes $z_c + z_d = \dfrac{200}{1.25} = 160 \quad ...(3)$

$$\frac{n_a}{n_d} = 12 = \frac{z_d}{z_c} \times \frac{z_b}{z_a}$$

$$\frac{z_d}{z_c} \times \frac{104}{24} = 12$$

$$z_d = 2.77$$

From Eq. (3), we get $z_c = 42.4 \cong 42$ and $z_d = 118$

$$\frac{n_a}{n_d} = \left(\frac{118}{2}\right) \times \left(\frac{104}{24}\right) = 12.17 \cong 12$$

■ **Example 10.3** Two shafts A and D are in the same line (axes in one line). They are geared together through an intermediate parallel shaft carrying wheels B and C which mesh with the wheels on A and B respectively, as shown in Fig. 10.9. Wheels A and B have a module of 4 mm and the wheels C and D have a module of 9 mm. The number of teeth on any wheel is

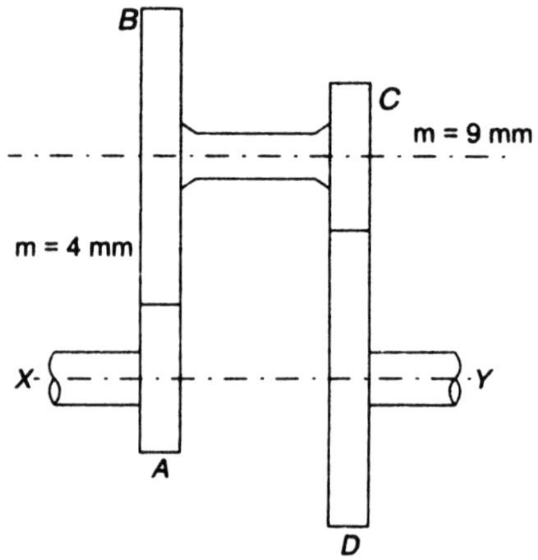

Fig. 10.9

not to be less than 15 and the speed of D is to be about, but not greater than 1/2 the speed of A and the ratio of each reduction is the same. Find (a) suitable number of teeth for the wheels, (b) the actual reduction, and (c) the distance of the intermediate shaft from the axes of the shafts A and D (centre distance), (d) indicate the configuration with a sketch, (e) How is addendum modification related to correction of gears and when are they used in practice? (Gear A is on shaft A and gear D on shaft D)

Solution : Given : $m_a = m_b = 4\text{ mm}, m_c = m_d = 9\text{ mm}, z_{min} = 15, \dfrac{n_a}{n_d} = 12$

Now $$\frac{n_a}{n_b} = \frac{n_c}{n_d}$$

But $$n_b = n_c$$

Therefore, $$n_b^2 = n_c^2 = n_a n_d \text{ or } n_b = n_c = \sqrt{n_a n_d} = \sqrt{12}\, n_d$$

$$\frac{n_c}{n_d} = \frac{n_b}{n_d} = \sqrt{12} = 3.464 \simeq 3.5$$

For same centre distance,

$$4(z_a + z_b) = 9(z_c + z_d)$$
$$(z_a + z_b) = 2.25(z_c + z_d)$$
$$z_a(1 + 3.5) = 2.25 z_c(1 + 3.5)$$
$$z_a = 2.25 z_c$$

Let $z_c = 1.5$, then $$z_a = 33.75 \cong 34$$

Now $\dfrac{z_d}{z_c} = 3.5$, $$z_d = 51.96 \cong 52$$

$$34 + z_b = 2.25(15 + 52)$$
$$z_b = 116.75 \simeq 117$$
$$\frac{n_A}{n_B} = (3.464)^2 = 12$$
$$c = 4\,(34 + 117) = 604\text{ mm}$$

■ **Example 10.4** In an epicylic gear train, as shown in Fig. 10.10, the number of teeth on wheels A, B and C are 50, 25 and 52 respectively. If the arm rotates at 420 rpm cw, find (a) speed of wheel C when A is fixed, and (b) speed of wheel A when C is fixed.

Solution : Given : $z_a = 50, z_b = 25, z_c = 52$

Operation	*Arm*	*Gear A, 50*	*Gear B, 25*	*Gear C, 52*
1. Arm fixed, + 1 revolution given to gear A, ccw	0	+ 1	$-\dfrac{za}{zb} = -\dfrac{50}{25} = -2$	$+\dfrac{za}{zc} = \dfrac{50}{52}$
2. Multiply by x	0	$+x$	$-2x$	$+\dfrac{50x}{52}$
3. Add y	y	$y + x$	$y - 2x$	$y + \dfrac{50x}{52}$

(a) When gear A is fixed, $x + y = 0$

Now $$y = 420\text{ rpm}$$

Therefore, $$x = -420\text{ rpm}$$

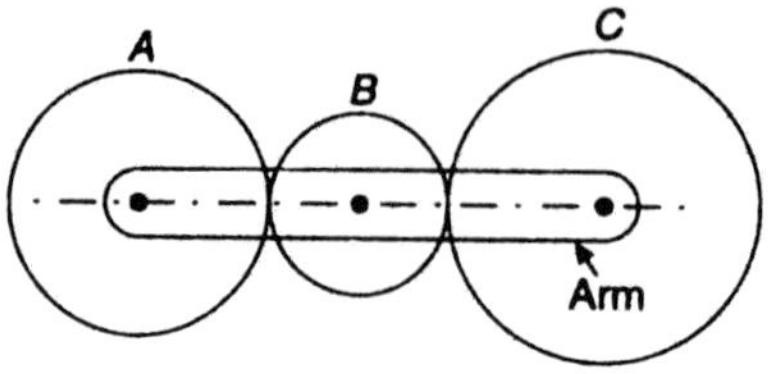

Fig. 10.10

$$n_c = \frac{50x}{52} = 420 - 420 \times \frac{50}{52} = 16.15 \text{ rpm}$$

(b) When C is fixed

$$n_c = y + x\frac{z_a}{z_c} = 0$$

or
$$420 + \frac{50x}{52} = 0$$

or
$$x = -436.8 \text{ rpm}$$

$$n_a = x = y = -436.8 + 420 = -16.8 \text{ rpm.}$$

■ **Example 10.5** An epicyclic gear train, as shown in Fig. 10.11, is composed of a fixed annular wheel A having 150 teeth. The wheel A is meshing with wheel B which drives wheel D through an idle wheel C, D being concentric with A. The wheels B and C are carried on an arm which revolves clockwise at 100 rpm about the axis of A and D. If the wheels B and D have 25 teeth and 40 teeth respectively, find the number of teeth on C and the speed and sense of rotation of C.

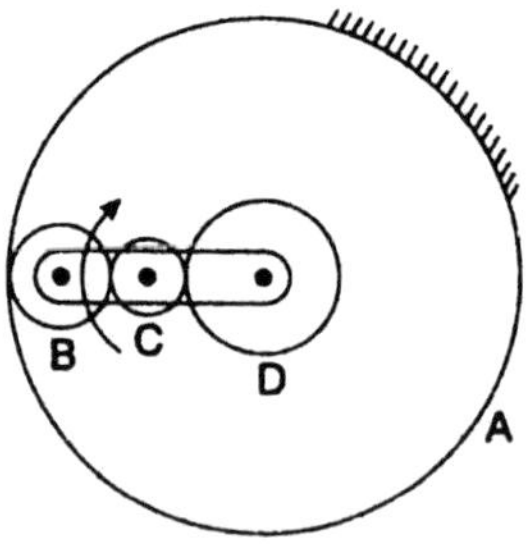

Fig. 10.11

Solution : Given $z_b = 25, z_b = 40$

Operation	Arm	Gear A, 150	Gear B, 25	Gear C	Gear D, 40
1. Arm fixed + 1 revolution given to gear A, ccw	0	+1	$+\frac{z_a}{z_b} = \frac{150}{25} = +6$	$-\frac{z_a}{z_c} = -\frac{150}{z_c}$	$+\frac{z_a}{z_d} = \frac{150}{40} = \frac{+15}{4}$
2. Multiply by x	0	$+x$	$+6x$	$\frac{-150x}{c}$	$\frac{-15x}{4}$
3. Add y	y	$y+x$	$y+6x$	$y - \frac{150x}{z_c}$	$y + \frac{15x}{4}$

For A to be fixed, $x + y = 0, y = -100$ rpm, $x = +100$ rpm

Now $$\frac{d_a}{2} = d_b + d_c + \frac{d_d}{2}$$

For same module, $$\frac{z_a}{2} = z_b = z_c + \frac{z_d}{2}$$

$$\frac{150}{2} = 25 + z_c + \frac{40}{2}$$

$$z_c = 30$$

$$n_c = -100 - 150 \times \frac{100}{30} = -600 \text{ rpm},$$

i.e., 600 rpm cw.

■ **Example 10.6** An epicyclic gear train, as shown in Fig. 10.12, has a sun wheel S of 30 teeth and two planet wheels P of teeth 50 each. The planet wheels mesh with the teeth of internal gear A. The driving shaft carrying the sun wheel transmits 6 kW at 300 rpm. The driven shaft is connected to an arm which carries the planet wheels. Determine the speed of the driven shaft and the torque transmitted, if the overall efficiency is 95%.

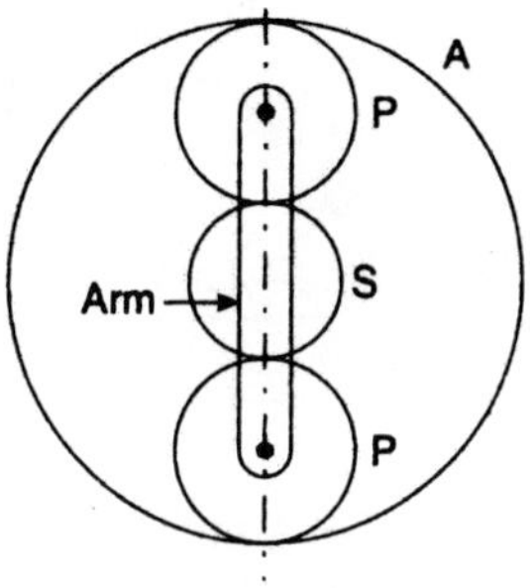

Fig. 10.12

Solution : Given : $z_s = 30, z_p = 50$, power transmitted = 6 kW, $N = 300$ rpm, efficiency = 0.95

Operation	*Arm*	*Sungear S, 30*	*Planet P, 50*	*Annular A*
1. Arm fixed + 1 revolution given to S, ccw	0	+ 1	$-\frac{z_s}{z_p} = \frac{-30}{50} = \frac{-3}{5}$	$\frac{-z_s}{z_a}$
2. Multiply by x	0	$+x$	$\frac{-3x}{5}$	$-x\frac{z_s}{z_a}$
3. Add y	y	$y + x$	$y - \frac{3x}{5}$	$y - x\frac{z_s}{z_a}$

$$n_s = x + y = 300 \quad \text{...(1)}$$

Now $$d_a = 2d_p + d_s$$

For same module, $$z_a = 2z_p + z_s = 100 + 30 = 130$$

For to be fixed, $$\frac{-zx_s}{z_a} + y = 0$$

$$= \frac{-3x}{13} + y = 0 \quad \text{...(2)}$$

From (1) and (2), we get

$$x = \frac{975}{4}, y = \frac{225}{4}$$

Speed of driven shaft (arm) = 56.25 rpm

$$T_s = 6 \times 10^3 \times \frac{50}{(2\pi \times 300)} = 190.986 \text{ N-m}$$

$$T_a = -T_s n_s \frac{\eta}{n_a} = -190.986 \times 300 \frac{0.95}{56.25} = -967.7 \text{ N-m}$$

■ **Example 10.7** A compound epicylic gear train is shown in Fig. 10.13. The gears *A, D* and *E* are free to rotate on the axis *P*. The compound gear *B* and *C* rotate together on the axis *Q* at the end of arm *F*. All the gears have equal module. The number of teeth on the gears *A, B* and *C* are 18, 45 and 21 respectively. The gears *D* and *E* are annual gears. The gear *A* rotates at 120 rpm ccw and the gear *D* rotates at 450 rpm cw. Find the speed and direction of the arm and the gear *E*.

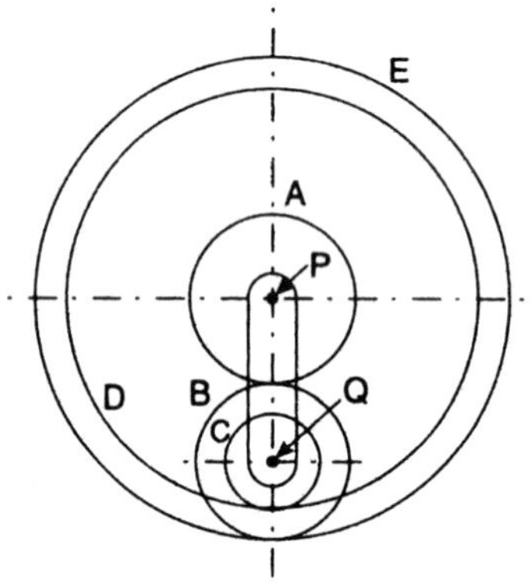

Fig. 10.13

Solution : Given : $z_a = 18, z_b = 45, z_c = 21, n_a = 120$ rpm ccw, $n_d = 450$ rpm cw

$$\frac{z_e}{2} = \frac{z_a}{2} + z_b$$

$$z_e = z_a + 2z_b = 18 + 90 = 108$$

$$z_d = z_a + z_b + z_c = 18 + 45 + 21 = 84$$

Operation	Arm	Gear A, 18	Gear B, C 45, 21	Gear D, 84	Gear E, 108
1. Arm fixed +1 revolution given to gear *A*. ccw	0	+1	$+\frac{z_a}{z_b} = \frac{-18}{45} = \frac{-2}{5}$	$\frac{-z_a}{z_b} \times \frac{z_c}{z_d} = \frac{-18}{45} \times \frac{21}{84} = \frac{-1}{10}$	$\frac{-z_a}{z_b} \times \frac{z_b}{z_e} = \frac{-18}{45} \times \frac{45}{108} = \frac{-1}{6}$
2. Multiply by *x*	0	$+x$	$\frac{-2x}{5}$	$\frac{-x}{10}$	$\frac{-x}{6}$
3. Add *y*	y	$y + x$	$y - \frac{2x}{5}$	$y - \frac{x}{10}$	$y - \frac{x}{6}$

$$x + y = 120$$

$$\frac{-x}{10} + y = -450$$

$$x = 518.2 \text{ rpm}, \; y = -395.2 \text{ rpm}$$

$$n_f = y - 398.2 \text{ rpm, } i.e., \text{ cw}$$

$$n_e = y - \frac{x}{6} = -398.2 - \frac{518.2}{6} = -484.57 \text{ rpm, } i.e., \text{ cw}$$

■ **Example 10.8** An epicyclic gear train shown in Fi.g 10.14, consists of two sun wheels *A* and *D* with 28 and 24 teeth respectively, engaged with a compound planet wheels *B* and *C* with 22 and 26 teeth. The sun wheel *D* is keyed to the driven shaft and the sun wheel *A* is a fixed wheel coaxial with the driven shaft. The planet wheels are carried on an arm *E* from the driving shaft which is coaxial with the driven shaft. Find the velocity ratio of gear train. If 1.2 kW is transmitted and input speed is 120 rpm, determine the torque required to hold the sun wheel *A*.

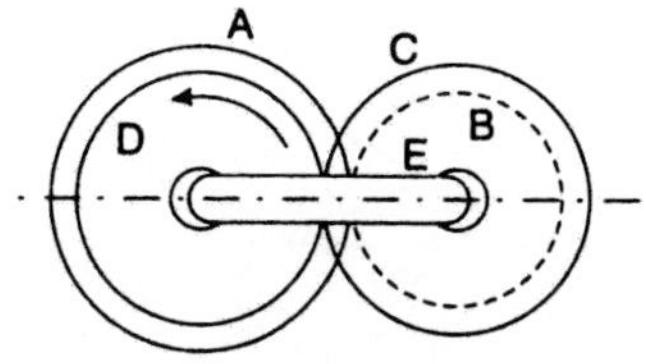

Fig. 10.14

Solution : Given : $z_a = 28, z_d = 24, z_b = 22, z_c = 26$, power transmitted = 1.2 kW, $N = 120$ rpm

Operation	Arm, E	Gear A, 28	Gear B, C, 22, 26	Gear D, 24
1. Arm fixed + 1 revolution given to gear *A*, ccw	0	+1	$-\frac{z_a}{z_b} = \frac{-28}{22} = \frac{-14}{11}$	$+\frac{z_a}{z_b} \times \frac{z_c}{z_d} = \frac{28}{22} \times \frac{26}{24} = \frac{91x}{66}$
2. Multiply by *x*	0	$+x$	$-\frac{14x}{11}$	$\frac{91x}{66}$
3. Add *y*	y	$y + x$	$y - \frac{14x}{11}$	$y + \frac{91x}{66}$

$$y = +120 \text{ rpm}$$

$$n_d = y + \frac{91x}{66} = 120 + \frac{91x}{66} = 0$$

$$x = -87 \text{ rpm}$$

$$n_a = x + y = -87 = 120 = 33 \text{ rpm}$$

Speed of driven shaft (gear A) = 33 rpm cw

$$\text{Speed ratio} = \frac{n_d}{n_a} = \frac{87}{33} = 2.637$$

$$M_1 = \frac{1.2 \times 10^3 \times 60}{2\pi \times 12)} = +95.5 \text{ N-m}$$

$$\omega_1 = \frac{2\pi \times 120}{60} = 12.57 \text{ rad/s} \quad ,$$

$$\omega_2 = \frac{2\pi \times 33}{60} = 3.456 \text{ rad/s}$$

$$M_2 = M_1 \times \frac{\omega_1}{\omega_2} = 95.5 \times \frac{12.57}{3.456} = 347.35 \text{ N-m}$$

Holding torque, $\quad M_3 = -M_1 - M_2 = -95.5 - 347.235 = -442.85\text{N-m}$

■ **Example 10.9** An epicyclic gear train is shown in Fig. 10.15. The main driving shaft *G* has a gear S_1 integrally mounted and driving the internal gear A_1 on the casing through two intermediate gears P_1 mounted on either side. The gears *P*, are free to revolve on arms *R*, which are integral with gear S_2 which in turn drives the internal gear A_2 on another casing through two gears P_2. The driven shaft H is integral with the casing carrying the internal gear A_1 and arms R_2 on which the gears P_2 are free to rotate. The casing A_2 and gear S_2 are free to rotate on shaft. *G*.

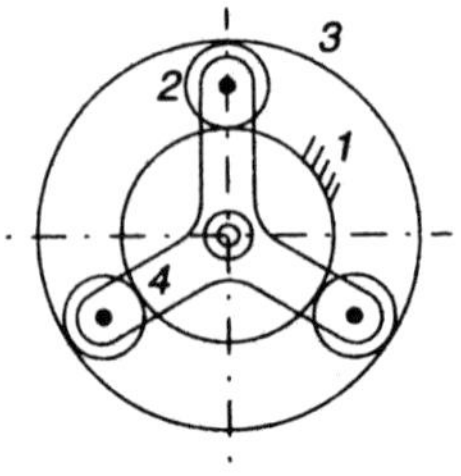

Fig. 10.15

Calculate the speed of shaft *H* when *G* rotates at 1000 rpm anti-clockwise when (a) A_2 is stationary; (b) A_2 rotates at 500 rpm clockwise.
The number of teeth on gears are : $S_1 = S_2 = 30, A_1 = A_2 = 90.$

Solution :

Operation	Arm, R	S_1, 30	P_1	A_1, 90	S_2, 30	P_2	A_2, 90
1. Arm R_1 fixed +1 revolution given to S_1, ccw	0	+1	$\frac{-30}{z_{p1}}$	$\frac{-30}{90} = \frac{-1}{3}$	$\frac{-1}{3}$	$\frac{-1}{3} \times \frac{30}{z_{p2}} = \frac{-10}{z_{p2}}$	$\frac{1}{3} \times \frac{1}{3} = \frac{1}{9}$
2. Multiply by *x*	0	+*x*	$\frac{-30x}{z_{p1}}$	$\frac{-x}{3}$	$\frac{-x}{3}$	$\frac{-10x}{z_{p2}}$	$\frac{x}{9}$
3. Add *y*	*y*	*y* + *x*	$y - \frac{30x}{z_{p1}}$	$y - \frac{x}{3}$	$y - \frac{x}{3}$	$y - \frac{10x}{z_{p2}}$	$y + \frac{x}{9}$

(a) A_2 fixed :

$$\frac{x}{9} + y = 0$$

$$x + y = 1000$$

$$y = -125$$

Speed of shaft $H = y = 125$ rpm cw

(b)
$$\frac{x}{9} + y = -1500$$
$$\frac{x}{9} + y = -500$$
$$x + y = 1000$$
$$x = \mathbf{1687.5\ rpm},\ y = -\mathbf{687.5\ rpm}$$

Speed of shaft $H = y =$ **687.5 rpm cw.**

■ **Example 10.10** Fig. 10.16 shows diagramatically a compound epicyclic gear train. Wheels *A*, *D* and *E* are free to rotate independently on spindle *C*, while *B* and *C* are compound and rotate together on spindle *P*, on the end of arm *OP*. All the teeth on different wheels have the same module. *A* has 12 teeth, *B* has 30 teeth and *C* has 14 teeth cut externally. Find the number of teeth on wheels *D* and *E* which are cut internally.

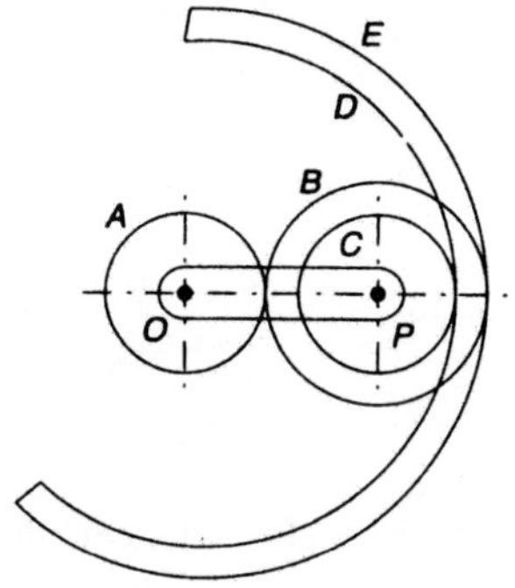

Fig. 10.16

If the wheel *A* is driven clockwise at 1 rps while *D* is driven clockwise at 5 rps, determine the magnitude and direction of the angular velocities of arm *OP* and wheel *E*.

Solution : Given : $z_a = 12, z_b = 30, z_c = 14, n_a = 1, n_d = 5$

$$z_e = z_a + 2z_b = 12 + 60 = 72$$
$$z_d = z_a + z_b + z_c = 12 + 30 + 14 = 56$$

Operation	Arm, OP	Gear A, 12	Gear B, C 30, 14	Gear D, 56	Gear E, 72
1. Arm fixed +1 revolution given to A, ccw	0	+1	$\frac{-12}{30} = \frac{2}{5}$	$\frac{-2}{30} \times \frac{14}{56} = \frac{-1}{10}$	$\frac{-24}{30} \times \frac{30}{72} = \frac{-1}{3}$
2. Multiply by x	0	$+x$	$\frac{-2x}{5}$	$\frac{-x}{10}$	$\frac{-x}{3}$
3. Add y	y	$y + x$	$y - \frac{2x}{5}$	$y - \frac{x}{10}$	$y - \frac{x}{3}$

$$x + y = -1$$
$$\frac{-x}{10} + y = -5$$
$$x = \frac{40}{11},\ y = -4.636\ \text{rpm}$$

Angular velocity of $OP = -2\pi \times \dfrac{4.636}{60} = -0.486$ rad/s cw

Angular velocity of wheel E:

$$y - \frac{x}{3} = 4.636 - \frac{3.636}{3} = -5.848 \text{ rpm}$$

$$\omega_e = 2\pi \times \frac{5.848}{60} = 0.612 \text{ rad/s cw}$$

■ **Example 10.11** A mechanism for recording the distance covered by the bicycle, as shown in Fig. 10.17, is as follows :

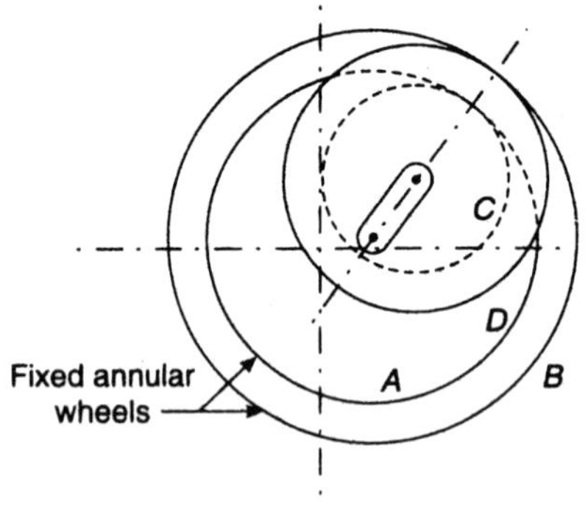

Fig. 10.17

There is a fixed annular wheel A of 22 teeth and another annular wheel B of 23 teeth, which rotates loosely on the axis of A. An arm driven by the bicycle wheel through gearing not described, also revolves freely on the axis of A and carries on a pin at its extremity two wheels C and D, which are integral with one another. The wheel C has 19 teeth and meshes with A and the wheel D with 20 teeth meshes with B. The diameter of the bicycle wheel is 0.7 m. What must be the velocity ratio between the bicycle wheel and the arm, if B makes one revolution per 1.5 km covered?

Solution : Given : $z_a = 22, z_b = 23, z_c = 19, z_d = 20$

Operation	*Arm, E*	*Gear A, 22*	*Gear C, D, 19, 20*	*Gear B, 23*
1. Arm fixed + 1 revolution given to gear A, ccw	0	+ 1	$-\dfrac{z_a}{z_c} = \dfrac{+22}{19}$	$\dfrac{+z_a}{z_c} \times \dfrac{z_d}{z_b} = \dfrac{+22}{19} \times \dfrac{20}{23} = \dfrac{+440x}{437}$
2. Multiply by x	0	$+x$	$\dfrac{+22x}{19}$	$\dfrac{+440x}{437}$
3. Add y	y	$y + x$	$y + \dfrac{22x}{19}$	$y + \dfrac{440x}{437}$

A fixed :

$$x + y = 0$$

$$\frac{440x}{437} + y = 1$$

$$x = \frac{437}{3} = 145.67 \text{ revolutions for every 1.5 km}$$

Revolutions made by bicycle wheel per 1.5 km

$$= 1000 \times \frac{1.5}{(\pi \times 1.5)} = 682$$

Velocity ratio of gearing

$$= \frac{\text{Total velocity rtio}}{\text{Velocity ratio provided by mechanism}}$$

$$= \frac{682}{145.67 \times 1} = 4.682$$

■ **Example 10.12** In an epicyclic gear train of the sun and planet type shown in Fig. 10.18, the pitch circle diameter of the annular wheel *A* is to be nearly equal to 216 mm, and the module is 4 mm. When the annular wheel is stationary, the spider which carries three planet gears *P* of equal size, has to make one revolution for every five revolutions of the driving spindle carrying *S* gear. Determine the number of teeth on all the wheels and also the exact pitch circle diameter of *A*.

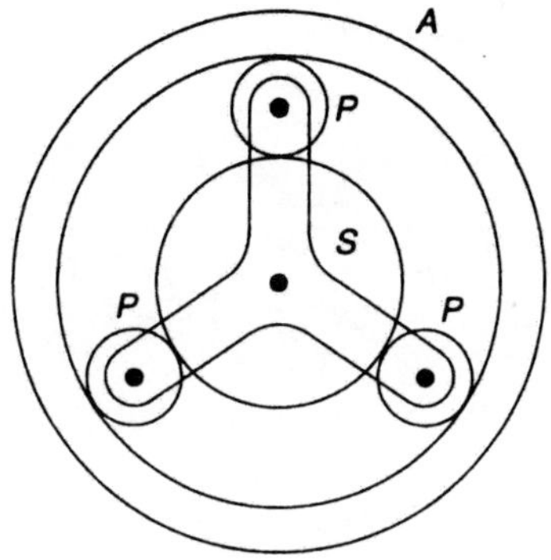

Fig. 10.18

Solution :

$$d_a = d_s + 2d$$

$$216 = 4(z_s + 2z_p)$$

$$z_s + 2z_p = 54$$

$$d_a = mz_a, z_a = \frac{216}{4} = 54 \qquad \text{...(1)}$$

Operation	*Spider*	*S*	*P*	*A*
1. Spider fixed +1 rev to *S*, ccw	0	+1	$\frac{-z_s}{z_p}$	$\frac{-z_s}{z_a}$
2. Multiply by *x*	0	+*x*	$\frac{-xz_s}{z_p}$	$\frac{-xz_s}{za}$
3. Add *y*	*y*	*y* + *x*	$y - \frac{xz_s}{z_p}$	$y - \frac{xz_s}{z_a}$

$$x + y = 5$$

$$y - \frac{xz}{z_p} = -1$$

Solving, we get

$$x\left(1+\frac{z_s}{z_p}\right)=6 \qquad \ldots(2)$$

Also $$y-\frac{xz_s}{z_a}=0$$

Thus $$x\left(1+\frac{z_s}{z_a}\right)=5 \qquad \ldots(3)$$

Solving Eqs. (1), (2) and (3), we get

$$z_s = 10, z_p = 22, d_a \; 216 \text{ mm.}$$

■ **Example 10.13** An epicyclic gear train as shown in Fig. 10.19, has a sun wheel S of 30 teeth and two planet wheels $P - P$ of 45 teeth. The planet wheels mesh with the internal teeth of a fixed annulus A. The driving shaft carrying the sun wheel transmits 4 kW at 360 rpm. The driven shaft is connected to an arm, which carries the planet wheels. Determine the speed of the driven shaft and the torque transmitted, if the overall efficiency is 95%.

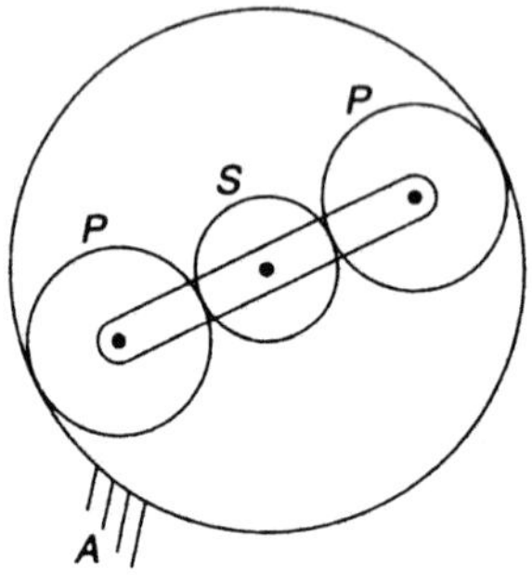

Fig. 10.19

Solution : Given : $z_s = 30, z_p = 45$

$$d_a = d_s + 2d_p$$
$$z_a = z_s + 2z_p$$
$$z_a = 30 + 90 = 120$$

Operation	Arm	S	P	A
1. Arm fixed +1 revolution given to S, ccw	0	$+1 = -\frac{2}{3}$	$\frac{-z_s}{z_p} = -\frac{30}{45} = -\frac{1}{4}$	$\frac{-z_s}{z_a} = \frac{-30}{120}$
2. Multiply by x	0	$+x$	$-\frac{2x}{3}$	$\frac{-x}{4}$
3. Add y	y	$y + x$	$y - \frac{2x}{3}$	$y - \frac{x}{4}$

A fixed : $$y-\frac{x}{4}=0$$

$$y + x = 360$$

$$x = 288 \text{ rpm}, \; y = 72 \text{ rpm}$$

Speed of driven shaft = 72 rpm

Input torque, $$T_1 = \frac{4 \times 10^3 \times 60}{2\pi \times 360} = 106.1 \text{ N-m}$$

$$T_1 n_1 + T_2 n_2 = 0$$

$$T_2 = -106.1 \times \frac{360}{72} = -530.5 \text{ N-m}$$

■ **Example 10.14** The diagram of Fig. 10.20 shows a single epicyclic gear. The wheel A is stationary. What is the speed of C when B rotates about its own axis at1800 rpm? The number of teeth on A and B are 120 and 145 respectively.

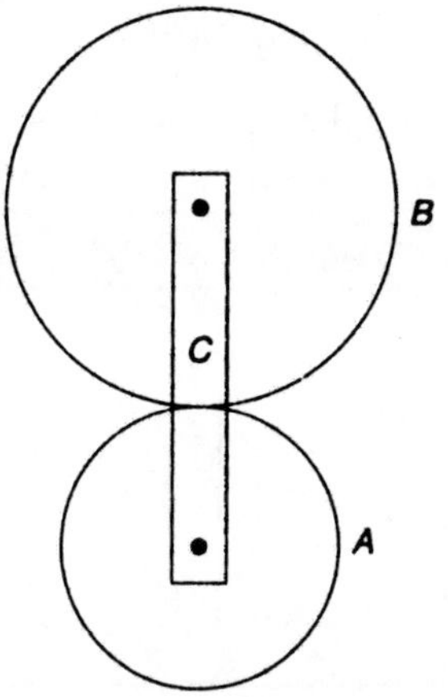

Fig. 10.20

Solution : Given : $n_b = 1800$ rpm, $z_a = 120, z_b = 145, n_c = ?$

Operation	Arm C	A	B
1. Arm fixed + 1 revolution given to S, ccw	0	+1	$-\frac{z_a}{z_b}$
2. Multiply by x	0	$+x$	$-x\frac{z_a}{z_b}$
3. Add y	y	$y+x$	$y-x\frac{z_a}{z_b}$

Now $$x + y = 0$$

$$y = -x$$

$$y - x\frac{z_a}{z_b} = 1800$$

$$-x - x\frac{120}{145} = 1800$$

$$-x\left(1 + \frac{120}{145}\right) = 1800$$

$$x = -984.9$$

$$n_c = y = -x = 984.9 \text{ rpm cw}$$

Speed of arm, $$C = 984.9 \text{ rpm.}$$

■ **Example 10.15** Fig. 10.21 shows an epicyclic gear train arrangement. Wheel E is fixed and gears C and D are integrally cost and mounted on openin. If the arm 'A' makes one revolution/sec counter clockwish, determine the speed and the direction of rotation of wheels B and F.

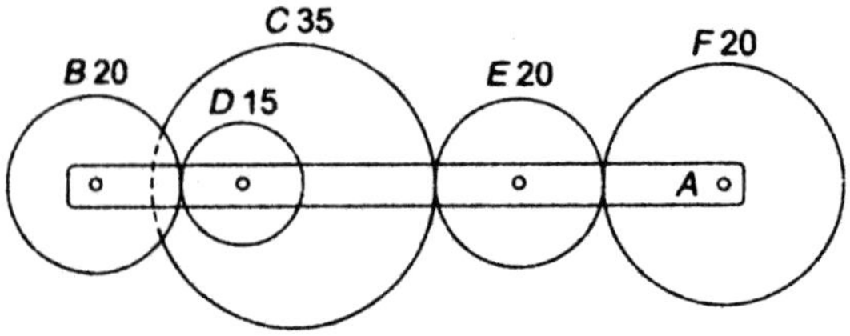

Fig. 10.21

Solution : Given : $z_b = 20, z_c = 35, z_d = 15, z_e = 20, z_f = 30$

$$n_a = 1 \text{ rps, } ccw, n_e = 0$$

Operation	*Arm*	**B**	**C, D**	**E**	**F**
1. Arm fixed + 1 revolution to B, cw	0	+1	$-\frac{z_b}{z_d}$	$+\frac{z_c}{z_e}$	$-\frac{z_e}{z_f}$
2. Multiply by x	0	$+x$	$-x\frac{z_b}{z_d}$	$x\frac{z_c}{z_e}$	$-x\frac{z_e}{z_f}$
3. Add y	y	$y+x$	$y-x\frac{z_b}{z_d}$	$y+x\frac{z_c}{z_e}$	$y-x\frac{z_c}{z_f}$

$$y = -1$$

$$y + x\frac{z_c}{z_e} = 0$$

$$-1 + x \times \frac{35}{20} = 0$$

$$x + \frac{4}{7}$$

$$n_b = \frac{4}{7} - 1 = \frac{-3}{7} - 0.428 \text{ or } 0.428 \text{ rps ccw}$$

$$n_f = -1 - \frac{4}{7} \times \frac{20}{30} = -0.952 \text{ or } 0.952 \text{ ccw}$$

■ **Example 10.16** In a reverted epicyclic gear train, as shown in Fig. 10.22, the arm A carries two gears B and C and compound gear D-E. The gear B meshes with gear D. The number of teeth on gears B, C and D are 75, 30 and 90 respectively. Find the speed and direction of gear C when gear B is fixed and the arm A makes 100 rpm clockwise.

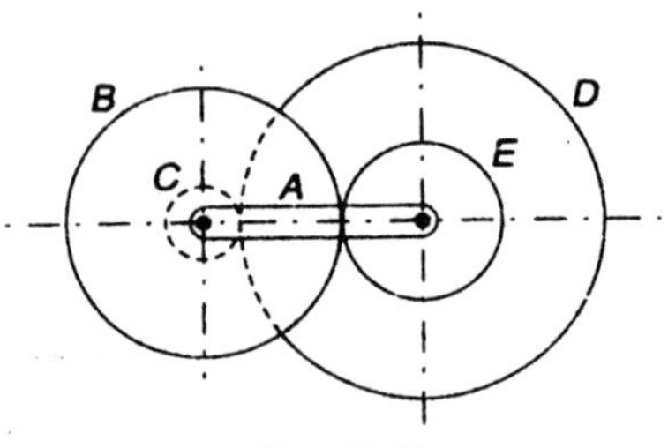

Fig. 10.22

Solution : Given : $z_b = 75, z_c = 30, z_d = 90$

$n_a = 100$ rpm cw, B is fixed

$$d_b + d_e = d_c + d_d$$

or
$$z_b + z_e = z_c + z_d$$

$$75 + z_e = 30 + 90$$

$$z_e = 45$$

Operation	*Arm*	*B*	*D, E*	*C*
1. Arm A fixed + 1 revolution given to A, cw	0	+ 1	$-\frac{z_b}{z_e}$	$+\frac{z_b}{z_e} \times \frac{z_d}{z_c}$
2. Multiply by x	0	x	$-x\frac{z_b}{z_3}$	$+x\frac{d_b}{z_e} \times \frac{z_d}{z_c}$
3. Add y	y	$y + x$	$y + x\frac{z_b}{z_e}$	$y + x\frac{z_b}{z_e} \times \frac{z_d}{z_c}$

$$n_a = 100 \text{ rpm}, \; y = 100 \text{ rpm}$$

B is fixed, $x + y = 0$

$$x = -100$$

$$n_c = y + x\frac{z_b}{z_e} \times \frac{z_d}{z_c} = 100 - 100 \times \frac{75}{45} \times \frac{90}{30}$$

$$= -400 \text{ rpm or } 400 \text{ rpm ccw}$$

■ **Example 10.17** Fig. 10.23 shows the gear train arrangement to drive a planning machine. The driving pulley is driven at 360 rpm by an electric motor keyed to the same shaft at this pulley is a 25 tooth pinion driving an 85 tooth gear wheel which is fixed to the same shaft 35 teeth pinion. This pinion drives a gear wheel that has 130 teeth, fixed to the same shaft as this gear wheel is a 30 teeth pinion which drives a bull wheel. This bull wheel has 75 teeth and meshes with a rack that is bolted to the underside of the table. The pitch of the rack teeth is 25 mm. Calculate the bull wheel speed and the table speed.**

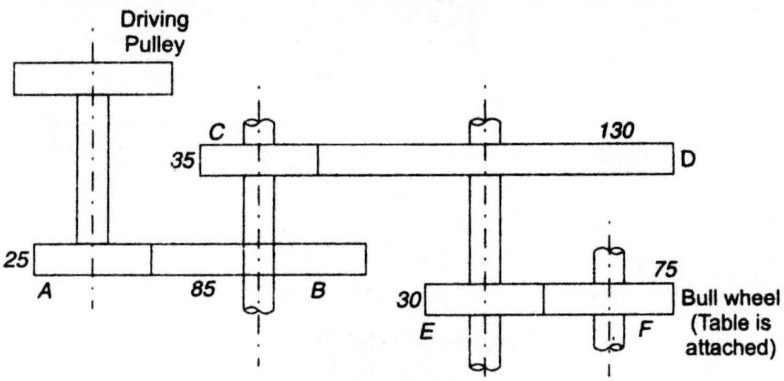

Fig. 10.23

Solution : $n_a = 360$ rpm, $z_a = 25$, $z_b = 85$, $z_c = 3$, $z_d = 130$, $z_e = 30$, $z_f = 75$, pitch of rack, $p = 25$ mm, $n_f = ?$

$$\text{Speed ratio} = \frac{\text{product of number of teeth of driving tears}}{\text{product of number of teeth of driven gears}}$$

$$= \frac{z_a \times z_c \times z_e}{z_b \times z_b \times z_f} = \frac{25 \times 35 \times 30}{85 \times 130 \times 75} = \frac{1}{31.57}$$

$$\text{Speed of bull wheel, } n_f = \frac{360}{31.75} = 11.4 \text{ rpm}$$

$$\text{Table speed} = pn_f = \frac{25 \times 11.4}{60} = 4.75 \text{ mm/s.}$$

SOLVED QUESTIONS FROM UNIVERSITY EXAMINATIONS

■ **Example 10.18** With the help of sketches, discuss the following types of gear trains and their application access as well as advantages :

(a) Simple gear trains, (b) Compound gear trains, and (c) Reverted gear trains.

[MDU, Dec., 2007]

Solution : Refer to art. 10.2.

■ **Example 10.19** Fig. 10.24 represents an epicyclic gear train. It has the following teeth on different gears :

Gear	*Number of teeth*
S	40
P	20
A	80

If sum gear *S* rotates at 150 rpm counter-clockwise and the annular (ring) clockwise at 400 rpm, find the arm speed.

[MDU, Dec., 2007]

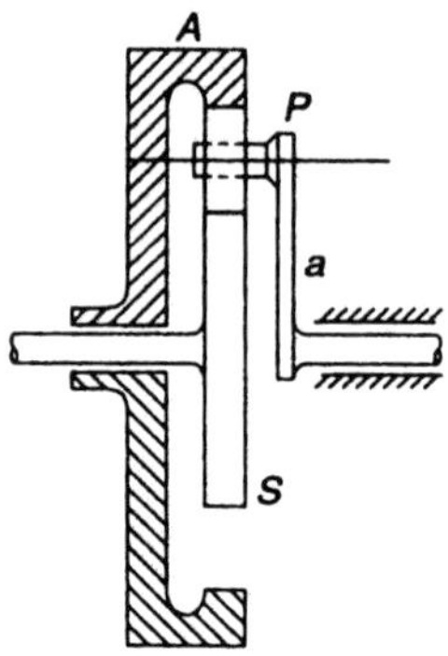

Fig. 10.24

Solution : Given $z_s = 40, z_p = 20, z_a = 80, N_s = 150$ rpm (ccw),

$$N_a = 400 \text{ rpm (cw)}$$

$$d_a = d_s + 2d_p$$

For same module,

$$z_a + z_s + 2z_p = 40 \times 2 \times 20 = 80 \text{ (given)}$$

Operation	Revolution of			
	Arm a	S	P	A
1. Arm fixed + 1 revolution given (ccw)	0	1	$-z_x/z_p$	$-\frac{z_s}{z_p}\times\frac{z_p}{z_a}=-\frac{z_s}{z_a}$
2. Multiply by x	0	$+x$	$-x\,z_s/z_p$	$-x\,z_s/z_a$
3. Add y	y	$x+y$	$y-x\,z_s/z_p$	$y-x\,z_s/z_a$

$$x + y = N_s = 150 \quad ...(1)$$

$$y - x \times \frac{40}{80} = N_a = -400$$

or

$$y - 0.5\,x = -400 \quad ...(2)$$

Eqs. (1), (2) gives,

$$1.5x = 550,\ x = 366.67 \text{ rpm}$$

$$y = 150 - 366.67 = -216.67\text{rpm}$$

or

$$216.67 \text{ rpm (cw)}$$

Arm speed = 216.67 rpm (cw)

■ **Example 10.20** What do you mean by gear train? Mention various types of gear trains. [MDU, May, 2008]

Solution : Refer to art. 10.2 and 10.2.

■ **Example 10.21** Name the different methods of obtaining the velocity ratio of epicyclic gear trains. Explain one method in detail. [MDU, May, 2009]

Solution : Refer to art.10.3.

■ **Example 10.22** In Fig. 10.25, an epicyclic gear train annular wheel A having 54 teeth meshes with a planet wheel B which gears with a sum wheel C, the wheels A and C being coadial. The wheel B is carried on a pin fixed on one end of arm.P wheel rotates about the axis of the wheels A and C. If the wheel A makes 20 rpm in a clockwise sense and the arm rotates at 100 rpm in the anti-clockwise direction and the wheel C has 24 teeth, determine rpm and sense of rotation of C. [MDU, Dec., 2005]

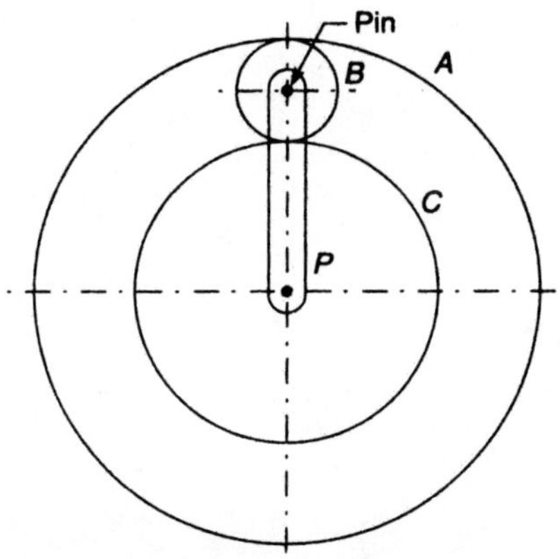

Fig. 10.25

Solution : Given : $z_A = 54, N_A = 20$ rpm (cw), $N_p = 100$ rpm (ccw), $z_C = 24, N_C = ?$

$$d_A = d_C + 2d_B$$

or
$$z_Z = z_C + 2z_B \text{ for same module.}$$

$$54 = 24 + 2z_B$$

$$z_B = \frac{30}{2} = 15$$

Operation	Revolution of			
	Arm P	Sun gear C	Planes gear B	Annular wheel A
1. Arm fixed + 1 revolution given to C (ccw)	0	+1	$-z_C/z_B$	$-\frac{z_C}{z_B} \times \frac{z_B}{z_A} = -\frac{z_C}{z_A}$
2. Multiply by x	0	x	$-x\frac{z_C}{z_B}$	$-x\frac{z_C}{z_A}$
3. Add y	y	$y + x$	$y - x\frac{z_C}{z_B}$	$y - x\frac{z_C}{z_A}$

$$y = +100 \qquad ...(1)$$

$$y - x\frac{24}{54} = -20$$

or
$$y - \frac{4}{9}x = -20 \qquad ...(2)$$

Eqs. (1) and (2) gives

$$\frac{4}{9}x = 120$$

$$x = 270 \text{ rpm}$$

$$y = -20 + \frac{4}{9} \times 270 = 100 \text{ rpm}$$

$$z_C = x + y = 270 + 100 = 370 \text{ rpm (ccw)}$$

■ **Example 10.23** Fig. 10.26 shows an epicyclic gear train. Pinion A has 15 teeth and is rigidly fixed to motor shaft. The wheel B has 20 teeth and gears with A and also with the annular fixed wheel D. Pinion C has 15 teeth and is integral with B (B, C being a compound gear

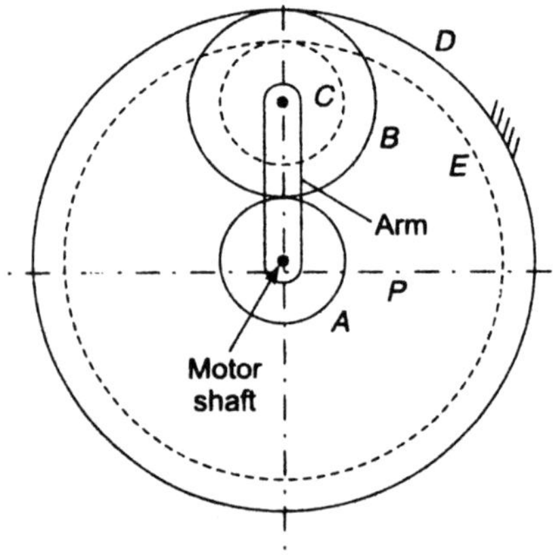

Fig. 10.26

wheel). Gear C meshes with annular wheel E, which is keyed to the machine shaft. The arm rotates about the same shaft on which A is fixed and carries the compound wheel B, C. If the motor runs at 1000 rpm, find the speed of the machine shaft. Find the torque exerted on the machine shaft if the motor develops a torque of 100 N-m. [MDU, May, 2009]

Solution : Given : $z_A = 15, z_B = 20, z_C = 15,\ N_m = N_A = 100 \text{ rpm},\ T_m = 100 \text{ N-m}$

$$d_D = d_A + 2d_B$$
$$d_E = d_D - (d_B - d_C)$$

For same module of all gears,

$$z_D = z_A + 2z_B = 15 + 2 \times 20 = 55$$
$$z_E = 55 - (20 - 15) = 50$$

Action	Revolution of				
	Arm A	Pinion A	Compound wheel B,C	Wheel E	Wheel D
1. Arm fixed + 1 revolution to A (ccw)	0	+1	$-\frac{z_A}{z_B}$	$-\frac{z_A}{z_B} \times \frac{z_C}{z_E}$	$-\frac{z_A}{z_B} \times \frac{z_B}{z_D} = -\frac{z_A}{z_D}$
2. Multiply by x	0	x	$-x\frac{z_A}{z_B}$	$-x\frac{z_A}{z_B} \times \frac{z_C}{z_E}$	$-x\frac{z_A}{z_D}$
3. Add y	y	$y + x$	$y - x\frac{z_A}{z_B}$	$y - x\frac{z_A z_C}{z_B z_E}$	$y - x\frac{z_A}{z_D}$

$$x + y = N_A = 1000 \qquad \ldots(1)$$

Wheel D is fixed,

$$y - x\frac{z_A}{z_D} = 0$$

$$y = x + \frac{15}{55} = \frac{3}{11}x \qquad \ldots(2)$$

$$x + \frac{3}{11}x = 1000$$

$$x + \frac{11}{14} \times 1000 = 785.71 \text{ rpm}$$

$$y = 1000 - 785.71 = 214.29 \text{ rpm}$$

Speed of machine shaft,

$$N_E = y - x\frac{z_A z_C}{z_B z_E}$$

$$= 214.29 - 785.7\left(\frac{15 \times 5}{20 \times 50}\right)$$

$$= 3.75 \text{ rpm (ccw)}$$

Assuming 100% efficiency,

$$T_E \cdot N_E + T_A \cdot N_A + T_D \cdot N_D = 0$$

or

$$T_E \cdot N_E + T_A \cdot N_A = 0 \qquad (\because N_D = 0)$$

$$T_E = -\frac{T_A N_A}{N_E} = -\frac{100 \times 1000}{37.5} = -2666.67 \text{ N-m}$$

■ **Example 10.24** In the epicyclic gear train shown in Fig. 10.27, the compound wheels A and B as well as internal wheels C and D rotate independently about the axis O. The wheels E and F rotate on the pins fixed to arm P. All the wheels are of same module. The number of teeth of each wheel are :

$$z_A = 52, z_B = 56, z_E = z_F = 36$$

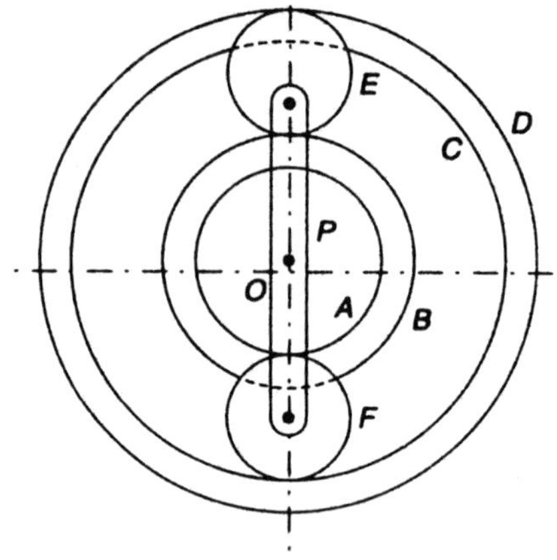

Fig. 10.27

Determine the speed of C if :

(i) The wheel D is fixed and arm P rotates at 200 rpm clockwise.

(ii) The wheel D rotates at 200 rpm counter-clockwise and rpm P rotates at 200 rpm counter-clockwise. [MDU, Dec., 2009]

Solution : Given : $z_A = 52, z_B = 56, z_E = z_F = 36, N_D = 200$ rpm, $N_P = 200$ rpm

$$d_D = d_B + 2d_E$$

or $$z_D = z_B + 2z_E = 56 + 2 \times 36 = 128$$

$$d_C = d_A + 2d_F$$

or $$z_C = z_A + 2z_F$$

$$52 + 2 \times 36 = 124$$

Action	Revolution of					
	Arm A	Wheels A, B	Wheels E	Wheels F	Wheel C	Wheel D
1. Arm P fixed, wheel A given +1 revolution (ccw)	0	+1	$-\frac{z_B}{z_E}$	$-\frac{z_A}{z_F}$	$-\frac{z_A}{z_F} \times \frac{z_F}{z_C} = -\frac{z_A}{z_C}$	$-\frac{z_B}{z_E} \times \frac{z_E}{z_D} = -\frac{z_B}{z_D}$
2. Multiply by x	0	x	$-x\frac{z_B}{z_E}$	$-x\frac{z_A}{z_F}$	$-x\frac{z_A}{z_C}$	$-x\frac{z_B}{z_D}$
3. Add y	y	$y + x$	$y - x\frac{z_B}{z_E}$	$y - x\frac{z_A}{z_F}$	$y - x\frac{z_A}{z_C}$	$y - x\frac{z_B}{z_D}$

(i) Wheel D is fixed,

$$\therefore \quad y - x\frac{z_B}{z_D} = 0$$

or $$y - x + \frac{56}{128} = 0$$

$$y - 0.4375x = 0 \quad \ldots(1)$$

$$y = N_P = -200 \quad \ldots(2)$$

$$\therefore \quad -200 - 0.4375x = 0$$

$$x = -457.14 \text{ rpm}$$

$$N_C = y - x\frac{z_A}{z_C} = -200 + 457.14 \times \frac{52}{124}$$

$$= -8.29 \text{ or } 8.29 \text{ rpm (cw)}$$

(ii) $N_D = 200$ rpm (ccw) and $N_p = 200$ rpm (ccw)

$$\therefore \quad y = 200\, y - x\frac{z_B}{z_D} = 200$$

$$200 - x \times \frac{56}{128} = 200$$

$$x = 0$$

$$N_C = y - x\frac{z_A}{z_C} = 200 - 0 = 200 \text{ rpm (ccw)}$$

■ **Example 10.25** A epicycligc gear-train is shown in Fig. 10.28.

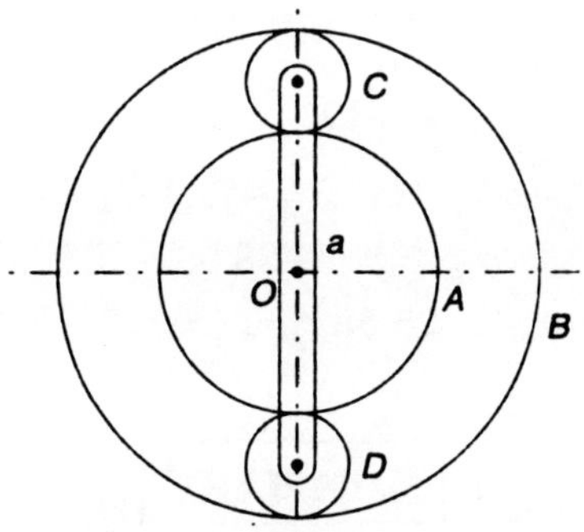

Fig. 10.28

The number of teeths on A and B are 80 and 200.
Determine the speed of arm 'a' if :
(i) A rotates at 100 rpm clockwise and B at 50 rpm in counter-clockwise.
(ii) A rotates at 100 rpm clockwise and B is stationary. [MDU, Dec., 2010]

Solution : Given : $z_A = 80, z_B = 200$

$$d_B = d_A + 2d_C \quad (\because d_C = d_D)$$

or $$z_B = z_A + 2z_C$$

$$200 = 80 + 2z_C$$

$$\therefore \quad z_C = 60$$

Action	Revolution of			
	Arm a	Gear A	Gear C and D	Gear B
1. Arm is fixed and + 1 revolution given to gear A	0	+1	$-\frac{z_A}{z_C}$	$-\frac{z_A}{z_C} \times \frac{z_C}{z_B} = -\frac{z_A}{z_B}$
2. Multiply by x	0	x	$-x\frac{z_A}{z_C}$	$-x\frac{z_A}{z_B}$
3. Add y	y	$x+y$	$y-x\frac{z_A}{z_C}$	$y-x\frac{z_A}{z_B}$

(i) $N_A = -100$ and $N_B = 50$

$$x + y = -100 \quad \text{...(1)}$$

$$y - x\frac{z_A}{z_B} = 50$$

$$y - x \times \frac{80}{200} = 50$$

or
$$y - 0.4x = 50 \quad \text{...(2)}$$

Eqs. (1)–(2) gives,

$$1.4x = -150, x = -107.14$$

$$y = -100 - x = -100 + 107.14 = 7.14$$

$\therefore$ Speed of arm, $N_a = y = 7.14$ rpm (ccw)

(ii) $N_A = -100, N_B - 0$

$$x + y = -100 \quad \text{...(1)}$$

$$y - x \times \frac{80}{200} = 0 \text{ or } y - 0.4x = 0 \quad \text{...(2)}$$

Eqs. (1) and (2) gives, $1.4x = -100, x = -71.42, y = -28.58$

$\therefore$
$$N_a = y = -28.58 \text{ rpm}$$

or
28.58 rpm (cw)

SUMMARY

1. Speed ratio of simple gear train

$$\frac{n_1}{n_m} = \frac{\text{angular velocity of the first gear in the train}}{\text{angular velocity of the last gear}}$$

2. Speed ratio of compound gear train

$$= \frac{\text{Speed of driven gear}}{\text{Speed of driving gear}} = \frac{\text{Product of teeth of driving gears}}{\text{Product of teeth of driven gears}}$$

3. Reverted gear train

Centre distance, $C = r_1 + r_2 + r_3 + r_4$

For same module of all gears, $z_1 + z_2 = z_3 + z_4$

4. **Speed ratio of planetary gear trains**

   The speed ratio of a planetary gear train can be determined by the following methods :

   1. Relative velocity method

   2. Algebraic or tabular method.

5. **Input torque, T_1 + Output torque, T_2 + Braking torque, $T_3 = 0$**

$$T_1 w_1 + T_2 w_2 + T_3 w_3 = 0$$

For $w_3 = 0$,

$$T_2 = -\left(\frac{w_1}{w_2}\right) T_1$$

$$T_3 = \left(\frac{w_1}{w_3}\right) T_1$$

$$T_3 = \left(\frac{w_1}{w_2} - 1\right) T_1$$

Short Answer Type Questions

1. What is a gear train?
2. List various types of gear trains.
3. What are epicyclic gears? What are their uses?
4. What is the formula of speed ratio of simple gear train?
5. What is the law for the speed ratio of a compound gear train?

Exercises

1. Two coaxial shafts A and B are geared together through an intermediate parallel shaft C. The wheels connecting A and C having a module of 2 mm and those connecting C and B have a module of 3.5 mm. Speed of B is less then 1/10th that of A. If the two pinions have each 20 teeth, find suitable number of teeth for the wheels, the actual velocity ratio, and corresponding centre distance of shafts C and A.

   [**Ans.** 92, 44; 10.12, 224 mm]

2. Two spur gears A and B of an epicyclic gear train shown in Fig. 10.29 have 25 and 35 teeth respectively. The arm C rotates at 105 rpm in clockwise direction. Find speed of gear B on its own axis, when gear A is fixed. If gear A rotates at 200 rpm in the counter-clockwise direction, what will be the speed of B?

   [**Ans.** 180 rpm cw, 322.86 rpm cw]

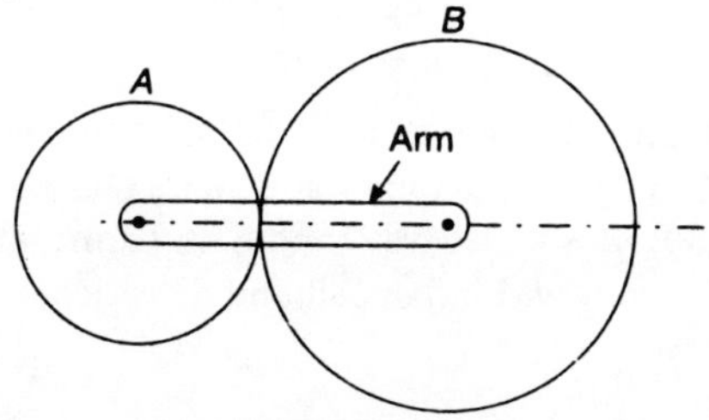

Fig. 10.29

3. In a epicyclic gear train shown in Fig. 10.30, the number of teeth on gears A, B and C are 50, 25, and 55 respectively. The arm rotates at 450 rpm clockwise. Calculate (a) speed of gear C when A is fixed, and (b) speed of gear A when C is fixed. [**Ans.** 40.9 rpm cw, 45 rpm ccw]

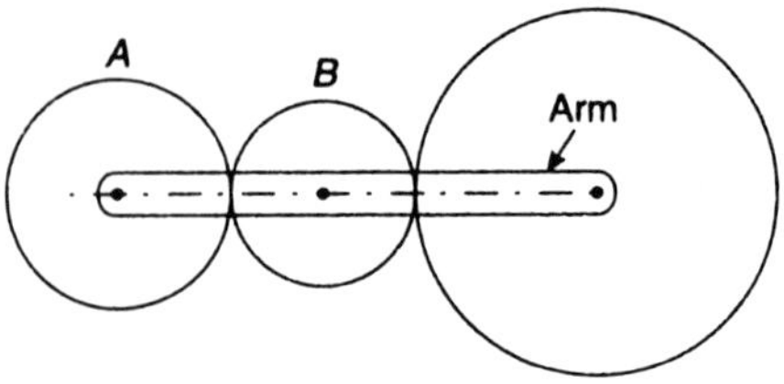

Fig. 10.30

4. In a epicyclic gear train shown in Fig. 10.31, the arm A is fixed to shaft S. The wheel B having 90 teeth rotates freely on shaft S and wheel F with 135 teeth is separately driven. If arm A runs at 240 rpm clockwise and wheel F at 120 rpm in the same direction, find (a) number of teeth on wheel C, and (b) speed of wheel B. [**Ans.** 23.421.3 rpm]

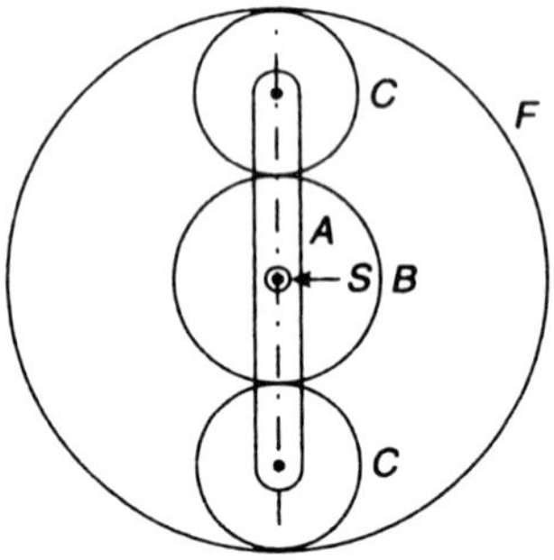

Fig. 10.31

5. In an epicyclic gear train shown in Fig. 10.32, the internal wheels A and F and compound wheel C-D rotate about the axis O. The wheel B and E rotate on pins fixed to arm L. The wheels have the same module and number of teeth are : $z_B = z_E = 20, z_c = 30$ and $z_D = 28$. If arm L makes 200 rpm cw, find the speed of F, when (a) wheel A is fixed, and (b) wheel a makes 20 rpm ccw. [**Ans.** 7.84 rpm cw, 11.37 rpm ccw]

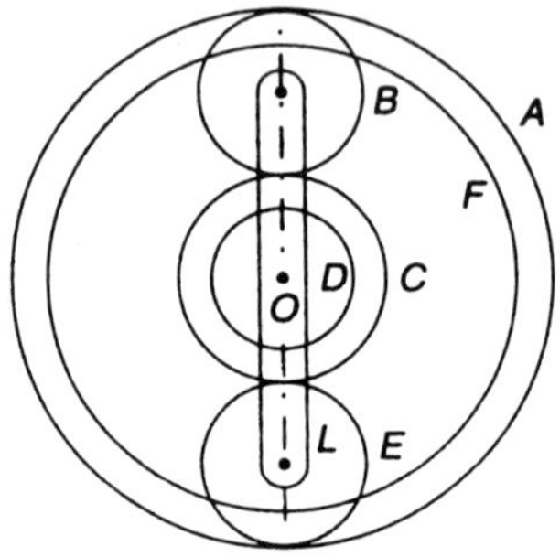

Fig. 10.32

6. A compound epicyclic gear train is shown in Fig. 10.33. The gears A, D and E are free to rotate on axis P. The compound gear B and C rotate together on axis Q at the end of arm F. All the gears have equal module, $z_A = 20, z_B - 50, z_C = 25$. Gears D and E and annular gears. Gear A rotates at 100 rpm cw and gear D rotates at 500 rpm cw. Find speed and direction of arm and gear E. [**Ans.** 328.6 rpm cw, 400 rpm cw]

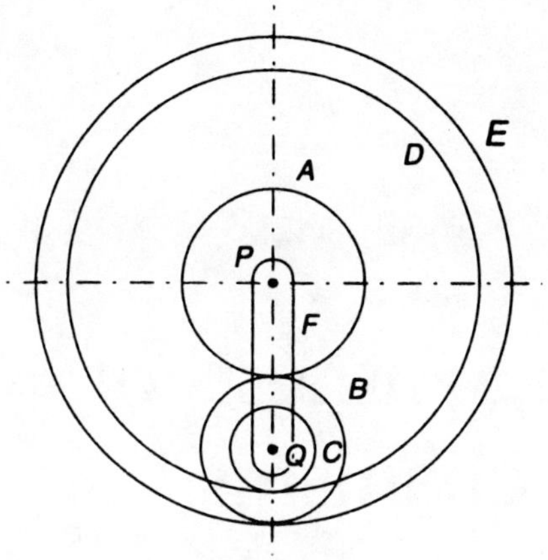

Fig. 10.33

7. In an epicyclic gear of the sun and planet type shown in Fig. 10.34, the annular gear A has 56 teeth and meshes internally. Here planet wheels of equal size mesh with annular gear A and sun wheel S. When gear A is stationary, the spider C which carries the planet wheels is to make one revolution for every 5 rotations of spindle carrying the sun wheel S. Calculate the number of teeth for all the wheels. **[Ans.** $z_S = 14, z_p = 21$]

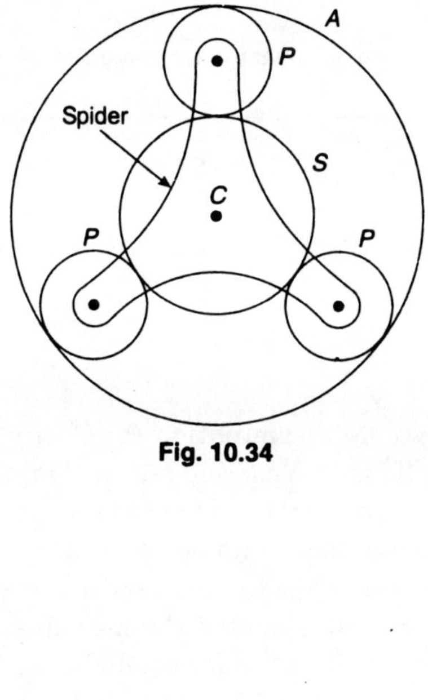

Fig. 10.34

❑❑❑

Chapter 11

Kinematic Synthesis of Planar Mechanisms

Inside This Chapter

11.1 INTRODUCTION

Kinematic analysis is the process of determination of velocity and acceleration of the various links of an existing mechanism. This was carried out in Chapters 2 and 3. On the other hand, kinematic synthesis deals with the determination of the lengths and orientation of the various lengths of the links so that a mechanism could be enolved to satisfy certain conditions.

Kinematic synthesis of a mechanisms requires the determination of the lengths of various links that satisfy the requirements of motion of the mechanism. The usual requirements are related to specified positions of the input and output links. It is simple to design a planar mechanism when the position of input and output links are known at fewer positions as compared to satisfying large number of positions.

11.2 MOVABILITY (OR MOBILITY) OR NUMBER SYNTHESIS

Movability of a mechanism means the number of degrees of freedom, which is equal to the number of independent coordinates required to specify the configuration in order to define its motion. The concept is sometimes known as *number synthesis*.

The Gruebler's or (Kutzbach) criterion for degrees of freedom of planar mechanisms is given by :

$$F = 3(n-1) - 2p - h \qquad ...(11.1)$$

where $p = \frac{1}{2}(2n_2 + 3n_3 + 4n_4 + ... in_i)$

= number of simple joints or lower pairs having one degree of freedom.

n_2 = number of binary links

n_3 = number of ternary links

n_4 = number of quaternary links, and so on

h = number of higher pairs having two degrees of freedom.

$n = n_2 + n_3 + n_4 + ... n_i$ = total number of links.

If $h = 0$

$$F = 3(n-1) - 2p$$

$$\therefore \quad F = 3[(n_2 + n_3 + n_4 + ... n_i) - 1] - (2n_2 + 3n_3 + 4n_4 + ... + in_i) \qquad ...(11.2)$$

Simplifying and re-arranging the equation, we get

$$n_2 = (F+3) + [n_4 + 2n_5 + 3n_6 + ... + (i-3)n_i] \qquad ...(11.3)$$

For a fully constrained mechanism, $F = 1$. Thus

$$n_2 = 4 + [n_4 + ... + (i-3)n_i] \qquad ...(11.4)$$

From Eq. (11.4), it is quite evident that the minimum number of binary links is equal to four. Therefore, the four-bar kinematic chain is the simplest mechanism.

For

$$F = 1, n_2 \geq 4$$

$$F = 2, n_2 \geq 5$$

$$F = 3, n_2 \geq 6$$

....

$$F = k, n \geq (k+3)$$

For a fully constrained motion, $F = 1$, so that

$$1 = 3(n-1) - 2p \qquad ...(11.5)$$

or

$$p = \frac{3n}{2} - 2$$

11.3 GRASHOF'S LAW

This law states that for a four-bar mechanism the sum of the lengths of the largest and the shortest links should be less than or equal to the sum of the lengths of the other links, that is,

$$(l + s) \leq (p + q) \qquad ...(11.6)$$

where l, s = lengths of the longest and the shortest links, respectively

p, q = lengths of the other two links.

Consider the four-bar chain shown in Fig. 11.1. Let the length of fixed link $O_2O_4 = d$, crank $O_2A = a$, coupler $AB = b$ and lever $BO_4 = c$. The following types of mechanisms are obtained by adjusting the lengths of various links :

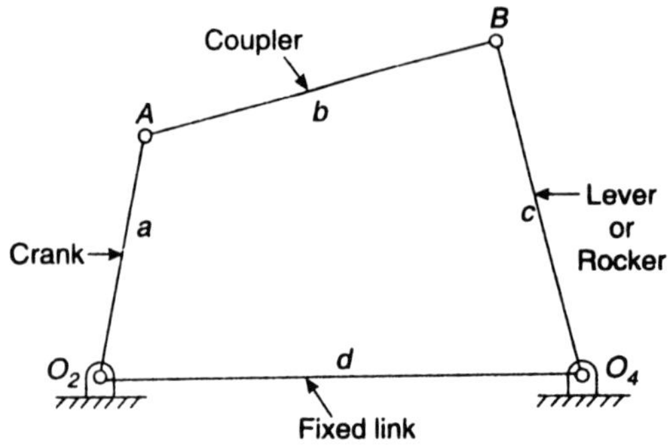

Fig. 11.1 Four-bar chain.

1. Crank-crank Mechanism

When the shortest link is fixed and $(l + s) \le (p + q)$ crank-crank or double crank mechanism is obtained, as shown in Fig. 11.2. Links O_2A or O_4B may be the inputs links or cranks. They are able to make complete rotations about points O_2 and O_4 respectively. Shortest link $O_4O_4 = d$ is fixed.

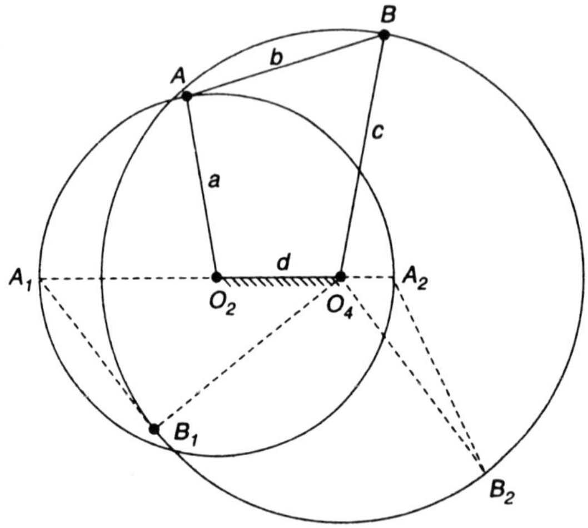

Fig. 11.2 Crank-crank mechanism.

2. Crank-Rocker (or lever) Mechanism

If any of the adjacent links of shortest link d, *i.e.*, 'a' or c is fixed, d can have full revolution and the link b opposite to it oscillates. In Fig. 11.3 (a), 'a' is fixed, d is the crank to rotate about O_2 and b oscillates, whereas in Fig. 11.3 (b), c is fixed, d is the crank to rotate about O_4 and b oscillates. Again $(l + s) \le (p + q)$ applies.

3. Rockr-rocker Mechanism

If the link b opposite to the shortest link d is fixed and shortest link d is made the coupler, the other two links, 'a' and c would oscillate, as shown in Fig. 11.4 (a). When $(l + s) \le (p + q)$, the linkage is known as a class-I four-bar linkage.

4. Class-II Four-bar Linkage

When $(l + s) > (p + q)$, the linkage is known as a class-II four-bar linkage. In such a mechanism, fixing of any of the links always results in rocker-rocker mechanism, as shown in Fig. 11.4 (b).

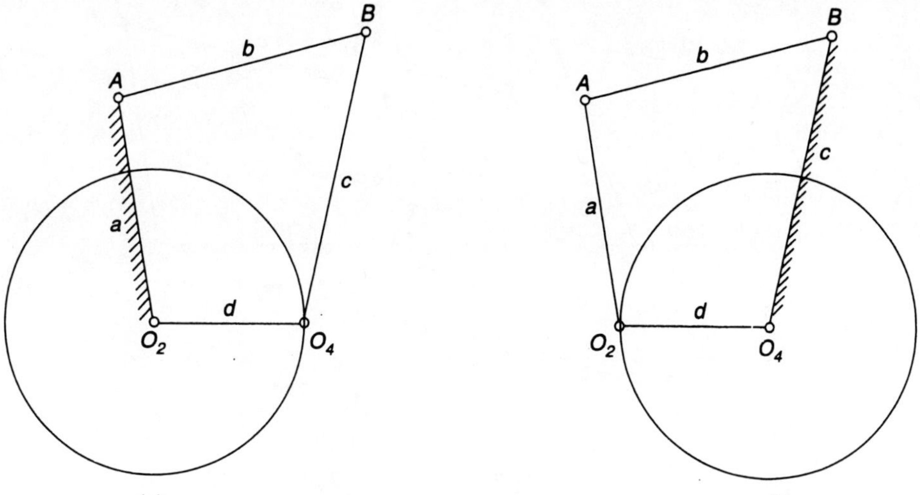

(a) (b)

Fig. 11.3 Crank-rocker mechanism.

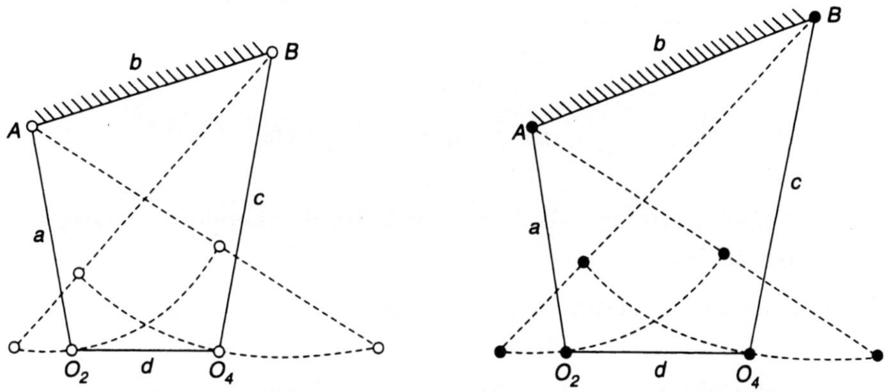

(a) $(l + s) < (p + q)$ **(b)** $(l + s) > (p + q)$

Fig. 11.4 Rocker-rocker mechanism.

■ **Example 11.1** Identify the nature of each mechanism shown in Fig. 11.5 (a) to (d).

Solution :

(a) $l = 10\,\text{cm}$, $s = 5\,\text{cm}$, $p + q = 7 + 9 = 16$ cm, $l + s = 15\,\text{cm}$

$$\therefore \quad (l + s) \angle (p + q)$$

The shortest link is fixed. Hence it is a crank-crank (or drag link) mechanism

(b) $l = 11\,\text{cm}$, $s = 6\,\text{cm}, l + s = 17\,\text{cm}$, $p + q = 9 + 10 = 19\,\text{cm}$

$$\therefore \quad (l + s) \angle (p + q)$$

The link adjacent to the shortest link is fixed. Therefore, it is a crank-rocker mechanism.

(c) $l = 12\,\text{cm}$, $s = 6\,\text{cm}$, $l + s = 18\,\text{cm}$, $p + q = 21\,\text{cm}$

$$\therefore \quad (l + s) \angle (p + q)$$

The link adjacent to the shortest link is fixed. Therefore, it is a crank-lever mechanism.

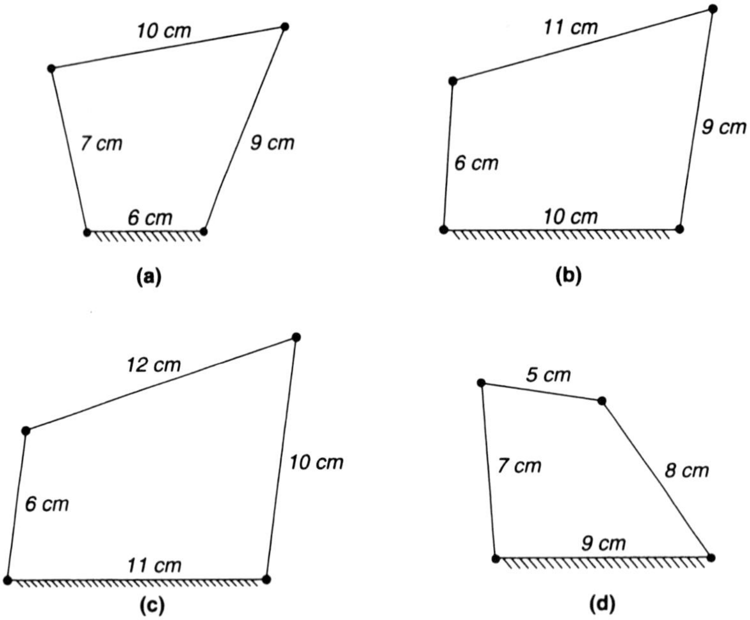

Fig. 11.5

(d) $l = 9\,\text{cm}$, $s = 5\,\text{cm}$, $l + s = 14\,\text{cm}$, $p + q = 7 + 8 = 15\,\text{cm}$

$\therefore$ $$(l + s) \angle (p + q)$$

The link opposite to the shortest link is fixed. Therefore, it is a rocker-rocker mechanism of class-I.

■ **Example 11. 2** Identify the nature of the mechanisms shown in Fig. 11.6 a) to (d).

Solution :

(a) $l = 800\,\text{mm}$, $s = 100\,\text{cm}$, $l + s = 400\,\text{mm}$, $p + q = 500\,\text{mm}$

$\therefore$ $$(l + s) \angle (p + q)$$

The shortest link is fixed. Therefore, it is a crank-crank mechanism.

(b) $l = 800\,\text{mm}$, $s = 100\,\text{mm}$, $l + s = 900\,\text{mm}$, $p + q = 1100\,\text{mm}$

$\therefore$ $$(l + s) \angle (p + q)$$

The link opposite to the shortest link is fixed. Hence, it is a rocker-rocker mechanism.

(c) $l = 800\,\text{mm}$, $s = 100\,\text{mm}$, $l + s = 900\,\text{mm}$, $p + q = 1100\,\text{mm}$

$\therefore$ $$(l + s) \angle (p + q)$$

The link adjacent to the shortest link is fixed. Hence, it is a crank-rocker mechanism.

(d) $l = 450\,\text{mm}$, $s = 250\,\text{mm}$, $l + s = 700\,\text{mm}$, $p + q = 650\,\text{mm}$

$\therefore$ $$(l + s) > (p + q)$$

Therefore, it is a rocker-rocker mechanism of class-II.

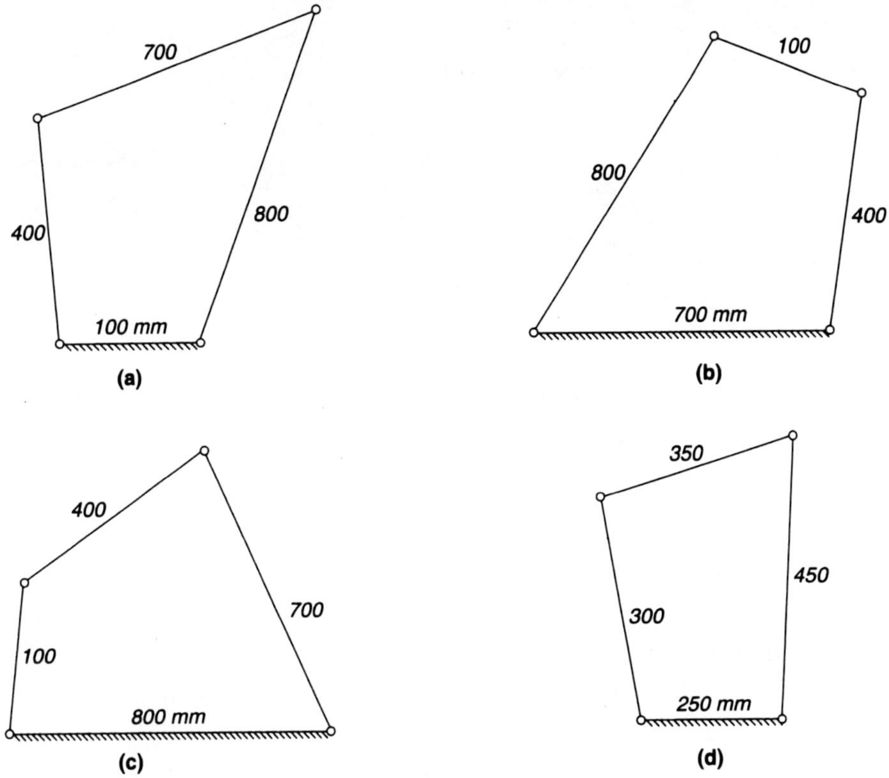

Fig. 11.6

11.4 DIMENSIONAL SYNTHESIS

It deals with the determination of actual dimensions of the mechanism to satisfy the specified motion characteristics. The actual dimensions could be the lengths of the links between adjacent hinge pairs, angles between the arms of a bell-crank lever, or cam contour dimensions, etc.

Synthesis of mechanism, may be carried out by graphical methods or analytical methods.

11.5 GRAPHICAL METHOD

11.5.1 Pole

Consider a four-bar mechanism $O_2\ ABO_4$ in two positions $O_2A_1B_1O_2$ and $O_2A_2B_2O_4$ as shown in Fig. 11.7. The coupler link ABhas moved from the position A_1B_1 to A_2B_2. The input link O_2A and output link O_4B have moved through angles θ_{12} and ϕ_{12} respectively in the clockwise direction. For the motion of the coupler $AB$from A_1B_1 to A_2B_2, P_{12} is its centre of rotation with respect to the fixed link. P_{12} lies at the point of intersection of the perpendicular bisectors of the coupler link ABin its two position A_1B_1 and $A_2B_2.P_{12}$ is called the pole. P_{12} also lies at the point of intersection of the mid-normals a_{12} and b_{12} of chords A_1A_2 and B_1B_2 respectively.

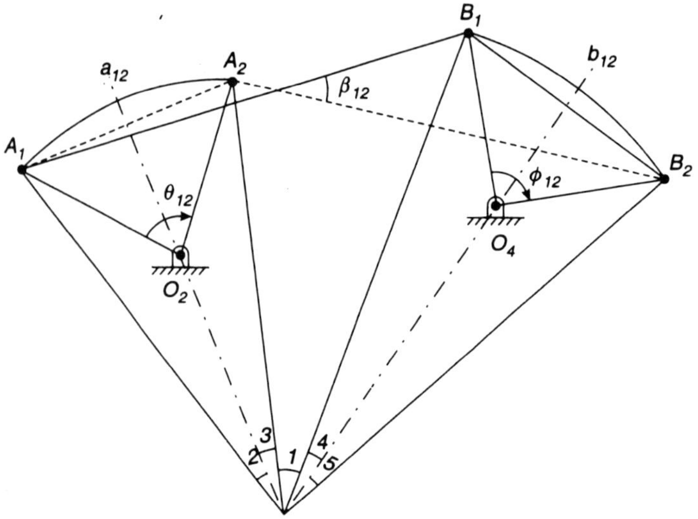

Fig. 11.7

Properties of Pole Point

1. Since $AB = A_1B_1 = A_2B_2$, the perpendicular bisectors of A_1A_2 and B_1B_2 pass through fixed centres O_2 and O_4.
2. Since coupler AB rotates about P_{12} from position A_1B_1 to A_2B_2, and therefore $\Delta A_1B_1P_{12} = \Delta A_2B_2P_{12}$,

   $\therefore \quad \angle 2 + \angle 3 + \angle 1 = \angle 1 + \angle 4 + \angle 5$

   or angle subtended by A_1B_1 at P_{12} = angle subtended by A_2B_2 at P_{12}
3. Since $\angle 2 + \angle 3 + \angle 1 = \angle 1 + \angle 4 + \angle 5$

   $\therefore \quad \angle 2 + \angle 3 = \angle 4 + \angle 5$

   Thus A_1A_2 and B_1B_2 subtend equal angle at P_{12}.
4. Since P_{12} lies on the perpendicular bisectors of A_1A_2 and B_1B_2.

   $\therefore \quad \angle 2 = \angle 3$ and $\angle 4 = \angle 5$
5. Since $\angle 2 + \angle 3 = \angle 4 + \angle 5$

   and $\quad \angle 2 = \angle 3$ and $\angle 4 = \angle 5$

   $\therefore \quad \angle 2 = \angle 4$ and $\angle 3 = \angle 5$

   Therefore, input and output links subtend equal angles at pole P_{12} as they move from one position to another.
6. Since $\angle 2 + \angle 3 + \angle 1 = \angle 1 + \angle 4 + \angle 5 = \angle 1 + \angle 4 + \angle 3$ ($\because \angle 5 = 3$)

   i.e., angle subtended by the coupler link AB = angle subtended by the fixed link O_2O_4.
7. The triangle $A_1B_1P_{12}$ moves as one link about P_{12} to the position $A_2B_2P_{12}$.

   Angular displacement of coupler A_1B_1 = angular displacement of $P_{12}B_1$

   i.e., $\quad \angle\beta_{12} = \angle 4 + \angle 5$

11.5.2 Relative Pole

The pole of a moving link is the centre of its rotation with respect to a fixed link. However, if the rotation of the link is considered relative to another moving link, the pole is called as the relative pole. To determine the relative pole, fix the link of reference and observe the motion of the other link in the reverse direction.

(a) Determination of Relative Pole for Four-Bar Chain

Consider the four-bar chain O_2ABO_4 in its two positions $O_2AB_1O_4$ and $O_2AB_2O'_4$ as shown in Fig. 11.8 with links O_2A fixed. The relative pole of AB relative to O_2A is at A. The relative pole of O_4B relative to O_2A is at R_{12}, which can be determined as follows :

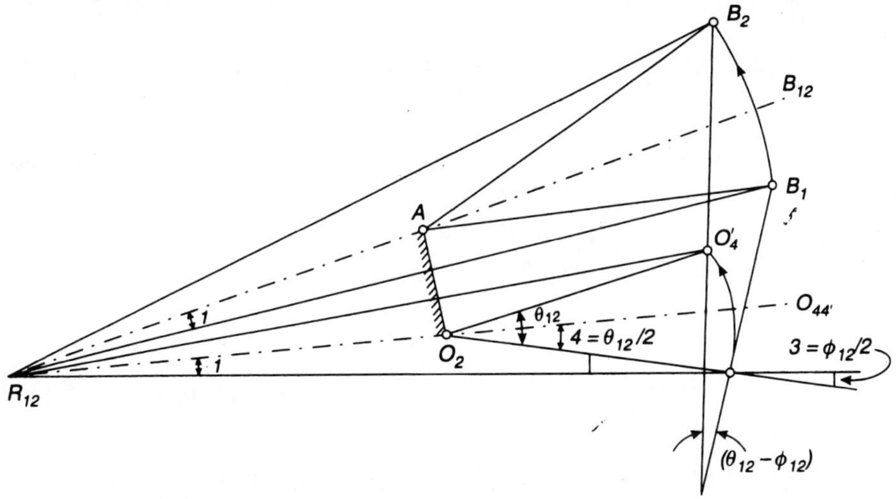

Fig. 11.8 Determination of relative pole of four-bar chain.

Let θ_{12} = angle of rotation of O_2A (clockwise)

ϕ_{12} = angle of rotation of O_4B (clockwise)

1. Assume O_2 and A as the fixed pivots and rotate O_2O_4 about O_2 through angle θ_{12} in the counter-clockwise direction (opposite to the direction of rotation of O_2A). Let O_4' be the new position after the rotation of O_2O_4.
2. Locate the point B_2 by drawing arc with centres A and O_4' radii equal to AB_1 and O_4B_1 respectively. Then $O_2AB_2O_4'$ is the inversion of $O_2AB_1O_4$.
3. Draw mid-normals of O_2O_4' and B_1B_2 passing through O_2 and A respectively to intersect at R_{12}, which is the required relative pole.

   $\phi_{12} - \theta_{12}$ = angle of rotation of the output link O_4B relative to the input link. The angle will be negative if $O_4B > O_2A$ and positive if $O_4B < O_2A$.

Angular displacement of $R_{12}O_4$ = angular displacement of O_4B_1.

$$\angle O_4R_{12}O'_4 = -(\angle\phi_{12} - \theta_{12}) \qquad \text{(assuming } O_4B > O_2A)$$

or
$$2\,\angle 1 = -(\angle\phi_{12} - \angle\theta_{12})$$

or
$$\angle 1 = \frac{1}{2}(\angle\phi_{12} - \angle\theta_{12})$$

In $\Delta O_2R_{12}O_4$,
$$\angle 4 = \angle 1 + \angle 2$$

or
$$\frac{1}{2}\angle\theta_{12} = -\frac{1}{2}(\angle\phi_{12} - \angle\theta_{12}) + \angle 3 \qquad (\because \angle 2 = \angle 3)$$

$$= \frac{1}{2} \angle\phi_{12} + \frac{1}{2} \angle\theta_{12} + \angle 3$$

$$\therefore \qquad \angle 3 = \frac{1}{2} \angle\phi_{12} \qquad ...(11.7)$$

Procedure :

1. Join O_2O_4 and extend it further. Rotate O_2O_4 about O_2 through an angle $\frac{1}{2}\theta_{12}$ in a direction opposite to that of O_2A.
2. Again rotate O_2O_4 about O_4 through an angle $\frac{1}{2}\phi_{12}$ in a direction opposite to that of O_4B.
3. The point of intersection of these two positions of O_2O_4 after rotation about O_2 and O_4, is the required relative pole R_{12}.

The angles subtended by O_4O_4' and B_1B_2 at R_{12} are the same

i.e., $$\angle O_4R_{12}O_4' = \angle B_1R_{12}B_2$$

or $$2\angle O_4R_{12}O_2 = 2\angle B_1R_{12}A$$

or $$\angle O_4R_{12}O_2 = \angle B_1R_{12}A$$

$\therefore$ Angle subtended by the first pivots (O_2 and O_4) at the relative pole = Angle subtended by the coupler AB.

(b) Determination of Relative Pole for Slider-Crank Mechanism

Consider the slider crank mechanism shown in Fig. 11.9. The point B on the slider reciprocates through a horizontal distance x. Its centre of rotation will lie at infinity on a vertical line where the point O_4 can also be assumed to lie. Then O_2O_4 will also be a vertical line through O_2. Rotating O_2O_4 about O_4 through $\frac{1}{2}\theta_{12}$ in the counter-clockwise direction. Rotating O_2O_4 about O_4 through $\frac{1}{2}\theta_{12}$ would mean a vertical line towards left of O_2, at a distance of $\frac{x}{2}$. The intersection of the two lines locates R_{12}.

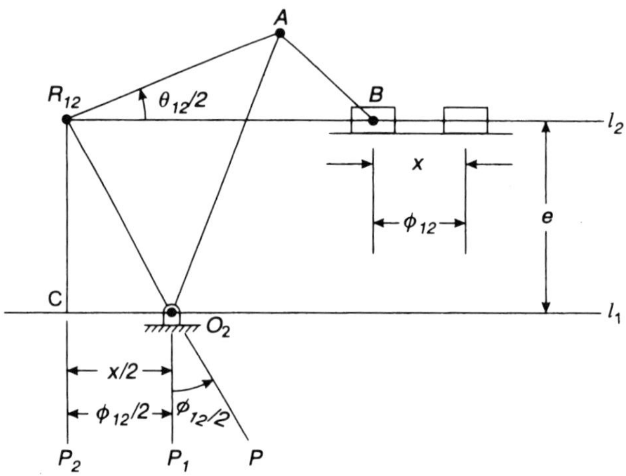

Fig. 11.9 Determination of relative pole of slider-crank mechanism.

The procedure to locate the relative pole of a slider-crank mechanism is as follows :

1. Draw two parallel lines l_1 and l_2 at a distance equal to the eccentricity e (if there is any eccentricity).
2. Select a line segment $O_2C = \frac{x}{2}$ on line l_1 such that C is measured in a direction opposite to the motion of the slider.
3. Draw perpendicular lines O_2P_1 and CP_2 to line l_1.
4. Make $\angle\frac{1}{2}\,\theta_{12}$ at a point O_2 with the line O_2P_1 in a direction opposite to the rotation of the input link.
5. The intersection of this line with the line CP_2 extended locates the relative pole R_{12}.

11.6 DESIGN OF MECHANISMS BY RELATIVE POLE METHOD

11.6.1 Four-Bar Mechanism

(a) Two position synthesis

Let for a four-bar mechanism, length of the fixed link O_2O_4 along with the angular displacement θ_{12} (between position 1 and 2) of the input link O_2A and angular displacement ϕ_{12} (between position 1 and 2) of the output link O_4B, are known.

To design the mechanism (Fig. 11.10), first locate the relative pole R_{12} as explained in art. 11.5.2 (a).

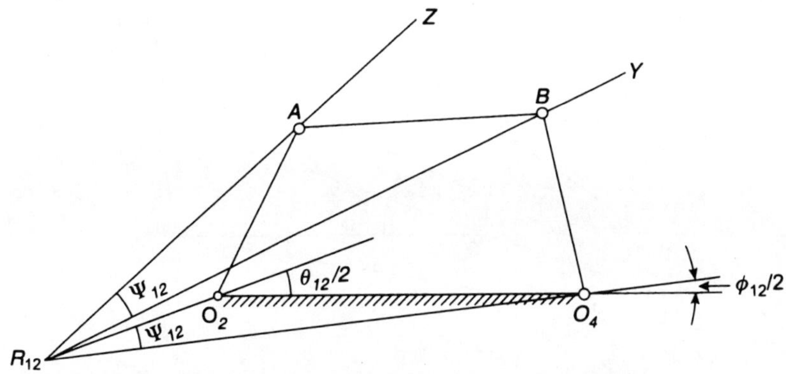

Fig. 11.10 Two-position method for four-bar mechanism.

Now, angle subtended by the coupler AB at R_{12}

$$= \text{angle subtended by the fixed link } O_2O_4 \text{ at } R_{12}$$

$$= \frac{1}{2}\angle\theta_1 - \frac{1}{2}\angle\phi_{12} \qquad \text{(Assuming } O_4B \angle O_2A\text{)}$$

$$= \angle\Psi_{12} \qquad ...(11.8)$$

Procedure :

1. Construct an angle Ψ_{12} at an arbitrary position R_{12}. Join any two points on the two arms of this angle to obtain the coupler link AB of the mechanism. Join A with O_2 and B with O_4 to get the input and output links O_2A and O_4B respectively.

2. Locate point B arbitrarily so that BO_4 is the output link. Construct $\angle BR_{12}Z = \Psi_{12}$. Take any suitable point A on $R_{12}Z$. Join AB and O_2A.
3. Instead of locating point B as above, locate point A arbitrarily so that O_2A is the input link. Construct $\angle AR_{12}Y = \Psi_{12}$. Select any point B on $R_{12}Y$. Join A and B to get AB. Join B with O_4 to get output link O_4B.

(b) Three Position Synthesis

Let three positions of input link θ_1, θ_2 and θ_3, and three positions of output link ϕ_1, ϕ_2 and ϕ_3 are known. The relative poles R_{12} and R_{13} can be determined considering.

$$\theta_{12} = \theta_2 - \theta_1 \text{ and } \phi_{12} = \phi_2 - \phi_1 \text{ to locate } R_{12}$$

and
$$\theta_{13} = \theta_3 - \theta_1 \text{ and } \phi_{13} = \phi_3 - \phi_1 \text{ to locate } R_{13}.$$

Then angle subtended at R_{12},

$$\Psi_{12} = \frac{1}{2}(\theta_{12} - \phi_{12}) \quad \text{...(11.9a)}$$

and at R_{13},

$$\Psi_{13} = \frac{1}{2}(\theta_{13} - \phi_{13}) \quad \text{...(11.9b)}$$

Construct the angles Ψ_{12} and Ψ_{13} at the points R_{12} and R_{13} respectively at arbitrary positions such that the arms of the angles intersect at A and B. Join A and B to get the coupler AB. Join A with O_2 and B with O_4 to get input and output links respectively, as shown in Fig. 11.11.

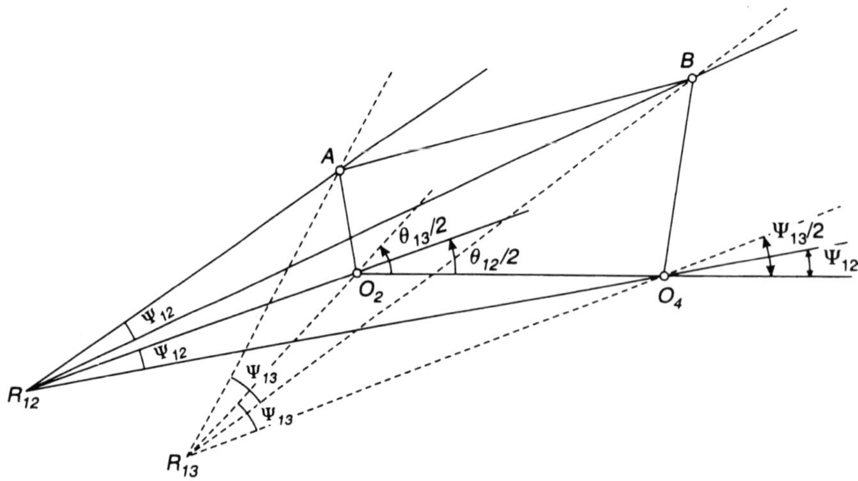

Fig. 11.11 Three-position method for four-bar mechanism.

11.6.2 Slider-Crank Mechanism

(a) Two-position Synthesis

Let

θ_{12} = angular displacement of input link O_2A (angle between θ_1 and θ_2) in cw direction.

x = linear displacement of slider to the right.

e = eccentricity

Locate the relative pole R_{12} as explained in art. 11.5.2 (b) and shown in Fig. 11.12.

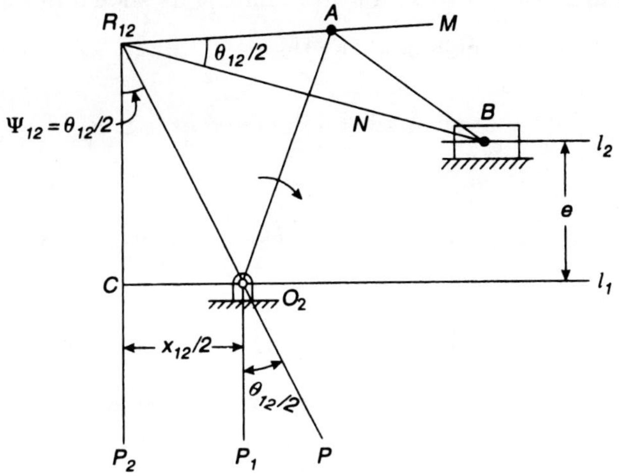

Fig. 11.12

Procedure :

1. Draw two parallel lines l_1 and l_2 at a distance equal to eccentricity *e*.
2. Construct an angle equal to $\theta_{12}/2$ at point R_{12}, chosen arbitrarily and at a convenient position. The intersection of an arm *N* of this angle with line l_2 gives the position of the slider at point *B*.
3. Select point *A* arbitrarily on the other arm *M* of the angle. Join *A* with O_2 to get the input link (crank) O_2A. Join *A* and *B* to get the coupler (connecting rod) *AB*.

(b) Three-position Synthesis

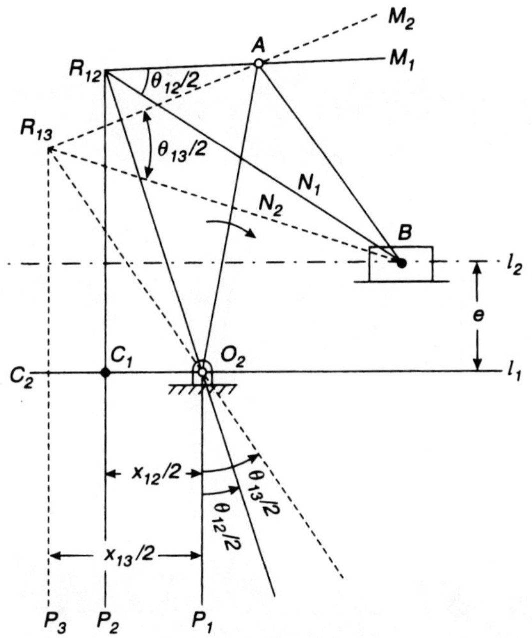

Fig. 11.13

Three position of input link θ_1, θ_2 and θ_3 are known along with corresponding slider positions x_1, x_2 and x_3 are also known. Find R_{12} and R_{13} as shown in Fig. 11.13 . Here

$$\frac{\theta_{12}}{2} = \text{angle made by the fixed link at } R_{12}.$$

$$\frac{\theta_{13}}{2} = \text{angle made by the fixed link at } R_{13}.$$

Procedure :

1. Construct angle $\frac{\theta_{12}}{2} = \angle M_1 R_{12} N_1$ at R_{12} in an arbitrary position with arm N_1 locating point B.
2. Draw angle $\frac{\theta_{13}}{2} = \angle M_2 R_{13} N_2$ at R_{13} with an arm N_2 along $R_{13}B$.
3. Intersection of the two arms M_1 at R_{12} and M_2 at R_{13} (not through B) of the two angles locates the point A. Join A with B to get the coupler AB.

11.7 ANALYTICAL METHOD

11.7.1 Function Generation

In function generation, the motion of input (or driver) link is correlated to the motion of the output (or follower) link. Let θ and ϕ be the angles of rotation of input and output links respectively. Let $y = f(x)$ be the function to be generated. The angle of rotation θ of the input link O_2A represents the independent variable x and the angle of rotation ϕ of the output link O_4B represents the dependent variable y, as shown in Fig. 11.14. The relation between x and θ and that between y and ϕ is generally assumed to be linear. Let θ_i and ϕ_i be the initial values of θ and ϕ representing x_i and y_i respectively. Then

$$\frac{\theta - \theta_i}{x - x_i} = r_x = \text{const.} = \frac{\theta_f - \theta_i}{x_f - x_i} \quad \text{...(11.10a)}$$

and

$$\frac{\phi - \phi_i}{y - y_i} = r_y = \text{const.} = \frac{\phi_f - \phi_i}{y_f - y_i} \quad \text{...(11.10b)}$$

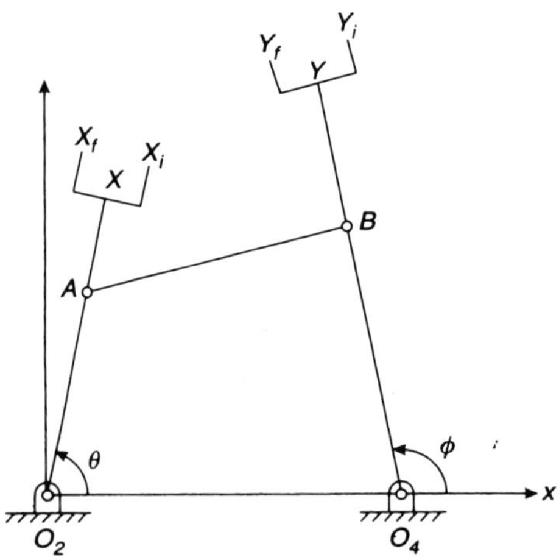

Fig. 11.14

where the constants r_x and r_y are called the *scale factors*. The subscripts i and $f$denote the initial and final values.

11.7.2 Chebyshev's Spacing for Precision Points

Let x_i and x_j be the initial and final values of variable x respectively. A function $f(x)$ is desired to be generated in the interval $x_i \le x \le x_f$. Let the generated function be $F(x, R_1, R_2, \ldots R_n)$ where $R_1, R_2, \ldots, R_n$ are the design parameters. The difference $E(x)$ between the desired function and generated function can be represented by,

$$E(x) = f(x) - F(x, R_1, R_2, \ldots, R_n) \qquad \ldots(11.11)$$

At precision points, say for $x = x_1, x_2, \ldots, x_n$, the desired and generated functions agree and $E(x) = 0$. At other points $E(x)$ will have some value, called the structural error. It is desirable that $E(x)$ should be minimum. Therefore, the spacing of precision points is very important. The precision points, according to Chebyshev's spacing, are given by :

$$x_m = a + b\cos\left[\frac{(2m-1)\pi}{2n}\right],\ m = 1, 2, 3 \qquad \ldots(11.12)$$

where $a = \dfrac{x_i + x_f}{2}$, $b = \dfrac{x_f - x_i}{2}$ and

n = number of precision points.

11.7.3 Graphical Method to Locate Precision Points

The Precision points can be obtained by the graphical method as explained in the following steps :

1. Draw a circle of radius 'b' and centre on the x-axis at a distance 'a' from point O.
2. Inscribe a regular polygon of side $2n$ in this circle such that the two sides are perpendicular to the x-axis.
3. Determine the locations of n accuracy points by projecting the vertices on x-axis as shown in Fig. 11.15. It is sufficient to draw semi-circles only showing inscribed polygon to get the values of precision points.

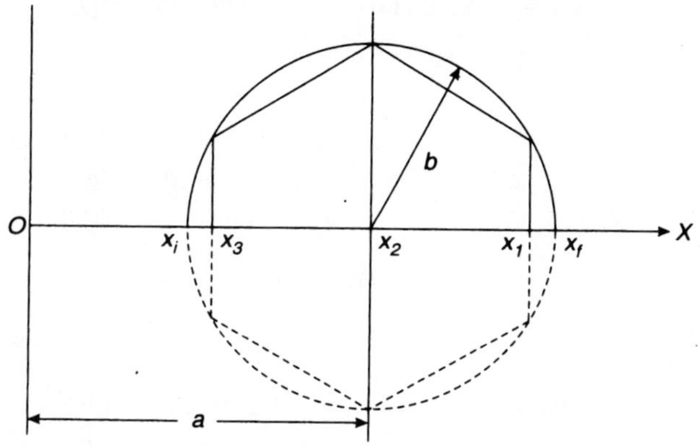

Fig. 11.15 Graphical method to determine precision points.

11.7.4 Freudenstein's Equation for Precision Points

Freudenstein's equation helps to determine the length of links of a four-bar mechanism. The displacement equation of a four-bar mechanism, shown in Fig. 11.16. is given by

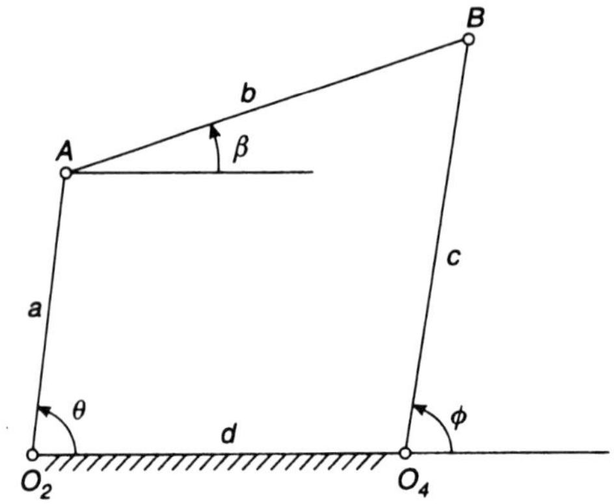

Fig. 11.16

$$2cd \cos\phi - 2ad \cos\theta + a^2 - b^2 + c^2 + d^2 = 2ac (\cos\theta \cos\phi + \sin\theta \sin\phi)$$

Dividing throughout by $2ac$ and rearranging, we have

$$\frac{d}{a} \cos\phi - \frac{d}{c} \cos\theta + \frac{a^2 - b^2 + c^2 + d^2}{2ac} = \cos(\theta - \phi)$$

Let $$k_1 = \frac{d}{a},\ k_2 = -\frac{d}{c} \text{ and } k_3 = \frac{a^2 - b^2 + c^2 + d^2}{2ac}$$

Then $$k_1 \cos\phi + k_2 \cos\theta + k_3 = \cos(\theta - \phi) \quad \text{...(11.13)}$$

Eq. (11.13) is known as the Freudenstein's equation.

Let the input and output are related by some function, such as, $y = f(x)$.

For three specified positions, let

$$\theta_1, \theta_2, \theta_3 = \text{three positions of input link}$$
$$\phi_1, \phi_2, \phi_3 = \text{three positions of output link}$$

Then substituting these values in Eq. (11.13), we have

$$k_1 \cos\phi_1 + k_2 \cos\theta_1 + k_3 = \cos(\theta_1 - \phi_1)$$
$$k_1 \cos\phi_2 + k_2 \cos\theta_2 + k_3 = \cos(\theta_2 - \phi_2)$$
$$k_1 \cos\phi_3 + k_2 \cos\theta_3 + k_3 = \cos(\theta_3 - \phi_3)$$

These equations can be written in the matrix form as :

$$\begin{bmatrix} \cos\phi_1 & \cos\theta_1 & 1 \\ \cos\phi_2 & \cos\theta_2 & 1 \\ \cos\phi_3 & \cos\theta_3 & 1 \end{bmatrix} = \begin{Bmatrix} k_1 \\ k_2 \\ k_3 \end{Bmatrix} = \begin{Bmatrix} \cos(\theta_1 - \phi_1) \\ \cos(\theta_2 - \phi_2) \\ \cos(\theta_3 - \phi_3) \end{Bmatrix}$$

These equations can be solved by any numerical technique.

Using Cramer's rule, let

$$A = \begin{vmatrix} \cos\phi_1 & \cos\theta_1 & 1 \\ \cos\phi_2 & \cos\theta_2 & 1 \\ \cos\phi_3 & \cos\theta_3 & 1 \end{vmatrix},\ A_1 = \begin{vmatrix} \cos(\theta_1 - \phi_1) & \cos\theta_1 & 1 \\ \cos(\theta_2 - \phi_2) & \cos\theta_2 & 1 \\ \cos(\theta_3 - \phi_3) & \cos\theta_3 & 1 \end{vmatrix}$$

$$A_2 = \begin{vmatrix} \cos\phi_1 & \cos(\theta_1 - \phi_1) & 1 \\ \cos\phi_2 & \cos(\theta_2 - \phi_2) & 1 \\ \cos\phi_3 & \cos(\theta_3 - \phi_3) & 1 \end{vmatrix}$$

$$A_3 = \begin{vmatrix} \cos\phi_1 & \cos\theta_1 & \cos(\theta_1 - \phi_1) \\ \cos\phi_2 & \cos\theta_2 & \cos(\theta_2 - \phi_2) \\ \cos\phi_3 & \cos\theta_3 & \cos(\theta_3 - \phi_3) \end{vmatrix}$$

Then $$k_1 = \frac{A_1}{A},\ k_2 = \frac{A_2}{A},\ k_3 = \frac{A_3}{A}$$

Knowing k_1, k_2 and k_3, the values of a, b, c, and d can be calculated. Value of either 'a' or 'd' can be assumed to be unity to obtain the proportionate values of other parameters.

■ **Example 11.3** Design a four-bar mechanism to coordinate three positions of the input and output links given by :

$$\theta_1 = 25°,\ \phi_1 = 30°;\ \theta_2 = 35°,\ \phi_2 = 40°,\ \theta_3 = 50°\ \phi_3 = 60°.$$

Solution : $\cos\theta_1 = \cos 25° = 0.9063$, $\cos\theta_2 = \cos 35° = 0.8191$,

$\cos\theta_3 = \cos 50° = 0.6428$

$\cos\phi_1 = \cos 30° = 0.8660$, $\cos\phi_2 = \cos 40° = 0.7660$,

$\cos\phi_3 = \cos 60° = 0.5000$

$$\cos(\theta_1 - \phi_1) = \cos(25° - 30°) = 0.9962$$

$$\cos(\theta_2 - \phi_2) = \cos(35° - 40°) = 0.9962$$

$$\cos(\theta_3 - \phi_3) = \cos(50° - 60°) = 0.9848$$

$$A = \begin{vmatrix} 0.8660 & 0.9063 & 1 \\ 0.7660 & 0.8191 & 1 \\ 0.5000 & 0.6428 & 1 \end{vmatrix}$$

$$= 0.8660\,(0.8191 - 0.6428) - 0.9063\,(0.7660 - 0.5000) + 1\,(0.7666 \times 0.6428 - 0.8191 \times 0.5000)$$

$$= -5.5652 \times 10^{-3}$$

$$A_1 = \begin{vmatrix} 0.9962 & 0.9063 & 1 \\ 0.9962 & 0.8191 & 1 \\ 0.9848 & 0.6428 & 1 \end{vmatrix}$$

$$= 0.9962\,(0.8191 - 0.6428) - 0.9063\,(0.9962 - 0.9848) + 1\,(0.9962 \times 0.6428 - 0.8191 \times 0.9848)$$

$$= -9.9408 \times 10^{-4}$$

$$A_2 = \begin{vmatrix} 0.8660 & 0.9062 & 1 \\ 0.7660 & 0.9962 & 1 \\ 0.5000 & 0.9848 & 1 \end{vmatrix}$$

$$= 0.8660\,(0.9962 - 0.9848) - 0.9962\,(0.7660 - 0.5000) + 1\,(0.7666 \times 0.9848 - 0.9962 \times 0.9848)$$

$$= -1.14 \times 10^{2}$$

$$A_3 = \begin{vmatrix} 0.8660 & 0.9063 & 0.9062 \\ 0.7660 & 0.8191 & 0.9962 \\ 0.5000 & 0.6448 & 0.9848 \end{vmatrix}$$

$$= 0.8660\ (0.8191 \times 0.9848 - 0.9962 \times 0.6428)$$
$$- 0.9063\ (0.7666 \times 0.9848 - 0.9962 \times 0.5000)$$
$$= - 5.746 \times 10^{-3}$$

$$k_1 = \frac{A_1}{A} = \frac{-9.9408 \times 10^{-4}}{-5.5652 \times 10^{-3}} = 0.1786 = \frac{d}{a}$$

Let $d = 1$ unit, then $a = 5.598$ units

$$k_2 = \frac{A_2}{A} = \frac{1.14 \times 10^{-3}}{5.5652 \times 10^{-3}} = - 0.2048$$
$$= -\frac{d}{c},\ c = 4.88 \text{ units}$$

$$k_3 = \frac{A_3}{A} = \frac{-5.7164 \times 10^{-3}}{5.5652 \times 10^{-3}} = 1.0272$$
$$= \frac{a^2 - b^2 + c^2 + d^2}{2ac}$$
$$= \frac{(5.598)^2 - b^2\ (4.88)^2 + 1}{2 \times 5.598 \times 4.88},\ b = 0.176 \text{ units}$$

The mechanism is shown in Fig. 11.17.

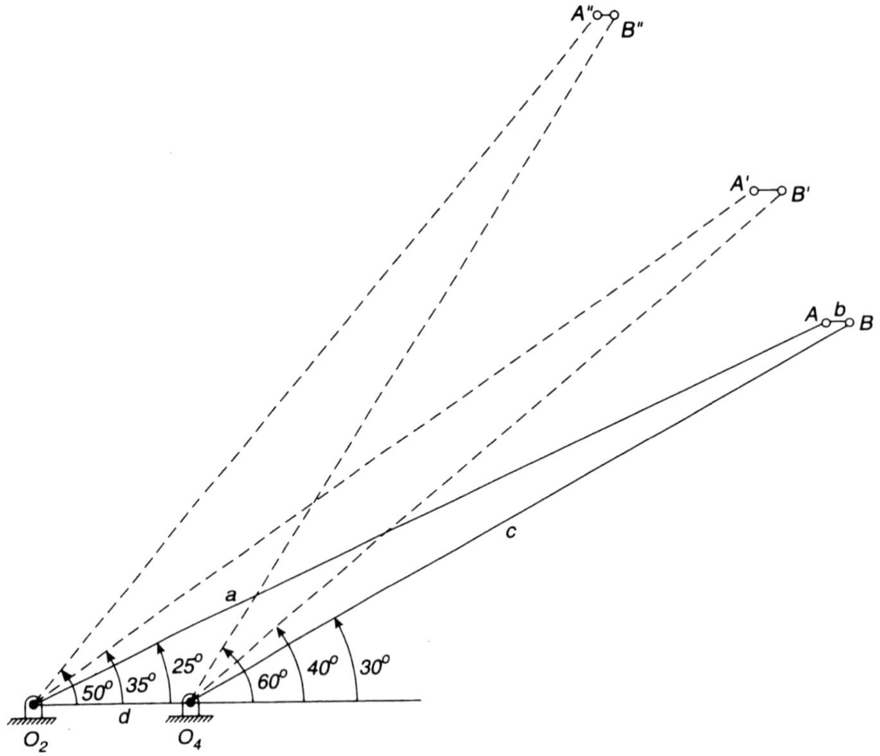

Fig. 11.17

■ **Example 11.4** Design a four-bar mechanism when the motions of the input and output links are governed by a function $y = 2x^2$ and x varies from 2 to 4 with an interval of 1. Assume θ to vary from 40° to 120° and ϕ from 60° to 132°.

Solution : The angular displacement of input link is governed by x whereas that of the output link by y. θ varies from 40° to 120° (*i.e.*, through 80°) and ϕ from 60° to 132° (*i.e.*, through 72°).

$x = 2, 3, 4$

The corresponding values of y are : $2 \times 2^2 = 8, 2 \times 3^2 = 18, 2 \times 4^2 = 32.$

$$r_x = \frac{\theta_f - \theta_i}{x_f - x_i} = \frac{120° - 40°}{4 - 2} = \frac{80}{2} = 40$$

$$\frac{\theta_2 - \theta_i}{x_2 - x_i} = r_x, \frac{\theta_2 - 40°}{3 - 2} = 40, \theta_2 = 80°$$

$$r_y = \frac{\phi_f - \phi_i}{y_f - y_i} = \frac{132 - 60}{32 - 8} = \frac{72}{24} = 3$$

$$\frac{\phi_2 - \phi_i}{y_f - y_i} = r_y = \frac{\phi_2 - 60}{18 - 8} = 3, \phi_2 = 90°$$

Precision point	*x*	*y*	*θ, deg*	*φ, deg*
1	2	8	40	60
2	3	18	80	90
3	4	32	120	132

cos 40° = 0.7660, cos 0.5000, cos (40° – 60°) = 0.9397

cos 80° = 0.1736, cos 90° = 00, cos (80° – 90°) = 0.9848

cos 120° = – 0.5, cos 132° = 0.6691, cos (120° – 132°) = 0.9781

$$A = \begin{vmatrix} 0.5000 & 0.7660 & 1 \\ 0.0000 & 0.1736 & 1 \\ -0.6691 & -0.5000 & 1 \end{vmatrix} = -0.0596$$

$$A_1 = \begin{vmatrix} 0.9397 & 0.7660 & 1 \\ 0.9848 & 0.1736 & 1 \\ -0.9791 & -0.5000 & 1 \end{vmatrix} = -0.03435$$

$$A_2 = \begin{vmatrix} 0.5000 & 0.9397 & 1 \\ 0.0000 & 0.9848 & 1 \\ -0.6691 & -0.9781 & 1 \end{vmatrix} = 0.0335,$$

$$A_3 = \begin{vmatrix} 0.5000 & 0.7660 & 0.9397 \\ 0.0000 & 0.1736 & 0.9848 \\ -0.6691 & -0.5000 & 0.9781 \end{vmatrix} = -0.0645$$

$$k_1 = \frac{A_1}{A} = 0.5763 = \frac{d}{a}, a = \frac{1}{0.5763} = 1.73$$

$$k_2 = \frac{A_2}{A} = -0.562 = \frac{-d}{c}, c = 1.78$$

$$k_3 = \frac{A_3}{A} = \frac{-0.0645}{-0.0596} = 1.082 = \frac{a^2 - b^2 + c^2 + d^2}{2ac}$$

$$= \frac{(1.73)^2 - b^2 + (1.78)^2 + 1}{2 \times 1.73 \times 1.78}, \; b = 0.7$$

The mechanism is shown in Fig. 11.18.

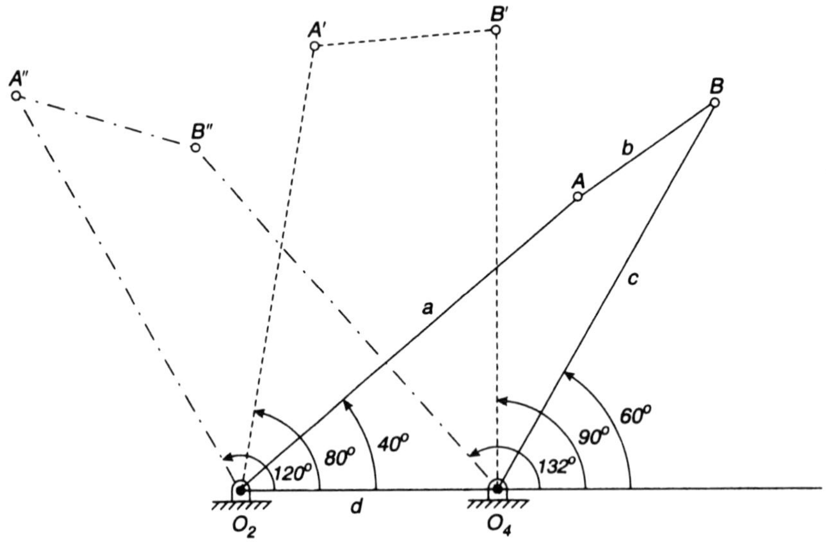

Fig. 11.18

■ **Example 11.5** Determine the lengths of all the four links of a four-bar mechanism to generate $y = \log x$ in the interval $1 \le x \le 11$ for three precision points. The length of the smallest link is 10 cm. The range of input angles is $45° \le \theta \le 105°$ and output angles $135° \le \phi \le 225°$.

Solution : $x_i = 1, x_f = 11, n = 3$

Using Chebyshev's precision points,

$$a = \frac{1}{2}(x_i + x_f) = \frac{1}{2}(1 + 11) = 6$$

$$b = \frac{1}{2}(x_f - x_i) = \frac{1}{2}(11 - 1) = 5$$

$$x_m = a + b \cos\left[\frac{(2m-1)\pi}{2}\right], \; m = 1, 2, 3$$

$$x_1 = a + b \cos\left[\frac{(2 \times 1 - 1)\pi}{6}\right] = a + b \cos\frac{\pi}{6} = 6 + 5\cos\frac{\pi}{6} = 10.33$$

$$x_2 = a + b \cos\frac{\pi}{2} = 6$$

$$x_3 = a + b \cos\left(\frac{5\pi}{6}\right) = 6 + 5\cos\left(\frac{5\pi}{6}\right) = 1.67$$

$$y_1 = \log x_1 = \log 10.33 = 1.014$$
$$y_2 = \log x_2 = \log 6 = 0.778$$
$$y_3 = \log x_3 = \log 1.67 = 0.223$$

$$y_i = \log x_i = \log 1 = 0$$
$$y_f = \log x_f = \log 1 = 1.0414$$

Scale factors are :

$$r_x = \frac{\theta_f - \theta_i}{x_f - x_i} = \frac{105 - 45}{11 - 1} = \frac{60}{10} = 6$$

$$r_y = \frac{\phi - \phi_i}{y_f - y_i} = \frac{225 - 135}{1.0414 - 0} = \frac{90}{1.0414} = 86.423$$

$$r_x = \frac{\theta - \theta_i}{x - x_i}, \frac{\theta_1 - \theta_i}{x_1 - x_i} = - \frac{\theta_1 - 45°}{10.33 - 1} = 6, \theta_1 + 100.98°$$

$$\frac{\theta_2 - 45°}{6 - 1} = 6, \theta_2 = 75°, \frac{\theta_3 - 45°}{1.67 - 1} = 6, \theta_3 = 49.02°$$

$$r_y = \frac{\phi - \phi_i}{y - y_i}, \frac{\phi_1 - \phi_i}{y_1 - y_i} = \frac{\phi - 135°}{1.014 - 0} = 86.423, \phi_1 = 222.63°$$

$$\frac{\phi_2 - 135°}{0.778 - 0} = 86.423, \phi_2 = 202.24°$$

$$\frac{\phi_3 - 135°}{0.223 - 0} = 86.423, \phi_3 = 154.27°$$

Precision point	x	y	ϕ, deg	$\cos\phi$	θ, deg	$\cos\theta$	$\theta - \phi$, deg	$\cos(\theta - \phi)$
1	10.33	1.014	222.63	−0.736	100.98	− 0.190	− 121.65	− 0.525
2	6	0.778	202.24	− 0.926	75	0.259	− 127.24	− 0.605
3	1.67	0.223	154.27	− 0.901	49.02	0.656	− 105.25	− 0.263

The Freudenstein's equations becomes.

$$\begin{bmatrix} -0.736 & -0.190 & 1 \\ -0.926 & 0.259 & 1 \\ -0.901 & 0.556 & 1 \end{bmatrix} \begin{Bmatrix} k_1 \\ k_2 \\ k_3 \end{Bmatrix} = \begin{Bmatrix} -0.525 \\ -0.605 \\ -0.263 \end{Bmatrix}$$

$$A = \begin{vmatrix} -0.736 & -0.190 & 1 \\ -0.926 & 0.259 & 1 \\ -0.901 & 0.656 & 1 \end{vmatrix} = -0.087$$

$$A_1 = \begin{vmatrix} -0.525 & -0.190 & 1 \\ -0.605 & 0.259 & 1 \\ -0.263 & 0.656 & 1 \end{vmatrix} = -0.185,$$

$$A_2 = \begin{vmatrix} -0.736 & -0.525 & 1 \\ -0.926 & 0.605 & 1 \\ -0.901 & 0.263 & 1 \end{vmatrix} = -0.063$$

$$A_3 = \begin{vmatrix} -0.736 & -0.190 & -0.525 \\ -0.926 & 0.259 & -0.605 \\ -0.9C1 & 0.656 & -0.263 \end{vmatrix} = -0.103$$

$$k_1 = \frac{A_1}{A} = \frac{-0.185}{-0.087} = 0.126 = \frac{d}{a}, \quad a = \frac{d}{2.126}$$

$$k_2 = \frac{A_2}{A} = \frac{-0.063}{-0.087} = 0.724 = -\frac{d}{c} \text{ or } \frac{d}{c} = -0.724,\ c = \frac{-d}{0.724}$$

$$k_3 = \frac{A_3}{A} = \frac{-0.103}{-0.087} = 1.184 = \frac{a^2 - b^2 + c^2 + d^2}{2ac}$$

$$\left(\frac{d}{2.126}\right)^2 - b^2 + \left(\frac{-d^2}{0.724}\right) + d^2 = 1.184 \times 2 \times \frac{d}{2.126} \times \left(\frac{-d^2}{0.724}\right)$$

$$\frac{d^2}{4.52} - b^2 + \frac{d^2}{0.524} + d^2 = -1.538\, d^2$$

$$d^2 = (0.221 + 1.908 + 1 + 1.3538) = b^2$$

$$4.667 d^2 = b^2$$

$$\frac{d}{b} = 0.463$$

$$\therefore \quad \frac{d}{a} = 2.126, \frac{d}{b} = 0.463, \frac{d}{c} = -0.724$$

$$\therefore \quad \frac{d}{a} > \frac{d}{c} > \frac{d}{b}$$

or

$$a \angle d \angle c \angle b$$

∴ Link a is the smallest. Thus $a = 10$ cm, $d = 21.26$ cm

$$b = \frac{21.26}{0.463} = 45.92 \text{ cm}, \quad c = \frac{-21.26}{0.724} = -29.36 \text{ cm}.$$

– ve sign indicates that the length c is to be drawn in the reverse direction.

■ **Example 11.6** Design a four-bar mechanism so that $\theta_{12} = 70°$ and $\phi_{12} = 50°$. Length of fixed link is 4 cm. Input and ouput links rotate anti-clockwise.

Solution : Refer to Fig. 11.19.

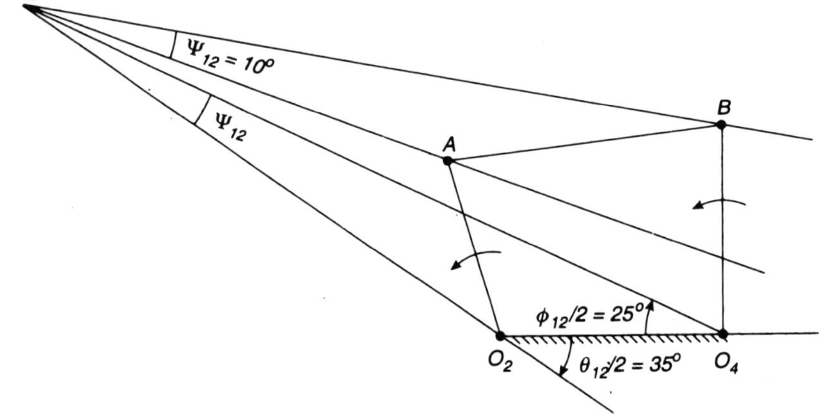

Fig. 11.19

1. Draw $O_2O_4 = 4$ cm. Rotate O_2O_4 about O_2 through $\theta_{12}/2 = 35°$ clockwise.
2. Rotate O_2O_4 about O_4 through $\phi_{12}/2 = 25°$ clockwise. The point of intersection of these two lines locates relative pole R_{12}.
3. Construct an angle $\Psi_{12} = \frac{1}{2}(\theta_{12} - \phi_{12}) = \frac{1}{2}(70° - 50°) = 10°$ at R_{12}. Join any two points on the two arms of this angle to obtain the coupler link AB. Join A with O_2 and B with O_4 to get the input and output links.
4. O_2ABO_4 is the desired mechanism.

■ **Example 11.7** Design a four-bar mechanism such that $\theta_{12} = 120°$, $\theta_{13} = 160°$ and $\phi_{12} = 70°$, $\phi_{13} = 100°$. Input link rotates clockwise and output link also rotates clockwise. Length of fixed link is 5 cm.

Solution : Refer to Fig.11.20

$$\theta_{12}/2 = \frac{120}{2} = 60°, \theta_{12}/2 = \frac{70}{2} = 35°, \Psi = \frac{1}{2}(\theta_{12} - \phi_{12}) = 25°$$

$$\theta_{13}/2 = \frac{160}{2} = 80°, \theta_{13}/2 = 50°, \Psi_{13} = \frac{1}{2}(\theta_{13} - \phi_{12}) = 30°$$

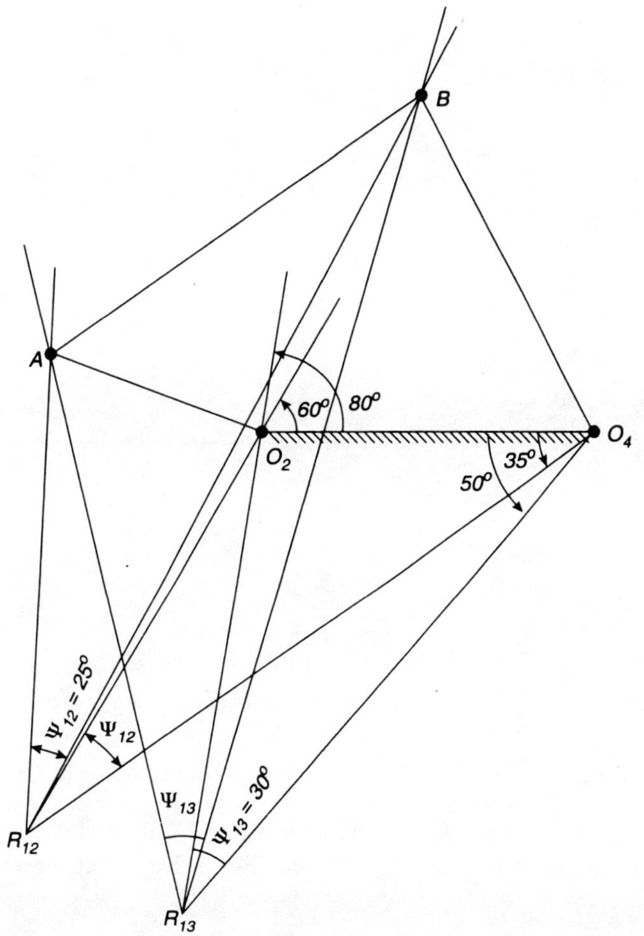

Fig. 11.20

1. Draw $O_2O_4 = 4$. Rotates R_{12} and R_{13}.
2. Construct angles Ψ_{13} and Ψ_{12} at R_{12} and R_{13} respectively such that the arms of the angles intersect at points A and B. Join AB to get the coupler.
3. Join O_2A and O_4B to get the input and output links respectively.
4. O_3ABO_4 is the required mechanism.

■ **Example 11.8** Design a four-bar mechanism such that $\theta_{12} = 120°$, $\theta_{13} = 160°$ and $\phi_{12} = 70°$, $\phi_{12} = 70°$. $\phi_{13} = 100°$. Input link rotates clockwise and output link rotates anti-clockwise. Length of fixed link is 10 cm.

Solution : Refer to Fig. 11.21

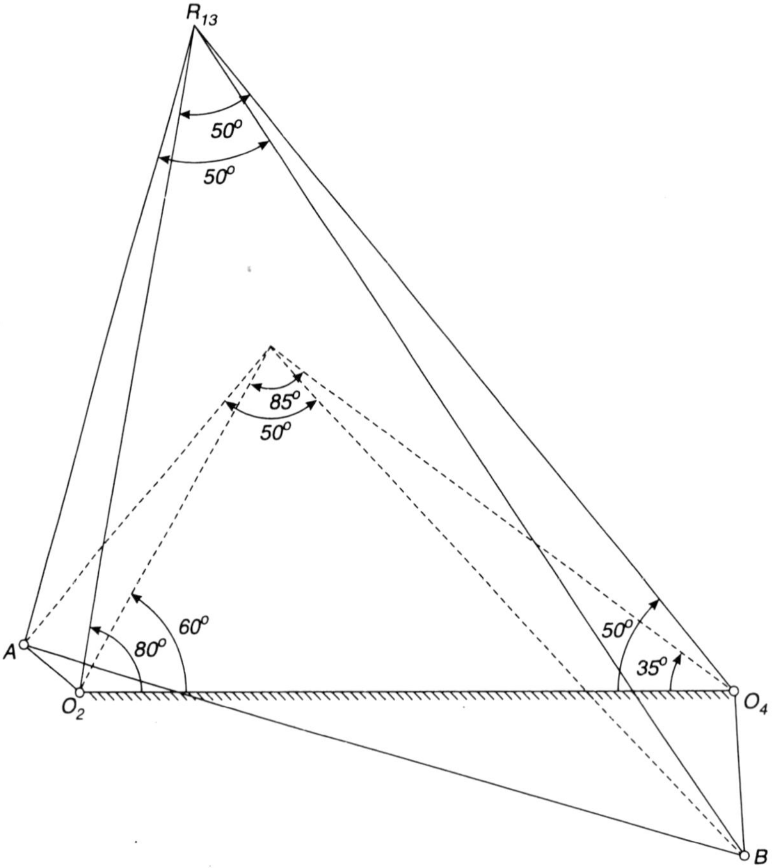

Fig. 11.21

$$\theta_{12}/2 = 60°,\ \phi_{12}/2 = 35°,\ \Psi_{12} = 180° = \frac{1}{2}(\theta_{12} + \phi_{12}) = 180° = 85°$$

$$\theta_{13}/2 = 80°,\ \phi_{13}/2 = 50°,\ \Psi = 180° - \frac{1}{2}(\theta_{13} - \phi_{13}) = 180° - (80 + 50) = 50°$$

1. Draw $O_2O_4 = 10\,\text{cm}$ with O_2 as centre rotate O_2O_4 anti-clockwise through 60°. With O_4 as center rotate O_2O_4 through 35° clockwise. Intersection of these two arcs given R_{12}.

2. With O_2 as centre rotate O_2O_4 anti-clockwise through 80°. With O_4 as centre rotate O_2O_4 through 50° clockwise. Intersection of these two arcs give R_{13}.
3. Construct angles Ψ_{12} and Ψ_{13} at the points R_{12} and R_{13} respectively such that the arms of the angles intersect at A and B. Join AB to get the coupler AB.
4. Join A with O_2 and B with O_4 to get the input and output links.

■ **Example 11.9** Synthesise a slider crank mechanism with eccentricity, $e = 1$ cm for the two input positions of input link, $\theta_{12} = 60°$ and output displacement, $x = 1.5$ cm to the right.

Solution : Refer to Fig. 11.22

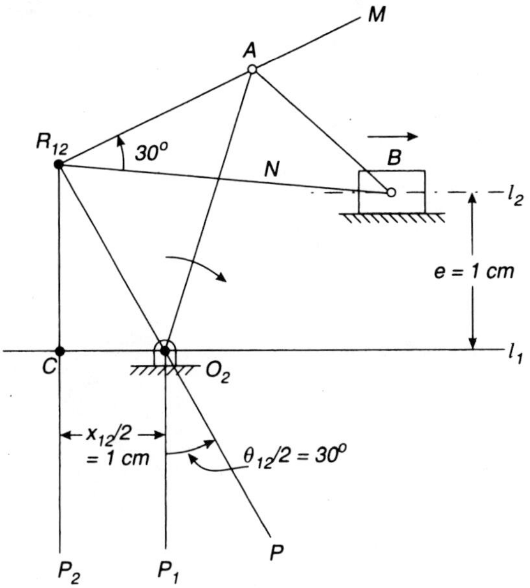

Fig. 11.22

1. Draw two parallel lines l_1 and l_2 at an eccentricity, $e = 1$ cm.
2. Select an arbitrary point O_2 on line l_1. Measure $CO_4 = \frac{x}{2} = \frac{1.5}{2} = 0.75$ cm the left.

   Draw CP_2 and O_2P_1 perpendicular to lines l_1.
3. Make $\angle\theta_{12}/2 = 30° = \angle P_1O_2P$ in the anti-clockwise direction. The intersection of P_1O_2 and P_2C produced locates the relative pole R_{12}. Join $R_{12}B$.
4. Make $\angle\theta_{12}/2 = 30°$ with BR_{12} in the anti-clockwise direction.
5. Choose an arbitrary point A on the arm of the angle other than the one passing through point B. Join AB to get the coupler. Also join AO_2 to get the input crank.

■ **Example 11.10** Synthesise a slider-crank mechanism for its three positions $\theta_{12} = 60°$ and $\theta_{13} = 100°$ of the input crank and three positions $x_{12} = 2$ cm and $x_{13} = 5$ cm of the output slider block.The eccentricity is 2 cm. The slider is moving outwards.

Solution : Refer to Fig. 11.23.

1. Draw two parallel lines l_1 and l_2 at eccentricity, $e = 2$ cm.
2. Select an arbitrary point O_2 on l_1. Measure $O_2C_1 = x_{12}/2 = 1$ cm and $O_2C_2 = x_{13}/2 = 2.5$ cm to the left of O_2. Draw O_2P_1, C_1P_2, C_2P_3 perpendicular on line l_1.

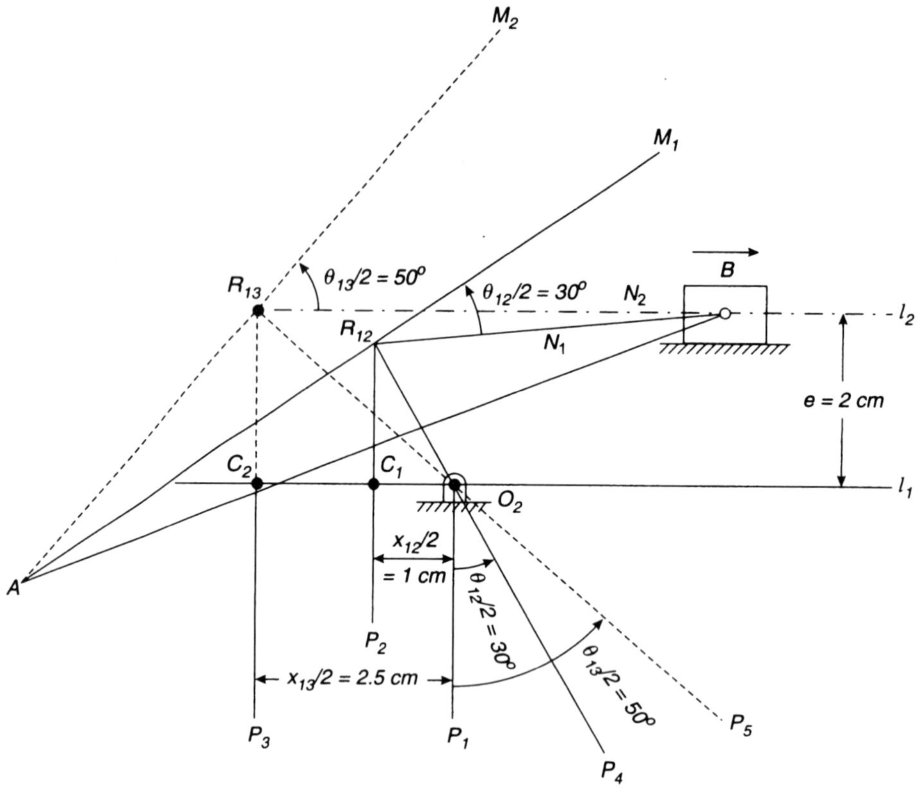

Fig. 11.23

3. Draw $\angle P_1O_2P_4 = \theta_{12}/2 = 30°$ and $\angle P_1O_2P_5 = \theta_{13}/2 = 50°$ in the anti-clockwise direction.
4. Produce O_2P_4 and O_2P_5 lines to meet the perpendiculars at C_1 and C_2 to locate the relative poles R_{12} and R_{13}.
5. Join $R_{12}B$ and $R_{13}B$.
6. Draw $\angle BR_{12}M_1 = \theta_{12}/2 = 30°$ and $\angle BR_{13}M_2 = \theta_{13}/2 = 50°$ in the anti-clockwise direction. Produce the lines $R_{12}M_1$ and $R_{13}M_2$ bookward to meet at A.
7. Join AB and AO_2 to get the coupler and input crank respectively.

■ **Example 11.11** For the four-bar linkage the following data are given :

$$\theta_2 = 60°, \theta_4 = 90°$$
$$\omega_2 = 3 \text{ rad/s}, \omega_4 = 2 \text{ rad/s}$$
$$\alpha_2 = -1 \text{ rad/s}^2, \alpha_4 = 0$$

Determine the link-length ratios.

Solution : The Freudenstein's equation for the four-bar mechanism shown in Fig. 11.24 is :

$$k_1 \cos\theta_4 + k_2 \cos\theta_2 + k_3 = \cos(\theta_2 - \theta_4) \quad ...(1)$$

where $k_1 = \dfrac{d}{a}$, $k_2 = -\dfrac{d}{c}$, $k_3 = \dfrac{a^2 - b^2 + c^2 + d^2}{2ac}$

Taking the first time derivative of Eq. (1), we have

$$k_1\omega_4 \sin\theta_4 + k_2\,\omega_2 \sin\theta_2 = (\omega_2 - \omega_4)\sin(\theta_2 - \theta_4) \quad ...(2)$$

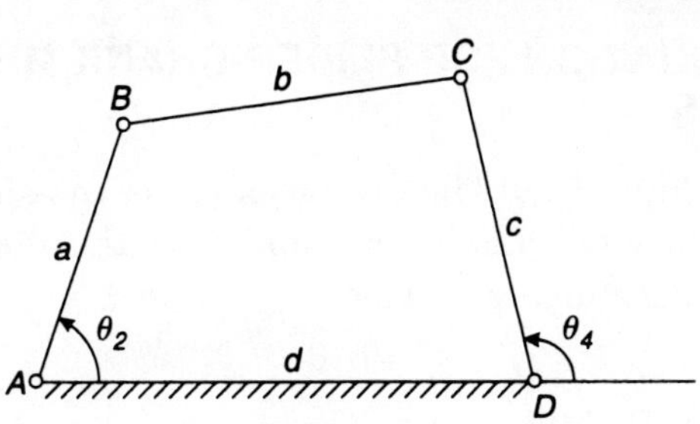

Fig. 11.24

Taking the second time derivative of Eq. (2)

$$k_1 (\omega_4^2 \cos \theta_4 + \alpha_4 \sin \theta_4) + k_2 (\omega_2^2 \cos \theta_2 + \alpha_2 \sin \theta_2)$$
$$= (\alpha_2 - \omega_4) \sin (\theta_2 - \theta_4) + (\omega_2 - \omega_4)^2 \cos (\theta_2 - \theta_4) \quad ...(3)$$

Substituting the values of various terms in Eqs. (1) to (3), we get

$$k_1 \cos 90° + k_2 \cos 60° + k_3 = \cos (60° - 90°)$$

or $$0.5\, k_2 + {}_3 = \frac{\sqrt{3}}{2} \quad ...(4)$$

or $$k_2 + 2k_3 = \sqrt{3}$$

$$k_1 \times 2 \times \sin 90° + k_2 \times 3 \sin 60° = (3 - 2) \sin (60° - 90°)$$

or $$2k_1 + \frac{3\sqrt{3}}{2} k_2 = -\frac{1}{2}$$

or $$4k_1 + 3\sqrt{3}k_2 = -1 \quad ...(5)$$

$$k_1 (4 \cos 90° + 0 \times \sin 90°) + k_2 (9 \times \cos 60° - 1 \times \sin 60°)$$
$$= (-1 - 0) \times \sin (60° \times 90°) + (3 - 2)^2 \times \cos (60° - 90°)$$

or $$k_2 \left(9 \times \frac{1}{2} - \frac{\sqrt{3}}{2} \right) = (-1) \times -\frac{1}{2} + 1 \times \frac{\sqrt{3}}{2}$$

$$k_2 (4.5 - 0.866) = \frac{1}{2} + \frac{\sqrt{3}}{2} \quad ...(6)$$

$$k_2 = 0.376$$

Solving Eqs. (4) to (6), we get

$$k_1 = -0.738, k_3 = 0.678$$

$$k_2 = \frac{d}{a}, a = \frac{-1}{0.738} = -1.355$$

$$k_2 = -\frac{d}{c}, c = \frac{-1}{0.376} = -2.659$$

$$k_3 = \frac{a^2 - b^2 + c^2 + d^2}{2ac}$$

$$0.678 \times 2 \times (-1.355) \times (-2.659) = (-1.355)^2 - b^2 + (-2.659)^2 + 1$$

$$b^2 = 5.0207, b = 2.24$$

Thus $$d = 1.0, a = -1.355, b = 2.24, c = -2.659$$

11.8 FREDENTEIN'S EQUATION FOR SLIDER-CRANK MECHANISM FOR THREE PRECISION POINTS

The mechanism is shown in Fig. 11.25. The displacement of the slider B has to be coordinated with the rotation of the crank OA. Let us assume that the displacement of the slider is proportional to the crank rotation expressed as :

$$s_f - s_i = c\,(\theta_f - \theta_i) \qquad ...(1)$$

where c = a constant of proportionality

s = distance of slider from the origin O.

θ = angle of rotation of crank from the line of stroke.

i, f = subscripts for initial and final values respectively.

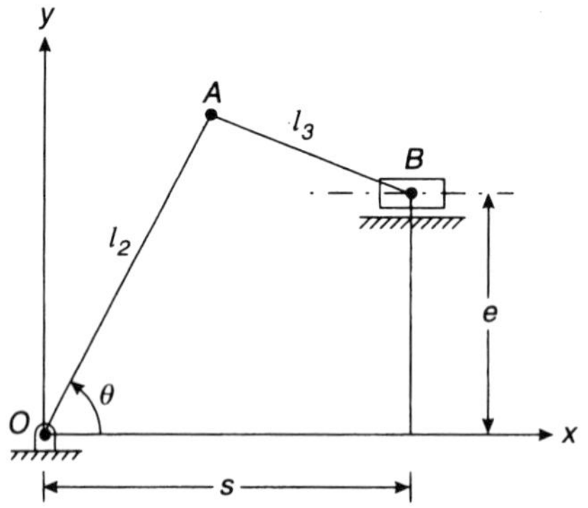

Fig. 11.25

The coordinates of points A and B are :

$$x_A = l_2 \cos\theta,\ y_A = l_2 \sin\theta$$

$$x_B = s,\ y_B = e$$

Also
$$AB^2 = (x_B - x_A)^2 + (y_A - y_B)^2$$

$$= (s - l_2 \cos\theta)^2 + (l_2 \sin\theta - e)^2$$

$$= s^2 + l_2^2 \cos^2\theta - 2\,s\,l_2 \cos\theta + l_2^2 \sin^2\theta + e^2 - l_2 e \sin\theta$$

$$l_3^2 = s^2 + l_2^2 + e^2 - 2s\,l_2 \cos\theta - l_2\,e \sin\theta$$

or
$$s^2 = (l_3^2 - l_2^2 - e^2) + 2sl_2 \cos\theta + 2l_2 e \sin\theta$$

Let $k_1 = 2l_2$, $k_2 = 2l_{2e}$, and $k_3 = l_3^2 - l_2^2 - e^2$

Then
$$s^2 = k_1 s \cos\theta + k_2 \sin\theta + k_3 \qquad ...(11.14)$$

Equation (11.14) is the Freudenstein's equation for the slider-crank mechanism.

Now
$$c = \frac{s_f - s_i}{\theta_f - \theta_i}$$

$\therefore$
$$s - s_i = c\,(\theta - \theta_i)$$

The precision points according to Chebyshve are given by,

$$\theta_m = a + b \cos\left[\frac{(2m-1)\pi}{2n}\right],\ m = 1, 2, 3$$

where $a = \frac{\theta_i - \theta_f}{2}$, $b = \frac{\theta_f - \theta_i}{2}$, $n = 3$

The following steps may be followed to determine lengths.

1. Obtain the three accuracy points by using Chebyshev's spacing, *i.e.*, θ_1, θ_2, and θ_3.
2. Calculate values of s_1, s_2 and s_3 from the following equation :
$$s_m - s_i = c(\theta_m - \theta_i), m = 1, 2, 3.$$
3. Obtain three equations in k_1, k_2 and k_3.
4. Solve for k_1, k_2 and k_3 using Cramer's rule.
5. Obtain the values of l_2, l_3 and e from the values of k_1, k_2 and k_3.

■ **Example 11.12** Design a slider-crank mechanism so that displacement of the slider is proportional to the crank rotation in the interval $30° \le \theta \le 100°$. Assume initial distance of the slider equal to 15 cm and final distance to be 10 cm.

Solution : Given : $(s - s_i) \propto (\theta - \theta_i)$

or
$$s - s_i = c(\theta - \theta_i)$$

where c = constant of proportinality.

Now $s = s_f = 10\,\text{cm}$ when $\theta = \theta_f = 100°$

and $s = s_i = 15\,\text{cm}$ when $\theta = \theta_i = 30°$

$$\therefore \quad s_f - s_i = c(\theta_f - \theta_i)$$

$$c = \frac{10 - 15}{100 - 30} = \frac{-5}{70} = -\frac{1}{14}$$

$$\therefore \quad s - s_i = \left(-\frac{1}{14}\right)(\theta - \theta_i)$$

The Chebyshve's precision spacing are given by :

$$\theta_m = a + b\cos\left[\frac{(2m-1)\pi}{2n}\right], m = 1, 2, 3$$

where $a = \frac{\theta_i + \theta_f}{2} = \frac{30 + 100}{2} = 65°$

$$b = \frac{\theta_f - \theta_i}{2} = \frac{1}{2}(100 - 30) = 35°$$

$$\theta_1 = 65 + 35\cos\frac{\pi}{6} = 95.3°$$

$$\theta_2 = 65 + 35\cos 90° = 65°$$

$$\theta_3 = 65 + 35\cos\frac{5\pi}{6} = 34.69°$$

Now
$$s_m - s_i = \left(-\frac{1}{14}\right)(\theta_m - \theta_i)$$

$$\therefore \quad s_1 - 15 = \left(-\frac{1}{14}\right)(98.3 - 30), s_1 = 10.336\,\text{cm}$$

$$s_2 = 15 = \left(-\frac{1}{14}\right)(65 - 30), s_2 = 12.50\,\text{cm}$$

$$s_3 - 15 = \left(-\frac{1}{14}\right)(34.69 - 30), s_3 = 14.665 \text{ cm}$$

Now
$$k_1 s \cos\theta + k_2 \sin\theta + k_3 = s^2$$

$$k_1 \times 10.336 \times \cos 95.3° + k_2 \sin 95.3° = (10.336)^2$$

$$-0.95474\, k_1 + 0.90631\, k_2 + k_3 = 106.833 \quad ...(1)$$

$$k_1 \times 12.50 \times \cos 65° + k_2 \sin 65° + k_3 = (12.50)^2$$

$$5.28273\, k_1 + 0.90631\, k_2 + k_3 = 156.25 \quad ...(2)$$

$$k_1 \times 14.665 \times \cos 34.69° + k_2 \sin 34.69° + k_3 = (14.665)^2$$

$$12.0582\, k_1 + 0.56914\, k_2 + k_3 = 215.062 \quad ...(3)$$

Subtracting Eqs. (1) and (2) from Eq. (3), we get

$$13.013\, k_1 - 0.4266\, k_2 = 108.229 \quad ...(4)$$

$$6.775\, k_1 - 0.3372\, k_2 = 58.812 \quad ...(5)$$

$$30.504\, k_1 - k_2 = 253.70 \quad ...(6)$$

$$20.092\, k_1 - k_2 = 174.41 \quad ...(7)$$

Subtracting Eq. (7) from Eq. (6), we get

$$10.412\, k_1 = 79.29$$

$\therefore$
$$k_1 = 7.615$$

Then
$$k_2 = -21.4$$

$$k_3 = 135.418$$

Now
$$k_1 = 2l_2,\ l_2 = \frac{7.615}{2} = 3.808 \text{ cm}$$

$$k_2 = 2l_2 e,\ e = \frac{-21.4}{7.615} = -2.81 \text{ cm}$$

$$k_3 = l_3^2 - l_2^2 - e^2$$

$$135.418 = l_3^2 - (3.808)^2 - (2.81)^2$$

$$l_3^2 - 157.815$$

$$l_3 = 12.562$$

Thus $l_2 = 3.808$ cm, $l_3 = 12.562$ cm, $e = -2.81$ cm

■ **Example 11.13** Design a slider crank mechanism so that the displacement of the slider is proportional to the square of the crank rotation in the interval $45° \le \theta \le 135°$. Use three points Chebyshev spacing.

Solution : Consider the slider-crank mechanism as shown in Fig. 11.26.

Let l_2 = length of crank

l_3 = length of connecting rod

s = distance of slider from the crank shaft

e = eccentricity of slider from crank shaft.

Then
$$\left.\begin{aligned} x_A &= l_2 \cos,\ y_A = l_2 \sin\theta \\ x_B &= s,\ y_B = e \end{aligned}\right\} \quad ...(1)$$

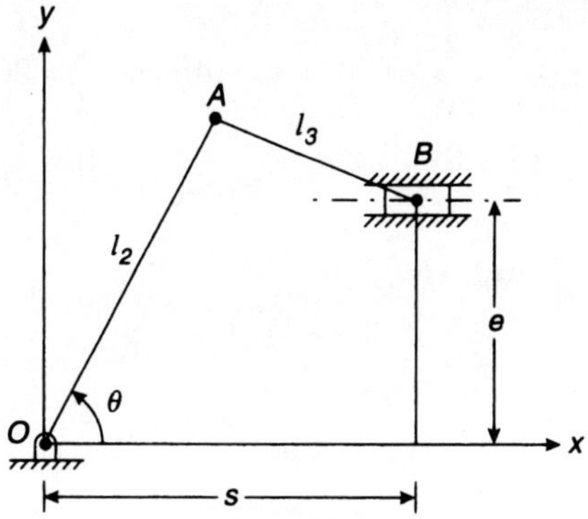

Fig. 11.26

$$\text{Now } AB^2 = (x_B - x_A)^2 + (y_A - y_B)^2$$
$$= (s - l_2 \cos\theta)^2 + (l_2 \sin\theta - e)^2$$
$$= s^2 + l_2^2 \cos^2\theta - 2sl_2 \cos\theta + l_2^2 \sin^2\theta + e^2 - 2l_2 e \sin\theta$$
$$l_3^2 = s^2 + l_2^2 + e^2 - 2sl_2 \cos\theta - 2l_2 e \sin\theta$$

or
$$s^2 = l_3^2 - l_2^2 - e^2 + 2sl_2 \cos\theta + 2l_2\, e \sin\theta$$

Let
$$A_1 = 2l_2,\ A_2 = 2l_2 e \text{ and } A_3 = l_2^2 - l_3^2 + e^2 \quad \text{...(2)}$$

Then
$$A_1 s \cos\theta + A_2 \sin\theta - A_3 = s^2 \quad \text{...(3)}$$

The displacement Eq. (3) has three variables A_1, A_2, and A_3.

Let
$$s_i = 10\,\text{cm and } s_f = 3\,\text{cm}$$

where subscripts i and f represent the initial and final values.

Given :
$$s_f - s_i = c\,(\theta_f - \theta_i)^2$$

or
$$c = \frac{s_f - s_i}{(\theta_f - \theta_i)} = \frac{3 - 10}{(135 - 45)^2} = \frac{-7}{(90)^2} = \frac{-7}{8100}$$

where c = constant of proportionality

Using Chebyshev's equation for precision spacing, we have
$$x_m = a - b\cos\left[\frac{(2m-1)\pi}{2n}\right],\ m = 1, 2, 3$$

where $a = \dfrac{x_i + x_f}{2}$, $b = \dfrac{x_f - x_i}{2}$, $n = 3$ number of precision points

Here $x_i = \theta_i$, the angular position of crank.
$$a = \frac{\theta_i + \theta_f}{2} = \frac{45 + 135}{2} = 90°$$
$$b = \frac{\theta_f + \theta_i}{2} = \frac{135 + 45}{2} = 45°$$
$$\theta_1 = 90 - 45\cos\left[\frac{(2 \times 1 - 1)\pi}{2 \times 3}\right] = 90 - 45\cos\frac{\pi}{6} = 51.03°$$

$$\theta_2 = 90 - 45 \cos\left[\frac{(2 \times 2 - 1)\pi}{6}\right] = 90 - 45 \cos\frac{\pi}{2} = 90°$$

$$\theta_3 = 90 - 45 \cos\left[\frac{(2 \times 3 - 1)\pi}{6}\right] = 90 - 45 \cos\frac{5\pi}{6} = 128.97°$$

The given relationship for s and θ is :

$$s - s_i = c(\theta - \theta_i)^2$$

$$\therefore \quad s_1 - 10 = \left(-\frac{7}{8100}\right)(51.03 - 45)^2$$

$$s_1 = 9.97 \text{ cm}$$

$$s_2 - 19 = \left(-\frac{7}{8100}\right)(90 - 45)^2$$

$$s_2 = 8.25 \text{ cm}$$

$$s_3 - 10 = \left(-\frac{7}{8100}\right)(128.97 - 45)^2$$

$$s_3 = 3.91 \text{ cm}$$

The precision points obtained are :

Precision point	***s, cm***	***θ, deg***
1	9.97	51.03
2	8.25	90
3	3.91	128.97

Substituting in Eq. (3), we get

$$A_1 \times 9.97 \cos 51.3° + A_2 \sin 51.03° - A_3 = (9.97)^2 \quad \ldots(4)$$

or

$$6.27\,A_1 + 0.78A_2 - A_3 = 99.4$$

$$A_1 \times 8.25 \cos 90° + A_2 \sin 90° - A_3 = (8.25)^2$$

or

$$A_2 - A_3 = 68.06 \quad \ldots(5)$$

$$A_1 \times 3.91 \cos 128.97° + A_2 \sin 128.97° - A_3 = (3.91)^2$$

$$-2.459A_1 + 0.78A_2 - A_3 = 15.29 \quad \ldots(6)$$

We have

$$\begin{bmatrix} 6.27 & 0.78 & -1 \\ 0 & 1 & -1 \\ -2.459 & 0.78 & -1 \end{bmatrix} \begin{Bmatrix} A_1 \\ A_2 \\ A_3 \end{Bmatrix} = \begin{Bmatrix} 99.4 \\ 68.06 \\ 15.29 \end{Bmatrix}$$

$$[B]\{A\} = \{C\}$$

$$|B| = -1.92$$

$$A_1 = \frac{\begin{vmatrix} 99.4 & 0.78 & -1 \\ 68.06 & 1 & -1 \\ 15.29 & 0.78 & -1 \end{vmatrix}}{-1.92} = \frac{-18.5}{-1.92} = 9.64$$

$$A_2 = \frac{\begin{vmatrix} 6.27 & 99.4 & -1 \\ 0 & 68.06 & -1 \\ -2.459 & 15.29 & -1 \end{vmatrix}}{-1.92} = \frac{-253.8}{-1.92} = 132.2$$

$$A_3 = \frac{\begin{vmatrix} 6.27 & 0.78 & 99.4 \\ 0 & 1 & 68.06 \\ -2.459 & 0.78 & 15.29 \end{vmatrix}}{-1.92} = \frac{-123.1}{-1.92} = 64.11$$

$$l_2 = \frac{A_1}{2} = 4.82\,\text{cm},\; e = \frac{A_2}{2l_2} = \frac{132.2}{2 \times 4.82} = 13.7\,\text{cm}$$

$$A_3 = l_2^2 - l_3^2 + e^2,$$

$$l_3^2 = l_2^2 + e^2 - A_3 = (4.82)^2 + (13.7)^2 - 64.11 = 146.81$$

$$l_3 = 12.12\,\text{cm}$$

■ **Example 11.14** Design a 4 bar mechanism to coordinate the input and output angles as follows :

Input angles	15°	30°	45°
Output angles	30°	40°	55°

Also draw the designed mechanism for the above values of input and output.

Solution : Freudenstein's equation for a four-bar chain in (Fig. 11.27)

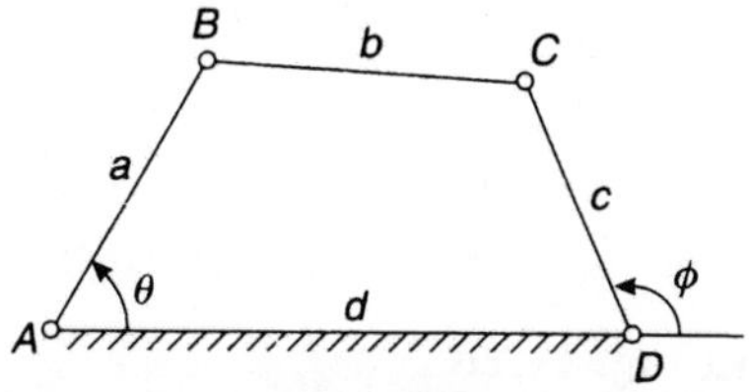

Fig. 11.27

$$k_1 \cos\phi + k_2 \cos\theta + k_3 = \cos(\theta - \phi)$$

where $k_1 = \dfrac{d}{a}$, $k_2 = -\dfrac{d}{c}$ and $k_3 = \dfrac{a^2 - b^2 + c^2 + d^2}{2ac}$

ϕ, deg	cos ϕ	θ, deg	cos θ	$\theta - \phi$, deg	cos $(\theta - \phi)$
30	0.866	15	0.966	– 15	0.966
40	0.766	30	0.866	– 10	0.985
55	0.573	45	0.707	– 10	0.985

$$\begin{bmatrix} 0.866 & 0.966 & 1 \\ 0.766 & 0.866 & 1 \\ 0.573 & 0.703 & 1 \end{bmatrix} = \begin{Bmatrix} k_1 \\ k_2 \\ k_3 \end{Bmatrix} = \begin{Bmatrix} 0.966 \\ 0.985 \\ 0.985 \end{Bmatrix}$$

$$|A| = \begin{vmatrix} 0.866 & 0.966 & 1 \\ 0.766 & 0.866 & 1 \\ 0.573 & 0.703 & 1 \end{vmatrix} = -3 \times 10^{-3}$$

$$A_1 = \begin{vmatrix} 0.966 & 0.966 & 1 \\ 0.985 & 0.866 & 1 \\ 0.985 & 0.703 & 1 \end{vmatrix} = -3.097 \times 10^{-3}$$

$$A_2 = \begin{vmatrix} 0.866 & 0.966 & 1 \\ 0.766 & 0.985 & 1 \\ 0.573 & 0.985 & 1 \end{vmatrix} = -3.667 \times 10^{-3}$$

$$A_3 = \begin{vmatrix} 0.866 & 0.966 & 0.966 \\ 0.766 & 0.866 & 0.985 \\ 0.573 & 0.703 & 0.985 \end{vmatrix} = -3.758 \times 10^{-3}$$

$$k_1 = \frac{|A_1|}{|A|} = \frac{-3.097}{-3} = \pm 1.032, k_2 = \frac{3.667}{-3} = -1.222,$$

$$k_3 = \frac{-3.758}{-3} = 1.253$$

$$k_1 = \frac{d}{a}, \text{Let } d = 1 \text{ unit}, a = \frac{d}{k_1} = \frac{1}{1.032} = 0.969 \text{ unit}$$

$$k_2 = -\frac{d}{c}, c = -\frac{d}{k_2} = \frac{-1}{-1.222} = 0.818 \text{ unit}$$

$$k_3 = \frac{a^2 - b^2 + c^2 + d^2}{2ac},$$

$$1.253 \times 2 \times 0.969 \times 0.818 = (0.969)^2 - b^2 + (0.818)^2 + 1$$

$$b^2 = 0.6217, b = 0.788 \text{ unit}$$

The mechanism is shown in Fig. 11.28.

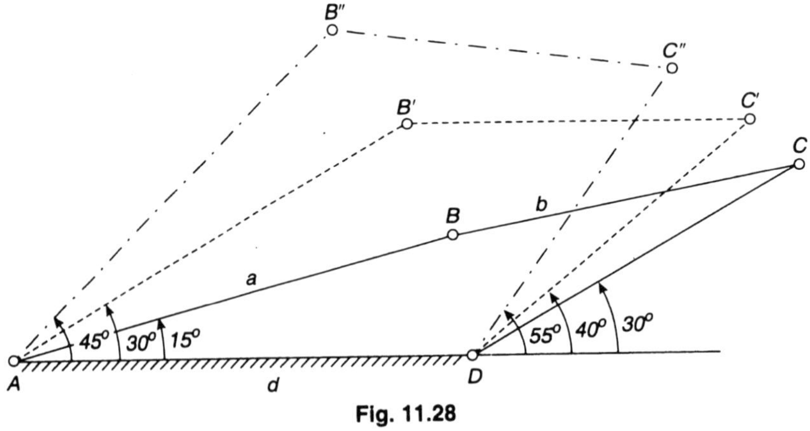

Fig. 11.28

■ **Example 11.15** Synthesize a four bar function generator to solve the equation $y = 2x^2 - 1$, $1 < x \le 2$. Use three precision points and Chebychev spacing. Take a $\Delta\phi = 60°$. $\Delta\Psi = 90°$, $\phi_0 = 30°$ and $\Psi_0 = 60°$, where $\Delta\phi$, $\Delta\Psi$ are ranges of input and output link rotations and Ψ_0, Ψ_0 are initial angular positions of input and output links respectively.

Solution : Chebyshev's precision points are given by,

$$x_m = a + b\cos\left[\frac{(2m-1)\pi}{2n}\right],\ m = 1, 2, 3$$

$n = 3, x_i = 1, x_f = 2^{2n}$

$$a = \frac{x_i + x_f}{2} = \frac{1+2}{2} = 1.5,\ b = \frac{x_f - x_i}{2} = \frac{2-1}{2} = 0.5$$

$$\therefore \quad x_i = 1.5 + 0.5\cos\left[\frac{(2m-1)\pi}{6}\right]$$

$$x_1 = 1.5 + 0.5\cos\frac{\pi}{6} = 1.933$$

$$x_2 = 1.5 + 0.5\cos\frac{\pi}{2} = 1.5$$

$$x_3 = 1.5 + 1.5\cos\frac{5\pi}{2} = 1.067$$

Now

$$y = 2x^2 - 1$$

$$y_1 = 2 \times (1.933)^2 - 1 = 6.473$$

$$y_2 = 2 \times (1.5)^2 - 1 = 3.5$$

$$y_3 = 2 \times (1.067)^2 - 1 = 1.277$$

Scale factors are :

$$r_x = \frac{\Delta\phi}{x_f - x_i} = \frac{60}{2-1} = 60$$

$$r_x = \frac{\Delta\Psi}{y_f - y_i}$$

Now

$$y_i = 2 \times 1^2 - 1 = 1$$

$$y_f = 2 \times 2^2 - 1 = 7$$

$$r_y = \frac{90}{7-1} = \frac{90}{6} = 15$$

$$r_x = \frac{\phi_m - \phi_i}{x_m - x_i},\ \phi_i = \phi_0 = 30°,\ x_i = 1$$

$$\therefore \quad \frac{\phi_1 - 30}{1.933 - 1} = 60,\ \phi_1 = 85.98°$$

$$\frac{\phi_2 - 30}{1.5 - 1} = 60,\ \phi_2 = 60°$$

$$\frac{\phi_3 - 30}{1.3067 - 1} = 60,\ \phi_3 = 34.02°$$

$$r_y = \frac{\Psi_m - \Psi_i}{y_m - y_i}, \phi_i = \phi_0 = 60°, y_1 = 34.02°$$

$$\frac{\Psi_1 - 60}{6.473 - 1} = 15, \Psi_1 = 142.1°$$

$$\frac{\Psi_2 - 60}{3.5 - 1} = 15, \Psi_2 = 97.5°$$

$$\frac{\Psi_3 - 60}{1.277 - 1} = 15, \Psi_3 = 64.155°$$

Precision point	*x*	*y*	Ψ, deg	cos Ψ	φ deg	cos φ	(φ – Ψ), deg	cos (φ – Ψ)
1	1.933	6.473	142.1	– 0.789	85.98	0.070	– 56.12	0.557
2	1.5	3.5	97.5	– 0.130	60	0.5	– 37.5	0.793
3	1.067	1.277	64.155	0.436	34.02	0.829	– 30.135	0.865

The Freudenstein's equations become :

$$\begin{bmatrix} -0.789 & 0.070 & 1 \\ -0.130 & 0.5 & 1 \\ -0.436 & 0.829 & 1 \end{bmatrix} = \begin{Bmatrix} k_1 \\ k_2 \\ k_3 \end{Bmatrix} = \begin{Bmatrix} 0.557 \\ 0.793 \\ 0.865 \end{Bmatrix}$$

$$[A]\{k\} = \{B\}$$

$$|A| = -0.02657$$

$$|A_1| = \begin{vmatrix} -0.557 & 0.070 & 1 \\ -0.793 & 0.5 & 1 \\ -0.865 & 0.829 & 1 \end{vmatrix} = 0.04668, k_1 = \frac{|A_1|}{|A|} = 1.757$$

$$|A_2| = \begin{vmatrix} -0.789 & 0.557 & 1 \\ -0.130 & 0.793 & 1 \\ -0.436 & 0.865 & 1 \end{vmatrix} = -0.57183, k_2 = \frac{|A_2|}{|A|} = 21.522$$

$$|A_3| = \begin{vmatrix} -0.789 & 0.070 & 0.557 \\ -0.130 & 0.5 & 0.793 \\ -0.436 & 0.865 & 1 \end{vmatrix} = -0.54377, k_3 = \frac{|A_3|}{|A|} = 20.465$$

$$k_1 = \frac{d}{a}, a = \frac{d}{k_1} = -\frac{1}{1.757} = -0.569$$

$$k_2 = -\frac{d}{c}, c = -\frac{d}{k_2} = \frac{1}{21.522} = -0.046$$

$$k_3 = \frac{a^2 - b^2 + c^2 + d^2}{2ac},$$

$$20.465 \times 2 \times (-0.569) \times (-0.046)^2 = (-0.569)^2 - b^2 + (0.046)^2 + 1$$

$$b^2 = 0.25457, b = 0.504$$

$$a = -0.569, b = 0.504, c = -0.046, d = 1$$

The sign indicates the reversed direction of length.

■ **Example 11.16** Synthesize, graphically a four-bar mechanism for three positions of input and output crank with $\Psi_{12} = 35°$, $\phi_{23} = 30°$, $\Psi_{12} = 40°$ and $\Psi_{23} = 60°$ where ϕ_{ij} and Ψ_{ij} are angular distance between position i and j of input and output links respectively.

Solution : Given $\phi_{12} = 35°, \phi_{23} = 30°; \Psi_{12} = 40°, \Psi_{23} = 60°$

$$\therefore \quad \phi_{13} = 35 + 30 = 65° \text{ and } \Psi_{13} = 100°]$$

$$\alpha_{12} = \frac{1}{2}(\phi_{12} - \Psi_{12}) = \frac{1}{2}(35 - 40) = -2.5°$$

$$\alpha_{13} = \frac{1}{2}(\phi_{13} - \Psi_{13}) = \frac{1}{2}(65 - 100) = -17.5°$$

Angle subtended by the coupler at the relative pole = angle subtended by the fixed link at the relative pole.

The four-bar mechanism with three positions may be synthesised as explained below : (Fig 11.29)

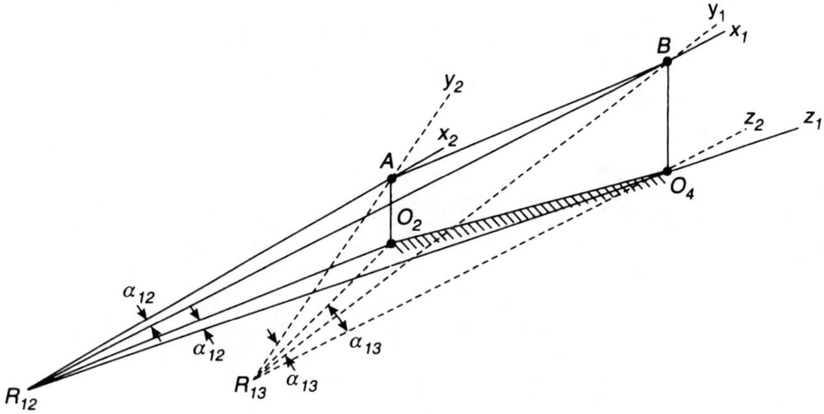

Fig. 11.29

1. Select relative poles R_{12} and R_{13} at positions arbitrarily chosen. Draw $\angle\alpha_{12} = 2.5°$ at R_{12} and $\angle\alpha_{13} = 17.5°$ at R_{13}.
2. Let x_2 and y_2 lines intersect at A and x_1 and y_1 lines at B. Join AB to get the coupler.
3. Select the fixed point O_2 of the fixed link O_2O_4 at an arbitrary position. Join O_2R_{12} and O_2R_{13}.
4. Draw $\angle O_2R_{12}Z_1 = \angle\alpha_{12} = 2.5°$ and $\angle O_2R_{13}Z_2 = \angle\alpha_{13} = 17.5°$. Let $R_{12}Z_1$ and $R_{13}Z_2$ lines intersect at O_4.
5. O_2A, O_2O_4 and O_4B. Then O_2ABO_4 is the required four-bar mechanism shown in Fig. 11.29.

■ **Example 11.17** Synthesize a four-bar mechanism to generate a function $y = \sin x$ for $0 \le x < 90°$. The range of the output crank may be chosen as 60° while that of input crank be 120°. Assume three precision point obtained from Chebyshev spacing. Assume fixed link to be 50 mm long and $\theta_1 = 105°$ and $\phi_1 = 60°$, where θ_j and ϕ_j are the angles made by input and output link at jth position.

Solution : $y = \sin x$ for $0 \le x \le 90°$

$$x_i = 0°, x_f = 90°$$

Cheybyshve's precision points are given by :

$$x_m = a + b\cos\left[\frac{(2m-1)\pi}{2n}\right]$$

$$a = \frac{1}{2}(x_i + x_f) = \frac{1}{2}(0 + 90) = 45°$$

$$b = \frac{1}{2}(x_f - x_i) = \frac{1}{2}(90 - 0) = 45°$$

$$x_m = 45 + 45\cos\left[\frac{(2m-1)\pi}{6}\right]$$

$$x_1 = 45 + 45\cos\frac{\pi}{6} = 82.97°,\ y_1 = 0.994$$

$$x_2 = 45 + 45\cos\frac{\pi}{2} = 45°,\ y_2 = 0.707$$

$$x_3 = 45 + 45\cos\frac{5\pi}{6} = 6.03°,\ y_3 = 0.105$$

Given $\Delta\theta = 120°$, $\Delta\phi = 60°$ (Fig. 11.30)

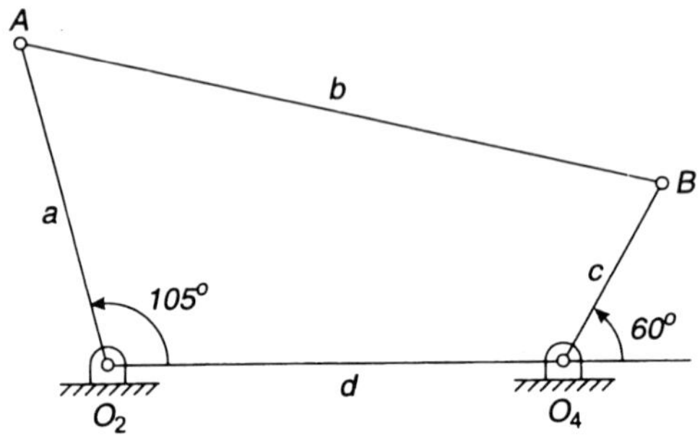

Fig. 11.30

Scale factors are :

$$r_x = \frac{\Delta\theta}{x_f - x_i} = \frac{120}{90-0} = \frac{120}{90} = \frac{4}{3}$$

$$y_i = \sin 0° = 0,\ y_f = \sin 90° = 1$$

$$r_y = \frac{\Delta\phi}{y_f - y_i} = \frac{60}{1-0} = 60$$

Now

$$r_x = \frac{\theta_m - \theta_i}{x_m - x_i}$$

$\therefore$

$$\frac{\theta_1 - 105}{83.97 - 0} = \frac{4}{3}$$

$$\theta_1 = 216.96°$$

$$\frac{\theta_2 - 105}{45 - 0} = \frac{4}{3},\ \theta_2 = 165°$$

$$\frac{\theta_3 - 105}{6.03 - 0} = \frac{4}{3},\ \theta_3 = 113.04°$$

$$r_y = \frac{\phi_m - \phi_i}{y_m - y_i}$$

$$\frac{\phi_1 - 60}{0.994 - 0} = 60, \phi_2 = 119.64°$$

$$\frac{\phi_2 - 60}{0.707 - 0} = 60, \phi_2 = 102.42°$$

$$\frac{\phi_3 - 60}{0.105 - 0} = 60, \phi_3 = 6.63°$$

Precision point	***x, deg***	***y***	***ϕ deg***	**cos ϕ**	***θ, deg***	**cos θ**	**(θ – ϕ), *deg***	**cos (θ – ϕ)**
1	83.97	0.994	119.654	– 0.494	216.96	– 0.8	197.32	– 0.955
2	45	0.707	102.42	– 0.215	165	– 0.966	62.58	0.460
3	6.03	0.105	6.63	0.993	113.04	– 0.391	– 106.41	– 0.282

The Freudenstein's equation is

$$k_1 \cos\phi + k_2 \cos\theta + k_3 = \cos(\theta - \phi)$$

where $k_1 = \frac{d}{a}, k_2 = \frac{-d}{c}, k_3 = \frac{a^2 - b^2 + c^2 + a^2}{2ac}$

Given $d_1 = 50$ mm

Substituting the values, we have

$$\begin{bmatrix} -0.494 & -0.8 & 1 \\ -0.215 & -0.966 & 1 \\ 0.993 & -0.391 & 1 \end{bmatrix} \begin{Bmatrix} k_1 \\ k_2 \\ k_3 \end{Bmatrix} = \begin{Bmatrix} -0.955 \\ 0.460 \\ -0.282 \end{Bmatrix}$$

$$[A]\{k\} = \{B\}$$

$$|A| = 0.36095, \; |A_1| = \begin{vmatrix} -0.955 & -0.8 & 1 \\ 0.460 & -0.966 & 1 \\ -0.282 & -0.391 & 1 \end{vmatrix} = -0.69045$$

$$|A_2| = \begin{vmatrix} -0.494 & -0.955 & 1 \\ -0.215 & -0.460 & 1 \\ 0.993 & -0.282 & 1 \end{vmatrix} = -1.91634$$

$$|A_3| = \begin{vmatrix} -0.494 & -0.8 & -0.955 \\ -0.215 & -0.966 & 0.460 \\ 0.993 & -0.391 & -0.282 \end{vmatrix} = -1.5367$$

$$k_1 = \left|\frac{A_1}{A}\right| = 1.913, k_2 = \left|\frac{A_2}{A}\right| = -5.309, k_3 = \left|\frac{A_3}{A}\right| = -4.2574$$

$$k_1 = \frac{d}{a}, a = \frac{d_1}{k_1} = \frac{50}{1.913} = 26.137 \text{ mm}$$

$$k_2 = \frac{-d}{c}, c = \frac{-d}{k_2} = \frac{-50}{-5.309} = 9.418 \text{ mm}$$

$$k_3 = \frac{a^2 - b^2 + c^2 + d^2}{2ac}$$

$$-4.2574 \times 2 \times 26.137 \times 9.418 = (26.137)^2 - b^2 + (9.418)^2 + 50^2$$

$$b^2 = 53.67.82, b = 73.265 \text{ mm}$$

Hence $a = 26.137$ mm, $b = 73.265$ mm, $c = 9.148$ m, $d = 50$ mm.

■ **Example 11.18** Synthesize a slider crank mechanism with an eccentricity of 9 mm for the two input positions of input link $\theta_{12} = 60°$ and output displacement of slider is 16 mm.

Solution : Given : $e = 9$ mm, $\theta_{12} = 60°$, $x_{12} = 16$ mm.

1. Draw two parallel lines l_1 and l_2 at an eccentricity, $e = 9$ mm (Fig. 11.31).

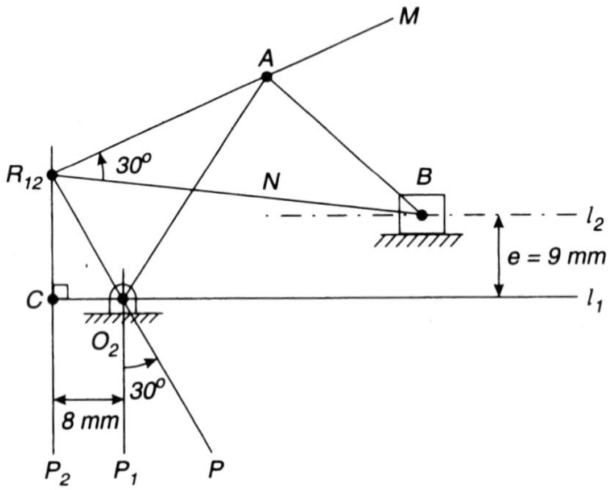

Fig. 11.31

2. Local point O_2 at a convenient position on line l_1. Draw $\angle P_1O_2P = \frac{1}{2}\theta_{12} = 30°$ opposite to the direction of crank rotation, *i.e.*, ccw.
3. Cut off $O_2C = \frac{1}{2}x_{12} = 8$ mm on line l_1 opposite to the direction of displacement of slider. Erect perpendicular P_2C at point C.
4. Extend lines PO_2 and P_2C to intersect at point R_{12}, to give the relative pole.
5. Construct an angle equal to $\frac{1}{2}\theta_{12} = 30°$ at R_{12}. One arm of this angle intersect line l_2 at B to locate the slider B.
6. Select point A arbitrarily and conveniently on the other arm of the above angle.
7. Join AB and AO_2 to give the connecting rod and crank respectively.

■ **Example 11.19** Design graphically a 4 bar mechanism such that

$$\theta_{12} = 120°; \theta_{13} = 160° \text{ and } \phi_{12} = 70°; \phi_{13} = 110°$$

Input moves anti-clockwise and output moves in clockwise direction.

Solution : Given : $\theta_{12} = 120°$, $\phi_{12} = 70°$; $\theta_1 3 = 160°$, $\phi_{13} = 110°$

$$\Psi = \frac{1}{2}(\theta_{12} - \phi_{12}) = \frac{1}{2}(120° - 70°) = 25°$$

$$\Psi_{13} = \frac{1}{2}(\theta_{13} - \phi_{13}) = \frac{1}{2}(160° - 110°) = 25°.$$

1. Select relative poles R_{12} and R_{13} at arbitrarily chosen positions. Draw $\angle\Psi_{12} = \angle x_1 R_{12} x_2 = 25°$ at R_{12} and $\angle\Psi_{13} = y_1 R_{13} y_2 = 25°$ at R_{13}.
2. Let x_2 and y_2 lines intersect at point A and x_1 and y_1 lines at point B. Join AB. Then AB is the coupler.
3. Select fixed point O_2 of fixed link O_2O_4 arbitrarily at a convenient place. Join R_{12} with O_2.
4. Draw $\angle O_2 R_{12} Z_1 = \angle\Psi_{12} = 25°$.
5. Join R_{13} with with O_2 and draw $\angle O_2 R_{13} Z_2 = \angle\Psi_{13} = 25°$. Let Z_1 and Z_2 lines intersect at O_4. Join O_2 and O_4. Then O_2O_4 is the fixed link.
6. Join O_2 with A and O_4 with B. Then O_2ABO_4 is the desired four-bar linkage shown in Fig. 11.32.

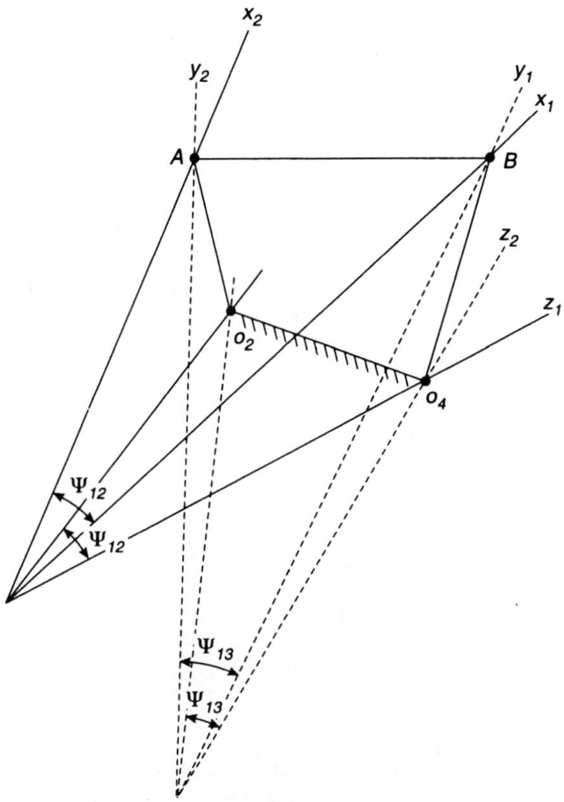

Fig. 11.32

■ **Example 11.20** For the four-bar linkage, the following data are given

$$\theta_2 = 60°; \theta_4 = 90°$$
$$\omega_2 = 3 \text{ rad/sec}; \omega_4 = 2 \text{ rad/sec}$$
$$\alpha_2 = -1 \text{ rad/sec}^2; \alpha_4 = 0$$

Determine the length of various link.

Solution : The Freudensein's equation for the displacement of a four-bar linkage (Fig. 11.33) is given by :

$$k_1 \cos\theta_4 + k \cos\theta_2 + k_3 = \cos(\theta_2 - \theta_4) \quad ...(1)$$

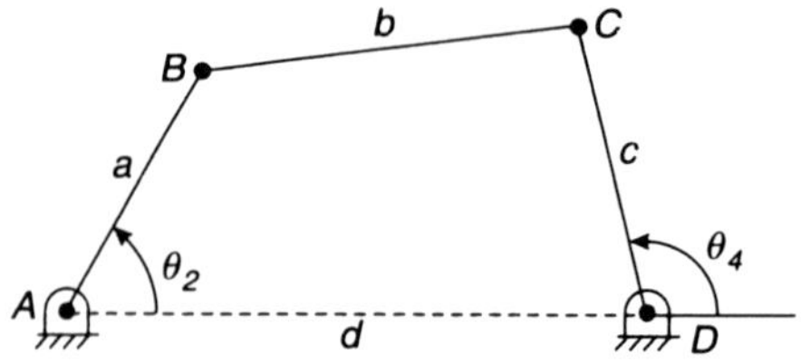

Fig. 11.33

where $k_1 = \frac{d}{a}, k_2 = -\frac{d}{c}, k_3 = \frac{a^2 - b^2 + c^2 + d^2}{2ac}$

Taking first time derivative of eq. (1), we have

$$k_1\omega_4 \sin\theta_4 + k_2\omega_2 \sin\theta_2 = (\omega_2 - \omega_4)\sin(\theta_2 - \theta_4) \quad ...(2)$$

Taking the second time derivative of eq. (2),we have

$$k_1(\omega_4^2 \cos\theta_4 + \alpha_4 \sin\theta_4) + k_2(\omega_2^2 \cos\theta_2 + \alpha_2 \sin\theta_2)$$

$$= (\alpha_2 - \alpha_4)\sin(\theta_2 - \theta_4) + (\omega_2 - \omega_4)^2 \cos(\theta_2 - \theta_4) \quad ...(3)$$

Substituting the values of various terms in eqs. (1) to (3), we get

$$k_1 \cos 90° + k_2 \cos 60° + k_3 = \cos(60° - 90°)$$

or
$$0.5\, k_2 + k_3 = \frac{\sqrt{3}}{2}$$

or
$$k_2 + 2k_3 = 1.732 \quad ...(4)$$

$$k_1 \times 2 \times \sin 90° + k_2 \times 3 \times \sin 60° = (3 - 1)\sin(60° - 90°)$$

or
$$2k_1 + \frac{3\sqrt{3}}{2} k_2 = -\frac{1}{2}$$

or
$$4k_1 + 5.196\, k_2 = -1$$

$$k_1(4\cos 90° + 0 \times \sin 90°) + k_2(9 \times \cos 60° - 1 \times \sin 60°)$$

$$= (-1 - 0)\sin(60° - 90°) + (3 - 2)^2 \cos(60° - 90°)$$

$$k_2\left(9 \times \frac{1}{2} - 1 \times \frac{\sqrt{3}}{2}\right) = \frac{1}{2} + 1 + \frac{\sqrt{3}}{2}$$

$$3.634 k_2 = 1.366$$

$$k_2 = 0.376$$

Substituting the values of k_2 in eqs. (5) and (4) gives,

$$k_1 = -0.738,\ k_3 = 0.678$$

Now
$$k_1 = \frac{d}{a},\ a = \frac{d}{k_1} = \frac{1}{-0.738} = -1.355$$

$$k_2 = -\frac{d}{c},\ c = -\frac{d}{k_2} = \frac{-1}{0.376} = -2.659$$

$$2\, ac\, k_3 = a^2 - b^2 + c^2 + a^2$$

$$2 \times (-1.355) \times (-2.659) \times 0.678 = (-1.355)^2 - b^2 + (-2.659)^2 + 1$$

$$b^2 = 5.0207,\ b = 2.24$$

Thus $a = -1.355$, $b = 2.24$, $c = -2.659$

$$d = 1.0$$

11.9 LEAST SQUARE TECHNIQUE

A four link mechanism can be designed precisely upto five positions of the input and the output links, provides θ and ϕ are measured from some arbitrary reference. In such cases, the synthesis equations become non-linear and cannot be solved by the Cramer's rule. It is also possible to design a mechnaism for more than five positions which gives least deviation from the specified positions and provides the average performance. The least-square technique is useful to synthesize such a mechanism.

Consider the Freudenstein's equation,

$$k_1 \cos \phi_i + k_2 \cos \theta_i + k_3 - \cos (\theta_i - \phi_i) = 0$$

Summing up for n values of θ and ϕ and defining,

$$S = \sum_{i=1}^{n} [k_1 \cos \phi_i + k_2 \cos \theta_i + k_3 - \cos (\theta_i - \phi_i)]^2$$

For S to be minimum, the conditions are :

$$\frac{\partial S}{\partial k_1} = 0, \frac{\partial S}{\partial k_2} = 0 \text{ and } \frac{\partial S}{\partial k_3} = 0$$

which gives,

$$\sum_{i=1}^{n} 2 [k_1 \cos \phi_i + k_2 \cos \theta_i + k_3 - \cos (\theta_i - \phi_i)] \cos \phi_i = 0$$

or $$k_1 \Sigma \cos^2 \phi_i + k_2 \Sigma \cos \theta_i \cos \phi_i + k_3 \Sigma \cos \phi_i = \Sigma \cos (\theta_i - \phi) \cos \phi_i \qquad ...(1)$$

Similarly,

$$k_1 \Sigma \cos \phi_i \cos \theta_i + k_2 \Sigma \cos^2 \theta_i + {}_3 \Sigma \cos \theta_i = \Sigma \cos (\theta_i - \phi) \cos \phi_i \qquad ...(2)$$

and $$k_1 \Sigma \cos \phi_i + k_2 \Sigma \cos \theta_i + k_3 \Sigma 1 = \Sigma \cos (\theta_i - \phi_i) \qquad ...(3)$$

Eqs. (1) and (3) are three simultaneous linear, non-homogeneous equations in three unknown k_1, k_2 and k_3. These can be solved by using the Cramer's rule. These equations can be written in matrix form as :

$$\begin{bmatrix} \Sigma \cos^2 \phi_i & \Sigma \cos \theta_i \cos \phi_i & \Sigma \cos \phi_i \\ \Sigma \cos \phi_i \cos \theta_i & \Sigma \cos^2 \theta_i & \Sigma \cos \theta_i \\ \Sigma \cos \phi_i & \Sigma \cos \theta_i & 1 \end{bmatrix} \begin{Bmatrix} k_1 \\ k_2 \\ k_3 \end{Bmatrix} = \begin{Bmatrix} \Sigma \cos(\theta_i - \phi_i) \cos \phi_i \\ \Sigma \cos (\theta_i - \phi_i) \cos \theta_i \\ \Sigma \cos (\theta_i - \phi_i) \end{Bmatrix} \qquad ...(11.14)$$

SOLVED QUESTIONS FROM UNIVERSITY EXAMINATIONS

■ **Example 11.21** Differentiate between function generation and path generation.

[KU, Dec. 2006, 2007]

Solution : Function generation : In function generation, the motion of input link is correlated to the motion of output link. If we want to synthesize a four-bar mechanism to generate a function $y = \sin x$, where x is the input crank angle, then the mechanism is designed so that the output link angle approximates the function y.

Path generation : In path generation, a point on the coupler link is constrained to describe a path with reference to a fixed frame. The common requirements are that a portion of the path may be a circular arc, an ellipse or a straight line.

■ **Example 11.22** Design a four link mechanism when the motions of input and output links are governed by the function $y = 2x^2$, x varies from 2 to 4 with an interval of 1. Sketch the mechanism indicating various dimensions. [KU, Dec., 2006]

Solution : Refer to Example 11.4

■ **Example 11.23** Distinguish between the following giving examples :

(i) Mechanism and Machine.

(ii) Analysis and synthesis of mechanism. [KU, Dec., 2008]

Solution : (i) Refer to art. 1.9.

(ii) Analysis of mechanism deals with the determination of velocity and acceleration of various links. Both analytical and graphical methods are used for analyzing a mechanism.

Synthesis of mechanism require the determination of lengths of various links that satisfy the requirement of motion of the mechanism. Generally the requirements are related to positions of input and output links.

■ **Example 11.24** What is Freudenstein's equation? How is it helpful in designing a four-link mechanism when three positions of the input $(\theta_1, \theta_2, \theta_3)$ and the output link (ϕ_1, ϕ_2, ϕ_3) are known? [KU, Dec., 2008]

Solution : Refer to art. 11.7.4.

■ **Example 11.25** Design a four-link mechanism when the motions of the input and the output links are governed by a function $y = x^2$ and x varies from 0 to 2 with an interval of 1. Assume θ to vary from 50° to 150° and ϕ from 80° to 160°. [KU, Dec., 2008]

Solution : The angular displacement of input link is governed by a x and that of the output link by y. θ varies from 50° to 150°, *i.e.*, through 100° and ϕ varies from 80° to 160°, *i.e.*, through 80°.

$$x = 0, 1, 2 \text{ with an interval of } 1.$$

Corresponding values of y are :

$$y = 0, 1, 4$$

Scale factors are :

$$r_x = \frac{\theta_f - \theta_i}{x_f - x_i} = \frac{150° - 50°}{2 - 0} = \frac{100}{2} = 50$$

$$\frac{\theta_2 - \theta_i}{x_2 - x_i} = r_x, \frac{\theta_2 - 50°}{1 - 0} = 50, \theta_2 = 50 + 50 = 100°$$

$$r_y = \frac{\phi_y - \phi_i}{y_f - y_i} = \frac{160° - 80°}{4 - 0} = \frac{80}{4} = 20$$

$$\frac{\phi_2 - \phi_i}{y_2 - y_i} = r_y, \frac{\phi_2 - 80°}{1 - 0} = 20, \phi_2 = 100°$$

Precision point	*x*	*y*	*θ, deg*	*ϕ, deg*
1	0	0	50	80
2	1	1	100	100
3	2	4	150	160

Freudenstein's equation is,

$$k_1 \cos\phi + k_2 \cos\theta_3 = \cos(\theta - \phi)$$

where $k_1 = \frac{d}{a}$, $k_2 = -\frac{d}{c}$, $k_3 = \frac{a^2 - b^2 + c^2 + d^2}{2ac}$

$$k_1 \cos 80° + k_2 \cos 50° + {}_3 = \cos(-30°) \quad \text{...(1)}$$

$$k_1 \cos 100° + k_2 \cos 100° + k_3 = \cos 0° \quad \text{...(2)}$$

$$k_1 \cos 160° + k_2 \cos 150° + k_3 = \cos(-10°) \quad \text{...(3)}$$

or

$$0.17365\text{k}_1 + 0.64278k_2 + k_3 = 0.86602 \quad \text{...(4)}$$

$$-0.17365\text{k}_1 - 0.17365k_2 + k_3 = 1 \quad \text{...(5)}$$

$$-0.93969\text{k}_1 - 0.86602k_2 + k_3 = 0.98480 \quad \text{...(6)}$$

Eq. (4) – Eq. (5) gives,

$$0.34730k_1 + 0.81643k_2 = -0.13398$$

or

$$k_1 + 2.35099\text{k}_2 = -0.38577 \quad \text{...(7)}$$

Eq. (5) – Eq. (6) gives,

$$0.76604k_1 + 0.69237k_2 = 0.0152$$

or

$$k_1 + 0.90383k_2 = 0.01984 \quad \text{...(8)}$$

Eq. (8) – Eq. (7) gives,

$$-1.44696k_2 = 0.40561$$

$$k_2 = -0.28032$$

From Eq. (8),

$$k_1 = 0.01984 - 0.90383 \times (-0.28032) = 0.2732$$

Eq. (4) gives,

$$0.17365 \times 0.2732 + 0.64278 \times (-0.28032) + k_3 = 0.86602$$

$$k_3 = 0.99876$$

Let $d = 5$ cm

$$k_1 = \frac{d}{a}, \quad a = \frac{5}{0.2732} = 18.3 \text{ cm}$$

$$k_2 = -\frac{d}{c}, \quad c = \frac{5}{0.28032} = 17.8 \text{ cm}$$

$$k_3 = \frac{(18.3)^2 - b^2 + (17.8)^2 + 5^2}{2 \times 18.3 \times 17.8} = 0.99876$$

$$676.73 - b^2 = 650.67$$

$$b = 5.1 \text{ cm}$$

■ **Example 11.26** State Grashof's law. State how it is helpful in classifying 4-bar link mechanism into different types. [MDU, Dec. 2009]

Solution : Refer art 11.3.

■ **Example 11.27** Explain kinematic synthesis of mechanisms. [MDU, May, 2008]

Solution : Refer to art. 11.4

■ **Example 11.28** Discuss the analytical method of syntheiss for slider-crank mechanism with appropriate sketches. [MDU, Dec., 2007]

Solution : Refer to art. 11.8.

■ **Example 11.29** Synthesize a four-bar mechanism such that $\theta_{12} = 50°, \theta_{23} = 40°$ and $\phi_{12} = 80°, \phi_{23} = 50°/$ [MDU, May 2008]

Solution : Refer to Example 11.7

■ **Example 11.30** Determine the proportions of four-bar mechanism by using three precision points to generate $y = x^{1.5}$, where x varies between 1 and 4. Assume $\theta_i = 30°, \Delta\theta = 90°$, $\phi_i = 90°, \Delta\phi = 90°$. Take length of fixed link as 25 mm. [MDU, May, 2009]

Solution : Chebyshev's Precision points are :

$$x_m = a + b\cos\left[\frac{(2m-1)\pi}{2n}\right], m = 1, 2, 3$$

Here $n = 3, x_i = 1, x_f = 4$

$$a = \frac{1}{2}(x_i + x_f) = \frac{1}{2}(1+4) = 2.5$$

$$b = \frac{1}{2}(x_f - x_i) = \frac{1}{2}(4-1) = 1.5$$

$$x_1 = a + b\cos\frac{\pi}{6} = 2.5 + 1.5 \times \cos\frac{\pi}{6} = 3.8$$

$$x_2 = a + b\cos\frac{\pi}{2} = 2.5 + 0 = 2.5$$

$$x_3 = a + b\cos\left(\frac{5\pi}{6}\right) = 2.5 + 1.5\cos\left(\frac{5\pi}{6}\right) = 1.2$$

$$y_1 = x_1^{1.5} = (3.8)^{1.5} = 7.41$$

$$y_2 = x_2^{1.5} = (2.5)^{1.5} = 3.95$$

$$y_3 = x_3^{1.5} = (1.2)^{1.5} = 1.31$$

$$y_i = x_i^{1.5} = 1, y_f = x_f^{1.5} = 4^{1.5} = 8$$

Scale factor are :

$$r_x = \frac{\Delta\theta}{x_f - x_i} = \frac{90}{4-1} = 30$$

$$r_y = \frac{\Delta\phi}{y_f - y_i} = \frac{90}{8-1} = \frac{90}{7} = 12.857$$

$$r_x = \frac{\theta - \theta_i}{x - x_i}, \frac{\theta_1 - \theta_i}{x_1 - x_i} = \frac{\theta_1 - 30}{3.8 - 1} = 30, \theta_1 = 114°$$

$$\frac{\theta_2 - \theta_i}{x_2 - x_i} = \frac{\theta_2 - 30}{2.5 - 1} = 30, \theta_2 = 75°$$

$$\frac{\theta_3 - \theta_i}{x_3 - x_i} = \frac{\theta_3 - 30}{1.2 - 1} = 30, \theta_3 = 36°$$

$$r_y = \frac{\phi - \phi_i}{y_1 - y_i}, \frac{\phi_1 - 90}{7.41 - 1} = 12.857, \phi_1 = 172.41°$$

$$\frac{\phi_2 - \phi_i}{y_2 - y_i} = \frac{\phi_2 - 90}{3.95 - 1} = 12.857, \phi_2 = 127.92°$$

$$\frac{\phi_3 - \phi_i}{y_3 - y_i} = \frac{\phi_3 - 90}{1.31 - 1} = 12.857, \phi_3 = 93.98°$$

Precision point	x	y	φ, deg	cos φ	θ, deg	cos θ	(θ – φ), deg	cos (θ – φ)
1	3.8	7.41	172.41	– 0.9912	114	– 0.4067	– 58.41	0.5238
2	2.5	3.95	127.92	– 0.6146	75	– 0.2588	– 52.92	0.6029
3	1.2	1.31	93.98	– 0.0694	36	0.8090	– 57.98	0.5302

The Freudenstein's equation for four-bar mechanism is,

$$k_1 \cos\phi + k_2 \cos\theta + k_3 = \cos(\theta - \phi)$$

$$-0.9912k_1 - 0.4067k_2 + k_3 = 0.5238 \quad ...(1)$$

$$-6146k_1 + 0.2588k_2 + k_3 = 0.6029 \quad ...(2)$$

$$-0.0694k_1 + 0.8090k_2 + k_3 = 0.5302 \quad ...(3)$$

Eq. (1) – (2) gives :

$$-0.3766k_1 - 0.6655k_2 = -0.0791$$

or

$$k_1 + 1.7671k_2 = 0.2100 \quad ...(4)$$

Eq. (2) – (3) gives :

$$-0.5452k_1 - 0.5502k_2 = 0.0727$$

or

$$k_1 + 0.9207k_2 = -0.1333 \quad ...(5)$$

Eq. (4)–(5) gives :

$$0.8486k_2 = 0.3433$$

$$k_2 = 0.4045$$

$$k_1 = 0.2100 - 1.7671 \times 0.4045 = -0.5048$$

$$k_3 = 0.5238 + 0.9912 \times (-0.5048) + 0.4067 \times 0.4045 = 0.1879$$

$$d = 25 \text{ mm (given)} = 0.1879$$

Now

$$\frac{d}{a} = k_1, a = \frac{25}{-0.5048} = -49.52 \text{ mm}$$

$$-\frac{d}{c} = k_2, c = -\frac{d}{k_2} = \frac{-25}{0.4045} = -61.80 \text{ mm}$$

$$\frac{a^2 - b^2 + c^2 + d^2}{2ac} = k_3,$$

$$\frac{(-49.52)^2 - b^2 + (-61.80)^2 + (25)^2}{2 \times (-49.52) \times (-61.80)} = 0.1879$$

$$2452.2 - b^2 = 3819.24 + 625 = 1150.07$$

$$b^2 = 5746.7$$

$$b = \pm 75.8 \text{ mm}$$

$$a = -49.52 \text{ m}, b = \pm 75.8 \text{ mm}, c = -61.80 \text{ mm}, d = 25 \text{ mm}$$

– ve sign indicates that length is to be measured in reverse direction.

■ **Example 11.31** Design a four-bar mechanism when the motion of the input and output links are governed by the function, $y = 2x^2$ and x varies from 2 to 4 with an interval of 1. Assume θ to very from 40° to 120° and ϕ from 60° to 132°. [MDU, Dec., 2009]

Solution : Refer to Example 11.4.

■ **Example 11.32** Determine the proportions of 4-bar mechanism by using three precision points to generate $y = x^{1.5}$ where x varies from 1 to 4. Assume θ to vary from 30° to 120° and ϕ from 60° to 130°. Take length of fixed link 30 cm. [MDU, Dec., 2010]

Solution : Refer to Example 11.30.

SUMMARY

1. Gruebler's criterion for degrees of freedom for planar mechanisms.
$$F = 3(n - 1) - 2p - h$$
where p = number of simple joints or lower pairs having one degree of freedom.
$$= \frac{1}{2}(2n_2 + 3n_3 + \ldots in_i)$$
$n_2, n_3 \ldots$ an so on are binary, ternary and so on links
h_2 = number of higher pairs having two degrees of freedom.
n = total number of links
$= n_2 + n_3 + \ldots$
For $h_2 = 0$ and $F = 1$,
$$n_2 = 4 + [n_4 + \ldots + (i - 3)n_i]$$
2. Grashof's law states that for a four-bar mechanism the sum of the lengths of the largest and the shortest links should be less than the sum of the lengths of the other two links.
$$(l + s) \le (p + q)$$
3. Number synthesis implies to determine the number of degree of freedom of a mechanism. It involves the determination of number of links together with number of type of joints required for a specified motion.
4. Number of degrees of freedom is equal to the number of independent coordinates required to specify its configuration in order to define its motion.
5. Dimensional synthesis deals with the determination of the significant dimensions of the mechanism to satisfy the specified motion characteristics. The significant dimensions could be link lengths, angles between the links, cam contour, etc.
6. Type synthesis refers to kind of mechanism selected such as gear combination, a belt-pulley combination or a cam mechanism.
7. The pole of the coupler link of a four-bar mechanism is its centre of rotation with respect to the fixed link.
8. The relative pole of the coupler link of a four-bar mechanism is its centre of rotation relative to other moving links.

9. Synthesis of four-bar mechanism :

(a) Two-position synthesis : θ_{12} and ϕ_{12} are known.

Angle subtended by coupler at R_{12},

$$\angle\Psi_{12} = \frac{1}{2}(\angle\theta_{12} - \angle\phi_{12})$$

(b) Three-position synthesis : $\theta_1, \theta_2, \theta_3$ and ϕ_1, ϕ_2, ϕ_3 are known.

$$\theta_{12} = \theta_2 - \theta_1. \theta_{13} = \theta_3 - \theta_1 \text{ and } \phi_{12} = \phi_2 - \phi_1, \phi_{13} = \phi_3 - \phi_1$$

Then

$$\angle\Psi_{12} = \frac{1}{2}(\angle\theta_{12} - \angle\phi_{12})$$

$$\angle\Psi_{13} = \frac{1}{2}(\angle\theta_{13} - \angle\phi_{13})$$

10. Synthesis of slider-crank mechanism

(a) Two-position synthesis

$$\theta_{12} \text{ and } x_{12} \text{ are known}$$

(b) Three-position synthesis

$$\theta_{12}, \theta_{13} \text{ and } x_{12}, y_{12} \text{ are known.}$$

11. In function generation, the motion of the input link is correlated to the motion of output link.

12. In path generation, a point on the coupler link is constrained to describe a path with reference to a fixed frame.

13. Scale factor are :

$$r_x = \frac{\theta - \theta_i}{x - x_i} = \frac{\theta_f - \theta_i}{x_f - x_i}$$

$$r_y = \frac{\phi - \phi_i}{y - y_i} = \frac{\phi_f - \phi_i}{y_f - y_i}$$

14. Chebyshev's spacing for precision points :

$$x_m = a + b\cos\left[\frac{(2m-1)\pi}{2n}\right], m = 1, 2, 3$$

where $a = \dfrac{x_i + x_f}{2}$, $b = \dfrac{x_f - x_i}{2}$ and n = number of precision points.

15. Freudenstein's equation for three precision points for a four-bar mechanism :

$$k_1 \cos\phi + k_2 \cos\theta + k_3 = \cos(\theta - \phi)$$

where $k_1 = \dfrac{d}{a}$, $k_2 = -\dfrac{d}{c}$ and $k_3 = \dfrac{a^2 - b^2 + c^2 + d^2}{2ac}$.

16. Freudenstein's equation for a slider-crank mechanism :

$$A_1 s \cos\theta + A_2 + \sin\theta - A_3 = s^2$$

where $A_1 = 2l_2$, $A_2 = 2l_2\, e$ and $A_3 = l_2^2 e$ and $A_3 = l_2^2 - l_3^2 + e^2$

SHORT ANSWER TYPE QUESTIONS

1. Differentiate between a pole and a relative pole of a coupler link of four-bar mechanism.
2. Define kinematic synthesis.
3. What do you understand by movability of a mechanism?
4. State Grashof's law.
5. When do we get double crank and crank-rocker mechanism?
6. What is a class-II double-rocker mechanism?
7. What are the properties of pole points?
8. What do you means by function generation?
9. What are Chebyshev's spacing for precision points?
10. Write the Freudenstein's equation of three precision points of a four-bar chain.

EXERCISES

1. Synthesis a four-bar mechanism so that $\theta_{12} = 45°$ and $\phi_{12} = 55°$. Both input and output cranks should move in the ccw direction.
2. Synthesis a four-bar linkage that will, in one of its positions, satisfy the following values for the angular velocities and accelerations :

$$y = x^{1.2} \text{ for } 1 \le x \le 5$$

using Chebyshve's spacing for three precision points. Take $\phi_0 = 30°$, $\Psi_o = 60°$ and $\Delta\phi = \Delta\Psi = 90°$ and $a = 10$ cm. [**Ans.** $a = 10$ cm, $b = 18.85$ cm, $c = 9.85$ cm, $d = -2.69$ cm]

3. Synthesize a four-bar mechanism to generate a function $y = \sin x$ for $0° \le x \le 90°$. The range of the output crank may be chosen as 60° while that of input crank be 120°. Assume three precision points which are to be form Chebychev's spacing. Length of fixed link = 52.5 mm, $\phi_1 = 105°$ and $\Psi_1 = 66°$.
4. Synthesize a four-bar linkage to generate $y = \log_{10} x$ in the interval $1 \le x \le 10$. The input crank length is to be 5 cm.The input crank is to rotate from 45° to 105° while the output crank moves from 135° to 225°. Use three accuracy points with Chebyshve's spacing. [**Ans.** $a = 5.9$ cm, $b = 22.11$ cm, $c = 5$ cm, $d = 10.05$ cm]
5. A four-bar mechanism is required such that the input and output angles are coordinated as given in the following table :

Input crank angle	30°	50°	80°
Output crank angle	0°	30°	60°

Synthesis the four-bar mechanism. [**Ans.** $a = 1, b = 2.3039, c = 1.2817, d = 1.8321$]

6. Synthesize a four-bar mechanism using Freudenstein's equation to generate the function $y = x^{1.5}$ for the interval $1 \le x \le 4$. The input crank is to start from $\theta_2 = 30°$ and is to have a range of 90°. The output crank angle is to vary from 90°. Take three accuracy points. [**Ans.** $a = 11.16$, $b = 7.76$, $c = 7.82$, $d = 1.0$]
7. Design a four-bar mechanism such that :

$$\theta_{12} = 120°, \theta_{13} = 170° \text{ and } \phi_{12} = 70°, \phi_{13} = 100°$$

Input moves anti-clockwise and output also moves anticlockwise.

8. Synthesize a slider cranks mechanism with eccentricity, $e = 0.9$ cm for the two input positions of input link $\theta_{12} = 56°$ and output displacement of slider $x_{12} = 1.6$ cm.
9. Synthesize a slider mechanism with eccentricity, $e = 0.9$ cm for the three input positions $\theta_{12} = 40°$, $\theta_{13} = 80°$ and output displacement of slider $x_{12} = 1.8$ cm and $x_{13} = 4.8$ cm.
10. Synthesize a four-bar linkage using Freudenstein's equation to satisfy the following specifications.

$$\theta_2 = 60°,\ \theta_4 = 90°$$
$$\omega = 5 \text{ rad/s},\ \omega_4 = 3 \text{ rad/s}$$
$$\alpha_2 = 2 \text{ rad/s}^2,\ \alpha_4 = 7 \text{ rad/s}^2$$

11. Synthesize a four-bar mechanism such that $\theta_{12} = 50°$, $\theta_{23} = 40°$ and $\phi_{12} = 80°$, $\phi_{23} = 50°$.
12. Determine the type of chains shown in Fig. 11.34

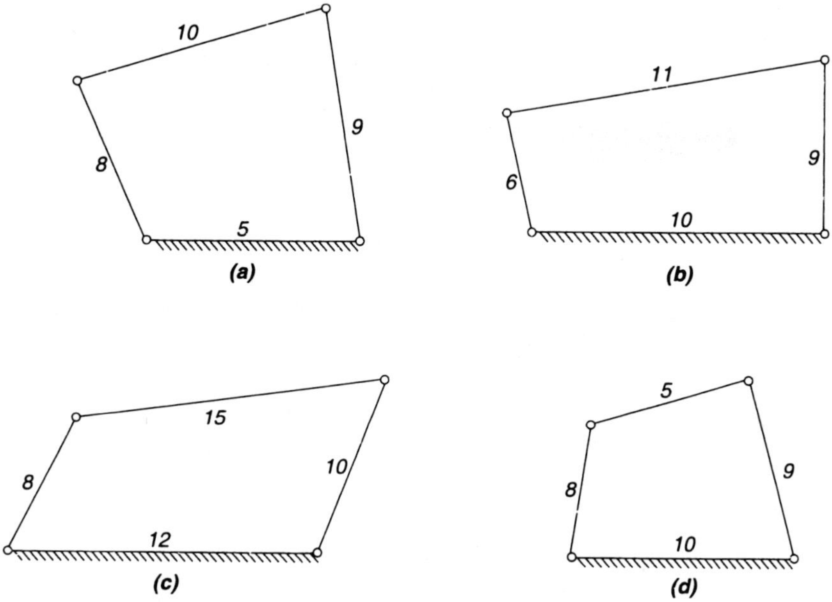

Fig. 11.34

13. Determine the type of chains shown in Fig. 11.35.

Fig. 11.35

INDEX

❑❑❑

Win Prizes

ATTENTION READERS

Win Prizes

We request for your frank assessment regarding various aspects of the book as given below :

KINEMATICS OF MACHINES

By: Dr. SADHU SINGH

Please Fill up the Sheet in Neat and Clean Writing

(*i*) What topics of your syllabus, important for your examination point of view, are not covered in this book ?

....

....

....

....

(*ii*) What are the chapters wherein the treatment of the subject matter is not systematic or organised or updated ?

....

....

....

....

(*iii*) Have you come across misprints/mistakes/factual inaccuracies in the book ? Please specify the chapters and the page numbers.

....

....

....

....

(*iv*) What other books on the same subject have you found/heard better than the present book ? Please specify in terms of price and quality.

....

....

....

....

(*v*) Further suggestions and comments for the improvement of this book.

....

....

....

....

(PS : If need be please attach a separate sheet to write your views)

OTHER DETAILS

(i) Who recommended you this book ? (Please tick (✓) in the box near the option relevant to you.

☐ Teacher ☐ Friends ☐ Bookseller

....

....

....

....

(ii) Name of the recommending teacher, his designation and address:

....

....

....

....

(iii) Name and address of the bookseller you purchased the book from :

....

....

....

....

(iv) Name and address of your institution (Please mention the University or Board as the case may be)

....

....

....

....

(v) Your name and complete address :

....

....

....

....

(vi) Course you are enrolled in :

....

....

....

....

The best assessment will be awarded each month : The award will be in the form of our publications as decided by the Editorial Board.

Please mail the filled up coupon at the vearliest to :

S.K. KATARIA & SONS®

4885/109, Prakash Mahal, Dr. Subhash Bhargav Lane,
Opposite Delhi Medical Association, Daryaganj, New Delhi-110002
Phone: +91-11-43551243 Mobile: +91-9871775858, 7678425209
email : katariabook@gmail.com, katariabooks@yahoo.com